Processing-Structure-Properties Relationships in Polymers

Processing-Structure-Properties Relationships in Polymers

Special Issue Editor

Roberto Pantani

MDPI • Basel • Beijing • Wuhan • Barcelona • Belgrade

MDPI

Special Issue Editor
Roberto Pantani
University of Salerno
Italy

Editorial Office
MDPI
St. Alban-Anlage 66
4052 Basel, Switzerland

This is a reprint of articles from the Special Issue published online in the open access journal *Polymers* (ISSN 2073-4360) from 2017 to 2018 (available at: https://www.mdpi.com/journal/polymers/special_issues/PSP_Relationships_in_Polymers).

For citation purposes, cite each article independently as indicated on the article page online and as indicated below:

LastName, A.A.; LastName, B.B.; LastName, C.C. Article Title. *Journal Name* **Year**, *Article Number*, Page Range.

ISBN 978-3-03921-880-6 (Pbk)
ISBN 978-3-03921-881-3 (PDF)

Contents

About the Special Issue Editor

Roberto Pantani was born in Nocera Inferiore (Salerno - Italy) in 1971. In 1995 he graduated "cum laude" in Chemical Engineering defending a thesis on shrinkage in injection molded PolyStyrene. He gained a PhD degree at the University of Palermo – Italy, in February 2000, defending a thesis entitled "Analysis of Shrinkage Development in Injection Molded Samples". From 1999 to 2001 he gained a research fellowship at University of Salerno, Dept. of Chemical Engineering, in "Processing and Characterization of Polymeric Materials". From September 2001 he has been an assistant professor (until 2007) and then Associate Professor (until 2017) of Transport Phenomena at University of Salerno, Dept. of Chemical Engineering; since October 2017 he is Full Professor at the Department of Industrial Engineering of the University of Salerno. He is a lecturer of Transport Phenomena and of Principles of Chemical Engineering and is currently the Director of the Department. His main research interests focus on the analysis and simulation of injection molding of thermoplastics, structure development in polymer processing, volume accuracy and stability in polymer processing, processing and degradation of biodegradable polymers. He collaborates with companies and research centers and is author of two patents, of more than 100 publications on international peer reviewed scientific journals, of several book chapters and of more than 200 publications on proceedings of international conferences.

Preface to "Processing-Structure-Properties Relationships in Polymers"

The extraordinary capacity of plastics to modify their properties according to a particular structure could be a difficulty, but also an opportunity, and it is one of the keys to the success of this class of materials. The same polymer can be transparent or opaque, rigid or flexible, permeable or impermeable, according to the spatial organization of its macromolecules or of a particular filler. Obviously, the key to taking profit of this peculiar capacity of plastics relies on our capacity of inducing, by means of a suitable processing, that specific spatial organization. This collection of research and review papers is aimed at depicting the state of the art on the possible correlations between processing variables, obtained structure and the special property, which this structure induces on the plastic part.

Roberto Pantani
Special Issue Editor

polymers

MDPI

Article

Structure–Properties Relations for Polyamide 6, Part 1: Influence of the Thermal History during Compression Moulding on Deformation and Failure Kinetics

Emanuele Parodi [1], Gerrit W. M. Peters [1] and Leon E. Govaert [1,2,*]

[1]   Department of Mechanical Engineering, Materials Technology Institute,
      Eindhoven University of Technology, P.O. Box 513, 5600 MB Eindhoven, The Netherlands;
      Emanuele.parodi@maxxistce.nl (E.P.); g.w.m.peters@tue.nl (G.W.M.P.)
[2]   Faculty of Engineering Technology, University of Twente, P.O. Box 217, 7500 AE Enschede, The Netherlands
*   Correspondence: l.e.govaert@tue.nl; Tel.: +31-040-247-2838

Received: 30 May 2018; Accepted: 20 June 2018; Published: 27 June 2018

Abstract: The deformation and failure kinetics of polyamide 6 samples prepared by several thermal histories were investigated by tests at different temperatures and relative humidities. PA6 samples were produced in quiescent condition and multiple cooling procedure. A characterization was performed to investigate the effect of the different thermal histories and the effect of hydration on both structures and glass transition temperature. The mechanical properties were investigated by tensile and creep tests at different temperatures and relative humidity. In order to describe the experimental results, the Ree–Eyring equation, modified with the "apparent temperature", was employed. In addition, the results of time-to-failure (creep tests) were described by the use of the "critical strain" concept. Eventually, a link between the Eyring theory and the structure evolution was made, i.e., a relation between the rate factors and the average lamellar thickness.

Keywords: polyamide 6; compression molding; polymorphism

1. Introduction

The processing of polymer products is a widely discussed topic in polymer science. The solidification procedure is a crucial element in processing because of its influence on the structures which often affect the product performance. The mechanical properties of polymers are often correlated to their yield stress. Yield stress is defined as the stress at which the material deforms mainly plastically. In the case of glassy polymers, the solidification procedure affects mainly the thermodynamic state, i.e., ageing; it was found that ageing increases as the cooling rate during solidification decreases, consequently the yield stress increases [1]. The case of semi-crystalline polymers is more complicated; pressure and cooling rate can affect strongly the material morphology, i.e., crystallinity and lamellar thickness [2–4]. The influence of morphology (structures) on properties was studied by several authors [5,6], and most of them concluded that the mechanical properties are highly dependent on structures. The yield kinetics of i-PP solidified upon different processing was investigated by [7,8]; moreover, the results were described by the use of a model based on the Ree–Eyring equation. It was found that lamellar thickness (or crystal thickness) is a key parameter in order to model the yield kinetics of i-PP produced by different processing.

This work focuses mainly on the effect of processing on structures and the relations between structures and mechanical properties of polyamide 6 (PA6) tested at different temperatures and different relative humidity. PA6 crystallize, by melt processing, in two forms: (a) α-phase for slow cooling ($\dot{T} < \approx 8$ °C), (b) γ-mesophase for intermediate cooling (≈ 8 °C $< \dot{T} < \approx 100$ °C); in case of quenching ($\dot{T} > \approx 100$ °C), a complete amorphous material can be obtained [9]. PA6 is hydrophilic [10], if exposed to a humid environment, it absorbs water. This is due to the polar character of its amide and carbonyl groups [11]. In dry conditions, the polar groups form hydrogen bonds between the polymer chains; these H-bonds give high strength to the material [12]. In case of hydration, part of the H-bonds between chains are broken and new H-bonds are made with the water molecules. This process, also known as plasticization, enhances the chain mobility and a decrease of glass transition temperature is obtained [13]. The hydration-induced depression of T_g has a strong impact on mechanical properties [14–17] as well as crystallographic properties [18].

In this work, the mechanical properties (both tensile and creep test) of polyamide 6 samples prepared with different thermal histories, and conditioned at different relative humidities, are investigated. A semi-empirical model, based on the Ree-Eyring equation, was implemented to predict the yield kinetics and time-to-failure.

2. State-of-the-Art

Observing the yield kinetics (i.e., the yield stress as a function of applied strain rate) of several semi-crystalline polymers, two different stress-dependences can be seen. These two dependences are generally attributed to an intra-lamellar deformation mechanism further referred to as processes I and an inter-lamellar mechanism, further referred to as processes II [19–21]. To describe the rate- and temperature-dependence of yield stress obtained by tensile test at a constant strain rate, the Eyring's activated flow theory [22], modified by Ree–Eyring [23], is used. In this theory, the two deformation processes are considered as independent and their stress contributions are additive. Consequently, the yield stress as a function of strain rate and temperature is calculated as follows:

$$\sigma_y(\dot{\varepsilon}, T) = \frac{kT}{V_I^*} \sinh^{-1}\left(\frac{\dot{\varepsilon}}{\dot{\varepsilon}_{0,I}} \exp\left(\frac{\Delta U_I}{RT}\right)\right) + \frac{kT}{V_{II}^*} \sinh^{-1}\left(\frac{\dot{\varepsilon}}{\dot{\varepsilon}_{0,II}} \exp\left(\frac{\Delta U_{II}}{RT}\right)\right),$$

(1)

where $\dot{\varepsilon}_{0,I}$, ΔU_I and V_I^* are rate factor, activation energy and activation volume related to process I and $\dot{\varepsilon}_{0,II}$, ΔU_{II} and V_{II}^* are related to the process II. As mentioned in Section 1, the mechanical properties of PA6 are strongly dependent on the glass transition which depends on the sample conditioning, i.e., the hydration level. In fact, PA6 can absorb water, which acts as a plasticizer; in other words, the absorption of water lowers the glass transition and therefore the mechanical properties. Thus, in order to include the effect of relative humidity onto the Ree–Eyring equation, the "apparent" temperature was introduced. This temperature, as explained in [24], is based on the principle that a humidity-induced reduction in glass transition temperature is subsequently regarded as an "apparent" increase in the ambient temperature. Consequently, we express this concept as follows:

$$\tilde{T} = T + (T_{g,dry} - T_{g,wet}),$$

(2)

where T is the actual experimental temperature, $T_{g,dry}$ is the glass transition temperature at the dry state and $T_{g,wet}$ is the T_g after conditioning. Subsequently, Equation (1) is rewritten:

$$\sigma_y(\dot{\varepsilon}, \tilde{T}) = \frac{k\tilde{T}}{V_I^*} \sinh^{-1}\left(\frac{\dot{\varepsilon}}{\dot{\varepsilon}_{0,I}} \exp\left(\frac{\Delta U_I}{R\tilde{T}}\right)\right) + \frac{k\tilde{T}}{V_{II}^*} \sinh^{-1}\left(\frac{\dot{\varepsilon}}{\dot{\varepsilon}_{0,II}} \exp\left(\frac{\Delta U_{II}}{R\tilde{T}}\right)\right).$$

(3)

Observing a creep test, i.e., a test with a constant applied load, three regimes can be found: (i) the primary creep in which the plastic flow rate decreases in time, (ii) the secondary creep where plastic flow rate is constant in time and (iii) the tertiary creep where plastic flow rate increases in time and finally failure occurs. Plotting, in a log–log scale plot, the plastic flow rate ($\dot{\varepsilon}_{pl}$) of the secondary regime as a function of time-to-failure, a linear trend with slope of −1 is found. This observation held for several polymers tested at different temperatures [25] and also different relative humidity [24]. Thus, the following relation is written:

$$\dot{\epsilon}_{pl}(\sigma) \cdot t_f(\sigma) = C, \tag{4}$$

where $\dot{\epsilon}_{pl}$ is the plastic flow rate in the secondary creep regime, t_f is the time-to-failure, σ is the applied stress and C is the constant (−1). It was observed that the steady state reached by test at constant strain rate (i.e., yield stress) and the steady state achieved by test at constant load are identical [26]. Eventually, the relation between the strain rate dependence of yield stress and the load dependence of time-to-failure is described by the critical strain concept [27,28]:

$$t_f(\sigma, T) = \frac{\epsilon_{cr}}{\dot{\epsilon}_{pl}(\sigma, T),} \tag{5}$$

where ϵ_{cr} is the critical strain that can be related to the amount of plastic deformation that the material would accumulate in the case in which the $\dot{\epsilon}_{pl}$ was constant all along the creep test.

3. Experimental

3.1. Materials

The material employed in this work was a polyamide 6 (Akulon K122) kindly provided by DSM ((Geleen, The Netherlands). This PA6 has a viscosity-average molar mass (Mv) of about 24.9 kg/mol.

3.2. Sample Preparation

Several different cooling procedures were performed in order to prepare sheets of 0.5 mm thickness with different structure parameters. After a drying procedure (1 night at 110 °C under vacuum), the pellets were placed in a "sandwich" consisting of two thick steel plates (about 3 mm), two thin aluminum foils (about 0.2 mm) and a 250 × 250 × 0.5 mm steel mold. The material was melted at 265 °C for 5 min, while a force of about 10 kN was applied. After this, different cooling procedures were applied (as summarised in Table 1):

- α-I, the hot press was switched off and the "sandwich" was left inside the hot press over night.
- α-II, the hot press was set to 180 °C, cooling was helped by a moderate flow of water, the "sandwich" was left inside during cooling and, once the set temperature was reached, it was removed after 5 min of isothermal.
- α-III, the same procedure of α-I was applied but an α-nucleating agent was added to the basic grade.
- γ-I, the "sandwich" was rapidly moved to a cold press set at 80 °C where the material was solidified in quiescent condition for 5 min.
- γ-II, the "sandwich" was rapidly moved to a cold press set at 110 °C where the material was solidified for 5 min.
- Q-I, the "sandwich" (only thin aluminum foils and mold) was rapidly moved to a bath of water with ice and salt (NaCl), water temperature around −14 °C.

According to the ISO527 type 1BA, dog-bone samples were prepared using a cutting die (main measures: width 5 mm, length 22 mm).

Table 1. Cooling protocols.

Sample	Method
α-I	slow cooled $\dot{T} \approx 0.5$ °C/s
α-II	isothermal at 180 °C
α-III	slow cooled $\dot{T} \approx 0.5$ °C/s—with nucleating agent
γ-I	isothermal at 80 °C
γ-II	isothermal at 110 °C

3.3. Sample Conditioning

In order to investigate the influence of hydration, the samples were stored at four different relative humidities, namely RH 0% (dry), RH 35%, RH 50% and RH 75%. For dry conditioning, samples were stored in a desiccators under vacuum at room temperature; for RH 50%, an environmental chamber was employed, while, in the case of RH 35% and RH 75%, two desiccators containing supersaturated salt solutions able to maintain a constant relative humidity in a close environment were employed. The supersaturated solutions were made of deionized water and two salts: sodium chloride and magnesium chloride hexahydrate for 75 and 35%, respectively.

3.4. Mechanical Tests

In order to perform uniaxial tensile and creep tests, a Zwick Unviversal Testing Machine (Ulm, Germany) provided with a 1 kN load-cell was employed. To control temperature and relative humidity, experiments were performed inside an environmental chamber. The tensile tests were repeated, at least, two times. Several conditions were investigated: a range of strain rates from 10^{-5} s^{-1} up to 3×10^{-2} s^{-1}, temperatures between 23 °C and 120 °C (dry) and relative humidity of 35, 50 and 75% (at 23 °C). A pre-load of 0.1 MPa was applied at a speed of 1 mm/min before each experiment. Creep was performed at three relative humidities: RH 35%, RH 50%, and RH 75%. The desired load was applied within 10 s, and kept constant up to failure. The time-to-failure was estimated as the time at which the strain reaches a fixed strain value of 40%, which was defined as strain at failure.

3.5. X-Ray Diffraction

Wide and small angle X-ray measurements were taken by a Ganesha X-ray instrument (Copenhagen, Denmark) equipped with a GeniX-Cu ultra low divergence source (Copenhagen, Denmark) ($l = 1.54$ Å) and a Pilatus 300 K silicon pixel detector (487 × 619 pixels of 172 × 172 μm^2). After normalization, the crystallinity was estimated by subtracting an amorphous halo (experimentally obtained) to the measured patterns. The degree of crystallinity is finally calculated by:

$$\chi_c = \frac{T_s - A}{T_s}, \tag{6}$$

where T_s is the total scattered intensity and I is the scattering from the amorphous halo. Moreover, a deconvolution analysis was performed. This was obtained by fitting Lorentzian functions, in proximity of each characteristic reflection. Eventually, all the functions and the amorphous halo were summed to check the fidelity of the fitting routine (green markers in Figure 1). Thus, the relative quantities $\chi_{c,\alpha}$ and $\chi_{c,\gamma}$ were calculated by the following:

$$\chi_{c,\alpha} = \left(\frac{A_l - A_\gamma}{A_m}\right) \quad and \quad \chi_{c,\gamma} = \left(\frac{A_l - A_\alpha}{A_m}\right), \tag{7}$$

where A_α and A_γ are the total area of the Lorentzian functions for the α and γ peaks, A_m is the total area of the measured pattern, and A_l is the sum off all the Lorentzian functions (α and γ). An example is given in Figure 1. As far as the small angle X-ray scattering (SAXS) experiments are concerned, Lorentz [29] and thermal density fluctuation [30] correction were applied. Thus, the peak position of the SAXS pattern (d^*) is used to define the long period (l_b):

$$L_b = \frac{2\pi}{d^*}, \tag{8}$$

which is used to estimate the average lamellar thickness:

$$l_c = \chi_{vol} \cdot l_b, \tag{9}$$

where l_b is the long period and χ_{vol} is the volumetric crystallinity, which is defined by the following:

$$\chi_{vol} = \frac{\dfrac{\chi_c}{\rho_c}}{\dfrac{\chi_c}{\rho_c} + \dfrac{100 - \chi}{\rho_a}},$$ (10)

where ρ_c is the density of the crystal (1.21 g/cm³ for α-phase [31], 1.16 g/cm³ for γ-mesophase [31]), ρ_a the density of the amorphous (1.09 g/cm³ [31]) and χ the mass crystallinity.

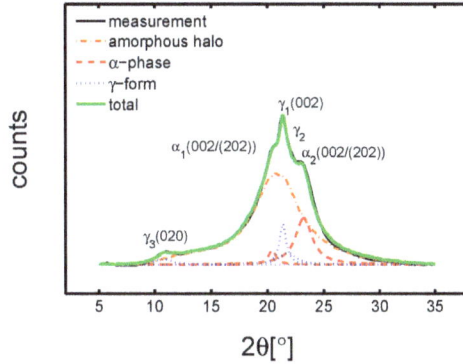

Figure 1. Example of wide angle x-ray diffraction (WAXD) pattern deconvolution analysis. The green line is the result of the deconvolution procedure, the orange line is the measured amorphous halo, and blue and red curves are the Lorentzian functions.

3.6. Dynamical Mechanical Thermal Analysis

Dynamical mechanical thermal analysis (DMTA) was employed to measure the glass transition temperature. The equipment was a TA instruments Q800 DMA (Asse, Belgium). The samples were films (rectangular shape) of about 5 mm width, 0.5 mm thickness. The sample were tested at a single frequency of 1 Hz and a temperature ramp (from −40 °C to 100 °C) with a heating rate of 3 °C/min. The T_g was defined as the maximum in tan(δ).

4. Results and Discussion

4.1. Samples Characterization

In order to understand the effect of different processing on structures, the first step of this study was a crystallographic characterization. It was performed by wide angle X-ray diffraction (WAXD) and small angle X-ray scattering (SAXS) on samples in the dry state. In Figure 2a,b, the integrated WAXD patterns and the results of deconvolution analysis are shown, respectively.

In Figure 2a, the integrated patterns show that the samples γ-I and γ-I have crystallized in the γ-form, which is recognizable by the characteristic central peak at around 2θ 21°and the secondary peak at 2θ 10°; the obtained crystallinity is around 30% and by deconvolution it is possible to state that, only in the case of γ-II, a small fraction of α-phase is obtained, as shown in Figure 2b. The samples α-I, α-II and α-III showed the two characteristic peaks of α-phase, at about 2θ 20°and 2θ 24°; the obtained crystallinity is around 40% in the case of α-III and α-I, while a slightly lower crystallinity (about 35%) is obtained for α-II. These three α-samples have crystallized in pure α-phase. To estimate the average lamellar thickness, small angle X-ray were performed on the dry samples. Examples of SAXS pattern are given in Figure 3a; accordingly to the procedure described in Section 3.5, the average lamellar thickness is estimated and plotted as a function of crystallinity (see Figure 3b).

In Figure 3a, the integrated patterns are reported, the peak position (red markers) is translated to the long period by Equation (8). The relationship between L_c and crystallinity is given in Figure 3b; the average lamellar thickness increases for increasing crystallinity with an asymptotic-like trend. It is remarked that, as stated in [32], crystallization of γ-mesophase at high under-cooling leads to the formation of non-lamellar morphology. Therefore, the values of l_c for the γ-samples should be intended as crystal thickness rather than lamellar thickness.

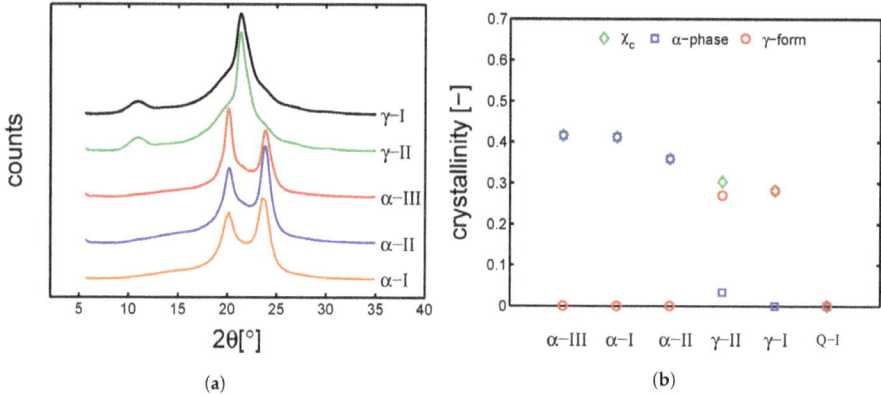

Figure 2. (**a**) wide angle X-ray diffraction integrated patterns; (**b**) deconvolution analysis of all the investigated samples with different thermal histories (dry state).

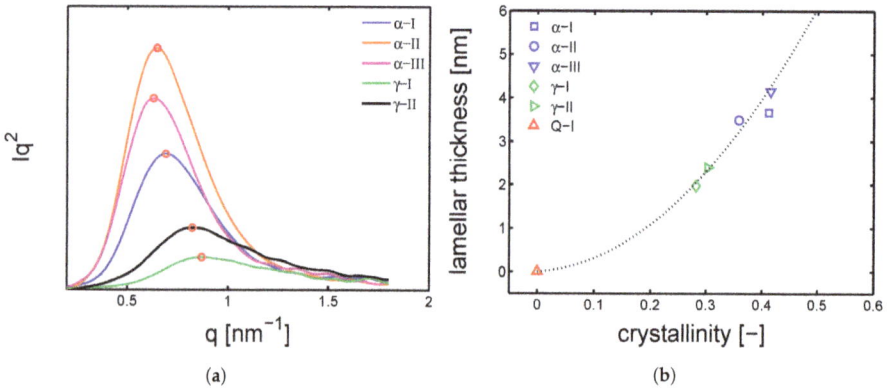

Figure 3. (**a**) small angle X-ray scattering integrated patterns in dry conditions; (**b**) lamellar thickness as a function of crystallinity in dry state. The dashed line is a guide to the eye.

4.2. Mechanical Properties

Next, the study can proceed with the investigation of the mechanical properties. This is initially done by tensile test at different temperatures, applying a range of strain rates (from 10^{-4} s^{-1} up to 3×10^{-2} s^{-1}) and the samples were kept in dry conditions. As a starting point for the mechanical properties' investigation, only three cases will be investigated: α-I for the polymorph α, γ-I for the polymorph γ and Q-I for the complete amorphous material.

As expected, the stress–strain response increases as the strain rate increases for both α-I and γ-I; while it decreases as temperature increases (see Figure 4a,b). In the case of α-I at 23 °C, the stress–strain response shows a very clear double yielding (see Figure 4a). This occurs because the amorphous and crystalline domains yield at different strains; indeed, the yield at low strain range (about 5–10%) is regarded as the contribution of amorphous domains, and the yield at higher strain range (about 15–35%) is considered to be the contributions of crystalline regions. This effect can be simply proven by observing the effect of temperature (for a fixed strain rate) on the stress–strain response: the first yield, well visible at room temperature, tend to disappear as temperature increases and clearly disappear when the temperature is above T_g (see Figure 4a green lines)—notwithstanding, also at temperatures higher than T_g, the double yielding would occur if very high strain rates were applied. The double yielding is less visible in the case of γ-I, which is likely due to a smaller contribution of the crystalline fraction, as demonstrated by the lower value of crystallinity estimated for γ-I ($\approx$30%) compared with the one of α-I ($\approx$40%). In order to study the yield kinetics, the yield stress is plotted as a function of the applied strain rate for different temperatures.

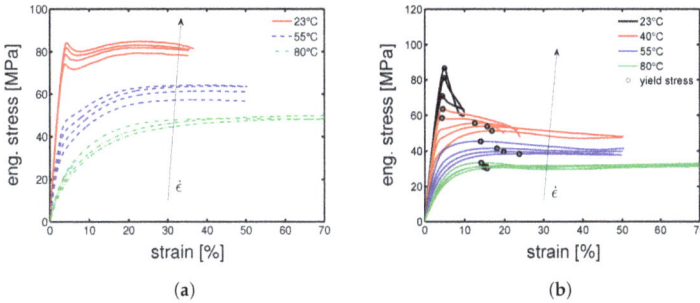

Figure 4. Stress-strain response at different temperatures and a range of strain rates for (**a**) α-I and (**b**) γ-I samples.

In Figure 5a,b, the yield kinetics are shown; in both α-I and γ-I cases, two different strain rate dependences are observed, as mentioned in Section 2. A steep slope is observed at low temperatures and (or) high strain rates, while a rather flat one is displayed at high temperatures and (or) low strain rates. In Figure 6, a schematic decomposition of the two processes is proposed.

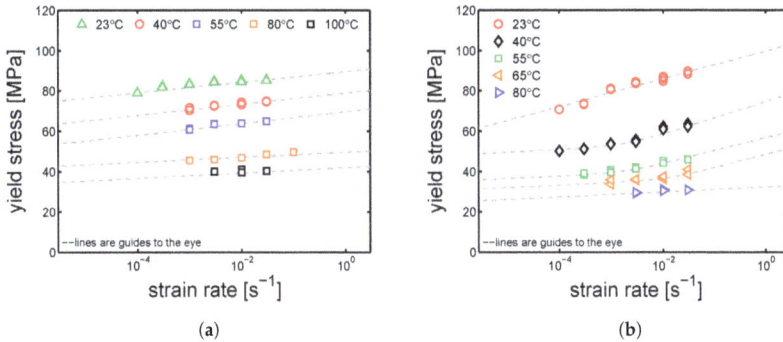

Figure 5. Yield kinetics (yield stress versus strain rate) at different temperatures for the (**a**) α-I and (**b**) γ-I samples. Lines are guides to the eye.

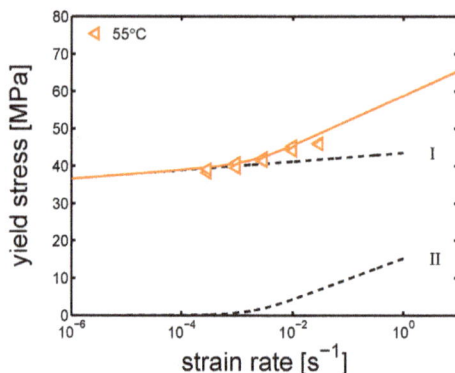

Figure 6. Example of two processes contributions. Yield stress versus strain rate at 55 °C; the black dashed lines are the two contributions separated (processes I and II).

In order to describe the yield kinetics obtained experimentally, Equation (1) is used and the parameters in Tables 2 and 3 were employed.

Table 2. Eyring parameters for α-I.

	V^* [m^3]	ΔU [J mol^{-1}]	$\dot{\varepsilon}_0$ [s^{-1}]
I	9e-27	1e6	1e108
II	6e-27	3.2e5	4e45

Table 3. Eyring parameters for γ-I.

	V^* [m^3]	ΔU [J mol^{-1}]	$\dot{\varepsilon}_0$ [s^{-1}]
I	9e-27	1e6	1e122
II	1.9e-27	3e5	2e45

In Figure 7a,b, it is shown that the model can describe well the results at different temperatures for both α-I and γ-I. Moreover, it is remarkable that the activation energy and activation volume employed for process I are the same for both α-I and γ-I. The parameters of process II do not match between the two different polymorphs; a plausible reason could be found introducing the concept of amorphous constraint. Looking at the mobility scenario in a semi-crystalline polymer, it is known that the crystalline regions have the lowest mobility, whereas the amorphous region should have the highest mobility. However, an elevated presence of crystalline regions may constrain the amorphous regions, with a consequent decrease of mobility. Thus, a difference in crystallinity may affect also the state (mobility) of the amorphous regions.

4.3. Influence of Temperature

To obtain a large overview about the effect of temperature on the yield stress, several tensile tests were performed at 10^{-2} s^{-1} and several temperatures were tested (see Figure 8). Observing the results in Figure 8a, it is possible to notice a clear double yielding behavior at temperature lower than T_g, while increasing the testing temperature, the first yield (at low strains) tends to disappear. The α-I samples show a predominantly crystalline contribution to yield, this is due to a high crystallinity index and a relatively high lamella thickness. Figure 8b shows a very different picture; at low temperature, yield takes place at low strains (≈5%) and it moves to about 15% strain when the temperature is increased above 50 °C. Comparing the strain at yield for both α-I and γ-I samples, we observed a very similar strain for the yield related to the amorphous domains (about 5%), whereas the contribution of the crystalline domains takes place at quite a different strain range, ≈15% for γ-I and about 30% for α-I.

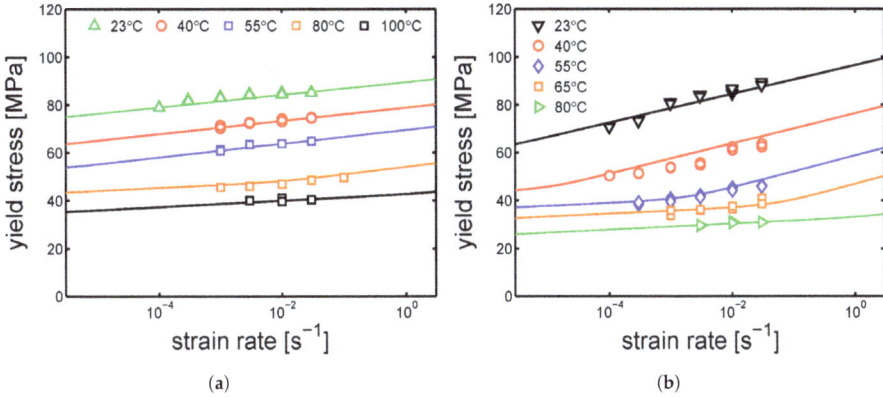

Figure 7. Stress–strain response at different temperatures (ranging from 23 °C to 100 °C) and a range of strain rate from 10^{-4} s^{-1} up to 3×10^{-2} s^{-1} for (**a**) α-I and (**b**) γ-I samples. Lines are the results of Equation (1).

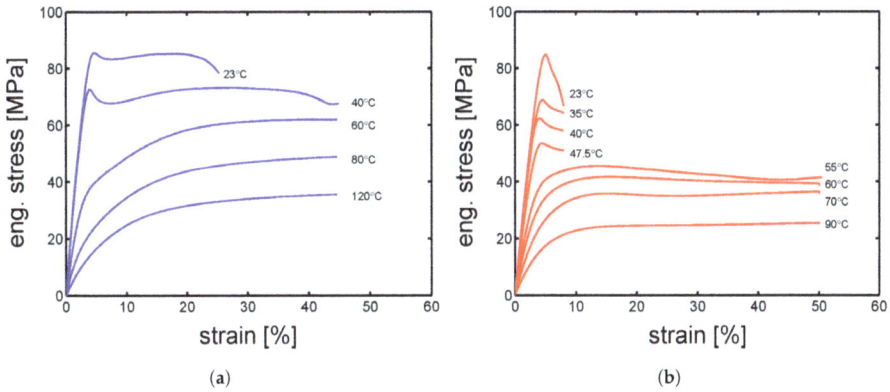

Figure 8. Stress-strain response at temperatures in a range from 23 to 120 °C, and strain rate of 10^{-2} s^{-1} for (**a**) α-I and (**b**) γ-I samples.

Figure 9 shows that α-I, γ-I and Q-I samples have three very different temperature dependencies. For the green markers, related to Q-I samples, no description (solid line) is provided. In fact, looking at the trend of yield stress as a function of temperature, three regions are found: (i) at low temperature, yield stress decreases drastically as temperature increases; (ii) at about 50 °C, the yield stress reaches a minimum; (iii) after which the material starts cold crystallizing and the yield stress increases and reaches a plateau up to 120°. The Q-I stress–strain behavior variation is due to an evolution of the material morphology, which is time- and temperature-dependent. For this reason, the Q-I samples will be taken out of this study. As far as the blue and red markers (α-I and γ-I, respectively) is concerned, they show very similar yield stress at 23 °C, but, as temperature increases, γ-I yield stress decrease more rapidly than α-I. In these two cases, the model is applied: in the case of α-I, the experimental results are matching the description; in the case of γ-I, the description matches the experimental results up to 90 °C. After this temperature, the experimental yield stress flattens. As already explained in [24], an evolution of crystallinity and/or lamellar thickness takes place at high temperatures; this phenomenon, called "annealing", is a time- temperature-activated process, it is governed by an enhancement of mobility that results in either cold crystallization and (or) lamellar thickening (perfectioning). Hence, higher crystallinity and (or) lamellar thickness results in an increase of yield stress, as shown in Figure 9. This effect is not observed in the case of α-I samples. It is hypothesized that "annealing" does not occur in α-I because of the already very high crystallinity and lamellar thickness obtained during processing.

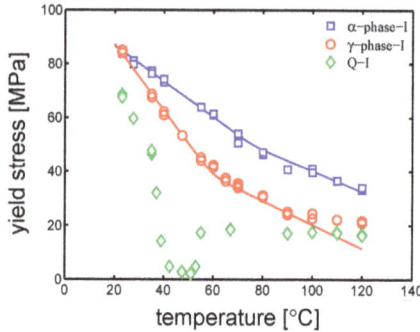

Figure 9. Yield stress versus testing temperature of samples tested with a strain rate of 10^{-2} s^{-1} and several temperatures. The lines are the results of Equation (1) in which the strain rate is a fixed value (10^{-2} s^{-1}) and the temperature is ranging from 23 to 120 °C.

The Ree–Eyring equation is applied with satisfactory results on the dry samples for both the two polymorphs, and two sets of parameters have been defined for the α-phase and γ-form, respectively; next, the influence of relative humidity and other thermal histories are investigated (see Table 1).

4.4. Influence of Humidity

As explained in Section 3.3, the samples were exposed to humid environments (relative humidity ranging from 35 to 75%) for a period long enough to allow the complete saturation. All the samples were conditioned at room temperature (23 °C). The absorbed water fraction was calculated by the following:

$$W\% = \left(\frac{W_i - W_0}{W_0}\right) \times 100, \tag{11}$$

where W_0 is the weight of the sample before conditioning and W_i is the weight at the time t_i.

The absorbed water fraction is plotted as a function of relative humidity (HR%) (see Figure 10a). The saturation level is different between the the samples because of a difference in crystallinity; as mentioned in Section 1, water can be absorbed only by the amorphous regions (because of their mobility state). In Figure 10b, the measured glass transition temperatures are plotted as functions of the relative humidity; a monotonic decrease of glass transition temperature is found by increasing the RH%. The results of the thermal-mechanical characterization, performed by DMTA, are shown in Figure 11a,b. The details about this technique are given in Section 3.6. Figure 11a shows the results of DMTA for the samples Q-I, α-I and γ-I after conditioning at different RH%; the markers are the defined T_gs. The Q-I is presented only in a dry condition because, upon conditioning, it crystallizes and therefore it changes its state drastically. Figure 12a shows the glass transition temperatures as functions of absorbed water fractions. At RH 0% (dry condition), all the investigated samples show a maximum in glass transition temperature; increasing the absorbed water fraction, a monotonic decrease of T_g is recorded. This is due to the plasticizing effect described in Section 1. Moreover, relying on the fact that only the amorphous region can absorb water, the glass transition temperature can be plotted as a function of normalized water fraction. The normalization is applied as follows:of

$$W_n\% = \frac{W\%}{1 - \chi_c} \times 100, \tag{12}$$

where χ_c is the crystallinity by WAXD and W% is the water fraction estimated experimentally (see Equation (11)). The results are shown in Figure 12b. As proposed, T_g follows a unique trend if plotted against a normalized water fraction. All of the T_g are reported in Table 4.

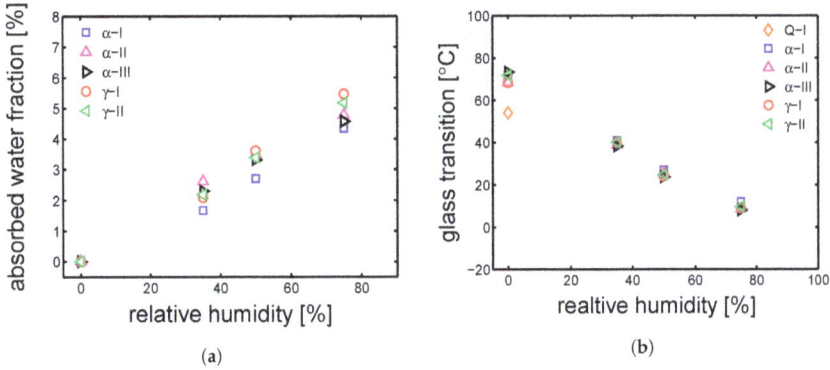

Figure 10. (**a**) absorbed water percentage at saturation in environments with different relative humidity at 23 °C. (**b**) glass transition temperature versus relative humidity.

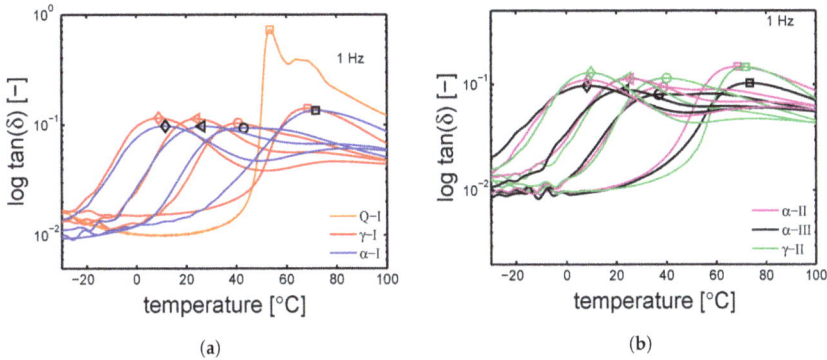

Figure 11. DMTA results, tan(δ) as a function of temperature for samples conditioned at different humidity; markers are the T_g at ($\square$) RH 0 %, ($\circ$) RH 35%, ($\triangleleft$) RH 50%, ($\diamond$) RH 75%. (**a**) samples α-I and γ-I; (**b**) α-II, α-III and γ-II.

Table 4. Glass transition temperature [°C].

Sample	Dry	35 RH%	50 RH%	75 RH%
α-I	72	43	27	12
α-II	68	39	26	9
α-III	73	38	24	8
γ-I	68	41	24	9
γ-II	72	40	25	10
Q-I	53	-	-	-

11

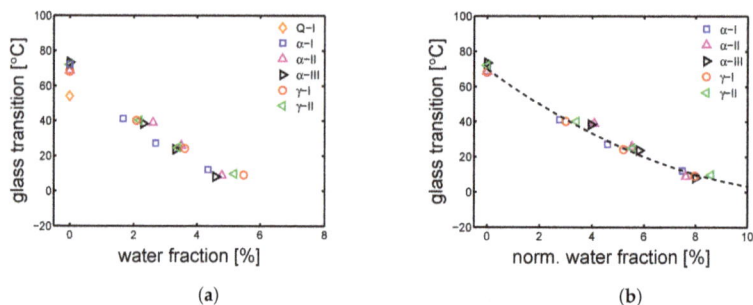

Figure 12. (**a**) glass transition temperatures (obtained by DMTA) as functions of the water fraction absorbed by the samples; (**b**) glass transition temperatures as functions of the normalized water fractions (line is a guide to the eye).

As already mentioned in Section 1, the glass transition can drop at temperatures even below room temperature. In this case, the polymer chains acquire the sufficient mobility needed to cold crystallize and (or) thickening the pre-existing crystals. The crystallinity can be plotted as a function of relative humidity for different samples as shown in Figure 13a.

Figure 13a shows that hydration has an actual effect on the crystallographic properties of PA6. In details, the crystallinity of Q-I samples rapidly increases for increasing relative humidity, γ-samples show a modest increase of crystallinity and γ-samples show a rather constant crystallinity. In order to investigate the process of "cold crystallization" taking place in the Q-I samples, the deconvolution analysis of the WAXD patterns was performed; results are shown in Figure 13b. The samples Q-I, starting from a completely amorphous material, are exposed to three different relative humidities, always at room temperature. The lowest RH% (35%) leads to crystallization of γ-form and α-phase with a balance slightly shifted towards the γ-form; at 50 RH%, the total χ_c increases and the balance γ-α is perfectly even; at the highest investigated relative humidity (75%), the crystallinity increases even further and the balance γ-α is slightly shifted towards the α-phase. In the case of γ-samples, the starting material is already semi-crystalline with a rather high crystalline index, thus secondary crystallization, phase transition and (or) lamellar perfectioning (thickening) are expected. In Figure 14a, the results of deconvolution analysis for γ-I are shown. The χ_c increases along the whole range of relative humidity, in particular a slight decrease of γ-form and a substantial increase of α-phase is detected. This effect can be related to a partial transformation γ to α, followed by a secondary crystallization of α-phase. In the case of α-I (see Figure 14b), the deconvolution reveals that no transformation takes place and only the crystallinity seems to increase a little (probably within an experimental error).

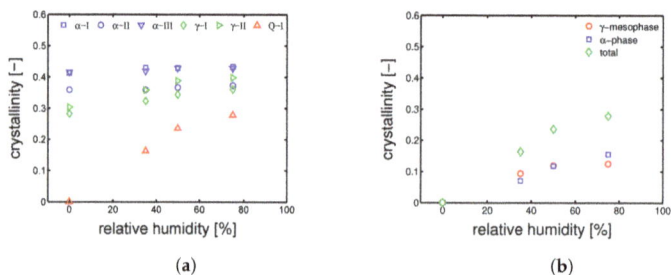

Figure 13. (**a**) overall crystallinity values as functions of relative humidity (conditioning performed at 23 °C); (**b**) evolution of the crystallographic phase contents as functions of relative humidity (conditioning performed at 23 °C) starting from Q-I (amorphous) sample.

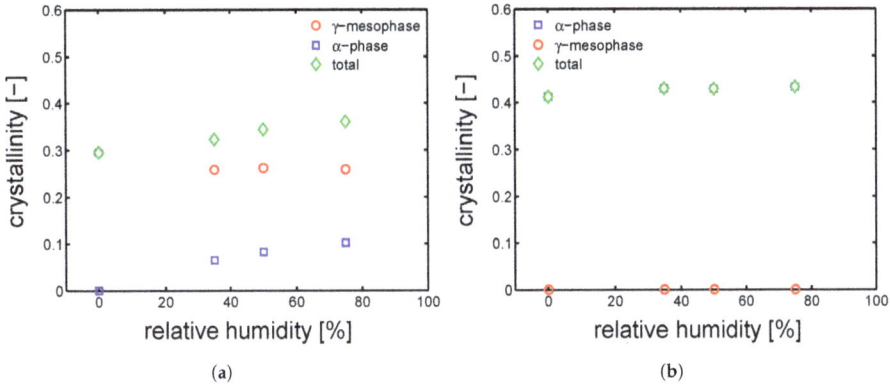

Figure 14. Evolution of the crystallographic phase contents as function of relative humidity (conditioning performed at 23 °C) starting from (**a**) γ-I and (**b**) α-I sample.

In order to capture the lamellar thickness evolution upon hydration, SAXS experiments were performed (as explained in Section 3.5). In Figure 15a, the results of lamellar thickness are proposed as a function of relative humidity. A similar trend to χ_c versus RH% are found. The lamellar thickness of Q-I samples increase quickly with relative humidity; in addition, the l_c of γ-samples increases although less rapidly than the amorphous samples, while the α-samples lamellar thickness are rather steady. These results can be plotted more intuitively as a function of the "apparent temperature". In Figure 15b, the l_c are plotted as a function of $\widetilde{T}$, in this way, the increase of l_c due to hydration can be easily regarded to a "cold crystallization" or "annealing" process, in which the samples are heated from the glassy state to a temperature above T_g.

Following this, the samples were tested by tensile test at different relative humidities and a range of strain rates (10^{-5} s^{-1} up to 3 × 10^{-2} s^{-1}). As it was done for yield kinetics in dry conditions, the aim is to also describe the results of test at different relative humidities by the Ree–Eyring equation (see Section 2, Equation (3)) employing the two set of parameters defined for the α-phase and γ-form (see Tables 2 and 3). Firstly, the model is applied to the samples α-I and γ-I (the results are shown in Figure 16a,b).

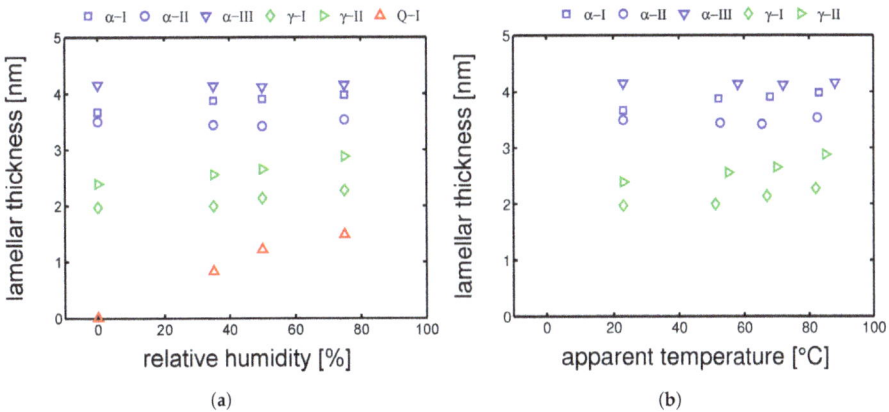

Figure 15. Lamellar thicknesses as functions of (**a**) relative humidity and (**b**) apparent temperature. All conditioning were performed at 23 °C.

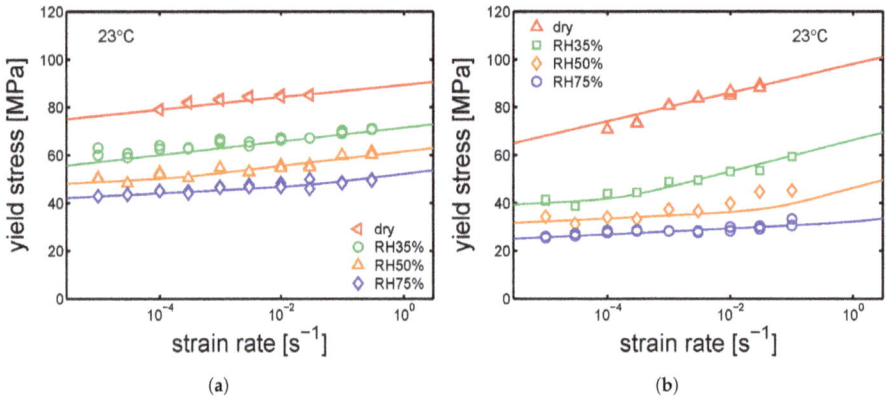

Figure 16. (**a**) yield stress as a function of strain rate for samples conditioned at different relative humidities, (**a**) α-I and (**b**) γ-I Testing temperature set at 23 °C.

It is shown that Equation (3) describes the experimental results obtained at different conditions for two polymorphs of PA6. As shown in Tables 2 and 3, the parameters employed for the α and γ polymorph differ mainly in the rate factors and the process II (inter-lamellar deformation). In Figure 17a,b, the yield kinetics of α-II, α-III and γ-II are shown, respectively. In order to describe these experimental results, the set of parameters used for α-I was employed for α-II and α-III, whereas the set used for γ-I was employed γ-II; in both cases, only the rate factors had to be changed. The parameters employed for these cases are listed in Tables 5–7.

Table 5. Eyring parameters for α-III.

	V^* [m^3]	ΔU [J mol^{-1}]	$\dot{\varepsilon}_0$ [s^{-1}]
I	9e-27	1e6	3e112
II	6e-27	3.2e5	3e48

Table 6. Eyring parameters for α-II.

	V^* [m^3]	ΔU [J mol^{-1}]	$\dot{\varepsilon}_0$ [s^{-1}]
I	9e-27	1e6	3e106
II	6e-27	3.2e5	8e47

Table 7. Eyring parameters for γ-II.

	V^* [m^3]	ΔU [J mol^{-1}]	$\dot{\varepsilon}_0$ [s^{-1}]
I	9e-27	1e6	1e123
II	1.9e-27	3e5	6e45

4.5. Time-to-Failure

Next, the influence of relative humidity on the PA6 lifetime was investigated. Several creep tests were performed at 23 °C and relative humidity ranging from 35% to 75%. All of the five sample series were tested at different applied loads. Subsequently, the $\dot{\varepsilon}_{pl}$ is estimated by the use of the Sherby–Dorn plot [33]; where the strain rate is plotted as a function of strain, and the $\dot{\varepsilon}_{pl}$ is defined as the minimum of the obtained curve. Finally, all the $\dot{\varepsilon}_{pl}$ are plotted as functions of the corresponding time-to-failure, as shown in Figure 18b. As mentioned in Section 2, the data plotted in a log–log graph show a slope of −1.

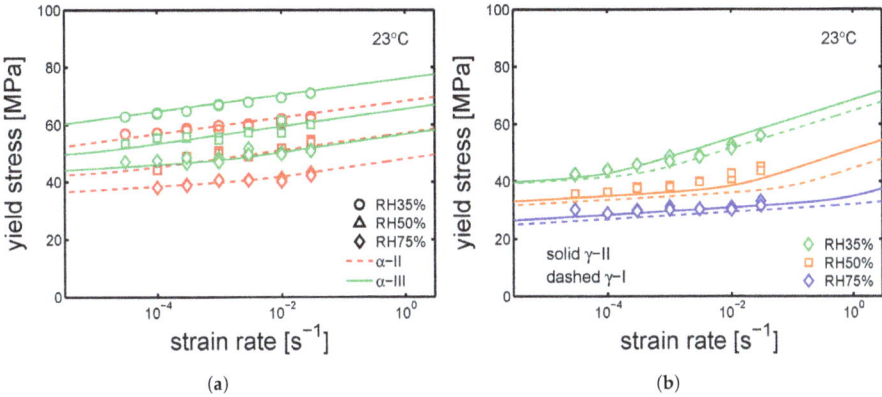

Figure 17. (**a**) yield stress as a function of strain rate for samples conditioned at different relative humidity, (**a**) α-II, α-III and (**b**) γ-II. Testing temperature set at 23 °C.

By extrapolating to $t_f = 1$ s, the results shown in Figure 18b, the ϵ_{cr} is estimated. Eventually, to describe the time-to-failure results obtained for the different samples series, the $\dot{\epsilon}_{pl}$ obtained by the Ree–Eyring equation (Equation (3)) modified to include the influence of relative humidity are combined with the ϵ_{cr} in Equation (5). In Figure 19a,b, the applied loads are plotted as functions of the time-to-failure. The lines are the results of Equation (5) and they are related to samples conditioned at different relative humidities (range from 35–75%) and 23 °C. The lines have the same stress dependence as the ones shown in Figure 16a,b but the slopes have opposite signs.

Figure 19a,b shows that a satisfactory prediction of time-to-failure is achieved for samples α-I and γ-I employing the set of parameters reported in Tables 2 and 3 for samples α-I and γ-I, respectively. By the use of the parameters listed in Tables 5–7, also samples α-II, α-III and γ-II are described by Equation (5); the results are shown in Figure 20a,b.

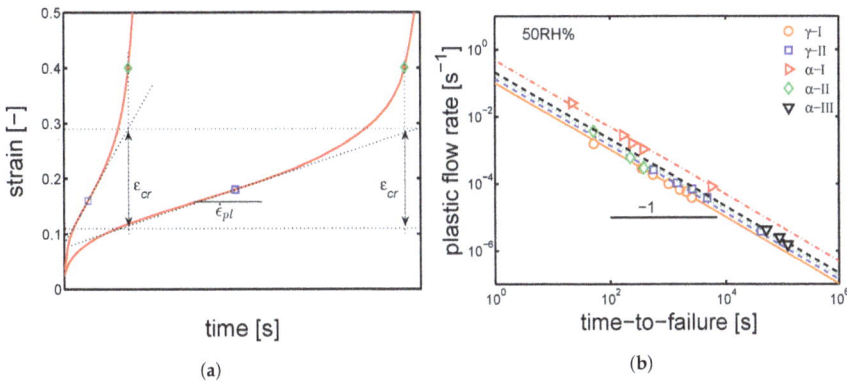

Figure 18. (**a**) examples of creep tests at constant applied load, the scheme shows the definition of $\dot{\epsilon}_{pl}$, ϵ_{cr} and t_f; (**b**) plastic flow rates as functions of time-to-failure for samples conditioned at RH 50% and 23 °C.

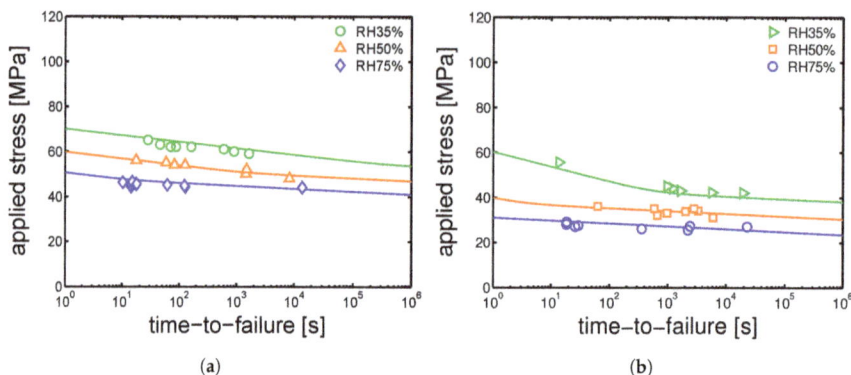

Figure 19. Creep tests for samples conditioned at RH 35%, RH 50%, RH 75% and 23 °C; applied loads as functions of time-to-failure for (**a**) α-I and (**b**) γ-I samples. Lines are the results of Equation (5).

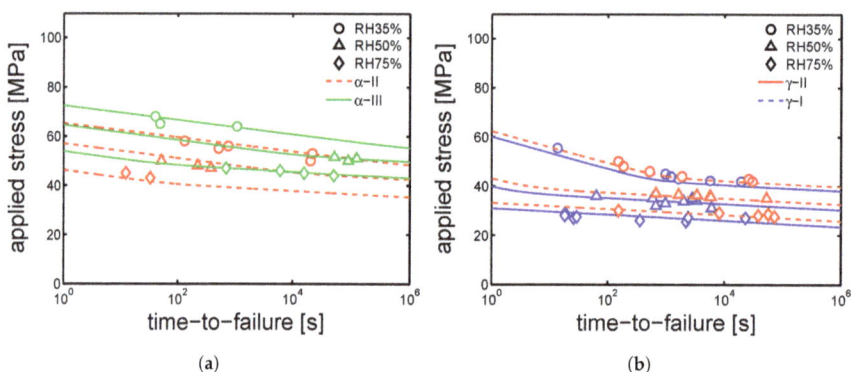

Figure 20. Creep tests for samples conditioned at RH 35%, RH 50% and RH75 % and 23 °C; applied loads as functions of time-to-failure for (**a**) α-II, α-III, (**b**) γ-I and γ-II samples. Lines are the results of Equation (5).

4.6. Structure–Properties Relations

For all the investigated PA6 samples, the temperature, relative humidity and stress dependent deformation kinetics were captured by the Ree–Eyring theory. To apply this theory, the characteristic parameters were defined, namely the activation volume (V^*), the activation energy (ΔU) and the rate factors ($\dot{\varepsilon}_0$). As explained in Section 4.2, PA6 shows two strain rate dependences that are related to two deformation mechanisms: an intra-lamellar deformation mechanism (also called process I) and an inter-lamellar mechanism (also called process II). As shown in Equation (1), each process needs one set of parameters.

The analysis has led to the conclusion that: (i) for process I, identical activation volume and activation energy can be used for all of the sample type, a part for the quenched samples whose structures, as explained in Section 4.3, are very dependent on temperature and relative humidity; (ii) for process II, V^* and ΔU are different for the two different polymorphs, i.e., α-phase and γ-mesophase; (iii) the rate factors were varied for each samples series, for both process I and process II. About the different V^* and ΔU determined for the two crystallographic phases in the case of process II, the author can only hypothesize that the reason might be in the different constriction level of the amorphous phase. In fact, as proposed in a previous study [24], process II is associated to the deformation of the amorphous phase. About the rate factors ($\dot{\epsilon}_{0,I}$ and $\dot{\epsilon}_{0,II}$), a rather clear correlation between the lamellar thickness (l_c) was found (see Figure 21a,b).

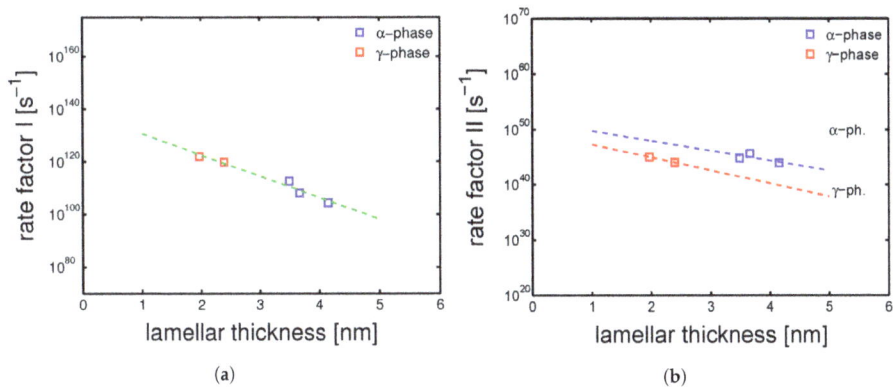

(a) (b)

Figure 21. (**a**) rate factor I and (**b**) rate factor II as functions of lamellar thickness. Lines are guides to the eye.

Figure 21a shows the relation between the rate factor I and the lamellar thickness; plotting the logarithm of $\dot{\epsilon}_{0,I}$ as functions of lamellar thickness for both α and γ samples, a linear trend is found. In the case of $\dot{\epsilon}_{0,II}$, two trends are found: one related to the α-samples and one for the γ-samples (see Figure 21b). Remember that also the V^* and the ΔU are different for process II, therefore there is no reason to expect that the trend of $\dot{\epsilon}_{0,II}$ for *alpha* matches the one for γ samples. This relation between the rate factors and the lamellar thickness was also proposed by other authors in the case of i-PP (isotactic polypropylene) [7,8]. The funding shown in Figure 21a,b are crucial for the prediction of the deformation kinetics and time-to-failure of PA6 processed with different histories. It is important to remark that, in the case of "real-life" applications, products are designed to work upon loads largely below the yield stress, hardly in dry conditions and often at high temperatures. Thus, the most frequent failure mode would be governed by intra-lamellar deformation, i.e., process I; consequently, the prediction would be governed by a V^* and ΔU, which are not dependent on the crystallographic phase.

5. Conclusions

In this work, the influence of thermal history on the structure of PA6 was investigated in regards to the mechanical properties (short and long term failure) of samples tested at different temperature and relative humidity. The investigation has led to several conclusions:

- the thermal history, i.e., the solidification procedure, has a crucial effect on the polymorphism and subsequently on the stress–strain response. By observing the three extreme cases (α-I, γ-I and Q-I), differences in stain rate- and temperature-dependence of the stress–strain response were found. The α-I samples has shown the least dependence on strain rate (see Figure 7a) and temperature (see Figure 9), while Q-I resulted in the most dependence on strain rate and temperature, and γ-I has shown a moderate dependence on strain rate and temperature.
- by quenching from the molten state, completely amorphous samples (Q-I) were obtained; these samples are extremely sensitive to temperature and relative humidity. After heating a Q-I sample above $\approx$50 °C, cold crystallization is obtained. Note that this phenomenon is time-temperature-dependent, therefore according to the temperature and time span of testing,

different stress–strain response is obtained. In addition, in case of exposition to moisture, cold crystallization occurs even at relative humidity of 35%.

- the exposition to moisture induced secondary crystallization, thickening of lamella and phase transformation (γ to α) for the γ-I and γ-II samples (see Figures 14a and 15a).

- the Ree–Eyring equation was successfully applied to describe the yield kinetics of samples α-I and γ-I. This was possible by employing two set of parameters, with identical V^* and ΔU for the process I and different V^* and ΔU for the process II.

- a modification of the Ree–Eyring equation to include the effect of relative humidity on the glass transition was applied successfully on all the investigated samples (a part for the quenched ones). Moreover, by the use of the critical strain concept, the time-to-failure of samples conditioned at different relative humidity were described.

- a correlation between the rate factors and the lamellar thickness was found. For a given activation volume and activation energy, the corresponding rate factor is dependent on the lamellar thickness (l_c).

- the effects of relative humidity on the crystallographic properties (i.e., lamellar thickness, crystallinity and crystal phase) are considered negligible at the scope of description of mechanical properties.

Author Contributions: E.P. performed the experiments, and wrote the manuscript. L.E.G. designed the modelling approach and assisted in the analysis. G.W.M.P. contributed to discussion and analysis.

Funding: This work is part of the Research Program of the Dutch Polymer Institute DPI, Eindhoven, The Netherlands, project number #786.

Conflicts of Interest: The authors declare no conflict of interest.

References

1. Govaert, L.; Engels, T.; Klompen, E.; Peters, G.; Meijer, H. Processing-induced properties in glassy polymers: Development of the yield stress in PC. *Int. Polym. Process.* **2005**, *20*, 170–177. [CrossRef]
2. Brucato, V.; Piccarolo, S.; Titomanlio, G. Crystallization kinetics in relation to polymer processing. *Makromol. Chem. Macromol. Symp.* **1993**, *68*, 245–255. [CrossRef]
3. Pesetskii, S.; Jurkowski, B.; Olkhov, Y.; Bogdanovich, S.; Koval, V. Influence of a cooling rate on a structure of PA6. *Eur. Polym. J.* **2005**, *41*, 1380–1390. [CrossRef]
4. La Carrubba, V.; Brucato, V.; Piccarolo, S. Influence of "controlled processing conditions" on the solidification of iPP, PET and PA6. *Macromol. Symp.* **2002**, *180*, 43–60. [CrossRef]
5. Popli, R.; Mandelkern, L. Influence of structural and morphological factors on the mechanical properties of the polyethylenes. *J. Polym. Sci. Part B Polym. Phys.* **1987**, *25*, 441–483. [CrossRef]
6. Elmajdoubi, M.; Vu-Khanh, T. Effect of cooling rate on fracture behavior of polypropylene. *Theor. Appl. Fract. Mech.* **2003**, *39*, 117–126. [CrossRef]
7. Van Erp, T.; Cavallo, D.; Peters, G.; Govaert, L. Rate-, temperature-, and structure-dependent yield kinetics of isotactic polypropylene. *J. Polym. Sci. Part B Polym. Phys.* **2012**, *50*, 1438–1451. [CrossRef]
8. Caelers, H.; Govaert, L.; Peters, G. The prediction of mechanical performance of isotactic polypropylene on the basis of processing conditions. *Polymer* **2016**, *83*, 116–128. [CrossRef]
9. Cavallo, D.; Gardella, L.; Alfonso, G.; Portale, G.; Balzano, L.; Androsch, R. Effect of cooling rate on the crystal/mesophase polymorphism of polyamide 6. *Colloid Polym. Sci.* **2011**, *289*, 1073–1079. [CrossRef]
10. Puffr, R.; Šebenda, J. On the structure and properties of polyamides. XXVII. The mechanism of water sorption in polyamides. *J. Polym. Sci. Part C Polym. Symp.* **2007**, *16*, 79–93. [CrossRef]
11. Murthy, N.; Stamm, M.; Sibilia, J.; Krimm, S. Structural changes accompanying hydration in nylon 6. *Macromolecules* **1989**, *22*, 1261–1267. [CrossRef]
12. Holmes, D.; Bunn, C.; Smith, D. The crystal structure of polycaproamide: Nylon 6. *J. Polym. Sci.* **1955**, *17*, 159–177. [CrossRef]
13. Ellis, T. Moisture-induced plasticization of amorphous polyamides and their blends. *J. Appl. Polym. Sci.* **1988**, *36*, 451–466. [CrossRef]

14. Boukal, I. Effect of water on the mechanism of deformation of nylon 6. *J. Appl. Polym. Sci.* **1967**, *11*, 1483–1494. [CrossRef]

15. Jia, N.; Fraenkel, H.; Kagan, V. Effects of moisture conditioning methods on mechanical properties of Injection molded Nylon 6. *J. Reinf. Plast. Compos.* **2004**, *23*, 729–737. [CrossRef]

16. Miri, V.; Persyn, O.; Lefebvre, J.M.; Seguela, R. Effect of water absorption on the plastic deformation behavior of nylon 6. *Eur. Polym. J.* **2009**, *45*, 757–762. [CrossRef]

17. Reimschuessel, H. Relationships on the effect of water on glass transition temperature and Young's modulus of nylon 6. *J. Polym. Sci. Polym. Chem. Ed.* **1978**, *16*, 1229–1236. [CrossRef]

18. Murthy, N.; Wang, Z.G.; Akkapeddi, M.; Hsiao, B. Isothermal crystallization kinetics of nylon 6, blends and copolymers using simultaneous small and wide-angle X-ray measurements. *Polymer* **2002**, *43*, 4905–4913. [CrossRef]

19. Boyce, M.; Parks, D.; Argon, A. Large inelastic deformation of glassy polymers. Part I: Rate dependent constitutive model. *Mech. Mater.* **1988**, *7*, 15–33. [CrossRef]

20. Kawai, H.; Hashimoto, T.; Suehiro, S.; Fujita, K.I. Dynamic X-ray diffraction studies of spherulitic poly-alpha-olefins in relation to the assignments of alpha and beta mechanical dispersions. *Polym. Eng. Sci.* **1984**, *24*, 361–372. [CrossRef]

21. Hiss, R.; Hobeika, S.; Lynn, C.; Strobl, G. Network stretching, slip processes, and fragmentation of crystallites during uniaxial drawing of polyethylene and related copolymers. A comparative study. *Macromolecules* **1999**, *32*, 4390–4403. [CrossRef]

22. Eyring, H. Viscosity, plasticity, and diffusion as examples of absolute reaction rates. *J. Chem. Phys.* **1936**, *4*, 283–291. [CrossRef]

23. Ree, T.; Eyring, H. Theory of Non-Newtonian Flow. I. Solid Plastic System. *J. Appl. Phys.* **1955**, *26*, 793–800. [CrossRef]

24. Parodi, E.; Peters, G.W.M.; Govaert, L.E. Prediction of plasticity-controlled failure in polyamide 6: Influence of temperature and relative humidity. *J. Appl. Polym. Sci.* **2018**, *135*, 45942. [CrossRef]

25. Kanters, M.; Remerie, K.; Govaert, L. A new protocol for accelerated screening of long-term plasticity-controlled failure of polyethylene pipe grades. *Polym. Eng. Sci.* **2016**, *56*, 676–688. [CrossRef]

26. Bauwens-Crowet, C.; Ots, J.M.; Bauwens, J.C. The strain-rate and temperature dependence of yield of polycarbonate in tension, tensile creep and impact tests. *J. Mater. Sci.* **1974**, *9*, 1197–1201. [CrossRef]

27. Visser, H.; Bor, T.; Wolters, M.; Engels, T.; Govaert, L. Lifetime assessment of load-bearing polymer glasses: An analytical framework for ductile failure. *Macromol. Mater. Eng.* **2010**, *295*, 637–651. [CrossRef]

28. Van Erp, T.; Reynolds, C.; Peijs, T.; Van Dommelen, J.; Govaert, L. Prediction of yield and long-term failure of oriented polypropylene: Kinetics and anisotropy. *J. Polym. Sci. Part B Polym. Phys.* **2009**, *47*, 2026–2035. [CrossRef]

29. Baltá-Calleja, F.; Vonk, C. *X-ray Scattering of Synthetic Polymers*; Number v. 8 in Polymer Science Library; Elsevier: Amsterdam, The Netherlands, 1989.

30. Ryan, A.; Bras, W.; Mant, G.; Derbyshire, G. A direct method to determine the degree of crystallinity and lamellar thickness of polymers: Application to polyethylene. *Polymer* **1994**, *35*, 4537–4544. [CrossRef]

31. Fornes, T.; Paul, D. Crystallization behavior of nylon 6 nanocomposites. *Polymer* **2003**, *44*, 3945–3961. [CrossRef]

32. Mileva, D.; Androsch, R.; Zhuravlev, E.; Schick, C. Morphology of mesophase and crystals of polyamide 6 prepared in a fast scanning chip calorimeter. *Polymer* **2012**, *53*, 3994–4001. [CrossRef]

33. Sherby, O.; Dorn, J. Anelastic creep of polymethyl methacrylate. *J. Mech. Phys. Solids* **1958**, *6*, 145–162.. [CrossRef]

polymers

MDPI

Article

Structure-Properties Relations for Polyamide 6, Part 2: Influence of Processing Conditions during Injection Moulding on Deformation and Failure Kinetics

Emanuele Parodi [1,2], **Gerrit W. M. Peters** [1] **and Leon E. Govaert** [1,*]

[1] Department of Mechanical Engineering, Materials Technology Institute, Eindhoven University of Technology, P.O. Box 513, 5600 MB Eindhoven, The Netherlands; Emanuele.parodi@maxxistce.nl (E.P.); g.w.m.peters@tue.nl (G.W.M.P.)

[2] Dutch Polymer Institute (DPI), P.O. Box 902, 5600 AX Eindhoven, The Netherlands

* Correspondence: l.e.govaert@tue.nl; Tel.: +31-402472838

Received: 30 May 2018; Accepted: 11 July 201; Published: 16 July 2018

Abstract: The effect of processing conditions during injection on the structure formation and mechanical properties of injection molded polyamide 6 samples was investigated in detail. A large effect of the mold temperature on the crystallographic properties was observed. Also the the effect of pressure and shear flow was taken in to consideration and analysed. The yield and failure kinetics, including time-to-failure, were studied by performing tensile and creep tests at several test temperatures and relative humidities. As far as mechanical properties are concerned, a strong influence of temperature and relative humidity on the yield stress and time-to-failure was found. A semi-empirical model, able to describe yield and failure kinetics, was applied to the experimental results and related to the crystalline phase present in the sample. In agreement with findings in the literature it is observed that for high mold temperatures the sample morphology is more stable with respect to humidity and temperature than in case of low mold temperatures and this effects could be successfully captured by the model. The samples molded at low temperatures showed, during mechanical testing, a strong evolution of the crystallographic properties when exposed to high testing temperature and high relative humidity, i.e., an increase of crystallinity or a crystal phase transition. This makes a full description of the mechanical behavior rather complicated.

Keywords: polyamide 6; injection molding; polymorphism; humidity; mechanical properties

1. Introduction

Injection molding is the most widely used technique to produce polymeric products. It is particularly preferred because of advantages such as fast production cycles, cheapness and the large versatility of product shapes. However, injection molding implies some challenges, interesting for users, engineers and polymer scientists. In fact, this technique involves high pressure, elevated shear flow and inhomogeneous transient temperature fields during solidification. The effect of the mold temperature on the mechanical properties is a widely studied topic in polymer engineering. In the case of amorphous polymers, the main influence of mold temperature on the yield stress is attributed to aging; a higher age (i.e., lifetime of a part) of the material correspond to higher yield stress [1]. In the case of semi-crystalline polymers, the topic becomes more complicated. It is well known that injection molded samples do not have a homogeneous morphology along the sample thickness and position [2–4]. Ideally, due to different conditions during solidification, three different morphologies can be detected in the sample thickness: (i) the outermost layer (also called skin layer) is subjected to fast cooling, which leads to amorphous or slightly crystalline material; (ii) the central layer (also called core layer) solidifies under high pressure and slow cooling rate, generally leading to high crystallinity; (iii) the inter-phase between the skin and core layer is called shear layer, which is subjected to high shear rates that can play an important role in the crystallization kinetics [5]. Moreover, because of the shear flow, a highly oriented morphology can be formed in this shear layer. The molecular orientation combined with the different position-dependent morphologies creates a strong mechanical anisotropy [4]. The morphology of injection molded nylon 6 was reported in literature [2,6]. These studies led to a common main conclusion: the metastable γ-mesophase (obtainable for moderate cooling rates [7]) is predominant near the surface of the sample, while the most stable α-phase (obtained by slow cooling [7]) takes over towards the center of the sample. The effect of this inhomogeneous morphology distribution on mechanical properties has barely been studied despite this being the most important factor for the end-users. Therefore, in this study, the effect of mold temperature on the mechanical properties of polyamide 6 is investigated. The samples for subsequent mechanical testing were conditioned at different temperatures and different relative humidities.

2. State of the Art

In previous studies, the authors have investigated the influence of structural properties (crystalline phase and lamellar thickness), temperature and relative humidity on the yield kinetics and time-to-failure of polyamide 6 processed under quiescent conditions. Firstly, the Ree-Eyring equation, normally used to describe the yield kinetics as a function of temperature and strain rate, was modified in order to include the effect of relative humidity [8]. To accomplish this, the "apparent" temperature equations was introduced:

$$\widetilde{T} = T + (T_{g,dry} - T_{g,wet}) \tag{1}$$

where T is the actual temperature, $T_{g,dry}$ is the glass transition temperature at the dry state and $T_{g,wet}$ is the T_g after conditioning. Nex, this "apparent" temperature is substituted in the Ree-Eyring equation:

$$\sigma_y(\dot{\epsilon}, \widetilde{T}) = \frac{k\widetilde{T}}{V_I^*} \sinh^{-1}\left(\frac{\dot{\epsilon}}{\dot{\epsilon}_{0,I}} \exp\left(\frac{\Delta U_I}{R\widetilde{T}}\right)\right) + \frac{k\widetilde{T}}{V_{II}^*} \sinh^{-1}\left(\frac{\dot{\epsilon}}{\dot{\epsilon}_{0,II}} \exp\left(\frac{\Delta U_{II}}{R\widetilde{T}}\right)\right) \tag{2}$$

where $\dot{\epsilon}_{0,I,II}$, $\Delta U_{I,II}$ and $V_{I,II}^*$ are the rate factor, the activation energy and the activation volume related to process I and process II, respectively. These processes are related to two different deformation mechanisms; process I to intra-lamellar deformation and process II to inter-lamellar deformation. Subsequently, this work focused on applying the model also in case of different crystal phases with varying lamellar thicknesses (l_c). For the two different polymorphs, two sets of parameters were defined, one for α-phase and one for γ-mesophase. Remarkably, the first process of both the polymorphs could be described with identical activation energy (ΔU_I) and activation volume (V_I^*) [9]. Also l_c, a relation between the lamellar thickness and the rate factors was found [9]. Thus, by selecting the right set of parameters accordingly to the present crystal phase, the yield kinetics of different samples with different l_c can be described. The parameters defined for α-phase and γ-mesophase are listed in Tables 1 and 2, respectively.

Moreover, by the use of the "critical strain" concept, the prediction made for the yield kinetics can be reconverted to predict for the time-to-failure [10]. The time-to-failure (t_f) is estimated by the equation:

$$t_f(\sigma, T) = \frac{\epsilon_{cr}}{\dot{\epsilon}_{pl}(\sigma, T)} \tag{3}$$

in which ϵ_{cr} is the critical strain and ϵ_{pl} is the plastic flow rate as function of load and temperature obtained by the yield kinetics, see Equation (2).

Table 1. Ree-Eyring parameters defined for α-phase.

	$V^* \, [\mathrm{m}^3]$	$\Delta U \, [\mathrm{J \, mol}^{-1}]$
I	9×10^{-27}	1×10^6
II	6×10^{-27}	3.2×10^5

Table 2. Ree-Eyring parameters defined for γ-mesophase.

	$V^* \, [\mathrm{m}^3]$	$\Delta U \, [\mathrm{J \, mol}^{-1}]$
I	9×10^{-27}	1×10^6
II	1.9×10^{-27}	3×10^5

3. Experimental

3.1. Material

The material employed in this work was a polyamide 6 (Akulon K122) kindly provided by DSM (Geleen, The Netherlands). This PA6 has a viscosity-average molar mass (Mv) of about 24.9 kg/mol, melting point 220 °C, density of the amorphous phase 1090 kg/m^3 and a viscosity number of 0.312 m^3/kg (ISO 307) [11].

3.2. Sample Preparation

To dry the pellets prior to processing, the material was placed in a vacuum oven at a temperature of 110 °C for 12 h. Next, the injection molding procedure was performed with the following parameters: temperature profile from the hopper to the nozzle, 70 °C, 230 °C, 240 °C, 250 °C, 245 °C and 240 °C; injection flow, 90 cm^3/s; maximum injection pressure, 500 bar; holding pressure, 500 bar; cooling time 30 s. Moreover, four mold temperature were used, such as 35, 85, 130 and 160 °C. The samples are 1 mm thick squared plates with side lengths of 70 mm. The injection gate was situated orthogonally to the plate plane, see Figure 1. Dog-bone shape samples were cut by the mean of a cutting die, according to the ISO527 type 1BA (main dimensions: width 5 mm, length 22 mm). The samples were cut in parallel and perpendicular direction compared to the flow, see Figure 1.

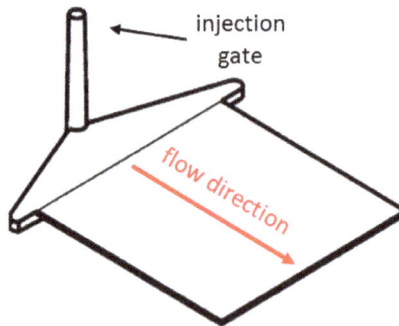

Figure 1. Schematic representation of a sample obtained by injection molding (70 × 70 × 1 mm).

As supporting experiments, also sheets with a thickness of 0.5 mm were prepared by compression molding (for a more extensive study on the effects of injection molding see [9]). The material was melted at 265 °C for 5 min, while a force of about 10 kN was applied, then it was rapidly moved to a cold press set at different temperatures, i.e., 80–120–140–160–180 °C where the material was solidified for 3 min.

3.3. Sample Conditioning

The samples were conditioned at room temperature (23 °C) at four different relative humidities: RH 0% (dry) using a vacuum chamber at room temperature, RH 35% and RH 75% by using chambers containing supersaturated solutions of sodium chloride and magnesium chloride hexahydrate, and RH 50% by using an environmental chamber. The samples were kept in the conditioning environment up to saturation. The absorbed water fraction was calculated with the following equation:

$$H_2O\% = (\frac{W_i - W_0}{W_0}) \times 100 \tag{4}$$

where W_0 is the weight of the sample in dry conditions and W_i is the weight at the time t_i. The saturation was identified as the level of water fraction after which a plateau is reached. Since water is mainly absorbed in the amorphous phase we define the normalized water fraction in which the water fraction is related to the the fraction of amorphous phase:

$$H_2O\%_N = \frac{H_2O\%}{1 - \chi_{vol}} \tag{5}$$

where χ_{vol} is the crystallinity.

3.4. Mechanical Tests

To investigate the yield and failure kinetics, uniaxial tensile and creep tests were performed using a Zwick Uniaxial Testing Machine (Ulm, Germany) equipped with a 1 kN load-cell and an environmental chamber with which temperature and relative humidity were controlled. Relative humidity ranged from RH 35% to RH 75%, strain rates from 10^{-5} s^{-1} up to 3×10^{-2} s^{-1} and temperatures from 23 to 120 °C. Each experiment was performed at least two times. As far as the the tensile tests is concerned, a pre-load of 0.1 MPa was applied prior the test with a speed of 1 mm/min. Subsequently the test was performed with constant strain rate up to a strain of ≈50%. The creep tests were performed at room temperature (23 °C) and three relative humidities (RH 35%, RH 50% and RH 75%). The load was applied within 10 s, then it was kept constant up to failure. Because of analytical issues, the time-to-failure was defined as the time to reach a strain of 35%.

3.5. X-ray Diffraction

To investigate the influence of different mold temperatures on structure, X-ray diffraction experiments were performed. First, wide angle X-ray diffraction (WAXD) measurements were done using a Ganesha X-ray instrument (Copenhagen, Denmark) equipped with a GeniX-Cu ultra low divergence source (l = 1.54 Å) and a Pilatus 300 K silicon pixel detector (487 × 619 pixels of 172 × 172 μm²). The patterns obtained were radially integrated and the weight percentage crystallinity was estimated by:

$$\chi_c = \frac{T - A}{T} \tag{6}$$

where T is the total scattered intensity and A is the scattering from the amorphous halo. The amorphous halo was retrieved performing WAXD on a completely amorphous sample obtained by quenching the material in water with ice and NaCl. However, especially in the case of injection molding, the material can crystallize in a mixture of α and γ. Thus the total crystallinity calculated by Equation (6) may actually consist of a fraction related to the α-phase and another fraction related to γ-mesophase. Therefore, a deconvolution analysis is performed. It consist of an analytical fitting of Lorentzian's functions to the characteristics peaks. Next, all the function areas are summed up to the amorphous halo area, and the resulting pattern is compared to the experimental result. The α-phase and γ-mesophase fractions follow from:

$$\chi_{c,\alpha} = \frac{A_\alpha}{A_m} \quad and \quad \chi_{c,\gamma} = \frac{A_\gamma}{A_m} \tag{7}$$

where A_α and A_γ are the total area of the Lorentzian functions for the α and γ peaks, and A_m is the total area of the measured pattern. Also, small angle X-ray scattering (SAXS) was performed on the same setup described for WAXD, only the samples to detector distance was increased. With SAXS, the difference of electronic density are detected, i.e., the distance covered by a lamella and an amorphous layer. This is called the long period (L_b). After applying Lorentz [12] and thermal density fluctuation [13] corrections, the long period is calculated with:

$$L_b = \frac{2\pi}{d*} \tag{8}$$

where $d*$ is the peak position expressed in inversed nanometers (nm^{-1}). Next, the lamellar thickness (l_c) is estimated by:

$$l_c = \chi_{vol} \cdot l_b \tag{9}$$

where χ_{vol} is the volumetric crystallinity percentage, which takes into account the different density of amorphous phase and the two polymorphs (α and γ):

$$\chi_{vol} = \frac{\dfrac{\chi_c}{\rho_c}}{\dfrac{\chi_c}{\rho_c} + \dfrac{100 - \chi}{\rho_a}} \tag{10}$$

where ρ_c is the density of the crystal (1.21 g/cm^3 for α-phase [11], 1.16 g/cm^3 for γ-mesophase [11]), ρ_a the density of the amorphous (1.09 g/cm^3 [11]) and χ the mass crystallinity. To investigate the influence of shear flow during crystallization, an azimuthal integration (180°) of the SAXS patterns was performed.

3.6. Dynamical Mechanical Thermal Analysis

To investigate the influence of processing and relative humidity on the glass transition temperature, a TA instruments Q800 (Asse, Belgium) was used to perform dynamical mechanical thermal analysis (DMTA). Flat rectangular samples (main sizes 0.5 × 5 mm) were tested performing a ramp of temperature from −40 °C to 120 °C with a heating speed of 3 °C/min and a frequency of 1 Hz. The glass transition temperature was defined ad the peak of the tan(δ) curve.

3.7. Dilatometry-PVT

A dilatometer (PVT) able to measure the specific volume of polymers as a function of cooling rate, shear flow, pressure and temperature, was employed [14]. It consists of a pressure cell which combines a traditional "piston-die" type dilatometer with a Couette geometry rheometer. The experiments were performed on ring-shaped samples produced by a mini injection molding machine (Babyplast, Molteno, Italy), with main dimensions: 22 mm outer diameter, 21 mm inner diameter and height of of 2.5 mm. To completely erase the thermal history, the sample was heated at 250° and kept at this temperature for 10 min. Then, the cooling procedure was performed for isobaric conditions. Two kind of cooling procedures were applied, ambient cooling ($\approx$0.1 °C/s) and air cooling ($\approx$1 °C/s). The pressure was varied in a range from 100 bar to 800 bar and the shear flow in range from 0 s^{-1} to 180 s^{-1}. The influence of shear flow was studied at 200 bar and the shear impulse was applied at 200 °C.

4. Results and Discussion

4.1. Samples Characterization

The first step of this study was a crystallographic characterization performed by WAXD and SAXS experiments (details about these techniques are reported in Section 3.5). WAXD experiments were performed on dry samples to understand the influence of mold temperature. In Figure 2, the integrated patterns are reported. It is observed that the mold temperature plays a crucial role on the crystallization: (i) at 35 °C an almost completely amorphous sample is obtained, (ii) at 85 °C a predominantly γ-mesophase sample is obtained, (iii) at 130 °C a mixture of α-phase and γ-mesophase are found and (iiii) at 160 °C a fully α-phase sample is obtained. In Figure 3b, the result of the deconvolution analysis is shown.

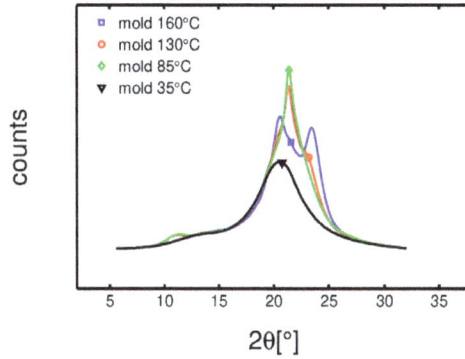

Figure 2. WAXD integrated patterns of samples molded at different temperatures; experiments performed at room temperature (23 °C) and dry condition.

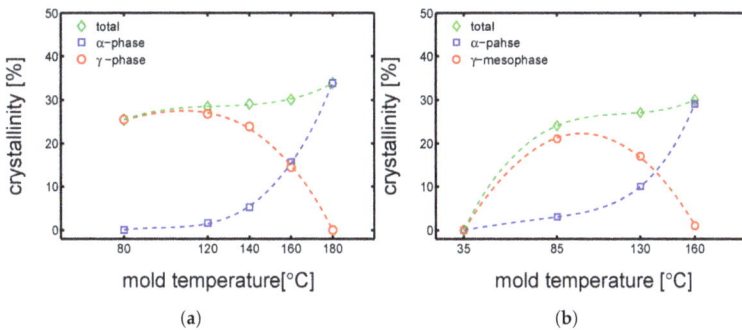

Figure 3. Deconvolution analysis of WAXD patterns. Crystallinity, α and γ fractions as functions of mold temperature. (**a**) Case: compression molding, (**b**) case: injection molding.

It is important to stress that, the results of X-ray analysis are all an average over the thickness. Normally, the cross section of an injection molded part shows a not homogeneous morphology. During processing, the material is exposed to very different conditions dependent on the history experienced in the mold. The material close to mold surface solidifies upon high cooling rate (generally called skin layer), the middle part solidifies relatively slowly under high pressure (called core layer), while the layer between the core and skin layer, solidifies under high shear (it is called shear layer). However, because of experimental limitations, it was chosen to present the results as an average over the thickness. In Figure 3a, the deconvolution analysis of samples compression molded at different temperatures are presented. Comparing the results shown in Figure 3a,b it possible to notice that in the case of injection molding, already a substantial fraction of α-phase is formed at a mold temperature of 130 °C, while in the case of compression molding mold temperature 130 °C leads to a sample with a rather small fraction of α-phase and a predominance of γ-mesophase. In order to study this effect, we have performed supporting PVT experiments.

4.1.1. Supporting Experiments-PVT

A dilatometer (PVT) was employed in order to perform cooling procedures that simulated processing conditions, i.e., crystallization during cooling at high pressures and subjected to shear flows. In Figure 4 an illustrative example of a PVT result is presented. The sample was cooled at ≈1 °C/s, with a constant pressure of 200 bar and no shear. The marker (circle) at about 172 °C (see Figure 4) is defined as the crystallization onset, which is the main outcome of this experiment. As far as the influence of pressure on the crystallization onset temperature is concerned, cooling at two different speeds (0.1 and 1 °C/s) were performed for several pressures (from 100 to 800 bar). The crystallization onset for these conditions is plotted as a function of applied pressures, see Figure 5a. The results show an increase of crystallization onset of about 17 °C from the minimum to the maximum applied pressure in both the investigated cooling rates. In the case of crystallization with applied shear flow and constant pressure (100 bar), the crystallization onset increases only few degrees. Therefore, it is concluded that the different crystallization behavior observed for the injection molding processing, compared to the one seen for compression molding, is mainly due to the effect of pressure which increases the crystallization temperature for a given cooling rate.

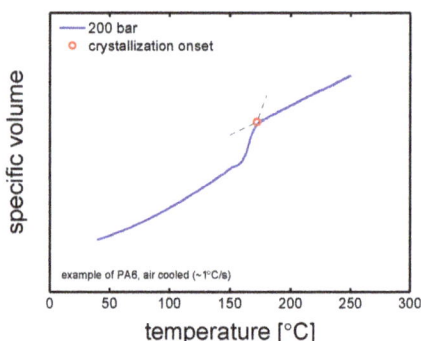

Figure 4. Dilatometry-PVT experiment, cooling at ≈1 °C/s and 200 bar with no shear flow. The circle indicates the crystallization onset temperature.

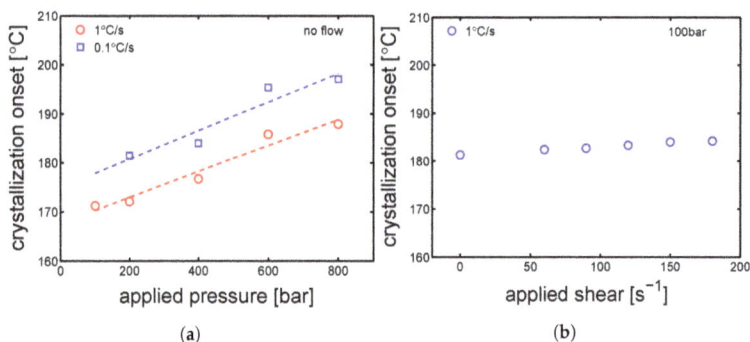

Figure 5. (**a**). Crystallization onset temperature obtained by cooling upon different pressures and cooling rates. Lines are just guides to the eye. (**b**) Crystallization onset temperature obtained by cooling upon different shear flow rates and 100 bar.

4.1.2. Effect of Flow on the Molecular Orientation

Next, SAXS experiments were performed on the dry samples processed at different mold temperatures. In Figure 6a, the results of radial integration of the SAXS patterns, which give the required information for to estimating the average long period (l_b) and, as explained in Section 3.5, the average lamellar thickness (l_c). Figure 6b shows the lamellar thickness values as functions of mold temperature; the highest l_c ($\approx$2.3 nm) is obtained with mold temperature 160 °C and the minimum ($\approx$1.6 nm) at mold temperature 85 °C.

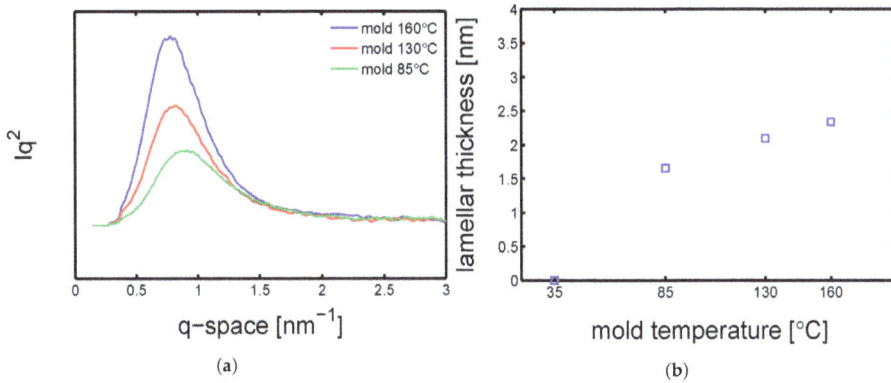

Figure 6. (**a**) SAXS integrated patterns of samples molded at different temperatures; experiments performed at room temperature (23 °C) and dry condition. (**b**) Lamellar thickness as a function of mold temperature; experiments performed at room temperature (23 °C) and dry condition.

As mentioned in the introduction, flow may lead to material orientation and this could affect the mechanical properties. To investigate the whether or not orientation was present, an azimuthal integration of the SAXS pattern was performed. In Figure 7 three examples of azimuthal integration of samples produced at different mold temperatures and conditioned at RH 0% (dry) at room temperature, are shown. A varying orientation is observed for all the three samples; in particular the mold temperature 85 °C shows two clear maximums at approximately 90° and 270°.

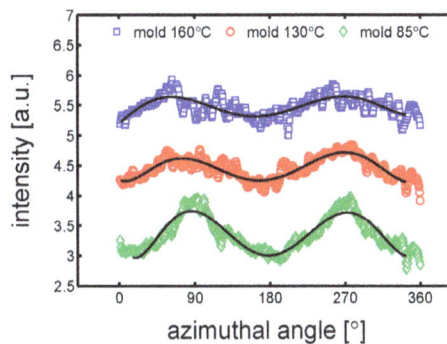

Figure 7. Azimuthal integration over a range from 0° to 180° of samples at dry conditions and room temperature. The solid lines are guide to the eye.

Therefore, the mechanical properties of samples cut in parallel and perpendicular direction compare to the flow direction, were tested by tensile tests at different strain rates. The stress-strain response of "parallel" and "perpendicular" sample molded at 160 °C are presented in Figure 8a. Astonishingly, both the "parallel" and "perpendicular" samples showed the same yield stress ($\pm$1 MPa), see Figure 8b. The same observations were made for the samples molded at 85 °C, see Figure 9a,b.

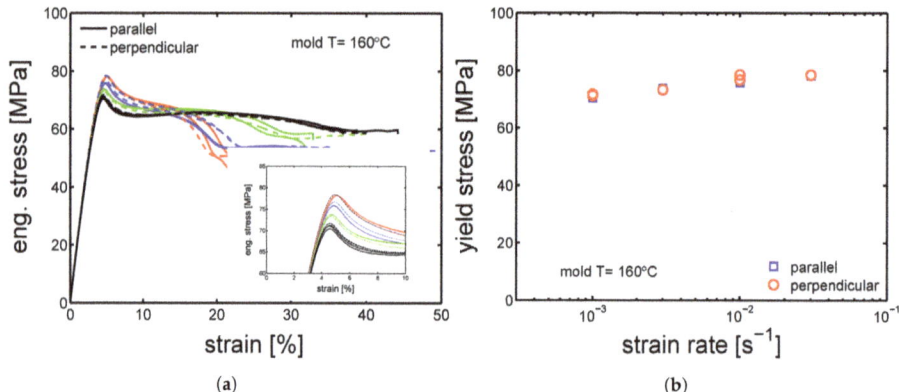

Figure 8. (**a**) Tensile tests at strain rate from 10^{-3} s^{-1} to 3×10^{-2} s^{-1} at 23 °C; comparison between samples cut in parallel (solid lines) and perpendicular (dashed lines) direction compare to the flow. (**b**) Yield stress as a function of strain rate. Samples molded at 160 °C.

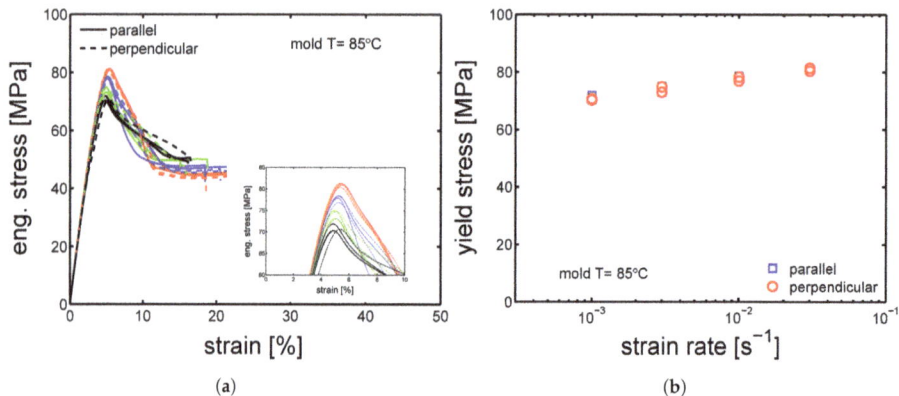

Figure 9. (**a**) Tensile tests at strain rate from 10^{-3} s^{-1} to 3×10^{-2} s^{-1} and 23 °C; comparison between samples cut in parallel (solid lines) and perpendicular (dashed lines) direction compare to the flow. (**b**) Yield stress as a function of strain rate. Samples molded at 85 °C.

4.2. Yield Kinetics

Because of the absence of effect of the orientation on the mechanical properties, the study continued on "parallel" samples only. The mechanical characterization continued with the investigation of yield kinetics for dry conditions. Uniaxial tensile test were performed in a range of temperatures from 23 °C to 80 °C and strain rates from 3×10^{-4} up to 3×10^{-2} s^{-1}. In Figure 10a,b and 11a,b, examples of stress-strain response at different temperatures are presented for mold temperature of 160 °C, 130 °C, 85 °C and 35 °C respectively. As expected, the yield stress increases for increasing strain rates and decreases for increasing temperature. However, this statement does not hold for the samples molded at 35 °C. For these, the increase of temperature from 23 °C to 47.5 °C leads rapidly to a dramatic drop of yield stress (from $\approx$70 MPa to $\approx$5 MPa), while increasing even further the temperature, the measured yield stress rises up to $\approx$20 MPa, see Figure 11b. This is a clear indication of the evolution of the sample state. Indeed, the samples molded at 35 °C are almost completely amorphous samples, see Figure 2. Heating an amorphous samples above its T_g causes cold crystallization and, consequently an enhancement of the yield stress is obtained. This effect is also visible in the DMTA experiments which are presented later on (see Figure 15a). The results for a mold temperature of 160 °C showed the weakest dependence on testing temperature and strain rate, while 85 °C showed the strongest (apart from the mold temperature 35 °C, which is a different case).

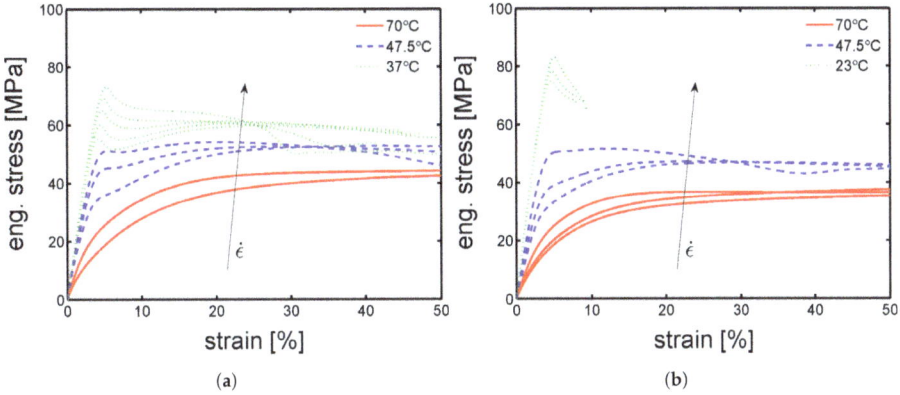

Figure 10. Stress-strain response at different temperatures and strain rates of samples molded at (a) 160 °C and (b) 130 °C.

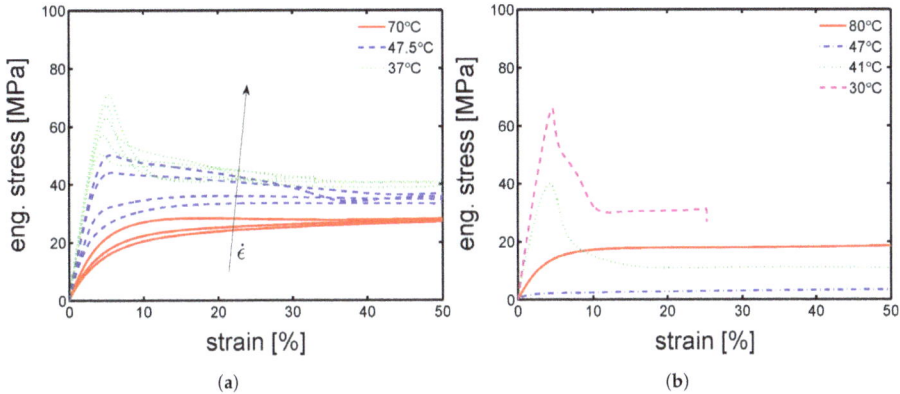

Figure 11. Stress-strain response at different temperatures and strain rates of samples molded at (a) 85 °C and (b) 35 °C.

In Figures 12a,b and 13a,b, the yield kinetics (yield stress as a function of strain rate) of dry samples are given. For the cases of samples molded at 160 °C and 130 °C, in which α-phase (160 °C) and γ-mesophase (130 °C) are predominant, predictions based on Equation (2) (in which $\tilde{T}$ is just the testing temperature) using parameters listed in Table 1 for mold 160 °C and Table 2 for 130 °C are given by the lines in Figure 12a,b. As explained in Section 2, the rate factors are defined accordingly to the lamellar thickness, see Tables 3–5. The lines match the experimental results rather well for these two cases. However, for a mold temperature of 85 °C, a clear mismatch between the prediction and the experimental results is found, see Figure 13a. In this case, the parameters related to γ-mesophase are employed. However, it is evident that the strain rate and temperature dependence of the yield stress is stronger then in the case of a mold temperature of 130 °C and 160 °C. This difference is even more clear for the results obtained at 37 °C, see Figure 13a. Figure 11b shows the stress-strain response of samples molded at 35 °C; the results obtained in this case are far from what found for a mold temperatures of 130 °C and 160 °C.

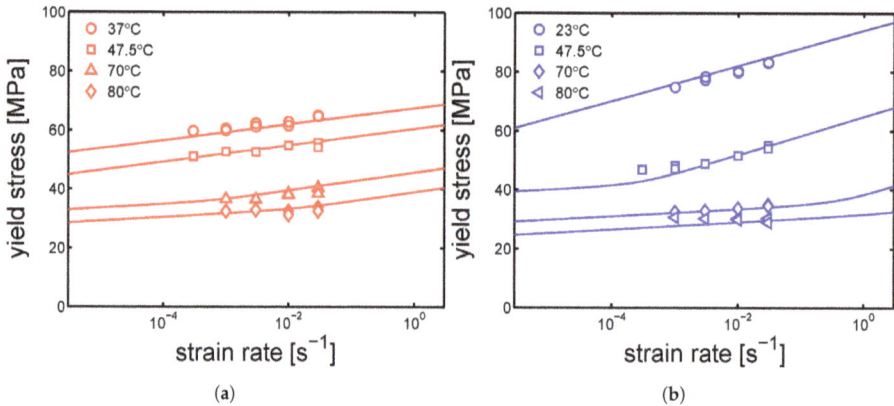

Figure 12. Yield kinetics (yield stress as a function of strain rate) of samples in dry condition, molded at (a) 160 °C (b) 130 °C. Lines are the the results of the Ree-Eyring equation.

Table 3. Ree-Eyring parameters: rate factors for samples molded at 160 °C.

	$\dot{\epsilon}_0$ [s^{-1}]
I	7×10^{119}
II	6×10^{45}

Table 4. Ree-Eyring parameters: rate factors for samples molded at 130 °C.

	$\dot{\epsilon}_0$ [s^{-1}]
I	1×10^{123}
II	4×10^{45}

Table 5. Ree-Eyring parameters: rate factors for samples molded at 85 °C.

	$\dot{\epsilon}_0$ [s^{-1}]
I	1×10^{130}
II	6×10^{45}

However, comparing results for a given mold temperature of 85 °C (Figure 13a) and a mold temperature 35 °C (Figure 13b), a very similar strain rate dependence (slope) of the yield stress is found at low temperatures, see Figure 14a. The reason of this behavior might be found by considering the samples homogeneity. As previously explained, injection molding samples show a non-homogeneous morphology along the thickness. It is likely that, in the case of mold temperature 85 °C, the skin layer (i.e., the outermost layer) is thicker than in the case of 130 °C and 160 °C, and thus a predominant contribution of the skin layer leads to a behavior closer to an amorphous sample rather then a semi-crystalline.

In Figure 14b, the yield stress obtained by tensile test at strain rate 10^{-2} s^{-1} is plotted as a function of temperature in a range from 23 °C to 120 °C. This figure shows clearly the behavior of the samples molded at 35 °C when tested at higher temperatures: at temperatures lower than ≈40 °C yield stress is slightly lower than the other samples molded at higher temperatures. Above ≈40 °C the yield stress decreases rapidly down to a minimum of about 5 MPa at 47.5 °C; increasing the temperature further, the yield stress increases again and at about 80 °C it reaches a plateau that continues up to the highest temperature investigated (120 °C). The samples molded at 130 °C and 160 °C show a similar behavior if plotted as function of temperature. The absolute value is slightly higher in the case of a mold temperature 160 °C and 130 °C. This is due to a larger lamellar thickness. At testing temperatures between 23 °C and ≈45 °C, the samples molded at 85 °C show an yield stress comparable with mold 130 °C and 160 °C; above ≈45 °C, the yield stress drops moderately till a minimum at is reached 120 °C, see Figure 14b.

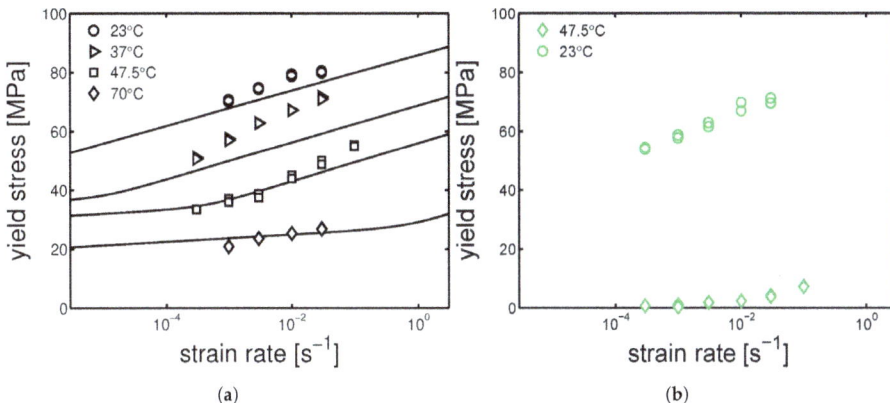

(a)   (b)

Figure 13. Yield kinetics (yield stress as a function of strain rate) of samples in dry condition, molded at (a) 85 °C (b) 35 °C. Lines are the the results of the Ree-Eyring equation.

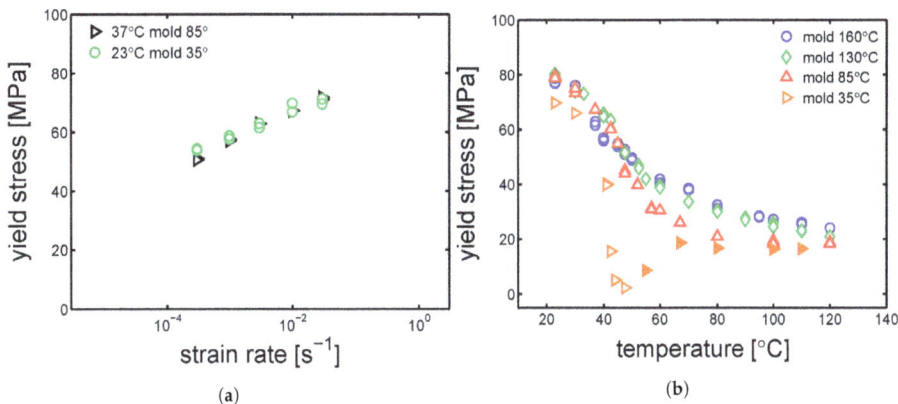

(a)   (b)

Figure 14. (a) Comparison between the yield kinetics of samples molded at 85 °C and 35 °C. (b) Yield stress as a function of temperature. In the case of the samples molded at 35 °C, the transition from unfilled to filled markers, is due to the fact that after ≈45 °C the samples start to cold-crystallize. Thus, the filled markers are not really representative of the samples molded at 35 °C but an evolution of those. Strain rate 10^{-2} s^{-1}.

4.3. Influence of the Conditioning Environment

The glass transition temperatures were measured by DMTA, after conditioning at a relative humidity 35%, 50% and 75% at room temperature (23 °C). The results are shown in Figures 15a,b and 16a,b. The T_g values are reported in Table 6. In Figure 15a, the cold crystallization previously mentioned, is clearly visible by observing the curve related to the dry sample at ≈60 °C.

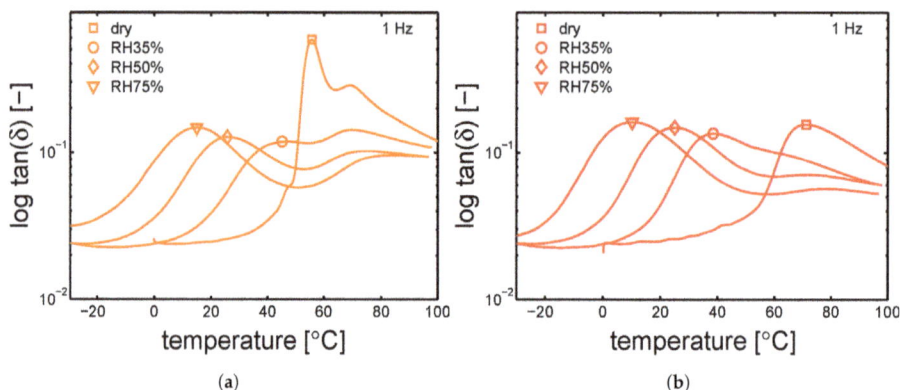

Figure 15. DMTA experiments for samples conditioned at different relative humidities, tan(δ) as a function temperature for samples molded at (**a**) 35 °C and (**b**) 85 °C. Markers indicate the measured T_g's.

The DMTA results for the samples that are conditioned at different humidities are vital in order to determine the "apparent" temperature (Equation (1)), see Section 2. The markers indicate the estimated glass transition temperatures.

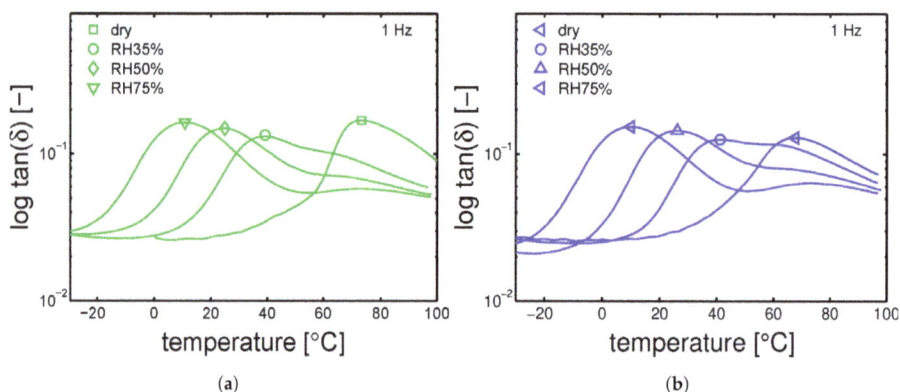

Figure 16. DMTA experiments for samples conditioned at different relative humidities, tan(δ) as a function temperature for samples molded at (**a**) 130 °C and (**b**) 160 °C. Markers are the selected T_g.

The glass transition temperatures are given as a function of the relative humidity (during conditioning), see Figure 17a. The largest differences in glass transition temperature are found in the case of a mold temperature 35 °C and dry condition. After conditioning, all the samples show a very similar T_g. The measured T_g are reported in Table 6. Finally, the glass transition temperatures can be also plotted as functions of normalized water fraction (see Equation (5)), which takes into account that only the amorphous fraction can absorb water [15]. Figure 17b shows that, when plotting T_g as a function of normalized water fraction, all the results are captured by a monotonic descending trend line.

Table 6. Glass transition temperature [°C].

Sample	Dry	35 RH%	50 RH%	75 RH%
mold 160 °C	68	41	26	10
mold 130 °C	73	39	25	11
mold 85 °C	71	38	25	10
mold 35 °C	56	45	26	15

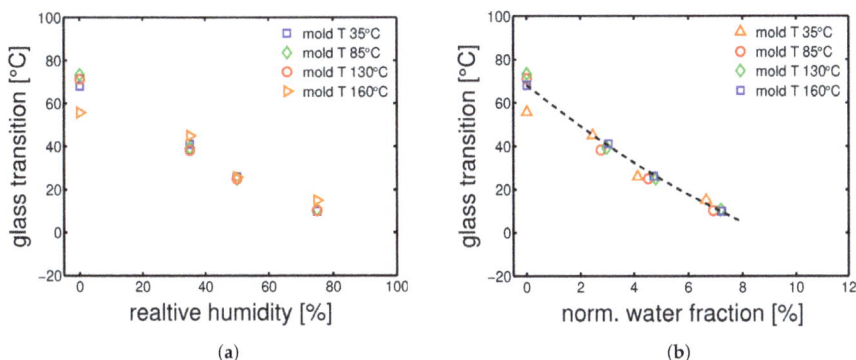

(a)

(b)

Figure 17. Glass transition temperatures as functions of (**a**) relative humidities and (**b**) normalized absorbed water fraction. The line is a guide to the eye.

4.3.1. Hydration-Induced Crystallographic Evolution

In order to check the influence of hydration on the crystallographic properties, WAXD experiments were carried out on conditioned samples. In Figures 18a, 19a, 20a and 21a, the integrated patterns are shown for the case of mold 160 °C, 130 °C, 85 °C and 35 °C; the corresponding deconvolution analysis of these patterns are given in figure Figures 18b, 19b, 20b and 21b. The drop of glass transition, due to water absorption, has different effects depending on the starting morphology. In the case of a mold temperature of 160 °C only a slight increase of crystallinity is recorded, most probably due to secondary crystallization. The samples molded at 130 °C show a partial transformation from γ to α-phase and a slight increase of the overall crystallinity. Similar behavior is seen in the case of mold temperature 85 °C. Finally, in the case of 35 °C, the cold crystallization leads, initially, to the crystallization of γ-mesophase (at RH 35%) and at higher relative humidity also α-phase is crystallized.

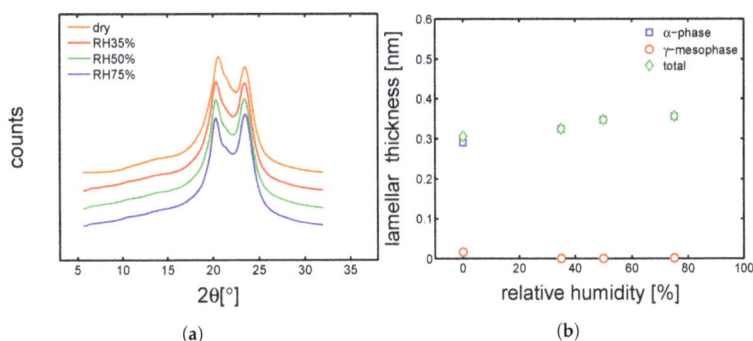

(a)

(b)

Figure 18. (**a**) Wide angle X-ray diffraction integrated patterns, samples molded at 160 °C and conditioned at different humidities. (**b**) Deconvolution analysis of the integrated WAXD patterns, crystalline, α and γ fraction as a function of relative humidity.

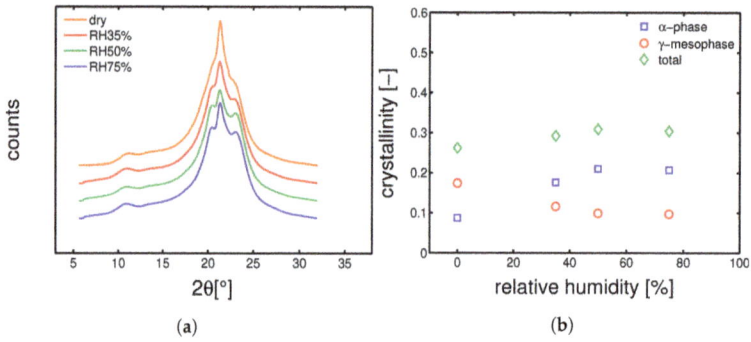

Figure 19. (**a**) Wide angle X-ray diffraction integrated patterns, samples molded at 130 °C and conditioned at different humidities. (**b**) Deconvolution analysis of the integrated WAXD patterns, crystalline, α and γ fraction as a function of relative humidity.

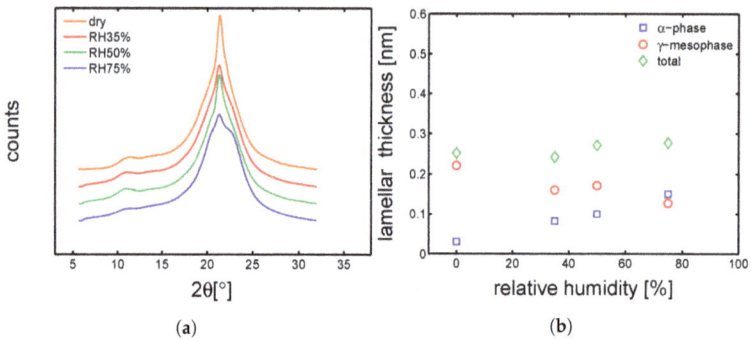

Figure 20. (**a**) Wide angle X-ray diffraction integrated patterns, samples molded at 85 °C and conditioned at different humidities. (**b**) Deconvolution analysis of the integrated WAXD patterns, crystalline, α and γ fraction as a function of relative humidity.

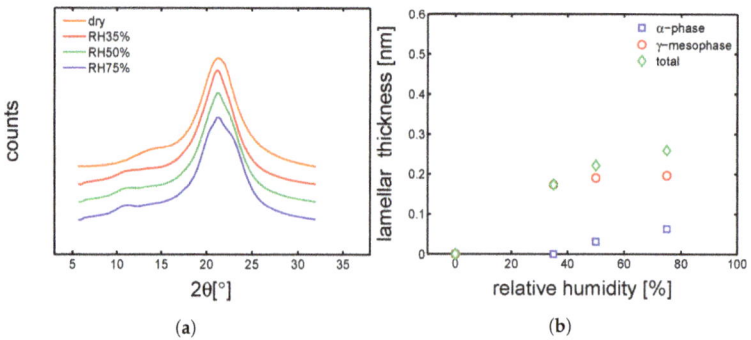

Figure 21. (**a**) Wide angle X-ray diffraction integrated patterns, samples molded at 35 °C and conditioned at different humidities. (**b**) Deconvolution analysis of the integrated WAXD patterns, crystalline, α and γ fraction as a function relative humidity.

4.3.2. Effect of Water Absorption on the Mechanical Response

After conditioning and determination of the glass transition temperature, uniaxial tensile tests were performed in environments with controlled temperature and relative humidity. The temperature was kept constant at 23 °C and three relative humidities were selected, RH 35%, RH 50% and RH 75%. In Figure 22a, an example of the effect of humidity on the stress-strain response of samples molded at 130 °C is presented. As already mentioned, the increase of relative humidity leads to a decrease of mechanical response. The yield stress values obtained at different relative humidity are plotted as function of the applied strain rate, see Figure 22b. The symbols are the values obtained experimentally, whereas the lines are the predictions based on the Equation (2) in which, in order to include the influence of RH% on the mechanical properties, the temperature was replaced by the "apparent" temperature equation. The agreement is excellent. The yield kinetics for samples molded at 160 °C and 85 °C are shown in Figure 23a,b, respectively. As in the case of dry samples, the predictions made for samples molded at 160 °C and 130 °C match quite well the experimental results. Even for the 85 °C mold temperature case, for which the agreement between model and experiment was not satisfactory for dry condition the description (lines) are not too far from the experimental results. Our explanation for this results is related to the effect of hydration on the crystallographic properties. As explained previously, because of the drop of glass transition due to hydration, cold crystallization is observed in the case of amorphous (quenched) samples. Thus, the amorphous skin layer of the samples molded at 85 °C are likely to cold-crystallize in γ-mesophase, see Figure 21b. Consequently, a decreased amorphous contribution on the mechanical properties is obtained and by the use of the parameters for γ-phase, a rather good description is obtained. Moreover, the hydration-induced drop of T_g decreases the mechanical contribution of the amorphous regions.

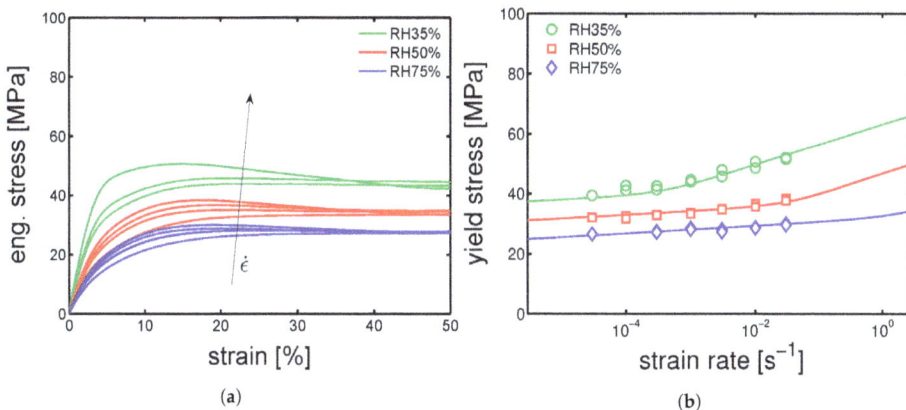

Figure 22. (a) Stress-strain response of samples conditioned at different relative humidities and tested at strain rate in a range from 3×10^{-5} s^{-1} to 3×10^{-2} s^{-1}. (b) Yield stress kinetics of samples molded at 130 °C.

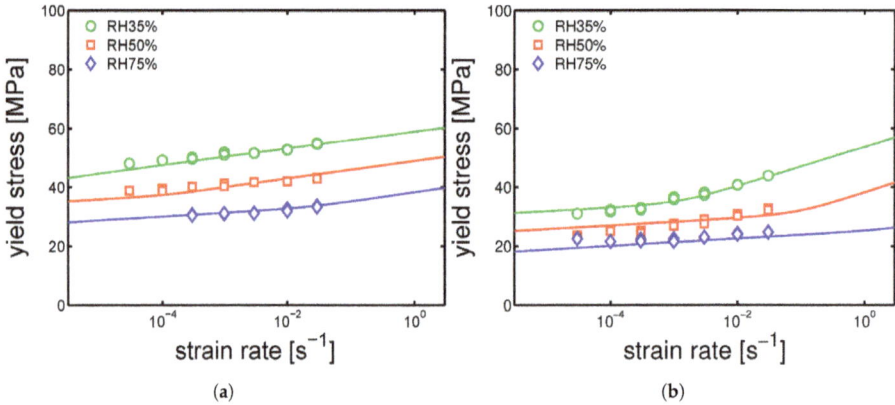

Figure 23. Yield stress kinetics of samples conditioned at 23 °C different relative humidities and molded at (**a**) 160 °C and (**b**) 85 °C.

4.4. Creep Tests

Finally, creep test at different relative humidity were performed. The samples were tested at 23 °C and different loads were applied. The time-to-failure (t-t-f) was defined as the time to reach a strain of 35%, and the plastic flow rate is calculated by selecting the minimum in the Sherby-Dorn plot [16]. Subsequently, from the plastic flow rates, plotted as functions of the corresponding measured time-to-failure in a bi-logarithmic plot, the critical strain (ϵ_{cr}) is determined, as explained in [10]. Next, using the predictions made for the yield kinetics, Equation (3) is applied. In Figures 24a,b and 25, the results of creep tests at different relative humidity and several applied load are shown for the samples injection molded at 160 °C, 130 °C and 85 °C, respectively.

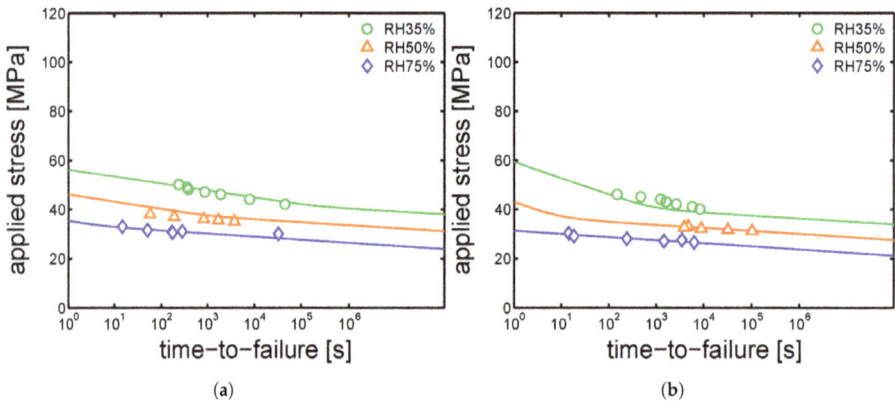

Figure 24. Creep tests, applied load as a function of time-to-failure for samples conditioned at different relative humidities and room temperature. The lines are the results of Equation (3). (**a**) Mold temperature 160 °C and (**b**) 130 °C.

The experimental results for mold temperatures of 160 °C and 130 °C are well described by the model (see Figure 24a,b). However, in the case of a mold temperature of 85 °C, the predictions made for the samples conditioned at relative humidity 35%, do not match the experimental values, see Figure 25. We can only speculate about this mismatch.

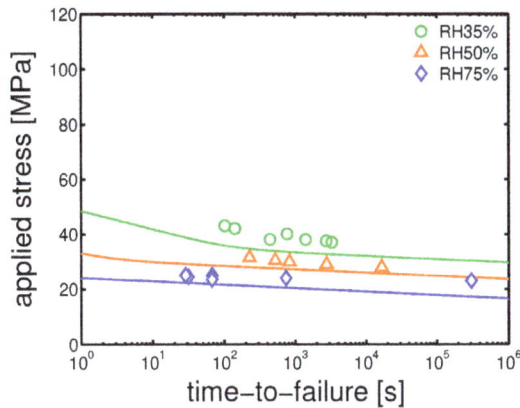

Figure 25. Creep tests, applied load as a function of time-to-failure for samples conditioned at different relative humidities and room temperature. The lines are the results of Equation (3). Mold temperature 85 °C.

4.5. Structures-Properties Relations for Modelling

As mentioned in Section 2, a previous study [9] on samples crystallized quiescently, a relation between the lamellar thickness and the rate factors ($\dot{\varepsilon}_{0,I,II}$) was proposed. In Figure 26a,b, the rate factors I and II defined for the samples processed by injection molding are plotted as functions of lamellar thickness together with the rate factors obtained for compression molding. In the case of the rate factor I, the injection molding markers are in good agreement with the trend obtained in the case of compression molding, see Figure 26a. Plotting the $\dot{\varepsilon}_{0,II}$ obtained for injection molding samples as a function of lamellar thickness, values are in line with trend found for compression molding, see Figure 26b. However, because of the mixture of α and γ crystals present in the sample molded by injection molding, the trends might results not perfectly in line with the polymorph division (see dashed lines in Figure 26b).

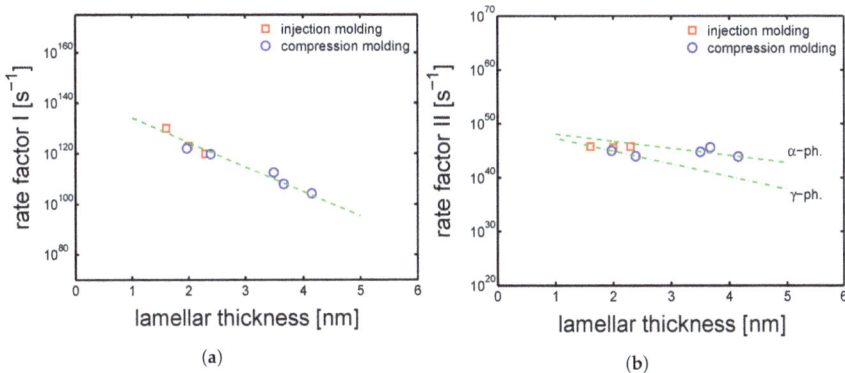

Figure 26. (**a**) Rate factor I and (**b**) rate factor II as functions of lamellar thickness. Lines are guides to the eye.

5. Conclusions

In this study, an industrial injection molding machine was used to produce polyamide 6 samples with different crystallographic properties. The mold temperature was varied in a range from 35 to 160 °C. A clear influence of mold temperature on the crystallization was detected. In the investigated range of temperatures, different crystallinity values, lamellar thicknesses and crystal polymorphs were obtained. Moreover, also the effect of shear flow and pressure during crystallization were studied by dilatometry (PVT). This led to the conclusion that the well known shift of the crystallization to higher temperature is dominant, while shear flow has a minor effect on crystallization kinetics. The effect of pressure during injection molding was particularly highlighted by a deconvolution analysis performed on WAXD patterns obtained for samples made with different mold temperatures. To confirm this finding, the deconvolution analysis performed on samples obtained by injection molding was compared with samples obtained by compression molding (in which pressure and shear flow effects are negligible). It was found that in the case of injection molding, even at a mold temperature of 160 °C, a fully α-phase sample was obtained; while in the case of compression molding at the same temperature, a mixture of α and γ was obtained. This effect was ascribed to the high pressure present during cooling in the case of injection molding. By azimuthal integration, the presence of molecular orientation was measured. Orientation was found but, surprisingly, the mechanical properties did not show an influence of the orientation. The mechanical properties were tested by uniaxial tensile tests and uniaxial creep tests at several temperature in dry condition and at room temperature with varying relative humidity. The Eyring's flow model, modified with the apparent temperature and combined with the concept of critical strain, were employed in order to describe the results obtained by tensile and creep tests. This was partially successful. In the case of samples molded at 160 °C and 130 °C, the model was successfully applied by the use of parameters previously determined as characteristic in the case of α-phase and γ-mesophase, respectively. In the case of mold temperature 85 °C, it was hypothesized that the inhomogeneous morphology (intrinsically induced by the molding technique) led to an excessive contribution of the amorphous layer (skin layer). Finally, in the case of mold temperature 35 °C, in which completely amorphous samples were obtained, a description was not achieved due to the continuous development of the micro-structure.

Author Contributions: E.P. performed the experiments, and wrote the manuscript. L.E.G. designed the modelling approach and assisted in the analysis. G.W.M.P. contributed to discussion and analysis.

Funding: This work is part of the Research Program of the Dutch Polymer Institute DPI, Eindhoven, the Netherlands, project number #786.

Acknowledgments: The authors thank the "Institute for Complex Molecular Systems" (ICMS) of Eindhoven, The Netherlands for providing the X-ray facilities and Marco M.R.M Hendrix for assisting with the experiments.

Conflicts of Interest: The authors declare no conflict of interest.

References

1. Govaert, L.E.; Engels, T.A.P.; Klompen, E.T.J.; Peters, G.W.M.; Meijer, H.E.H. Processing-induced properties in glassy polymers: Development of the yield stress in pc. *Int. Polym. Process.* **2005**, *20*, 170–177. [CrossRef]
2. Russell, D.P.; Beaumont, P.W.R. Structure and properties of injection-moulded nylon-6. *J. Mater. Sci.* **1980**, *15*, 197–207. [CrossRef]
3. Ghiam, F.; White, J.L. Phase morphology of injection-molded blends of nylon-6 and polyethylene and comparison with compression molding. *Polym. Eng. Sci.* **1991**, *31*, 76–83. [CrossRef]
4. Van Erp, T.B.; Govaert, L.E.; Peters, G.W.M. Mechanical performance of injection-molded poly(propylene): Characterization and modeling. *Macromol. Mater. Eng.* **2013**, *298*, 348–358. [CrossRef]
5. Schrauwen, B.A.G.; Breemen, L.C.A.; Spoelstra, A.B.; Govaert, L.E.; Peters, G.W.M.; Meijer, H.E.H. Structure, deformation, and failure of flow-oriented semicrystalline polymers. *Macromolecules* **2004**, *37*, 8618–8633. [CrossRef]
6. Yalcin, B.; Valladares, D.; Cakmak, M. Amplification effect of platelet type nanoparticles on the orientation behavior of injection molded nylon 6 composites. *Polymer* **2003**, *44*, 6913–6925. [CrossRef]
7. Cavallo, D.; Gardella, L.; Alfonso, G.C.; Portale, G.; Balzano, L.; Androsch, R. Effect of cooling rate on the crystal/mesophase polymorphism of polyamide 6. *Colloid Polym. Sci.* **2011**, *289*, 1073–1079. [CrossRef]
8. Parodi, E.; Peters, G.W.M.; Govaert, L.E. Prediction of plasticitycontrolled failure in polyamide 6: Influence of temperature and relative humidity. *J. Appl. Polym. Sci.* **2018**, *135*, 45942. [CrossRef]

9. Parodi, E.; Peters, G.W.M.; Govaert, L.E. Structure-properties relations for polyamide 6, part 1: Influence of the thermal history during compression moulding on deformation and failure kinetics. *Polymers* **2018**, *10*, 19. [CrossRef]
10. Kanters, M.J.W.; Remerie, K.; Govaert, L.E. A new protocol for accelerated screening of long-term plasticity-controlled failure of polyethylene pipe grades. *Polym. Eng. Sci.* **2016**, *56*, 676–688. [CrossRef]
11. Fornes, T.D.; Paul, D.R. Crystallization behavior of nylon 6 nanocomposites. *Polymer* **2003**, *44*, 3945–3961. [CrossRef]
12. Baltá-Calleja, F.J.; Vonk, C.G. *X-ray Scattering of Synthetic Polymers*; Number v. 8 in Polymer Science Library; Elsevier: New York, NY, USA, 1989.
13. Ryan, A.J.; Bras, W.; Mant, G.R.; Derbyshire, G.E. A direct method to determine the degree of crystallinity and lamellar thickness of polymers: Application to polyethylene. *Polymer* **1994**, *35*, 4537–4544. [CrossRef]
14. Forstner, R.; Peters, G.W.M.; Meijer, H.E.H. A novel dilatometer for pvt measurements of polymers at high cooling -and shear rates. *Int. Polym. Process.* **2009**, *24*, 114–121. [CrossRef]
15. Murthy, N.S. Hydrogen bonding, mobility, and structural transitions in aliphatic polyamides. *J. Polym. Sci. Part B Polym. Phys.* **2006**, *44*, 1763–1782. [CrossRef]
16. Sherby, O.D.; Dorn, J.E. Anelastic creep of polymethyl methacrylate. *J. Mech. Phys. Solids* **1958**, *6*, 145–162. [CrossRef]

MDPI

Article

Hydrophobicity Tuning by the Fast Evolution of Mold Temperature during Injection Molding

Sara Liparoti [1], Roberto Pantani [1], Andrea Sorrentino [2], Vito Speranza [1,*] and Giuseppe Titomanlio [1]

[1] Department of Industrial Engineering, University of Salerno, via G. Paolo II 132, 84084 Fisciano (SA), Italy; sliparoti@unisa.it (S.L.); rpantani@unisa.it (R.P.); gtitomanlio@unisa.it (G.T.)

[2] Institute for Polymers, Composites and Biomaterials (IPCB-CNR), via Previati n. 1/E, 23900 Lecco, Italy; andrea.sorrentino@cnr.it

* Correspondence: vsperanza@unisa.it; Tel.: +39-089-964-145

Received: 2 February 2018; Accepted: 9 March 2018; Published: 15 March 2018

Abstract: The surface topography of a molded part strongly affects its functional properties, such as hydrophobicity, cleaning capabilities, adhesion, biological defense and frictional resistance. In this paper, the possibility to tune and increase the hydrophobicity of a molded polymeric part was explored. An isotactic polypropylene was injection molded with fast cavity surface temperature evolutions, obtained adopting a specifically designed heating system layered below the cavity surface. The surface topology was characterized by atomic force microscopy (AFM) and, concerning of hydrophobicity, by measuring the water static contact angle. Results show that the hydrophobicity increases with both the temperature level and the time the cavity surface temperature was kept high. In particular, the contact angle of the molded sample was found to increase from 90°, with conventional molding conditions, up to 113° with 160 °C of cavity surface temperature kept for 18 s. This increase was found to be due to the presence of sub-micro and nano-structures characterized by high values of spatial frequencies which could be more accurately replicated by adopting high heating temperatures and times. The surface topography and the hydrophobicity resulted therefore tunable by selecting appropriate injection molding conditions.

Keywords: hydrophobicity; contact angle; polypropylene; atomic force microscopy; injection molding; mold temperature evolution

1. Introduction

Natural hydrophobicity and superhydrophobicity is grabbing the interest of many researchers and has inspired mimetic attempts in recent years [1–3]. The surfaces are characterized by high contact angle show self-cleaning [4,5], water repellence capabilities [6] and corrosion inhibition [7], and thus, thus they are very interesting for biomedical and pharmaceutical fields too [8–10].

Artificial super-hydrophobicity are obtained through the hierarchical structuring of a surface, namely by the formation of both micro and nano-structures on the surface [11–13]. The technologies adopted for the production of micro and nano-structured surfaces are classified as bottom-up or top-down methods. The bottom-up techniques take advantages of the self-assembling capabilities of the molecule thus the formation of a structured surface takes place layer by layer [14]. These methods can be adopted for a limited range of materials and are not suitable for mass production [15]. The top-down approaches consist in etching away bulk material to achieve the required smaller structural architectures [16], or in replicating master geometries [17]. These methods can be applied to a wide range of material, and are suitable for mass production of micro and nano-structured surfaces [15,18]. In particular, the "replica molding" techniques consist in the replication of a master, produced by lithography-based techniques, by hot embossing or injection molding [16,19]. The latter surely presents the advantage of cheaper and faster production [20–24]. Yamaguchi et al. [25]

proposed to replicate a master having nano-periodic structures, obtained by a femtosecond laser, layered on the cavity surface. They studied the effect of nano-structures replica on the wettability of acrylonitrile-ethylene-styrene moldings and found that the apparent contact angle increases with respect to the surface obtained on a smoother cavity. Puukilainen et al. [26] adopted a cavity with micro-pillars in micro-injection molding of polypropylene. They found that the distance between the pillars is the key parameter in tuning the surface wettability. Yoo et al. [27] adopted injection molding process to replicate nano-structured masters, having nano-pillars, on polypropylene and polycarbonate objects. They found that the replication accuracy increases with the mold temperature and, more important, that the hydrophobicity also increases with the mold temperature. The same research group [28] obtained similar results on polyethylene copolymer. All these studies highlight that the topography of a surface and, in particular, the quality of the replication of micro and nano-structures, directly influences the wettability of a surface. On their turn, all the processing parameters that control the accuracy of the replication influence also the wettability. Thus, it is possible to tune the wettability by controlling the same processing conditions that influence the replication accuracy. However, the replication of micro and nano-structures by injection molding process is not an easy task. The formation of a frozen layer on the polymeric surface, due to the contact of the molten polymer with the cold cavity surface, prevents an accurate replication [29]. It is necessary to increase the mold temperature, but this induces an increase of the processing time. The methods adopted to obtain an evolution of the mold temperature require additional tool cost and cause a significant increase in processing time [23,30–34].

In this paper, a fast evolution of mold temperature was obtained adopting a thin heating device, made of carbon black loaded poly(amide-imide), that is able to remarkably increase the temperature of a selected area on the cavity surface in few seconds, and to control this temperature level for a given time. This innovative heating device, as a result of its reduced thickness, 50 μm, also allows for a fast decrease in temperature soon after its deactivation. The fast evolution of the cavity surface temperature allows for a shorter processing time. The injection molding process coupled with such a heating system was adopted to obtain a better replication of a commercial steel layer. This paper aims to demonstrate that the control of the replication of a random structured surface by injection molding allows for tuning the hydrophobicity of polypropylene surfaces.

2. Materials and Methods

An isotactic polypropylene (iPP, Basell T30G, Ferrara, Italy), having an average molecular weight of M_w of 376,000 g·mol^{-1}, a polydispersity index M_w/M_n of 6.7, and a meso pentads content of 87.6% was adopted. A complete mechanical, rheological and thermal characterization of this polymer can be found elsewhere [35,36].

Injection molding tests were carried out by adopting a melt temperature of 220 °C, a mold temperature of 28 °C and a holding pressure of 26 MPa. Table 1 summarizes all the other operating conditions.

Table 1. Operative injection molding conditions (T_{melt} = melt temperature; T_{mold} = mold temperature; $P_{holding}$ = holding pressure; T_{level} = temperature reached on the cavity surface thanks to the activation of the heating device).

Test	T_{melt} [°C]	T_{mold} [°C]	$P_{holding}$ [MPa]	Holding time [s]	T_{level} [°C]	Heating time [s]
CIM	220	28	26	2	28	0
80-07	220	28	26	2	80	0.7
120-07	220	28	26	2	120	0.7
160-07	220	28	26	2	160	0.7
80-18	220	28	26	2	80	18
120-18	220	28	26	2	120	18
140-18	220	28	26	2	140	18
160-18	220	28	26	2	160	18

A rectangular cavity having a length of 70 mm, a width of 20 mm, and a thickness of 1 mm was adopted. A thin heater device, made of poly(amide-imide) loaded carbon black, was layered under the cavity surface. The heater was electrically insulated from the mold by poly(amide-imide) layers, having a thickness of 0.020 mm. An additional insulating layer 0.120 mm thickness was located between the heater and the mold to reduce the heat loss. A protective steel layer (0.100 mm thickness) covering all the length of the cavity was adopted to avoid the direct contact of the melt with the heating device. Supplementary information about the heating device is reported elsewhere [37]. The heating device adopted allows for heating a selected area on the cavity surface. Figure 1 shows a sketch of the heating device layered in the mold insert and a sketch of the cavity with the pressure transducer positions. In the configuration adopted in this work and depicted in Figure 1a, the cavity surface was only partially heated (for the first 35 mm adjacent to the gate).

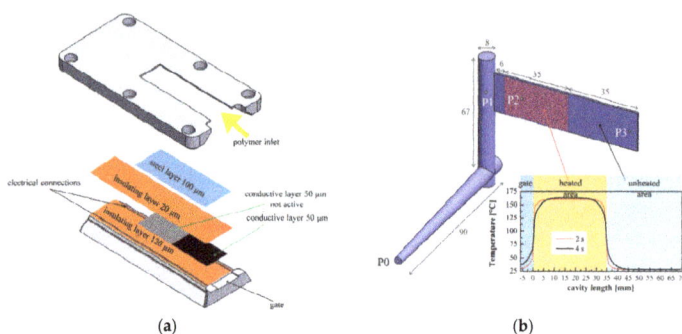

(a)   (b)

Figure 1. (a) Schematic drawing of the mold insert adopted to layer the heater below the cavity surface; (b) Sketch of the cavity adopted for the injection molding experiments (measures are in mm), pressure transducer positions (P0–P3) are indicated. A sketch of the surface temperature profile along the flow path is also reported.

As a reference, injection molding experiments were carried out in conventional injection molding conditions (CIM). In these cases, the temperature of the cavity surface at the beginning of the experiment corresponds to the temperature of the whole mold (28 °C). All the other experiments were performed adopting the heating device to increase the temperature of the cavity surface (see Figure 1b). The activation of the heating device on the half cavity length allows for comparing the contact angle of a surface obtained with high cavity temperature with the contact angle of a surface obtained with low cavity surface temperature, directly on the same sample. The heating devices were automatically activated with constant electrical power 0.5 s after the start of the injection, which corresponds to about 2 s before the contact with the polymer in position P2. The temperature reached on the cavity surface, which depends on the adopted electrical power, after 10 s heating time is denoted as T_{level} in the Table 1. 80, 120 and 160 °C were selected as T_{level}. Different heating times, 0.7 s (corresponding to the cavity filling time), and 18 s were adopted. In the following, each test was coded by sequencing the cavity surface temperature and the heating time (i.e., 160-18 refers to an experiment obtained with 160 °C cavity surface temperature and 18 s heating time).

Figure 1b also shows a sketch of the temperature profile along the flow path. It is possible to observe that the boundaries of the heated area are characterized by a temperature that is below the selected temperature, 160 °C in this case.

The surface of the injection molded samples was observed by Atomic Force Microscope (AFM). The area, 60 × 60 μm², for the AFM acquisitions was selected in order to include all the representative structures. Micro-graphs in air at room temperature were collected with a Dimension 3100 coupled with a Bruker Nanoscope V controller (Bruker, Billerica, MA, USA), operating in tapping mode. For each position along the flow direction, a square area of 3600 μm² was investigated with a resolution of 512 points/line and a scan rate of 0.5 line/s. Commercial probe tips (mod. OTESPA, Bruker, Billerica, MA, USA) with nominal spring constants of 42 N/m, resonance frequency of 300 kHz, tip with radius of 7 nm and height of 7–15 μm were adopted. AFM topographic maps were acquired with NanoScope software version 7.30 (Bruker, Billerica, MA, USA). To remove tilt or bow from the acquired surface a flatten procedure was applied by the NanoScope Analysis software version 1.80 (Bruker, Billerica, MA, USA); to remove noise a low pass filter with a cutoff frequency of 2 μm^{-1} was applied. The NanoScope Analysis software also allowed for performing the roughness and the frequency analysis of the acquired patterns.

The processed AFM acquisitions were analyzed by the software MountainsMap 7.1 (by DigitalSurf, Besançon, France) to obtain the 3D topographic parameters S_{da} (average close dale area), S_{ha} (average close hill area) and S_{dr} (developed area ratio) according to the standards defined by EUR15178N and ISO25178. The dale area, S_{da}, and hill area, S_{ha}, are calculated as the mean area of all individual motifs used to segment the surface acquired by AFM. The considered areas are the horizontal areas of the motifs projected onto the horizontal plane [38]. The developed area ratio S_{dr} is calculated by summing the local areas when following the surface curvature and it represents the area excess with respect to the projected area. It is expressed as the percentage of the projected area [38].

Contact angle measurement is an accurate method for determining the interaction energy between a liquid and a solid. Static contact angle measurements were carried out at room temperature with distilled water. A water drop (5 μL) was placed on the sample and photographed once a second for 30 s. The contact angle was determined mathematically by fitting a Young-Laplace curve around the drop. Values recorded between 6 and 30 s were averaged to obtain the contact angle for each measurement. The measurements were repeated on a series of three equivalent samples and mediated to calculate the contact angle value on each point of the molded sample. The contact angles were measured along the flow path.

3. Results

3.1. The Process

Figure 2 shows, for two representative experiments, the pressure in three positions inside the cavity, along the flow path, P0, P2 and P3 (see Figure 1), and the temperature evolutions measured in position P2. Figure 2a shows pressure and temperature evolutions recorded during CIM, the experiment performed without the heating device. Figure 2b shows the pressure and the temperature evolutions recorded during the experiment named 160-18.

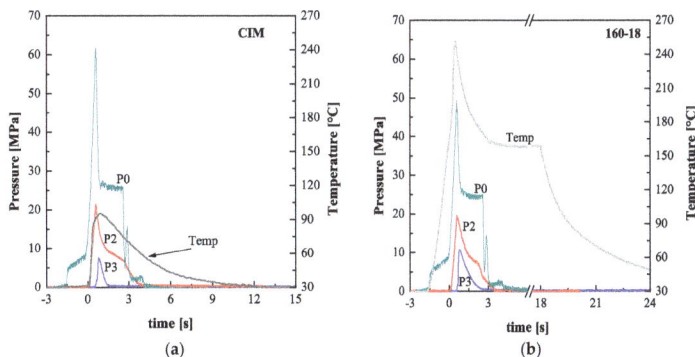

Figure 2. Pressure and temperature evolutions measured during the injection molding experiments conventional injection molding (CIM) (**a**) and 160-18 (**b**).

The pressure in position P0 was recorded in the injection chamber and represents the pressure imposed on the melt by the machine; it reached a maximum at the end of the filling stage ($t = 0.7$ s). This pressure is affected by the cavity surface temperature: as the cavity surface temperature increases the pressure necessary to fill the cavity at the set flow rate decreases. After the pressure peak, for both cases reported in Figure 2, the pressure in position P0 reaches the value set as holding pressure, 26 MPa, and is kept constant for 2 s, after that it decreases down to zero. When the polymer comes in contact with the cavity walls ($t = 0$ s), the pressure in position P2 starts to increase. It reaches a maximum value of about 20 MPa at the end of the cavity filling. The final filling pressure in position P2, at $t = 0.7$ s, is also affected by the adopted cavity surface temperature: it decreases as the cavity surface temperature increases. After the filling end, the pressure in position P2 decreases down in two steps: during the first step, the density increase (induced by the polymer cooling and crystallization) is counteracted by the packing flow; during the second step, the pressure decrease is faster due to the pressure release in position P0 [39]. When the melt contacts the cavity in position P3, the filling is almost complete; thus, the pressure evolution in this position is mainly driven by the packing flow. The pressure drop, determined by the packing flow, and the polymer viscosity are both smaller with higher cavity temperature, thus, the pressure in position P3 is higher during the experiment 160-18. The pressure decay in position P3, is faster with respect to the pressure decay in positions P2, for both cases shown in Figure 2, because the packing flow does not efficiently counteract the cooling. Figure 2 also shows the temperature evolutions in position P2. When the heating device is not present (Figure 2a) the temperature reaches 90 °C at the first contact of the hot melt with the cavity wall, afterward, the temperature decreases due to the contact with the cold cavity walls. Figure 2b shows that, at the contact with the melt, the surface temperature, which had already raised because device heater activation, undergoes an additional immediate increase and, soon after (see Figure 2b), a fast decrease toward a constant value denoted as T_{level}. At the heater deactivation, the temperature decreases down to the value of the whole mold (28 °C).

3.2. Contact Angle Distribution

In the literature, it has been demonstrated that the cavity temperature influences the hydrophobicity of the molded objects [27]. This paper aims to analyze the effect of the temperature evolution, namely the cavity surface temperature and the heating time, on the molded part surface hydrophobicity. Figure 3 shows the contact angle distributions along the flow path of the samples obtained with different cavity surface temperatures and 0.7 s heating time. The contact angle distribution for the experiment named CIM is also reported for comparison.

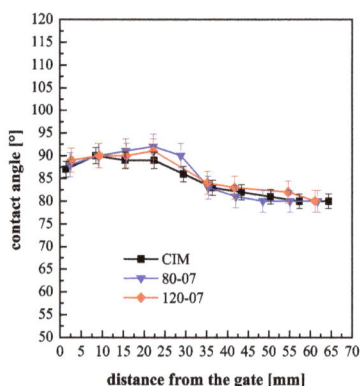

Figure 3. Contact angles of the molded samples obtained with CIM and different cavity surface temperature, keeping active the heating device only during the filling.

For all the considered samples, the contact angles show a similar distribution. When a heating time equal to the filling time, 0.7 s in these cases, was adopted, the contact angles showed values that have a negligible dependence on the temperature set on the cavity surface, within the measurement reproducibility. In particular, the contact angle distributions of the samples obtained with different cavity surface temperatures are similar to the one obtained for the CIM sample. These values show a slight decrease on increasing the distance from the gate, in agreement with the fact that both pressures and temperatures decrease on increasing the distance from the gate.

Figure 4a shows the contact angle distributions along the flow path, measured for the samples obtained with different cavity surface temperatures and 18 s heating time. The contact angle distribution along the flow path for the sample CIM is also reported for comparison. Figure 4b shows the plot of contact angles measured at 20 mm distance from the gate (namely at the center of the heated area) vs. the T_{level}.

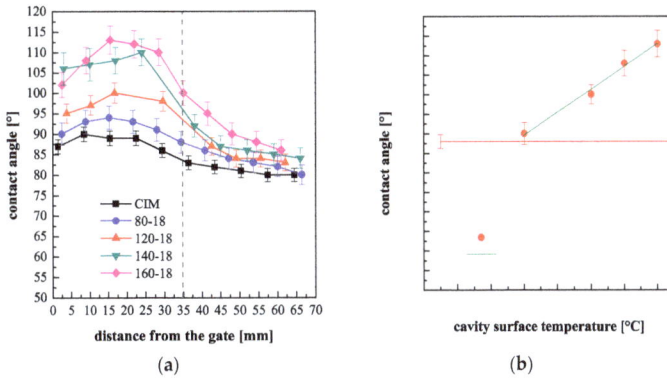

Figure 4. (**a**) Contact angles of the molded samples obtained with CIM and different T_{level}, keeping active the heating devices for 18 s. The vertical dashed line identifies the location of the heating element; (**b**) Contact angle dependence on the cavity surface temperature (the contact angles were measured at 20 mm distance from the gate on the samples obtained with 18 s heating time).

In these cases, since the heating device is kept active for long times, the cavity surface temperature has a strong influence on the contact angle: the contact angles show higher values, with respect to the CIM sample, especially at the central part of the heated area. The distributions of the contact angle present a maximum, for all the considered cases, because, as shown in Figure 1b, the temperature in the first 10 mm downstream the gate is smaller than T_{level}, due to the contact with a cold gate. On plotting the contact angle measured at 20 mm downstream the gate versus the T_{level}, a linear dependence is obtained, as shown in Figure 4b, in the whole range of set temperature explored (80–160 °C).

It can be concluded that the increase in the cavity surface temperature up to 160 °C, allows for a contact angle increase up to 28% with respect to the contact angle measured for the CIM sample. Expensive masters with specially designed structures were proposed in the literature to obtain a similar increase of the contact angle by injection molding process. In this work, this result was achieved adopting a commercial and cheap steel layer and by an evolution of the cavity surface temperature.

3.3. Analysis of the Topography of the Molded Surface

In the literature, many authors correlated the hydrophobicity of a surface to its topography [25,40]. The presence of a hierarchical structure on a surface induces an increase of the hydrophobicity with respect to a smooth surface [41]. This was assessed by replicating masters with complex surfaces on polymeric surfaces. In this work, a commercial steel layer is adopted as a master. The AFM acquisitions, shown in the uppermost of Figure 5, confirm that the steel layer is characterized by a complex surface, being composed of both micro and nano-structures, tightly packed. In particular, the components on different scales were highlighted applying high pass filters to the flattened maps, with cut off frequencies of 0.1 and 0.2 μm^{-1}. These frequencies attenuate by 50% and 70%, respectively, the spectral component of the signal acquired on the steel layer.

Since the steel layer is composed of structures on different scales, the hydrophobicity is expected to increase by the accurately replicate the steel layer surface on the molded object. Figure 5 shows the topographies acquired on the sample 160-18 at different positions, 5, 20 and 50 mm, downstream the gate, where differences in contact angles were detected.

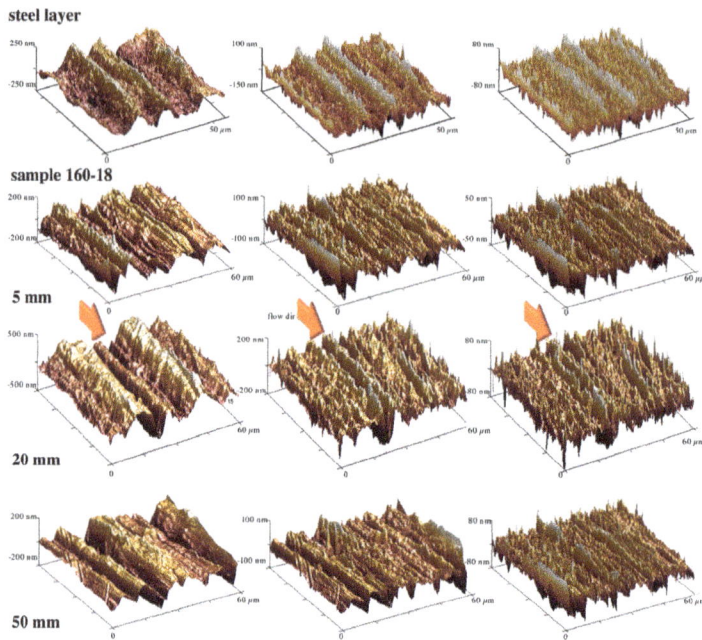

Figure 5. Topographies of the steel layer and of the sample 160-18 at different distances from the gate. From left to right are reported the atomic force microscopy (AFM) maps processed by applying a flatten procedure and two high pass horizontal filters, 0.1 and 0.2 μm^{-1} cut off frequency.

Figure 5 shows that also the surfaces of the 160-18 samples are characterized by sub-micro and nano-structures in all the positions along the flow path. To compare the structures and determine the accuracy of the replication, a roughness analysis was carried out and summarized in Table 2.

Table 2. Roughness of the steel layer and of the molded samples CIM and 160-18 at 20 mm distance from the gate.

Test	Filter	5 mm	20 mm	55 mm
Steel layer	flattened		164 ± 30 nm	
	0.1 μm^{-1}		64 ± 10 nm	
	0.2 μm^{-1}		25 ± 5 nm	
CIM	flattened	113 ± 25 nm	139 ± 10 nm	116 ± 10 nm
	0.1 μm^{-1}	48 ± 3 nm	38 ± 5 nm	48 ± 3 nm
	0.2 μm^{-1}	26 ± 3 nm	20 ± 2 nm	25 ± 2 nm
160-18	flattened	77 ± 5 nm	136 ± 15 nm	106 ± 12 nm
	0.1 μm^{-1}	25 ± 5 nm	40 ± 6 nm	24 ± 2 nm
	0.2 μm^{-1}	23 ± 5 nm	27 ± 3 nm	23 ± 2 nm

Table 2 shows that the roughness of the molded samples is nearly the same for both CIM and 160-18 samples and that the differences in the roughness values along the flow path are within the reproducibility of the measurements. Thus, the roughness is not sufficient in describing the replication accuracy of the surfaces. A deeper analysis of the profiles is necessary.

Figure 6 shows some profiles, obtained by the AFM maps acquired on the samples CIM and 160-18 (20 and 50 mm distance from the gate). The profile of the steel layer is also reported for comparison. Three patterns are reported for each sample: the profile obtained from the flattened maps and two additional profiles obtained applying high pass filters with cutoff frequencies of 0.1 and 0.2 µm^{-1} to the AFM maps.

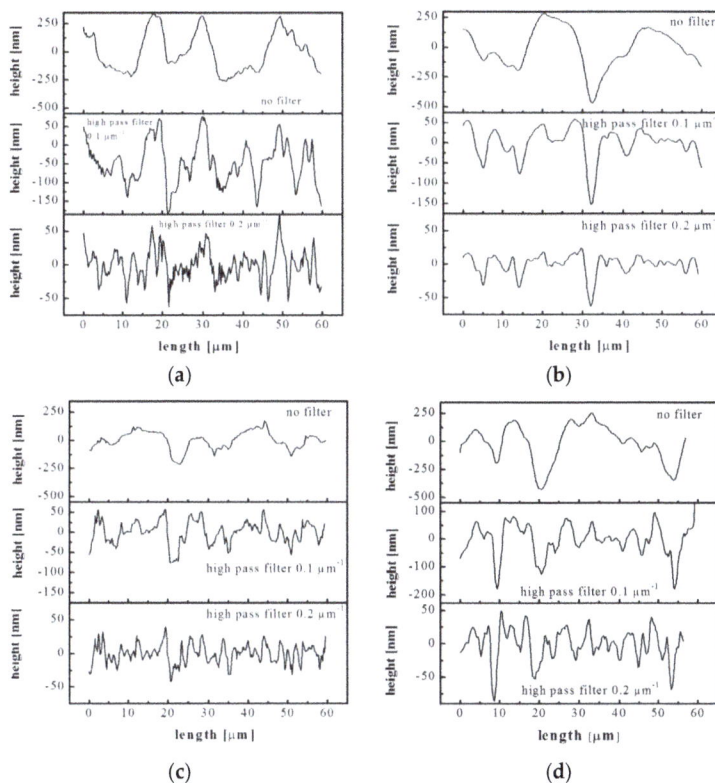

Figure 6. AFM patterns obtained from the maps acquired on the (**a**) steel layer; (**b**) CIM sample (20 mm distance from the gate); (**c**) 160-18 sample (20 mm distance from the gate) and (**d**) 160-18 sample (50 mm distance from the gate). Three patterns are reported for each sample: the flattened pattern, and the patterns obtained applying 0.1 and 0.2 µm^{-1} high pass filters.

Figure 6a, related to the steel layer, shows that the profile obtained from the flattened map is composed of structures on micro and nanoscale. Figure 6b shows that the pattern of CIM sample is smoother than the corresponding pattern of the steel layer and also with respect to the 160-18 sample obtained in the heated area (Figure 6c). This characteristic of the CIM sample is evident especially if comparing the patterns obtained after the application of 0.2 µm^{-1} high pass filter. The patterns of the sample 160-18 acquired at 50 mm distance from the gate (Figure 6d), appear smoother than the patterns acquired on the same sample but in the heated area (20 mm distance from the gate, Figure 6c). The mold temperature increase induces a better replication of the structures that are present on the cavity surface, in both micro and nano-metric scales [23,24,42]. This is the case of the 160-18 sample at 20 mm distance from the gate. Frequency analyses were performed to quantify the replication accuracy of the random surface of the steel layer. Figure 7 shows the harmonic content related to the flattened patterns of the steel layer and of the samples CIM and 160-18, the latter at two distances from the gate, 20 mm and 50 mm.

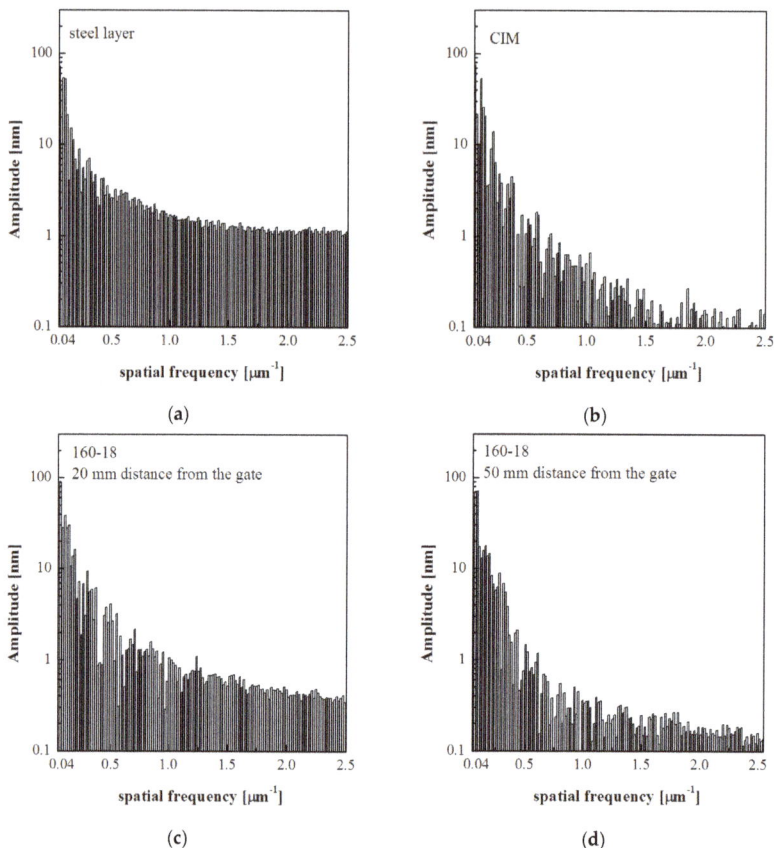

Figure 7. Spatial frequency analysis of the AFM acquired signal for (**a**) the steel layer and for the molded samples (**b**) CIM (20 mm distance from the gate), (**c**) 160-18 (20 mm distance from the gate) and (**d**) 160-18 (50 mm distance from the gate).

The frequency analysis of the AFM patterns shows that the steel layer is composed by significant contributes with high frequencies, up to 2.5 μm^{-1}, which correspond to structures of smaller dimensions. The CIM sample is characterized by a harmonic content completely different with respect to the steel layer, with significant contributes at the lower frequencies, and negligible contributes at the higher frequencies. The acquisition carried out at 20 mm distance from the gate on the sample 160-18 is characterized by a harmonic content similar to that one observed for the steel layer. The acquisition of the 160-18 sample at 50 mm distance from the gate shows a behavior similar to the CIM sample. This is consistent with a poor replication accuracy and also with the lower values of the contact angles. The frequency analysis performed on the samples considered in this paper show that the behavior became more similar to the one depicted in Figure 7c increasing T_{level}. Thus, on the basis of the frequency analysis, it is possible to conclude that the 160-18 sample, at 20 mm distance from the gate, replicates more accurately the steel layer. Being this latter characterized by a considerable portion of structures having high spatial frequencies, namely short distances between structures, its accurate replication imparts characteristics of hydrophobicity to the polymeric samples.

3.4. Determination of Relevant Sample Surface Parameters

The differences observed in the topography of the samples obtained with different evolutions of the cavity surface temperatures are responsible for the different hydrophobicity. Two models are considered to correlate the hydrophobicity to the surface topography: Wenzel [43] and Cassie–Baxter [44]. In the Wenzel regime, the hydrophobicity of a rough surface is enhanced by the increase of the solid-liquid contact area. In the Cassie–Baxter regime, wetted surface is composed by the solid and by the air, which results to be trapped between the droplet and the solid, in this case, the hydrophobicity increased by the decrease of the solid-liquid contact area. Figure 8 schematizes the two models applied to a random surface.

(a)   (b)

Figure 8. (a) Scheme of the Wenzel model and (b) Cassie–Baxter model applied to a random surface.

Wenzel and Cassie–Baxter models are reported in the Equations (1) and (2), respectively.

$$\cos\theta = r\cos\theta_0 \tag{1}$$

$$\cos\theta = f\cos\theta_0 - (1 - f) \tag{2}$$

The θ_0 is the Young contact angle corresponding to an ideal smooth surface. In this paper, the Young contact angle was measured on moldings obtained in the conditions named CIM, in which the cavity surface had a nominal roughness of 15 ± 2 nm. The molding roughness, being of 10 ± 2 nm, can be considered smooth [45], and its contact angle is $80°$, according to the smallest contact angle found in the moldings (see Figure 3).

The r is the ratio among the real rough surface area and the projected smooth surface, and it takes into account that the increase in contact angle is due to an increase of the wetted surface (which determines an increase of free energy) with respect to the projected one; f is a factor that enables taking into account that the droplet contacts a composed surface made of solid (on an area fraction f) and air (on an area fraction $1 - f$). r and f can be calculated directly from the 3D topographic parameters, as suggested by Vogler et al. [46], S_{dr}, S_{da} and S_{ha} mentioned in the method section [41,46]:

$$r = \frac{S_{dr} + 100}{100} \tag{3}$$

$$f = \frac{S_{ha}}{S_{ha} + S_{da}} \tag{4}$$

The topographic parameters were evaluated for all the considered AFM topographic maps to relate these parameters to the measured contact angles.

Figure 9 shows the contact angles, for the CIM sample and the samples obtained with different cavity surface temperature evolutions, as function of the parameters S_{dr} and f, respectively. The parameters S_{dr} and f were taken from the flattened AFM acquisitions. For each considered sample, the AFM acquisitions were taken in three different positions along the flow path.

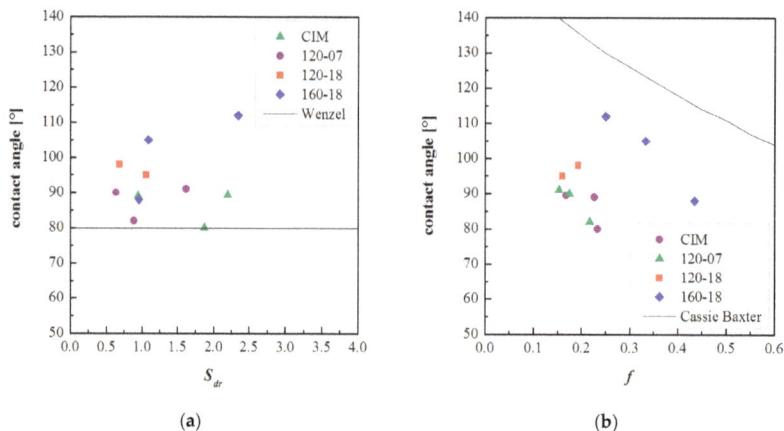

Figure 9. Contact angles vs. the parameters S_{dr} (**a**) and f (**b**) calculated from the AFM flattened maps. The Wenzel and the Cassie–Baxter models were also reported in each plot.

Figure 10 also shows the contact angles as function of the parameters S_{dr} and f but these parameters were calculated on the AFM acquisitions filtered with 0.2 μm^{-1} high pass cutoff frequency. Figures 9 and 10 reports also the Wenzel and the Cassie–Baxter model as black lines.

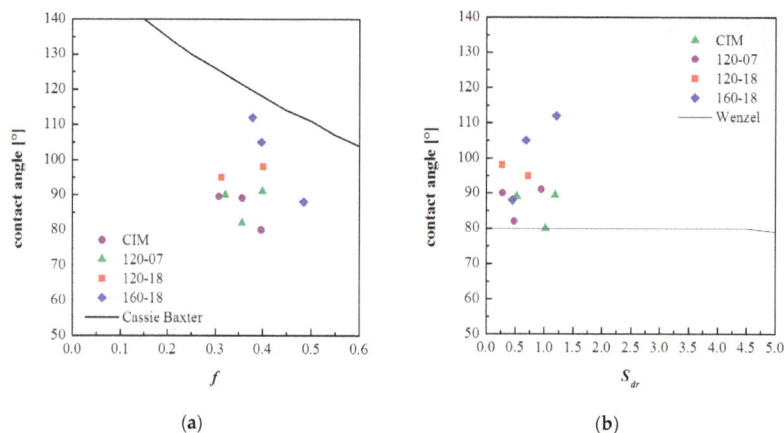

Figure 10. Contact angles vs. the parameters S_{dr} (**a**) and f (**b**) calculated on the AFM maps where a high pass filter of 0.2 μm^{-1} was applied. The Wenzel and the Cassie–Baxter models were also reported in each plot.

Figures 9a and 10a show that there is not a relationship between the contact angles and the surface parameter S_{dr}. Thus, the Wenzel model does not fit the experimental data.

The Cassie–Baxter model does not fit the experimental data; however, this model appears to be more suitable, especially for the sample 160-18, in describing the relationship between the measured contact angles and the surface characteristics. Figures 9b and 10b show that the parameter f has to be higher with respect to the calculated ones. The reason for this discrepancy could be due to the method adopted for calculating f, which does not take into account the overall complexity of the surface. The real surfaces are composed of structures on different scales, both micro and nano-metrical, with a complex harmonic content (see Figure 7). This aspect is not considered in the calculation of f, since S_{da} and S_{ha} are referred to projected areas; furthermore, they have averaged values that could be not representative of the real surface topography. The real surface is characterized by structures with a certain height and by different frequencies of dales and hills, this information has to be taken into account together in the calculation of the surface parameters.

4. Discussion

The steel layer adopted as cavity surface in this work is characterized by randomly dispersed structures (see Figures 5 and 6). These structures present micro and nano-metrical heights, i.e., roughness, and spatial distances between structures down to the nano-metrical level. The roughness is obviously not a good measure of the surface hydrophobicity, since surfaces showing similar roughness present quite different contact angles. The frequency analysis of the acquired patterns provides instead a method to check the replication accuracy both in the structure height and spatial distribution (Figure 7). The replication is enhanced by the temperature and heating time [47,48].

It is necessary to operate with long heating times to reach an accurate replication both in height and in the spatial distance, since the replication of structures is due to the filling of micro and nano-cavities by a pressure-driven mechanism. The temperature has to be high during the filling of the micro or nano-cavities to prevent solidification and crystallization that inhibit an accurate replication.

The moldings that show the most accurate replication of structures also show the highest hydrophobicity. In particular, the hydrophobicity increases for the surfaces showing the replication of structures characterized by high spatial frequencies, namely nano-metrical distances between structures.

The two models considered, Wenzel and Cassie–Baxter, identify relationships between the surface topography and the hydrophobicity. The latter model appears to better describe the behavior of the contact angles with respect to the surface topography. The discrepancy between the Cassie–Baxter model and the experimental findings might be due to the method adopted to evaluate the fraction f of the liquid in contact with the solid rather than with trapped gas, through the topographic parameters S_{da} and S_{ha}, which might be not suitable to describe a real liquid wetted surface. In particular, the trapped air prevents the complete filling of the dales by the liquid. While the topography of the surfaces proposed in the literature [26,41], being formed by isolated reliefs uniformly spaced on the surface, allows for a partial escaping of the trapped air, a random surface with closed dales could present limited escape routes for air. Furthermore, on a real surface, the liquid filled levels depend also on the pressure that the trapped air has to withstand, since its volume decreases with pressure. In addition, the topographic parameters S_{da} and S_{ha}, evaluated in this work according to the standards, are not representative of the real surface topography, being simply referred to the averaged projected areas of dales and hills; thus, neglecting the effect of their heights. Following the standard and also the literature [41], all these aspects were neglected in the calculation of the topographic parameters, and thus in the calculation of f; this may have determined a relevant part of the discrepancy between the Cassie–Baxter model predictions and the experimental findings, as shown in Figure 9.

5. Conclusions

In this work, a very efficient and new technique was adopted to obtain the evolution of the temperature on the cavity surface and to tune hydrophobicity of polyolefins. The heating device adopted to achieve the fast evolution of the cavity temperature allows for tuning the temperature locally, and it is also possible to keep constant the selected temperature for the desired time. Being limited to a small thickness below the cavity surface, the cooling after the heater deactivation results to be very fast, with a rate comparable with the cooling rate of the conventional injection molding process. This technique is very efficient in replicating structures, both in the height and shape, in a wide dimensional range, micro and nano-metric. By adopting this technique, the effect of the operating conditions on the hydrophobicity of molded iPP samples was analyzed. Furthermore, the correlation among the hydrophobicity and the surface topography of the moldings was explored. The replication of a random surface, composed by both sub-micro and nano-structures, was analyzed under different injection molding conditions. The presence of sub-micro and nano-structures characterized by high values of spatial frequency was detected and highlighted only in the molded samples obtained with high cavity surface temperatures.

The static water contact angles were measured on all molded samples at different distances from the gate along the flow path. The contact angles were found to be higher in the heated area and lower in the unheated area. A significant increase of the contact angle was measured when high cavity surface temperatures and long heating times were adopted during injection molding. Small heating times, comparable with the cavity filling time, were not sufficient to induce significant increases of the contact angles.

The experimental contact angles were plotted versus the main surface parameters, calculated from AFM acquisitions, to find correlations between sample surface morphology and the increase of the contact angle with respect to the Young value. Both the Wenzel and the Cassie–Baxter models were compared with the experimental data. It was found that the behavior of the contact angle with respect to the surface parameter f (which estimates the fraction of surface where the liquid is in contact with the solid rather than with trapped gas) was close to the behavior described by Cassie–Baxter model, especially for the sample obtained with the highest cavity surface temperature and the longest heating time. The residual discrepancy between the theoretical model and the experimental data could be attributed to the complexity of the description of both the sample surface topography and the split of water droplet contacts between the polymer and the entrapped air, which could not be adequately described into the parameter f.

Acknowledgments: Mariarosaria Marcedula (IPCB-CNR) is kindly acknowledged for their technical support. Andrea Sorrentino acknowledges financial support from the "VINMAC" project, ID 139455, D. R. Lombardia n. 9559 del 02/08/2017, CUP: E67H16000980009.

Author Contributions: Sara Liparoti and Giuseppe Titomanlio planned the experiments; Sara Liparoti performed the experiments; Vito Speranza analyzed the AFM data; Andrea Sorrentino measured the contact angles; Roberto Pantani analyzed the data and related them to the mathematical models; Sara Liparoti and Vito Speranza wrote the paper.

Conflicts of Interest: The authors declare no conflict of interest.

References

1. Marmur, A. The Lotus Effect: Superhydrophobicity and Metastability. *Langmuir* **2004**, *20*, 3517–3519. [CrossRef] [PubMed]
2. Tian, X.; Verho, T.; Ras, R.H.A. Moving superhydrophobic surfaces toward real-world applications. *Science* **2016**, *352*, 142–143. [CrossRef] [PubMed]
3. Kojevnikova, S.; Marmur, A. Multi-scale roughness and the Lotus effect: Discontinuous liquid-air interfaces. *Colloids Surf. A Physicochem. Eng. Asp.* **2017**, *521*, 78–85. [CrossRef]
4. Wang, H.; Zhu, Y.; Hu, Z.; Zhang, X.; Wu, S.; Wang, R.; Zhu, Y. A novel electrodeposition route for fabrication of the superhydrophobic surface with unique self-cleaning, mechanical abrasion and corrosion resistance properties. *Chem. Eng. J.* **2016**, *303*, 37–47. [CrossRef]
5. Quan, Y.-Y.; Zhang, L.-Z.; Qi, R.-H.; Cai, R.-R. Self-cleaning of Surfaces: The Role of Surface Wettability and Dust Types. *Sci. Rep.* **2016**, *6*, 38239. [CrossRef] [PubMed]
6. Förstner, R.; Barthlott, W.; Neinhuis, C.; Walzel, P. Wetting and Self-Cleaning Properties of Artificial Superhydrophobic Surfaces. *Langmuir* **2005**, *21*, 956–961. [CrossRef] [PubMed]
7. Liu, Y.; Yao, W.; Yin, X.; Wang, H.; Han, Z.; Ren, L. Controlling Wettability for Improved Corrosion Inhibition on Magnesium Alloy as Biomedical Implant Materials. *Adv. Mater. Interfaces* **2016**, *3*, 1500723. [CrossRef]
8. Falde, E.J.; Yohe, S.T.; Colson, Y.L.; Grinstaff, M.W. Superhydrophobic materials for biomedical applications. *Biomaterials* **2016**, *104*, 87–103. [CrossRef] [PubMed]
9. Shin, S.; Seo, J.; Han, H.; Kang, S.; Kim, H.; Lee, T. Bio-Inspired Extreme Wetting Surfaces for Biomedical Applications. *Materials* **2016**, *9*, 116. [CrossRef] [PubMed]

10. Zhou, J.; Frank, M.A.; Yang, Y.; Boccaccini, A.R.; Virtanen, S. A novel local drug delivery system: Superhydrophobic titanium oxide nanotube arrays serve as the drug reservoir and ultrasonication functions as the drug release trigger. *Mater. Sci. Eng. C* **2018**, *82*, 277–283. [CrossRef] [PubMed]

11. Shen, Y.; Liu, S.; Zhu, C.; Tao, J.; Wang, G. Facile fabrication of hierarchical structured superhydrophobic surface and its ultra dynamic water repellency. *Chem. Eng. J.* **2017**, *313*, 47–55. [CrossRef]

12. Gray-Munro, J.; Campbell, J. Mimicking the hierarchical surface topography and superhydrophobicity of the lotus leaf on magnesium alloy AZ31. *Mater. Lett.* **2017**, *189*, 271–274. [CrossRef]

13. MacGregor-Ramiasa, M.N.; Vasilev, K. Questions and Answers on the Wettability of Nano-Engineered Surfaces. *Adv. Mater. Interfaces* **2017**, *4*, 1700381. [CrossRef]

14. Rogers, J.A.; Lee, H.H. *Unconventional Nanopatterning Techniques and Applications*; Wiley: Hoboken, NJ, USA, 2008; ISBN 9780470099575.

15. Cesca, F.; Limongi, T.; Accardo, A.; Rocchi, A.; Orlando, M.; Shalabaeva, V.; Di Fabrizio, E.; Benfenati, F. Fabrication of biocompatible free-standing nanopatterned films for primary neuronal cultures. *RSC Adv.* **2014**, *4*, 45696–45702. [CrossRef]

16. Limongi, T.; Tirinato, L.; Pagliari, F.; Giugni, A.; Allione, M.; Perozziello, G.; Candeloro, P.; Di Fabrizio, E. Fabrication and Applications of Micro/Nanostructured Devices for Tissue Engineering. *Nano-Micro Lett.* **2017**, *9*, 1. [CrossRef]

17. Yang, C.; Huang, H.-X.; Castro, J.M.; Yi, A.Y. Replication characterization in injection molding of microfeatures with high aspect ratio: Influence of layout and shape factor. *Polym. Eng. Sci.* **2011**, *51*, 959–968. [CrossRef]

18. Limongi, T.; Schipani, R.; Di Vito, A.; Giugni, A.; Francardi, M.; Torre, B.; Allione, M.; Miele, E.; Malara, N.; Alrasheed, S.; et al. Photolithography and micromolding techniques for the realization of 3D polycaprolactone scaffolds for tissue engineering applications. *Microelectron. Eng.* **2015**, *141*, 135–139. [CrossRef]

19. Hwa, K.-Y.; Chang, V.H.S.; Cheng, Y.-Y.; Wang, Y.-D.; Jan, P.-S.; Subramani, B.; Wu, M.-J.; Wang, B.-K. Analyzing polymeric matrix for fabrication of a biodegradable microneedle array to enhance transdermal delivery. *Biomed. Microdevices* **2017**, *19*, 84. [CrossRef] [PubMed]

20. Calaon, M.; Tosello, G.; Hansen, H.N.; Nøregaard, J. Influence of process parameters on edge replication quality of lab-on-a-chip micro fluidic systems geometries. In Proceedings of the Annual Technical Conference—ANTEC Conference Proceedings, Cincinnati, OH, USA, 22–24 April 2013; Volume 2, pp. 1084–1088.

21. Baruffi, F.; Calaon, M.; Tosello, G. Effects of micro-injection moulding process parameters on accuracy and precision of thermoplastic elastomer micro rings. *Precis. Eng.* **2017**. [CrossRef]

22. Calaon, M.; Tosello, G.; Garnaes, J.; Hansen, H.N. Injection and injection-compression moulding replication capability for the production of polymer lab-on-a-chip with nano structures. *J. Micromech. Microeng.* **2017**, *27*, 105001. [CrossRef]

23. Menotti, S.; Hansen, H.N.; Bissacco, G.; Calaon, M.; Tang, P.T.; Ravn, C. Injection molding of nanopatterned surfaces in the sub-micrometer range with induction heating aid. *Int. J. Adv. Manuf. Technol.* **2014**, *74*, 907–916. [CrossRef]

24. Speranza, V.; Liparoti, S.; Calaon, M.; Tosello, G.; Pantani, R.; Titomanlio, G. Replication of micro and nano-features on iPP by injection molding with fast cavity surface temperature evolution. *Mater. Des.* **2017**, *133*, 559–569. [CrossRef]

25. Yamaguchi, M.; Sasaki, S.; Suzuki, S.; Nakayama, Y. Injection-molded plastic plate with hydrophobic surface by nanoperiodic structure applied in uniaxial direction. *J. Adhes. Sci. Technol.* **2015**, *29*, 24–35. [CrossRef]

26. Puukilainen, E.; Rasilainen, T.; Suvanto, M.; Pakkanen, T.A. Superhydrophobic polyolefin surfaces: Controlled micro- and nanostructures. *Langmuir* **2007**, *23*, 7263–7268. [CrossRef] [PubMed]

27. Yoo, Y.E.; Kim, T.H.; Choi, D.S.; Hyun, S.M.; Lee, H.J.; Lee, K.H.; Kim, S.K.; Kim, B.H.; Seo, Y.H.; Lee, H.G.; et al. Injection molding of a nanostructured plate and measurement of its surface properties. *Curr. Appl. Phys.* **2009**, *9*, e12–e18. [CrossRef]

28. Puukilainen, E.; Koponen, H.K.; Xiao, Z.; Suvanto, S.; Pakkanen, T.A. Nanostructured and chemically modified hydrophobic polyolefin surfaces. *Colloids Surf. A Physicochem. Eng. Aspects* **2006**, *287*, 175–181. [CrossRef]

29. Liparoti, S.; Sorrentino, A.; Speranza, V.; Titomanlio, G. Multiscale mechanical characterization of iPP injection molded samples. *Eur. Polym. J.* **2017**, *90*, 79–91. [CrossRef]

30. Chen, S.-C.; Lin, Y.-W.; Chien, R.-D.; Li, H.-M. Variable mold temperature to improve surface quality of microcellular injection molded parts using induction heating technology. *Adv. Polym Technol.* **2008**, *27*, 224–232. [CrossRef]

31. Lin, H.L.; Chen, S.C.; Jeng, M.C.; Minh, P.S.; Chang, J.A.; Hwang, J.R. Induction heating with the ring effect for injection molding plates. *Int. Commun. Heat Mass Transf.* **2012**, *39*, 514–522. [CrossRef]

32. Yao, D.; Kimerling, T.E.; Kim, B. High-frequency proximity heating for injection molding applications. *Polym Eng. Sci.* **2006**, *46*, 938–945. [CrossRef]

33. Su, Q.; Zhang, N.; Gilchrist, M.D. The use of variotherm systems for microinjection molding. *J. Appl. Polym Sci.* **2016**, *133*. [CrossRef]

34. Kim, B.M.; Niemeyer, M.F. Insulated Mold Structure for Injection Molding of Optical Disks. U.S. Patent 5,458,818A, 31 August 1993.

35. Pantani, R.; Speranza, V.; Titomanlio, G. Simultaneous morphological and rheological measurements on polypropylene: Effect of crystallinity on viscoelastic parameters. *J. Rheol.* **2015**, *59*, 377–390. [CrossRef]

36. Pantani, R.; Speranza, V.; Titomanlio, G. Evolution of iPP Relaxation Spectrum during Crystallization. *Macromol. Theory Simul.* **2014**, *23*, 300–306. [CrossRef]

37. Liparoti, S.; Landi, G.; Sorrentino, A.; Speranza, V.; Cakmak, M.; Neitzert, H.C. Flexible Poly(Amide-Imide)-Carbon Black Based Microheater with High-Temperature Capability and an Extremely Low Temperature Coefficient. *Adv. Electron. Mater.* **2016**, *2*, 1600126. [CrossRef]

38. Leach, R. *Characterisation of Areal Surface Texture*; Leach, R., Ed.; Springer: Berlin/Heidelberg, Germany, 2013; ISBN 978-3-642-36457-0.

39. Vietri, U.; Sorrentino, A.; Speranza, V.; Pantani, R. Improving the predictions of injection molding simulation software. *Polym Eng. Sci.* **2011**, *51*, 2542–2551. [CrossRef]

40. Wu, P.H.; Cheng, C.W.; Chang, C.P.; Wu, T.M.; Wang, J.K. Fabrication of large-area hydrophobic surfaces with femtosecond-laser-structured molds. *J. Micromech. Microeng.* **2011**, *21*. [CrossRef]

41. Belaud, V.; Valette, S.; Stremsdoerfer, G.; Bigerelle, M.; Benayoun, S. Wettability versus roughness: Multi-scales approach. *Tribol. Int.* **2015**, *82*, 343–349. [CrossRef]

42. Lucchetta, G.; Fiorotto, M.; Bariani, P.F. Influence of rapid mold temperature variation on surface topography replication and appearance of injection-molded parts. *CIRP Ann. Manuf. Technol.* **2012**, *61*, 539–542. [CrossRef]

43. Wenzel, R.N. Resistance of Solid Surfaces to Wetting by Water. *Ind. Eng. Chem.* **1936**, *28*, 988–994. [CrossRef]

44. Cassie, A.B.D.; Baxter, S. Wettability of porous surfaces. *Trans. Faraday Soc.* **1944**, *40*, 546. [CrossRef]

45. Erbil, H.Y.; Demirel, A.L.; Avci, Y.; Mert, O. Transformation of a Simple Plastic into a Superhydrophobic Surface. *Science* **2003**, *299*, 1377–1380. [CrossRef] [PubMed]

46. Vogler, E.A. Structure and reactivity of water at biomaterial surfaces. *Adv. Colloid Interface Sci.* **1998**, *74*, 69–117. [CrossRef]

47. Zitzenbacher, G.; Längauer, M.; Holzer, C. Modeling Temperature and Time Dependence of the Wetting of Tool Steel Surfaces by Polymer Melts. *Int. Polymer Process.* **2017**, *32*, 245–252. [CrossRef]

48. Rytka, C.; Opara, N.; Andersen, N.K.; Kristiansen, P.M.; Neyer, A. On The Role of Wetting, Structure Width, and Flow Characteristics in Polymer Replication on Micro- and Nanoscale. *Macromol. Mater. Eng.* **2016**, *301*, 597–609. [CrossRef]

polymers

MDPI

Article

Tube Expansion Deformation Enables In Situ Synchrotron X-ray Scattering Measurements during Extensional Flow-Induced Crystallization of Poly L-Lactide Near the Glass Transition

Karthik Ramachandran [1], Riccardo Miscioscia [2], Giovanni De Filippo [3], Giuseppe Pandolfi [2], Tiziana Di Luccio [1,2] and Julia A. Kornfield [1,*]

[1] Division of Chemistry and Chemical Engineering, California Institute of Technology, Pasadena, CA 91125, USA; kramacha@caltech.edu (K.R.); tidilu@caltech.edu (T.D.L.)
[2] Division of Sustainable Materials, ENEA, Centro Ricerche Portici, 80055 Portici, Italy; riccardo.miscioscia@enea.it (R.M.); giuseppe.pandolfi@enea.it (G.P.)
[3] Division of Photovoltaics and Smart Networks, Innovative Device Unit, Centro Ricerche Portici, 80055 Portici, Italy; giovanni.defilippo@enea.it
* Correspondence: jak@cheme.caltech.edu; Tel.: +1-626-395-4138

Received: 11 February 2018; Accepted: 6 March 2018; Published: 8 March 2018

Abstract: Coronary Heart Disease (CHD) is one of the leading causes of death worldwide, claiming over seven million lives each year. Permanent metal stents, the current standard of care for CHD, inhibit arterial vasomotion and induce serious complications such as late stent thrombosis. Bioresorbable vascular scaffolds (BVSs) made from poly L-lactide (PLLA) overcome these complications by supporting the occluded artery for 3–6 months and then being completely resorbed in 2–3 years, leaving behind a healthy artery. The BVS that recently received clinical approval is, however, relatively thick (~150 μm, approximately twice as thick as metal stents ~80 μm). Thinner scaffolds would facilitate implantation and enable treatment of smaller arteries. The key to a thinner scaffold is careful control of the PLLA microstructure during processing to confer greater strength in a thinner profile. However, the rapid time scales of processing (~1 s) defy prediction due to a lack of structural information. Here, we present a custom-designed instrument that connects the strain-field imposed on PLLA during processing to in situ development of microstructure observed using synchrotron X-ray scattering. The connection between deformation, structure and strength enables processing–structure–property relationships to guide the design of thinner yet stronger BVSs.

Keywords: PLLA; bioresorbable vascular scaffolds; stretch blow molding; biaxial elongation; SAXS; WAXS

---

1. Introduction

Coronary heart disease (CHD) results in obstructed blood flow to the heart due to the deposition of plaque on arterial walls. It is one of the leading causes of death in the world and claims over seven million lives each year [1–3]. Bioresorbable vascular scaffolds (BVSs) are emerging as a promising alternative to permanent metal stents for the treatment of CHD. These devices are referred to as "scaffolds" as opposed to "stents" owing to their transient nature in the implanted artery. In contrast to permanent metal stents, BVSs support the occluded artery for 3–6 months but are completely resorbed in 2–3 years [4], restoring vasomotion in the artery and minimizing the risk of fatal complications such as late stent thrombosis [5–7]. The first FDA-approved BVS [8] is made from the semicrystalline and biocompatible polymer poly L-lactide (PLLA), which hydrolyzes to form L-lactic acid, a metabolic product processed by the body [9–11]. However, PLLA is notorious for being a brittle polymer [12,13], making it a surprising choice for a device that must withstand crimping and deployment. PLLA blends [14,15] and copolymers [16,17] have superior ductility but these materials are not suitable for coronary implants as they prematurely lose strength due to rapid hydrolysis [18,19]. Therefore, processing conditions play a key role in overcoming PLLA's inherent brittleness by providing the BVS with a microstructure that facilitates deployment and confers lasting radial strength.

Bioresorbable vascular scaffolds are processed from extruded PLLA preforms in the following sequence: "tube expansion" biaxially stretches the initially amorphous preform and transforms it into a semicrystalline tube; "laser-cutting" converts the expanded tube into an as-cut scaffold that has a lattice network of struts; and "crimping" radially compresses the as-cut scaffold onto a balloon catheter for deployment in the artery [20]. The scaffold experiences severe local deformation during crimping (strain > 50%), yet shows no sign of failure upon deployment [20]. The strength of the scaffold in the deployed state has a counter-intuitive relationship with strength in the expanded tube. Tube expansion that is equibiaxial (200% strain in θ and 200% strain in z) creates a strong expanded tube; however, after laser-cutting and crimping, it performs poorly upon deployment (>40 cracks per scaffold when over-deployed) [20]. In contrast, tube expansion that is predominantly uniaxial (400% strain in θ and 20% strain in z) does not provide as strong an expanded tube, yet it produces scaffolds that perform well upon deployment (<10 cracks per scaffold when over-deployed) [20]. Microdiffraction results reconciled this disconnect in strength between the expanded tube and the deployed scaffold by elucidating the role of crimping; the predominantly uniaxial elongation process yields an expanded tube that develops a beneficial morphology in the crimped state that facilitates deployment without fracture [20]. The interplay of structural transformations that occur during tube expansion and crimping govern the strength of the deployed BVS, motivating investigation of structure development during the tube expansion process.

The promising success of current 150 μm thick BVSs motivates the design of thinner scaffolds (~80 μm, similar to metal stents [21,22]) to extend the benefits of resorbable implants to a broader patient population. The key to a thinner BVS lies in careful control of its microstructure during processing so that an 80 μm BVS has strength comparable to the current 150 μm FDA-approved BVS. Tube expansion, the first step in the manufacture of a vascular scaffold, plays a central role in the design of thinner scaffolds. Similar to stretch blow molding, tube expansion subjects a PLLA preform to biaxial elongation near the glass transition inside an outer mold [23]. The rapid deformation induces oriented crystallization, which in turn influences the macroscopic strain and the wall thickness of the resulting BVS. The process of tube expansion has been applied with excellent reproducibility to manufacture scaffolds with uniform wall thickness that meet clinical standards. However, conventional methodology is unable to provide structural insight on tube expansion at time scales relevant to processing (>400% strain in a matter of seconds). As a result, tube expansion is largely based on trial-and-error; expanded tubes are subjected to mechanical characterization and process parameters are iteratively adjusted to optimize strength.

Here, we discuss the implementation of a novel custom-built instrument that can subject polymeric preforms to tube expansion with acquisition of synchrotron X-ray scattering data at time scales relevant to processing (~1 s). To the best of our knowledge, the data acquired from this instrument are the first of their kind and relate deformation in the cylindrical geometry to the microstructure of the expanding preform. PLLA is subjected to tube expansion between 70 to 120 °C [24], relatively close to its glass transition (Tg ~55 °C) and well below its melting temperature (Tm ~170 °C). However, it is challenging to probe flow-induced crystallization in this temperature interval as PLLA is known to have poor thermal stability at elevated temperatures [25] and undergoes quiescent crystallization in the vicinity of 100 °C [26]. Therefore, a critical design requirement for our instrument is rapid heating—we are able to heat the preform above 100 °C (heating rate ~70 °C/min) and achieve inflation in less than 100 s; as a result, we avoid thermal degradation and isolate the impact of flow on the microstructure of the expanded tube. The primary emphasis of this article is on the morphology of PLLA, but the instrument and the insight gained from this approach can be broadly applied to study polymers (e.g., poly (ethylene terephthalate), PET) that are processed using stretch-blow molding.

2. Materials and Methods

2.1. Design Considerations for the Tube Expansion Instrument

The design of our instrument builds upon the production protocol described in the patent literature for bioresorbable vascular scaffolds, specifically the tube expansion process described above [27]. Tube expansion uses compressed air to inflate a PLLA preform (inner diameter (ID): 0.64 mm, outer diameter (OD): 1.5 mm) inside a glass mold (ID: 3.9 mm, OD: 6.0 mm) at a desired temperature; the inflation transforms the initially amorphous preform into an oriented, semicrystalline tube. The mold dimensions are selected to limit the maximum expansion to an OD of 160% relative to the preform's OD; in the present experiments, the expansion is self-arresting when the OD reaches approximately 150% of the preform's OD. Infrared heating (IR) is used to provide a rise time of approximately 50 s from 35 to 100 °C, which is fast enough to avoid quiescent crystallization in the PLLA preform [26]. In contrast to hot air convection or restive heating, which both act on the mold, IR directly heats the preform. The choice of IR radiation motivates the use of borosilicate glass (Pyrex) for the mold as Pyrex is mostly transparent to IR wavelengths that are strongly absorbed by PLLA (see Supplementary Materials: Figure S1 defines the orientation of lamps relative to the mold and Figure S2 describes the energy absorbed by the preform relative to the mold). The instrument is designed to selectively vary processing parameters such as the expansion temperature (T^e), the crystallization temperature (T^x), the imposed pressure inside the tube during expansion (only one value is examined here, 6.8 bar) and the annealing conditions to isolate their impact on the resulting PLLA microstructure.

2.2. Temperature Control

The PLLA preform is heated above its glass transition temperature (Tg ~55 °C) using two 500 W Philips T3 halogen lamps (OD: 10 mm, length: 118 mm, Figure 1a) that are placed 25 mm on either side of the preform (Figure 1a). The lamps are operated using a 48 V power supply for user safety, which results in an effective power of ~100 W per lamp. The lamps are oriented parallel to the preform as this geometry minimizes gradients in temperature along the axial direction (~10° over 30 mm, Figure S1a). The lamps are surrounded by curved reflectors that improve the uniformity of heating and direct stray radiation towards the preform. Numerical simulations performed in Zemax indicate that the parallel orientation of lamps minimizes axial gradients in the absorbed energy (heat flux varies by ~10% along the 60 mm length of the preform, Figure S2) and that the reflectors enhance the rate of heating (the preform absorbs >200% more energy with the curved reflectors, Figure S2). The control box (described below) continuously records the temperature of the mold using a K-type thermocouple connected to a Maxim MAX6675 cold-junction compensated controller, which adjusts the power of the lamps to match the set heating rate and temperature.

As the preform is expanded inside the Pyrex mold, it is challenging to experimentally probe the temperature of the preform. Therefore, we turn to numerical simulations that capture the geometry of the stretching apparatus to estimate the temperature of the PLLA preform. The volumetric heat flux provided by Zemax (~2.3 mW/mm^3 for the mold and ~4.5 mW/mm^3 for the preform, Figure S2) is incorporated in the finite element software Abaqus to predict spatial gradients in temperature for the mold and the preform as a function of time. The simulations predict a heating rate of ~1.1 °C/s for the mold and ~1.75 °C/s for the preform (Figure S3); the simulated heating rate of the mold is similar to experimental thermocouple data acquired on the mold (~1.2 °C/s, see Figure S3b, left for comparison). Furthermore, the simulations show that spatial gradients in temperature in the preform are ~1 °C after 50 s of heating (Figure S3a, right); a uniform temperature profile of the preform is consistent with the uniform thickness of the expanded tube (~140 μm). The higher temperature of the preform relative to the mold is corroborated by supporting experimental data, which also reveal that the preform temperature does not increase during the annealing step (Figure S4).

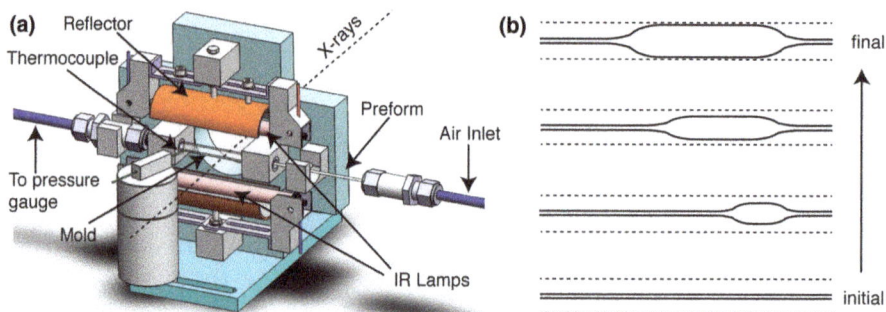

Figure 1. (**a**) A 3D diagram of the tube expansion instrument defines the positions of the IR lamps and reflectors above and below the mold, the thermocouple on the mold surface, the connection to the pressure transducer and the air inlet, and illustrates how the preform is inserted into the mold and oriented relative to the path of the X-rays. (**b**) Schematic diagram of the inflation process, which progresses from the side of tube attached to the air inlet to the opposite end.

2.3. Mechanical Assembly

The instrument core is comprised of a metal bracket that holds the IR lamps, the mold and the preform in position using Teflon supports (Figure 1a). The holders for the IR lamps are equipped with reflectors to direct stray radiation towards the mold-preform assembly (Figure 1a). A circular cutout in the metal bracket permits a path for X-rays to interact with the expanding preform (Figure 1a). Swagelok connectors secure one end of the preform to a source of compressed air (6.8 bar) and the opposite end to a Honeywell HSCDANN010BGAA5 pressure sensor (range 0–10 bar). The control box operates a pressure valve to trigger tube expansion, which propagates from the side of the tube attached to the air inlet towards the side connected to the pressure transducer (Figure 1b).

2.4. Control Box

The control box is comprised of an Arduino Mega 2560 board (microcontroller unit, MCU) manufactured by Arduino, Italy that operates at a 16 MHz clock speed to acquire data from the sensors and to drive the actuators (Figure 2). The MCU uses a 5 V relay to operate the IR lamps and a numerically controlled dimmer to apply proportional-integral-derivative (PID) control to regulate the power of the lamps in accord with the set heating rate and temperature (Figure 2). A Parker Lucifer 341 L91 07 coil pressure valve is operated using a 24 V DC power supply and is controlled by the MCU via a 5 V relay to trigger expansion at user-defined setpoints (Figure 2). A Java-based desktop application issues commands to the control box and is responsible for plotting and logging data.

2.5. Materials

Clinical grade poly L-lactide (PLLA) preforms (ID: 0.64 mm, OD: 1.5 mm) were provided by Abbott Vascular, Santa Clara, CA, USA.

2.6. In Situ Structural Characterization

The instrument was implemented at APS beamline 5-ID-D at the Argonne National Labs for the acquisition of simultaneous wide (WAXS) and small-angle X-ray scattering (SAXS) data. The incident X-ray beam is aligned parallel to the *r*-direction of the mold-preform assembly (Figure S5); diffraction patterns are acquired every 1 s at an exposure of 0.5 s using X-rays of wavelength = 0.7293 Å. WAXS and

SAXS images were acquired on Rayonix CCD detectors with a sample to detector distance of 200.83 mm and 8.502 m, respectively. The wavevector q is calibrated using a spinning silicon diffraction grid. A subtraction method is implemented to isolate the scattering of PLLA (thickness of expanded tube ~140 μm) from that of the Pyrex mold (~1 mm thick) in the WAXS patterns (described in detail in the Supplementary Materials, Figures S6–S10). At small angles, there is little to no scattering from the Pyrex mold and the in situ SAXS patterns are presented as is (no subtraction). Video recordings from the beamline camera (Figure S5) permit estimation of the strain and strain-rate experienced by the PLLA preform during expansion.

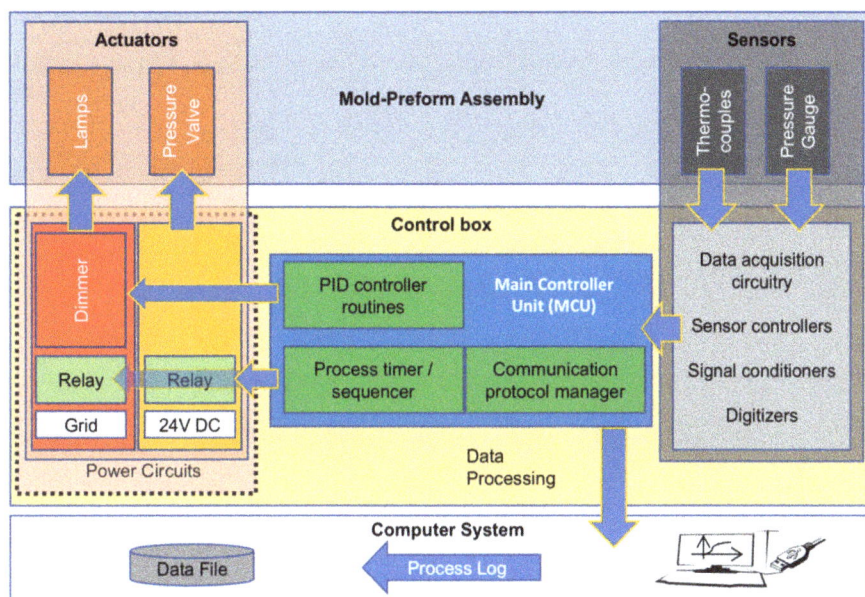

Figure 2. Block diagram describing the interface between the sensors, actuators, control unit and the data acquisition systems.

3. Results

Poly L-lactide (PLLA) cylindrical preforms are subjected to expansion using two modes of operation of the instrument. In the first mode, the pressure in the preform is not increased until a temperature of interest is reached during heating to a selected crystallization temperature: here, we use the temperature of the mold (T^e_{mold} of 80, 90 and 100 °C, where superscript "e" denotes "expansion") to trigger the elongation by opening a valve to 6.8 bar air, and use the mold temperature for feedback control during the isothermal crystallization process ($T^a_{mold} = 100$ °C, where superscript "a" denotes "annealing") for 5 min, followed by cooling under ambient conditions (Figure 3a). In the second mode of operation, the pressure in the preform is imposed prior to heating and is maintained throughout the process: here, we examine the effect of temperature during isothermal crystallization (T^x_{mold} of 80, 90 and 100 °C, where superscript "x" denotes "crystallization") for 5 min, followed by cooling under ambient conditions (Figure 3b). Numerical simulations predict that mold temperatures of 80, 90 and 100 °C correspond to PLLA preform temperatures of ~105, 115 and 125 °C, respectively (Figures S2 and S3). These two modes of operation were selected to isolate the impact of the expansion temperature (T^e) from that of the crystallization temperature (T^x) on the morphology of the expanded tube; expansion in the first mode is triggered at a set mold temperature while expansion in the second mode is reliant on the material properties of the preform.

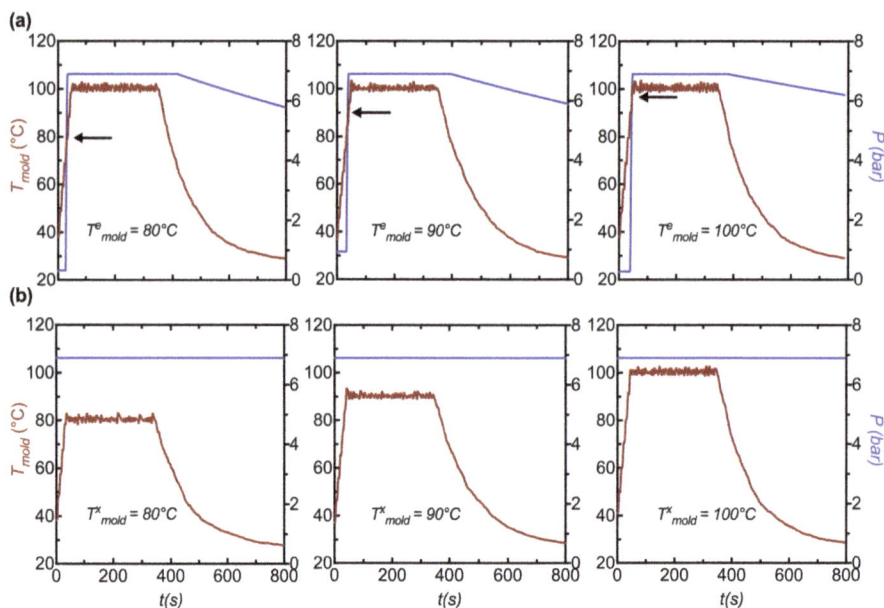

Figure 3. The transient temperature (red traces) and pressure (blue traces) profiles are presented in (**a**) for the first mode of operation and in (**b**) for the second mode of operation. In (**a**), inflation is triggered at a set mold temperature ($T^e{}_{mold}$ = 80, 90 and 100 °C) and the tube is subsequently annealed at a mold temperature of 100 °C for 5 mins prior to cooling. In (**b**), the tubes are pressurized prior to heating (note the constant pressure trace) and instead the tubes are annealed at three different mold temperatures ($T^x{}_{mold}$ = 80, 90 and 100 °C) for 5 mins prior to cooling. Numerical simulations described in the SI (Figures S2 and S3) predict that mold temperatures of 80, 90 and 100 °C translate to preform temperatures of ~105, 115 and 125 °C respectively.

Video recordings are used to estimate the transient outer diameter (OD) of the preform, which changes in a complex manner with time (Figure 4). While some elongation may occur along the axial direction, it is much less than the hoop elongation. Here, we make the approximation of neglecting the axial elongation and analyze the outer diameter to estimate the strain at the outer surface. When the internal pressure is abruptly imposed at a prescribed temperature (Figure 4a, i), the initial deformation progresses at a similar rate (~0.025 s^{-1} at the OD, Figure 4a, ii) till an inflection point at an OD strain of 40–50% (marked by black arrows in Figure 4a, ii). After the inflection, the strain rate increases dramatically, to approximately 0.07 s^{-1} (Figure 4a, ii), until the deformation rate slows down when the OD strain approaches ~150%. The rates of deformation are similar in the first mode of operation ($T^e{}_{mold}$ = 80, 90 and 100 °C) as the mold temperature has reached a steady state (100 °C) during inflation.

When the internal pressure is imposed before heating and maintained during heating (Figure 4b, i), the onset of deformation occurs when the PLLA preform reaches a temperature at which it can no longer withstand the pressure (Figure 4b, ii). This occurs approximately when the temperature of the mold reaches 80 °C. In the case of $T^x{}_{mold}$ = 80 °C, the mold has already reached a steady temperature when the PLLA begins to deform. On the other hand, for $T^x{}_{mold}$ = 90 and 100 °C, the temperature rises rapidly during the first several seconds of deformation, giving rise to progressively faster initial strain rates (the strain rate increases from 0.015 s^{-1} at $T^x{}_{mold}$ = 80 °C to 0.07 s^{-1} at $T^x{}_{mold}$ = 100 °C, t < 50 s in Figure 4b, ii). The initial deformation slows, giving a plateau prior to a steep increase in strain rate; the plateau OD strain is approximately 30–40% for all three values of $T^x{}_{mold}$, while the duration of the plateau decreases strongly with an increase in T^x_{mold} (compare strain profiles in Figure 4b, ii). The rapid deformation at the end of the plateau rolls off and the final OD strain is approximately 150% for all three T^x_{mold} cases. We now relate the macroscopic change in temperature and strain to the microstructure of the tube using in situ synchrotron X-ray scattering.

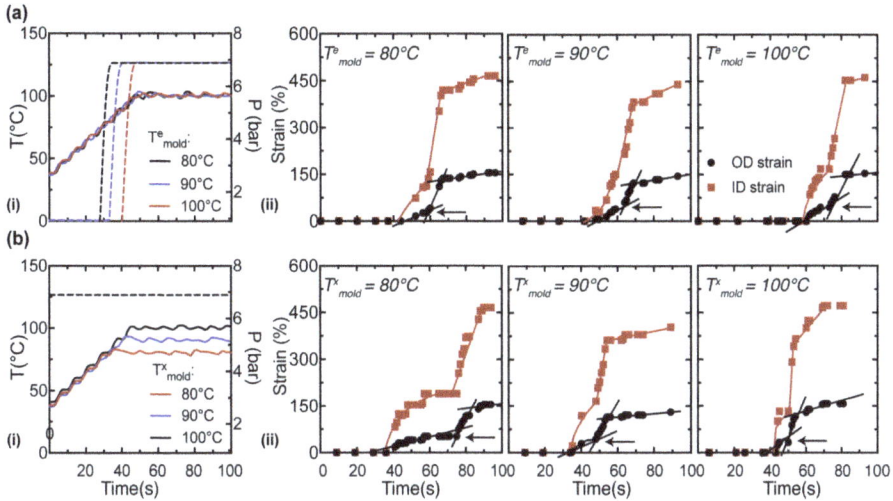

Figure 4. Plot of (**a, i**) temperature (solid lines) and pressure traces (dashed lines) for the first mode of operation; the pressure valve is triggered in ~5 s intervals to achieve expansion at T^e_{mold} = 80, 90 and 100 °C. The corresponding strain fields are presented in (**a, ii**). Plot of (**b, i**) temperature (solid lines) and pressure traces (dashed lines) for the second mode of operation; tubes expanded at T^x_{mold} = 80, 90 and 100 °C are pressurized to 6.87 bar prior to heating. The corresponding strain fields are presented in (**b, ii**). The strain at the outer diameter (OD, black circles in **a,b, ii**) of the preform is determined from video recordings (error in strain ~5%) and the strain at the inner diameter (ID, red squares in **a,b, ii**) is calculated from the OD strain assuming incompressibility (the difference in density between amorphous and crystalline PLLA is ~3% [28]). The slope of the straight black lines provides an estimate of the strain rate at the OD surface; the black arrows mark the inflection point when the strain rate abruptly increases. The error in synchronization between the video recordings and the synchrotron data (temperature, pressure and X-ray patterns) is ~5 s.

Here, we compare PLLA preforms stretched at T^e_{mold} = 100 °C and T^x_{mold} = 100 °C (in situ data for the other 4 experiments are presented in Figures S11–S14) to illustrate the impact of the onset of inflation (deformation occurs ~20 s later in T^e_{mold} = 100 °C relative to T^x_{mold} = 100 °C, Figure 4a,b, ii) on the morphology of the expanded tube. A selection of 12 diffraction patterns are presented for T^e_{mold} = 100 °C and T^x_{mold} = 100 °C that capture the transient expansion of the preform. Prior to expansion, we observe strong diffuse scattering from the amorphous phase and the presence of faint (110)/(200) peaks in the preform (compare Figures 5 and 6a, 0 s); the azimuthal position of these peaks suggests that a small population of crystallites have their c-axis oriented parallel to the z-direction of the preform. It is likely that these z-oriented crystallites are created in the extrusion process used to prepare the preforms. The initial phase of inflation occurs at a relatively slow strain rate (~0.025 s^{-1}) before the inflection point, Figure 4a, ii) and is characterized by a decrease in the amorphous intensity due to the thinning of preform; however, there is no indication of oriented crystallization in this interval (t < 70 s for T^e_{mold} = 100 °C and t < 45 s for T^x_{mold} = 100 °C, Figures 5 and 6a). After the initial inflection in the strain field (black arrows in Figure 4a,b, ii), the strain rate suddenly increases to ~0.07 s^{-1} and is accompanied by distinct (110)/(200) peaks with their c-axis oriented parallel to the θ-direction of the preform (at ~68 s for T^e_{mold} = 100 °C and ~44 s for T^x_{mold} = 100 °C, Figures 5 and 6a). The intensity of the newly formed θ-oriented (110)/(200) peaks increases rapidly over the next 10 seconds (68 to 77 s for T^e_{mold} = 100 °C and 44 to 55 s for T^x_{mold} = 100 °C; Figures 5 and 6a), a time interval that corresponds to a > 50% increase in OD strain (compare the strain field in Figure 4a,b with WAXS data in Figures 5 and 6a,b). Beyond 100 s, the expansion is nearly complete and amorphous material is steadily transformed into crystallites during annealing (Figures 5 and 6a, 350 s).

The azimuthal width of the (110)/(200) peaks (Figures 5 and 6a, ii) provides a measure of crystallite orientation in a material. In the first mode of operation, expansion is triggered in ~5 s intervals at a mold temperature of 80, 90 and 100 °C (the pressure pulse requires ~10 s to reach 6.8 bar in the tube, see Figure 4a, i). However, the bulk of the expansion occurs when the mold temperature has reached 100 °C. Therefore, we observe similar rates of expansion and consequently, the azimuthal width of the oriented PLLA crystallites are similar (~8° for T^e_{mold} = 80 °C and ~10° for both T^e_{mold} = 90 and 100 °C, Figure 7d, left). In the second mode of operation, the tubes are pressurized prior to heating (Figure 4b, i) and it appears that the azimuthal width of the (110)/(200) peaks decreases monotonically as the crystallization temperature increases (~11° for T^x_{mold} = 80 °C, ~9.5° for T^x_{mold} = 90 °C and ~8.5° for T^x_{mold} = 100 °C; Figure 7d, right). This seems counter-intuitive at first because the onset of inflation occurs at approximately the same time and temperature for all cases in the second mode of operation (Figures 3b and 4b). However, expansion progresses isothermally in the second mode of operation; therefore, the tube is expanding at a lower mold temperature in T^x_{mold} = 80 °C relative to T^x_{mold} = 100 °C and, consequently, the material deforms more rapidly at T^x_{mold} = 100 °C (compare the strain rate for T^x_{mold} = 80 °C vs. T^x_{mold} = 100 °C for t < 60 s, Figure 4b, ii) and develops a microstructure with greater crystallite orientation.

Comparison between simultaneously acquired WAXS and SAXS patterns indicate that WAXS is more sensitive than SAXS to the onset of expansion (SAXS features appear a few seconds after the occurrence of (110)/(200) peaks in WAXS, compare Figures 5 and 6a,b) [29]. The SAXS intensity continues to increase post expansion and adopts a bimodal distribution with peaks located along the meridian (parallel to the θ-direction) and the equator (parallel to the z-direction) of the patterns (see Figures 5 and 6b, 350 s). The SAXS peaks along the meridian are more than twice as intense as the equatorial peaks, suggesting a dominant oriented microstructure along the θ-direction of the tube. The SAXS intensity decreases during the cooling step in accord with densification of the amorphous interlamellar space (t > 350 s, Figures S15–S20). Interestingly, both the equatorial and the meridional peaks shift towards higher-q during annealing, indicating a steady decrease in the long period (note the shift in peak intensity towards higher-q is more apparent in Figures 5 and 6b, i; we return to this in the Discussion). The greater intensity of SAXS peaks in T^x_{mold} = 100 °C relative to T^e_{mold} = 100 °C (>15%, compare Figures 5 and 6a,b, 350 s and Figure 8a,b, left) suggests that the former introduces a greater population of regular spaced lamellar stacks (>25 nm, Figure 8a,b, right) along the θ-direction of the expanded tube.

At the end of each experiment, the tubes are extracted from the instrument and are stored for ex situ studies. Polarized light micrographs of ~50 μm thick sections of the expanded tubes reveal a gradient in morphology from the inner (ID) to the outer diameter (OD): the retardance at the ID is ~1300 nm (third order bluish-green, Figure 9) and decreases steadily towards the OD (~800 nm, second order yellow, Figure 9). This gradient in retardance is not observed in the extruded tube and is a consequence of the tube expansion process, which imposes greater strains at the ID (>400%) relative to the OD (~150%). The greater orientation of PLLA chains at the ID relative to the OD manifests in the observed gradient in retardance.

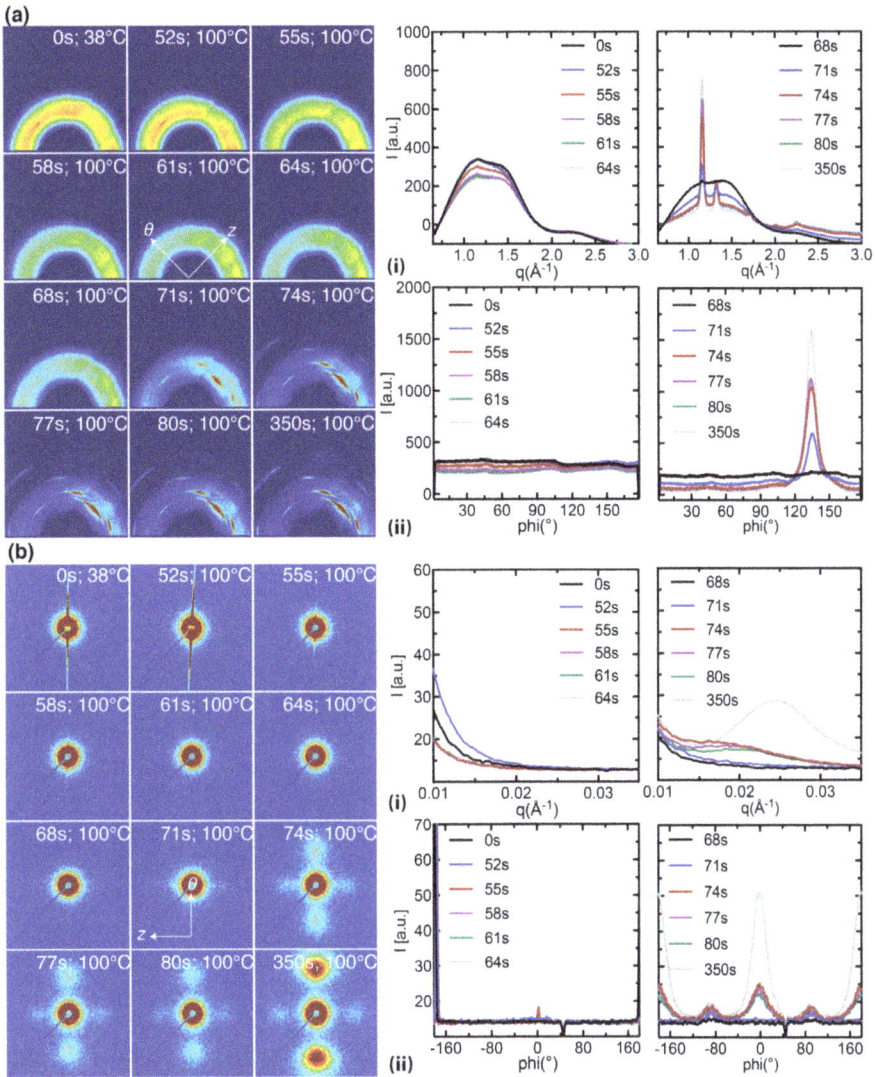

Figure 5. In situ (**a**) WAXS and (**b**) SAXS data acquired on a PLLA preform stretched at $T^e_{mold} = 100\,^\circ\text{C}$. The diffraction patterns capture the transient heating and annealing steps ($t < 350$ s, $T^a_{mold} = 100\,^\circ\text{C}$) and correspond to the temperature and strain profiles in Figures 3 and 4a. Diffraction patterns acquired during the cooling step ($t > 350$ s) can be found in Figure S17. The WAXS and SAXS data are presented as: 2D patterns; (**a,b, i**) azimuthally averaged intensity $I(q)$; and (**a,b, ii**) radially averaged intensity $I(\varphi)$.

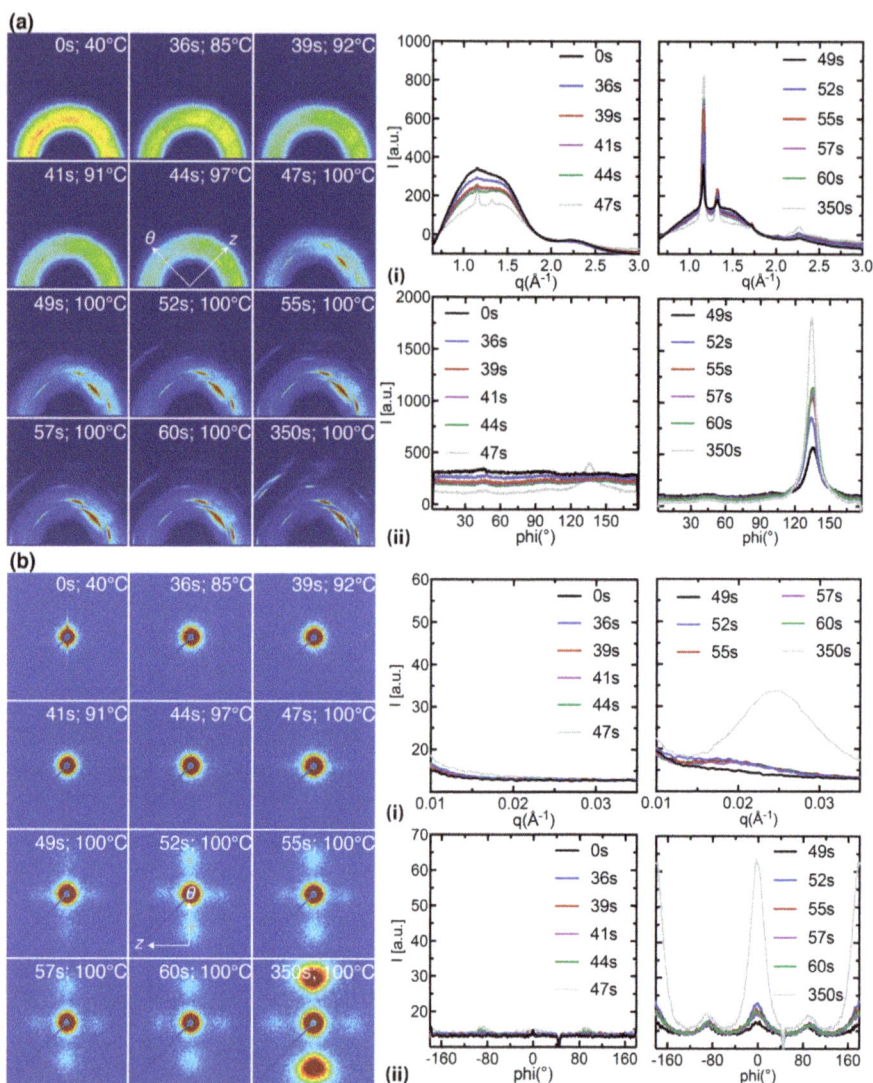

Figure 6. In situ (**a**) WAXS and (**b**) SAXS data acquired on a PLLA preform stretched at $T^x{}_{mold} = 100\ °C$. The diffraction patterns capture the transient heating and annealing steps (t < 350 s, $T^a{}_{mold} = 100\ °C$) and correspond to the temperature and strain profiles in Figures 3 and 4b. Diffraction patterns acquired during the cooling step (t > 350 s) can be found in Figure S20. The WAXS and SAXS data are presented as: 2D patterns; (**a,b, i**) azimuthally averaged intensity I(q); and (**a,b, ii**) radially averaged intensity I(φ).

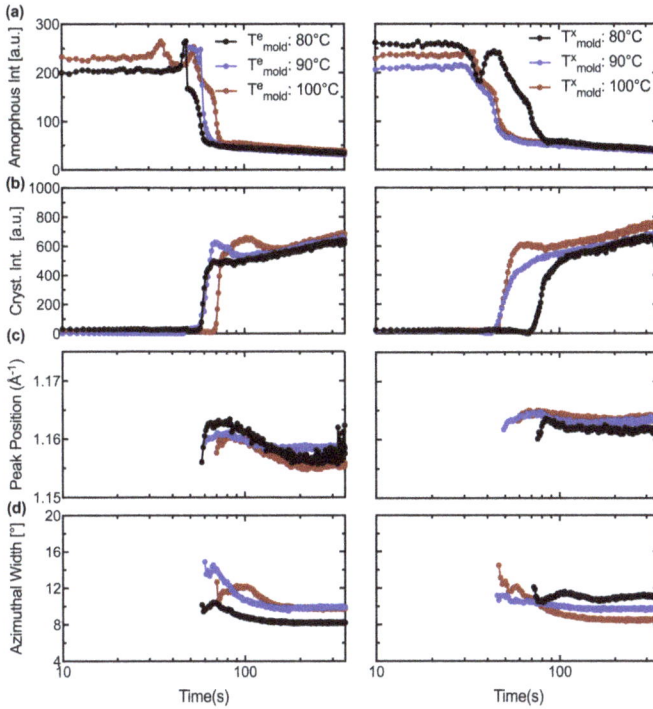

Figure 7. Quantitative characteristics of 1D WAXS profiles for expansion performed at: (**left**) T^e_{mold} = 80, 90 and 100 °C; and (**right**) T^x_{mold} = 80, 90 and 100 °C. The variation in: (**a**) amorphous content; (**b**) crystallinity; (**c**) peak position of the (110)/(200) peaks; and (**d**) full width at half maximum of the (110)/(200) peaks is presented during the heating and the annealing steps (t ≤ 350 s); quantitative analysis for the cooling step is presented in the Supplementary Materials (Figure S21).

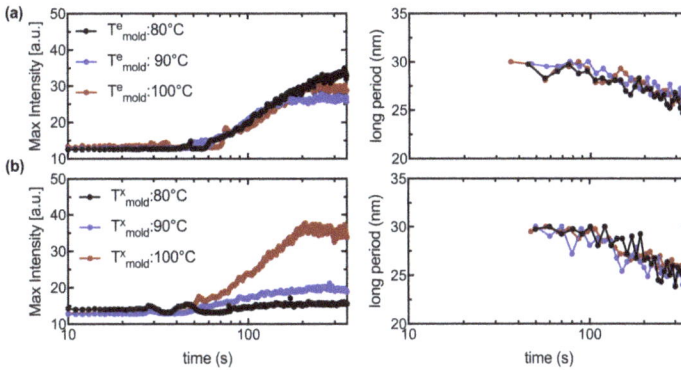

Figure 8. Quantitative characteristics of 1D SAXS profiles for expansion performed at: (**a**) T^e_{mold} = 80, 90 and 100 °C; and (**b**) T^x_{mold} = 80, 90 and 100 °C. The variation in: (**a,b**, **left**) the maximum intensity of the meridional SAXS peaks; and (**a,b**, **right**) the interlamellar spacing is presented during the heating and the annealing steps (t ≤ 350 s); quantitative analysis for the cooling step is presented in the Supplementary Materials (Figure S22).

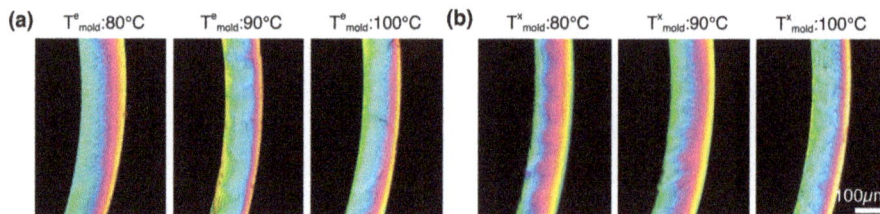

Figure 9. Expanded tubes stretched at: (**a**) T^e_{mold} = 80, 90 and 100 °C (Figures 3 and 4a); and (**b**) T^x_{mold} = 80, 90 and 100 °C were extracted from the instrument and microtomed parallel to the (r,θ) plane to yield consecutive 50 μm thick circular cross sections. The 50 μm sections were sandwiched between glass slides and imaged through linear crossed polarizers. Note that the sequence of Michel Levy colors from blue green to purple to pink to orange to yellow unambiguously shows that the retardance decreases from ID to OD.

4. Discussion

The tube expansion instrument provides structural information about PLLA during processing that is not captured by other techniques. The in situ WAXS patterns are very sensitive to the onset of inflation and probe changes in amorphous content, crystallinity, crystal morph, and crystallite orientation with a 1 s resolution. In all six experiments (Figures 3 and 4), the deformation of the material occurs in two stages. During the first one, the material remains predominantly amorphous (Figure S11, 45–56 s; Figure S12, 49–58 s; Figure 5, 52–64 s; Figure S13, 45–65 s; Figure S14, 30–41 s; and Figure 6, 36–44 s). The second stage begins with an abrupt increase in crystallinity (Figure S11, 59 s; Figure S12, 62 s; Figure 5, 71 s; Figure S13, 75 s; Figure S14, 47 s; and Figure 6, 49 s), and a decrease in the amorphous halo (Figure 7a). This step increases in crystalline diffraction accounts for approximately half of the total increase in diffraction that occurs during deformation and subsequent annealing (Figure 7b). During the second process, strongly-oriented diffraction peaks grow with the chain axis along the θ-direction. These oriented WAXS and SAXS patterns suggest the presence of a "shish-kebab" morphology with shish along the θ-direction of the tube, as expected due to the large elongational strain (in excess of 400% at the inner diameter) imposed during tube expansion. This morphology is characterized by regularly spaced lamellar stacks called "kebabs" that decorate a central stem of oriented precursors called "shish" [30]. The sharp WAXS peaks are detectable earlier than the corresponding oriented SAXS peaks, with the delay being shorter when the sample is heated prior to imposing stress and with increasing temperature, becoming negligible for both T^e_{mold} = 100 °C and T^x_{mold} = 100 °C. Together, these features indicate that the initial process is dominated by glassy deformations and suggest that the transition to melt flow enables the combination of stretching, orientation and organization of the polymer chains that allows rapid growth of oriented crystals. This sequence accords with observations in the literature on poly (ethylene terephthalate) (PET) [29].

Specifically, our experiments are close enough to the glass transition that the initial deformation occurs at a strain rate that is faster than the rate of long range conformational rearrangement of the chains. In accord with Mahendrasingam [29], when only local segmental reorientation occurs, crystallization cannot. The second stage occurs on longer timescales, consistent with the expectation that relaxation of the chains allows orientation and stretch of submolecules that are long enough to participate in crystallization (dozens of consecutive units along the backbone). This corresponds to Mahendrasingam's window of strain rate in which oriented crystals form during deformation, i.e., the strain rate is slow enough to permit chain retraction, but much faster than terminal relaxation by reptation. The transition between the two regimes correlates with the transition from glassy to rubbery mechanical properties.

Temperature affects the transition through its effect on the relaxation time, which is particularly strong for the glassy modes. The transition from glassy to rubbery relaxation offers a coherent explanation for the correlation between the abrupt increase in oriented crystallization (sharp WAXS peaks) and the abrupt increase in the rate of deformation (strain rates of ~0.07 s^{-1} or more after the inflection point, marked by black arrows in Figure 4a,b, ii). Experiments that use different crystallization temperatures suggest that the relaxation of the glassy modes requires tens of seconds at 80 °C, approximately 10–15 s at 90 °C and approximately 3–5 s at 100 °C (Figure 4b, ii). The rubbery relaxation is less sensitive to temperature, which explains the relatively similar rates of strain during the sudden tube expansion. It may also offer an explanation for the small effects of temperature on the final strain, manifested in the thickness of the expanded tube. There is precedent for flow-induced crystallization having such strong effects that the usual temperature dependence of crystallization is dramatically reduced [31]. Perhaps similar strain rates result in similar rates of oriented crystallization and, consequently, similar total strain when oriented crystallization brings deformation to a halt.

An interesting and perhaps technologically important observation is the pronounced difference in the degree of orientation between PLLA expanded with the $T^e{}_{mold}$ = 80 °C and $T^x{}_{mold}$ = 80 °C protocols (cf. black symbols in Figure 7, left vs. right); the azimuthal width of the (110)/(200) peaks (Figures 5 and 6a, ii) provides a measure of crystallite orientation. In the first mode of operation (Figure 7d, left), the preform is exposed to stress when the mold reaches $T^e{}_{mold}$ = 80 °C and heating continues until $T^a{}_{mold}$ = 100 °C. The glassy deformation stage is, consequently, brief and the rapid growth of oriented "kebabs" at 100 °C proceeds with strong correlation to the orientation of the "shish" created during elongation (see Figure S11 for X-ray data). In the second mode of operation (Figure 7d, right), the temperature at which expansion begins is dictated by the material properties as the stress is imposed on the tube throughout the heating process. In the $T^x{}_{mold}$ = 80 °C experiment (see Figure S13 for X-ray data), the sample begins to deform when the temperature of the mold reaches 80 °C and the deformation and crystallization processes take place isothermally at 80 °C. Due to the relatively low deformation temperature, there is a prolonged period of glassy deformation (approximately 20 s, Figure 4b,ii) when point-like precursors and threadlike precursors could form; and due to the relatively low crystallization temperature, the growth of crystallites from these precursors is slower and less correlated to the shish (Figure 7b, compare the slope of the main crystallization process for the black symbols). The dearth of oriented lamellar stacks for $T^x{}_{mold}$ = 80 °C is confirmed by SAXS results that show strong peaks indicative of shish-kebabs for $T^e{}_{mold}$ = 80 °C (black symbols in Figure 8a, left and Figure S11), but not for $T^x{}_{mold}$ = 80 °C (black symbols in Figure 8b, left and Figure S13). These results suggest that using a relatively low temperature during the tube expansion process could moderate the degree of anisotropy in the expanded tube, which is important for achieving balanced properties and ductile behavior in the BVS.

The intensity of the SAXS peaks increases during the first half of the annealing step (<200 s Figure 8a,b), then levels off. The increase in SAXS intensity occurs later and more gradually than the increase in WAXS (compare Figure 7b to Figure 8, left). This disconnect suggests that the increase in SAXS is due to a reorganization of previously formed crystals into increasingly coherent lamellae and, possibly, with a reduction of interlamellar density in favor of thicker crystallites (see Supplementary Materials for data on the cooling process, t > 350 s Figures S15–S20, where the SAXS intensity decreases). In conjunction with this reorganization, the interlamellar spacing monotonically decreases with time (Figure 8a,b, right). When this shift in long spacing occurs with an increase in SAXS intensity, it suggests that some interlamellar material is being pulled into the lamellae allowing the crystal to become wider. In some cases, the shift in long spacing occurs at a time when the SAXS intensity is decreasing, which suggests the growth of secondary lamellae in the space between previously formed lamellae. During the cooling step, the SAXS intensity of the present samples decreases due to the densification of amorphous interlamellar material and, possibly, growth of fringed micelles in the interlamellar space. By the time the sample temperature reaches ambient temperature, there is little to no evidence of lamellar superstructures (the SAXS pattern vanishes, see Supplementary Materials, t = 700 s in Figures S15–S20). Therefore, the transient structure reveals aspects of the present PLLA semicrystalline morphology that are concealed in the final state of the expanded tube.

The tube expansion apparatus has several features that enabled these experiments. The device is small and relatively robust, making it well suited for transporting to a synchrotron. The device requires only a few connections, which enables rapid setup with minimal loss of beamtime. Infrared (IR) heating is rapid relative to sample ovens or heater blocks. Therefore, the samples reach the desired test temperature before deforming under their weight, enabling experiments relevant to stretch-blow molding. In the case of materials that undergo chemical reactions at elevated temperature, rapid heating can be used to minimize changes in the material prior to measurements.

The most significant area for improvement is in the synchronization of strain measurements with the in situ X-ray scattering data. Synchronization of video images could be accomplished by displaying a flash of light at selected time intervals. Strain measurements during intervals when the IR lamps are on could be accomplished by providing the camera with a high-pass optical filter (blocking red and infrared wavelengths). Improved quantification of the strain field could be accomplished by placing markings on the preform, provided the markings do not significantly absorb or scatter infrared light. Implementation of improved strain measurements would open the way to a whole space of biaxial elongations. Here, we were limited to constant width elongation, which can be evaluated from the transient radius of the tube. The apparatus already has provisions for superimposing axial elongation and azimuthal elongation. Faster strain rates can be examined if a faster camera is used. A broader range of stresses can be examined by varying the imposed pressure inside the preform or by simply modifying the dimensions of the preform. The instrument could be used with any of a number of in situ structural measurements (e.g., birefringence or light scattering) with appropriate modifications to offset lens effects of the curved interfaces of the mold and the samples. Therefore, the tube-expansion geometry may prove useful for the broad community of scientists who are investigating deformation induced structure in polymers and other soft matter.

5. Conclusions

Bioresorbable vascular scaffolds (BVSs) are manufactured from a nominally brittle polymer (PLLA) but gain strength and ductility through processing. Tube expansion plays a critical role in the manufacture of a BVS but defies prediction due to the rapid time scales of deformation. The selection of tube expansion conditions determines the final thickness of the scaffold and influences the response of PLLA during crimping and deployment. Therefore, a detailed understanding of the morphology created in the expanded tube holds the key to a thinner yet stronger scaffold. In this report, we discuss the fabrication of an instrument that can probe the structure of PLLA in real time at conditions relevant to the processing of vascular scaffolds.

Video recordings and real time X-ray scattering data indicate that deformation in the PLLA preform progresses in two stages. During the first stage, PLLA exhibits glassy behavior and stretches at a relatively slow strain rate ($0.03\ \mathrm{s}^{-1}$) until the strain at the outer diameter (OD) reaches an inflection at ~50%. In the second stage, PLLA exhibits rubbery behavior (OD strain > 50%) and the strain rate abruptly increases ($0.07\ \mathrm{s}^{-1}$) till an OD strain of ~130%. The in situ WAXS data indicate that the material is predominantly amorphous during the first stage of deformation but develops oriented crystallites at the onset of the second stage of deformation. The corresponding SAXS data suggest that the rapid inflation induces growth of shish-kebabs along the θ-direction of the tube. By varying process parameters such as the expansion temperature, (T^{e}_{mold}), the crystallization temperature (T^{x}_{mold}), and the annealing conditions, we demonstrate that the microstructure of the expanded tube can be controlled to vary the crystallinity, crystallite orientation and the population of shish-kebabs to obtain desired mechanical properties.

The data presented in this report highlight the ability of a tube expansion instrument to provide in situ X-ray scattering data during and after elongational deformation. Future studies will probe the impact of total strain by using molds of different dimensions and strain-rate by varying the inflation pressure. The combination of in situ structural data with the transient strain field may prove broadly useful for understanding processing–structure–property relationships in polymers, as illustrated by

Polymers **2018**, *10*, 288

the example of PLLA relevant to bioresorbable vascular scaffolds that may improve treatment of cardiovascular disease.

Supplementary Materials: The following are available online at http://www.mdpi.com/2073-4360/10/3/288/s1, Figure S1: Selection of lamp orientation relative to the preform, Figures S2 and S3 & Table S1: Numerical simulations of the temperature profile of the preform, Figure S4: Experimental data used to validate the simulations, Figure S5: Implementation of the instrument at an X-ray beamline, Figures S6–S10: X-ray scattering analysis, Figures S11–S20: Summary of in situ WAXS and SAXS data, Figures S21 and S22: Quantitative analysis of in situ WAXS and SAXS data.

Acknowledgments: This research used resources of the Advanced Photon Source (APS), a US Department of Energy (DOE) Office of Science User Facility operated for the DOE Office of Science by Argonne National Laboratory under Contract DE-AC02-06CH11357. The authors acknowledge all the staff at beamline 5-ID-D DND-CAT of the Advanced Photon Source (APS) at the Argonne National Laboratories, especially Steven Weigand and James Rix for their support before and during the synchrotron experiments. The authors acknowledge Mary Beth Kossuth at Abbott Vascular for providing the preforms. KR and TDL are very thankful to Giuseppe Nenna at ENEA for useful discussions on ray tracing calculations. This project has received funding from the European Union's Horizon 2020 research and innovation program under the Marie Skłodowska-Curie grant agreement No 691238, the Jacobs Institute for Molecular Engineering for Medicine at the California Institute of Technology, and the National Heart, Lung, and Blood Institute of the National Institutes of Health under Award Number F31HL137308.

Author Contributions: Karthik Ramachandran and Julia A. Kornfield designed research; Karthik Ramachandran, Riccardo Miscioscia, Giovanni De Filippo, Giuseppe Pandolfi and Tiziana Di Luccio performed research; Karthik Ramachandran, Riccardo Miscioscia, Tiziana Di Luccio and Julia A. Kornfield analyzed data; and Karthik Ramachandran, Riccardo Miscioscia, Tiziana Di Luccio and Julia A. Kornfield wrote the paper

Conflicts of Interest: The authors declare no conflict of interest.

References

1. World Health Organization. *Global Atlas on Cardiovascular Disease Prevention and Control*; WHO: Geneva, Switzerland, 2010.
2. Centers for Disease Control and Prevention. *Health, United States*; CDC: Atlanta, GA, USA, 2016.
3. Nichols, M.; Townsend, N.; Scarborough, P.; Rayner, M. Cardiovascular disease in Europe 2014: Epidemiological update. *Eur. Heart J.* **2014**, *35*, 2950–2959. [CrossRef] [PubMed]
4. Kossuth, M.B.; Perkins, L.E.L.; Rapoza, R.J. Design principles of bioresorbable polymeric scaffolds. *Interv. Cardiol. Clin.* **2016**, *5*, 349–355. [CrossRef] [PubMed]
5. Ormiston, J.A.; Serruys, P.W.; Onuma, Y.; Van Geuns, R.J.; De Bruyne, B.; Dudek, D.; Thuesen, L.; Smits, P.C.; Chevalier, B.; McClean, D.; et al. First serial assessment at 6 months and 2 years of the second generation of ABSORB everolimus-eluting bioresorbable vascular scaffold a multi-imaging modality study. *Circ. Cardiovasc. Interv.* **2012**, *5*, 620–632. [CrossRef] [PubMed]
6. Serruys, P.W.; Ormiston, J.; van Geuns, R.-J.; de Bruyne, B.; Dudek, D.; Christiansen, E.; Chevalier, B.; Smits, P.; McClean, D.; Koolen, J.; et al. A polylactide bioresorbable scaffold eluting everolimus for treatment of coronary stenosis: 5-year follow-up. *J. Am. Coll. Cardiol.* **2016**, *67*, 766–776. [CrossRef] [PubMed]
7. Iqbal, J.; Onuma, Y.; Ormiston, J.; Abizaid, A.; Waksman, R.; Serruys, P. Bioresorbable scaffolds: Rationale, current status, challenges, and future. *Eur. Heart J.* **2014**, *35*, 765–776. [CrossRef] [PubMed]
8. Rizik, D.G.; Padaliya, B.B. Early U.S. experience following FDA approval of the ABBOTT vascular bioresorbable vascular scaffold: Optimal deployment technique using high resolution coronary artery imaging. *J. Interv. Cardiol.* **2016**. [CrossRef] [PubMed]
9. Gorrasi, G.; Pantani, R. Hydrolysis and Biodegradation of Poly(lactic acid). In *Synthesis, Structure and Properties of Poly(lactic acid)*; Springer: New York, USA, 2017.
10. Tibbitt, M.W.; Rodell, C.B.; Burdick, J.A.; Anseth, K.S. Progress in material design for biomedical applications. *Proc. Natl. Acad. Sci. USA* **2015**, *112*. [CrossRef] [PubMed]
11. Farokhzad, O.C.; Cheng, J.; Teply, B.A.; Sherifi, I.; Jon, S.; Kantoff, P.W.; Richie, J.P.; Langer, R. Targeted nanoparticle-aptamer bioconjugates for cancer chemotherapy in vivo. *Proc. Natl. Acad. Sci. USA* **2006**, *103*, 6315–6320. [CrossRef] [PubMed]
12. Renouf-Glauser, A.C.; Rose, J.; Farrar, D.F.; Cameron, R.E. The effect of crystallinity on the deformation mechanism and bulk mechanical properties of PLLA. *Biomaterials* **2005**, *26*, 5771–5782. [CrossRef] [PubMed]

13. Nakafuku, C.; Takehisa, S.Y. Glass transition and mechanical properties of PLLA and PDLLA-PGA copolymer blends. *J. Appl. Polym. Sci.* **2004**, *93*, 2164–2173. [CrossRef]
14. Hu, Y.; Rogunova, M.; Topolkaraev, V.; Hiltner, A.; Baer, E. Aging of poly(lactide)/poly(ethylene glycol) blends. Part 1. Poly(lactide) with low stereoregularity. *Polymer* **2003**, *44*, 5701–5710. [CrossRef]
15. Broz, M.E.; VanderHart, D.L.; Washburn, N.R. Structure and mechanical properties of poly(D,L-lactic acid)/poly(ε-caprolactone) blends. *Biomaterials* **2003**, *24*, 4181–4190. [CrossRef]
16. Rathi, S.; Chen, X.; Coughlin, E.B.; Hsu, S.L.; Golub, C.S.; Tzivanis, M.J. Toughening semicrystalline poly(lactic acid) by morphology alteration. *Polymer* **2011**, *52*, 4184–4188. [CrossRef]
17. Grijpma, D.W.; Pennings, A.J. (Co)polymers of L-lactide, 2. *Mechanical properties. Macromol. Chem. Phys.* **1994**, *195*, 1649–1663. [CrossRef]
18. Li, S.; Garreau, H.; Vert, M. Structure-property relationships in the case of the degradation of massive poly (α-hydroxy acids) in aqueous media, Part 3: Influence of the morphology of poly (L-lactic acid). *J. Mater. Sci. Mater. Med.* **1990**, *1*, 198–206. [CrossRef]
19. Li, S. Hydrolytic degradation characteristics of aliphatic polyesters derived from lactic and glycolic acids. *J. Biomed. Mater. Res.* **1999**, *48*, 342–353. [CrossRef]
20. Ailianou, A.; Ramachandran, K.; Kossuth, M.B.; Oberhauser, J.P.; Kornfield, J.A. Multiplicity of morphologies in poly (L-lactide) bioresorbable vascular scaffolds. *Proc. Natl. Acad. Sci. USA* **2016**, *113*, 11670–11675. [CrossRef] [PubMed]
21. Oberhauser, J.P.; Hossainy, S.; Rapoza, R.J. Design principles and performance of bioresorbable polymeric vascular scaffolds. *EuroIntervention* **2009**, *5*, 15–22. [CrossRef] [PubMed]
22. Ding, N.; Pacetti, S.; Tang, F.; Gada, M.; Roorda, W. XIENCE V (TM) stent design and rationale. *J. Interv. Cardiol.* **2009**, *22*, S18–S27. [CrossRef]
23. Menary, G.H.; Tan, C.W.; Harkin-Jones, E.M.A.; Armstrong, C.G.; Martin, P.J. Biaxial deformation and experimental study of PET at condtions applicable to stretch blow molding. *Polym. Eng. Sci.* **2012**, *52*, 671–688. [CrossRef]
24. Glauser, T.; Gueriguian, V.; Steichen, B.; Oberhauser, J.P.; Gada, M.; Kleiner, L.; Kossuth, M.B. Controlling Crystalline Morphology of a Bioabsorbable Stent 2015. U.S. Patent Appl No. 13/734,879, 15 December 2015.
25. Speranza, V.; De Meo, A.; Pantani, R. Thermal and hydrolytic degradation kinetics of PLA in the molten state. *Polym. Degrad. Stab.* **2014**, *100*, 37–41. [CrossRef]
26. Di Lorenzo, M.L. Crystallization behavior of poly(L-lactic acid). *Eur. Polym. J.* **2005**, *41*, 569–575. [CrossRef]
27. Glauser, T.; Gueriguian, V.; Steichen, B.; Oberhauser, J.P.; Gada, M.; Kleiner, L.; Kossuth, M.B. Controlling Crystalline Morphology of a Bioabsorbable Stent 2013. U.S. Patent Appl No. 12/559,400, 6 August 2013.
28. Aou, K.; Kang, S.; Hsu, S.L. Morphological study on thermal shrinkage and dimensional stability associated with oriented poly (lactic acid). *Macromolecules* **2005**, *38*, 7730–7735. [CrossRef]
29. Mahendrasingam, A.; Blundell, D.J.; Martin, C.; Fuller, W.; Mackerron, D.H.; Harvie, J.L.; Oldman, R.J.; Riekel, C. Influence of temperature and chain orientation on the crystallization of poly (ethylene terephthalate) during fast drawing. *Polymer* **2002**, *41*, 7803–7814. [CrossRef]
30. Kimata, S.; Sakurai, T.; Nozue, Y.; Kasahara, T.; Yamaguchi, N.; Karino, T.; Shibayama, M.; Kornfield, J.A. Molecular basis of the shish-kebab morphology in polymer crystallization. *Science* **2007**, *316*, 1014–1017. [CrossRef] [PubMed]
31. Kumaraswamy, G.; Kornfield, J.A.; Yeh, F.; Hsiao, B.S. Shear-enhanced crystallization in isotactic polypropylene. 3. Evidence for a kinetic pathway to nucleation. *Macromolecules* **2002**, *35*, 1762–1769.

polymers

MDPI

Article

Processing and Thermal Response of Temperature-Sensitive-Gel(TSG)/Polymer Composites

Jin Gong [1,2,*], Eiichi Hosaka [1], Kohei Sakai [2], Hiroshi Ito [1,*], Yoshikazu Shibata [3], Kosei Sato [3], Dai Nakanishi [3], Shinichiro Ishihara [3] and Kazuhiro Hamada [3]

[1]   Department of Polymer Science & Engineering, Graduate School of Organic Materials Science, Yamagata University, 4-3-16 Jonan, Yonezawa, Yamagata 992-8510, Japan; txy33515@st.yamagata-u.ac.jp
[2]   Department of Mechanical Systems Engineering, Graduate School of Science and Engineering, Yamagata University, 4-3-16 Jonan, Yonezawa, Yamagata 992-8510, Japan; tmf62483@st.yamagata-u.ac.jp
[3]   Kohjin Film & Chemicals Co., Ltd., 1-1 Koukokumachi, Yatsushiro, Kumamoto 866-8686, Japan; y_shibata@kohjin.co.jp (Y.S.); k_sato@kohjin.co.jp (K.S.); d_nakanishi@kohjin.co.jp (D.N.); s_ishihara@kohjin.co.jp (S.I.); kazuhiro_hamada@kohjin.co.jp (K.H.)
*   Correspondence: jingong@yz.yamagata-u.ac.jp (J.G.); ihiroshi@yz.yamagata-u.ac.jp (H.I.); Tel.: +81-238-26-3135 (J.G.); Tel: +81-238-26-3081 (H.I.)

Received: 4 April 2018; Accepted: 28 April 2018; Published: 1 May 2018

Abstract: Temperature-sensitive gels (TSGs) are generally used in the fields of medical, robotics, MEMS, and also in daily life. In this paper, we synthesized a novel TSG with good thermal durability and a lower melting temperature below 60 °C. We discussed the physical properties of he TSG and found it provided excellent thermal expansion. Therefore, we proposed the usage of TSG to develop a strategic breathable film with controllable gas permeability. The TSG particles were prepared firstly and then blended with linear low-density polyethylene/calcium carbonate (LLDPE/CaCO$_3$) composite to develop microporous TSG/LLDPE/CaCO$_3$ films. We investigated the morphology, thermal, and mechanical properties of TSG/LLDPE/CaCO$_3$ composite films. The film characterization was conducted by gas permeability testing and demonstration temperature control experiments. The uniformly porous structure and the pore size in the range of 5–40 μm for the TSG/LLDPE/CaCO$_3$ composite films were indicated by SEM micrographs. The demonstration temperature control experiments clearly proved the effect of the controllable gas permeability of the TSG and, more promisingly, the great practical value and application prospects of this strategic effect for the temperature-sensitive breathable film was proved.

Keywords: temperature sensitive; gel; controllable gas permeability; breathable film; polymer composite; processing

---

1. Introduction

Gels are a promising material having good compatibility with biological tissues. The three-dimensional cross-linked network of the gel, like collagen in tissues, is the framework structure contributing to the mechanical strength. Temperature-sensitive gels (TSGs) exhibit physical changes in response to changes in temperature. One research field in which the authors are interested is the synthesis and application of high-performance crystalline gels. Crystalline gels are gels with high crystallinity polymerized by introducing crystalline side-chains into the three-dimensional structure. Many crystalline gels have been synthesized [1–7]. TSGs are a kind of crystalline gel, and they are expected to apply in the fields of medical, robotics, as well as daily life [1,5,8]. The melting of side-chain crystals leads to the changing behavior in the physical properties, especially in viscosity, volume, and mechanical strength. The physical property changes provide the temperature sensitivity of TSGs. TSGs become viscous liquids and change their volume at the transfer temperature (melting temperature) [3,4,7].

Generally, volume change, especially thermal expansion, is considered one of the major drawbacks in polymer applications. Many researchers have been placing a great deal of effort in modifying the thermal expansion of polymers and polymer composites [9–16]. Usually it is the other way around, so we propose to make full use of the thermal expansion inversely.

Functional porous membranes/films, which can control the transfer of air, humidity, light, and temperature, are important and have attracted much attention due to their wide use in fields such as food packaging, electronics, the energy sector, and daily necessities [17–28]. Most of these porous membranes/films are environment- or signal-responsive. Even though environment-responsive pore size adjustment of porous membranes/films is desirable, environment-responsive pore size regulation remains challenging. In this paper, we report a temperature-responsive pore film that achieves the opening/closing of pores. The temperature-responsive pore adjustable film was fabricated with hybrid materials of TSGs within LLDPE/CaCO₃ composites. The TSG used here was designed to have high thermal expansion.

2. Materials and Methods

2.1. Materials

Two monomers containing vinyl groups ($-CH=CH_2$) were used as monomers. One monomer (M1) consists of a vinyl group and a long alkane chain with more than sixteen methylene units. The other monomer (M2) has a pendent group of acetoxy ($-COOCH_3$). N,N'-methylenebisacrylamide (MBAA) used as a cross-linker, and benzophenone (BP) used as a UV polymerization initiator were obtained from Wako Pure Chemical Industries Ltd., Osaka, Japan. LLDPE/CaCO₃ composite were supplied by Kohjin Film and Chemicals Co., Ltd., Yatsushiro, Japan. All the above chemicals are used as received.

2.2. Preparation of TSG Particles

Firstly, the TSG sheets are synthesized via bulk photopolymerization. The synthesis procedure is as follows: Into a 50 mL snap vial were placed monomer M1, monomer M2, and cross-linker MBAA. The mixture was stirred for 30 min and then initiator BP was added. The molar ratio of M1: M2: MBAA: BP was 1.00:3.00:0.02:0.04. The nitrogen bubbling was performed during the above process to eliminate the oxygen, which prevents the radical polymerization. Then the solution was poured into a self-made mold and irradiated under a UV lamp at 30 °C for 9–20 h. The obtained TSG sheets were frozen with liquid nitrogen and finely ground into particles using a dancing mill (ALM90DM, Nitto Kagaku Co., Ltd., Nagoya, Japan). The prepared TSG particles are a white powder.

2.3. Preparation of TSG Films

With these frozen, ground TSG particles, Uniform TSG films were prepared by a vacuum heating press machine (Imoto machinery Co., Ltd., Kyoto, Japan) for thermal mechanical analysis (TMA) testing. Hot pressing was conducted for 2 min at 140 °C under 8 MPa pressure between non-stick sheets. This ensured the complete melting/fusion of TSG particles during film formation. A force of 4 MPa was applied and maintained for 1 min during the cooling period until room temperature was reached. The thickness of the resultant TSG films was in the range of 200 to 300 μm.

2.4. Preparation of TSG/LLDPE/CaCO₃ Composites and Films

The pellets of TSG/LLDPE/CaCO₃ composites were prepared by a twin-screw extruder (LABO PLASTOMILL 4C150, Toyo Seiki Seisaku-sho, Ltd., Tokyo, Japan), set to a speed of screw rotation of 60 rpm and a speed of feeder rotation of 10 rpm. The extruder barrel temperature profile was 190/200/200 °C from the feeder to the die end. The mixing weight ratio of TSG to LLDPE/CaCO₃ was 10 to 90 wt %.

The flat films of TSG/LLDPE/CaCO₃ were extruded from the T-die twin-screw extruder (LABO PLASTOMILL MT100B type, 2D15W, Toyo Seiki Seisaku-sho, Ltd., Tokyo, Japan) using the resultant pellets, set to a speed of screw rotation of 9 rpm, a speed of feeder rotation of 10 rpm, and a temperature program of 190/200/200 °C. To further uniformize the film thickness, a vacuum heating press machine (Imoto machinery Co., Ltd., Kyoto, Japan) was used. Hot pressing was conducted at 200 °C for 1 min under 20 MPa pressure, then cooling pressing was conducted for 1 min under 10 MPa pressure at room temperature. The hot-pressed uniform films were used for biaxial stretching, and also punched out to produce dumbbell-shaped specimens for tensile testing.

2.5. Characterization

2.5.1. Differential Scanning Calorimetry (DSC)

The melting and crystallization behavior of TSG particles were studied using a differential scanning calorimeter (DSC) (Q-200, TA instruments Japan Inc., Tokyo, USA) operating under a nitrogen flow. Samples of about 5 mg were heated to 70 °C at a heating rate of 10 °C/min and held for 1 min to eliminate the thermal prehistory. Next, samples were cooled to 10 °C at a cooling rate of 5 °C/min and held for 1 mi, and finally reheated again to 70 °C at the same rate. Premium hermetic pans (TA Instruments T zero #901683.901) were used for the measurements.

2.5.2. Wide-Angle X-Ray Scattering (WAXS)

To determine the crystallinity of TSG particles, wide-angle X-ray scattering (WAXS) was performed on a X-ray diffractometer (Ultima IV, Rigaku Corporation, Akishima, Japan) with nickel-filtered Cu Kα radiation at a scanning rate of 5 $°\cdot min^{-1}$. The degree of crystallinity (W_c, %) was evaluated according to the following formula reported by W. Ruland [29]:

$$W_C = \frac{I_c}{I_c + I_a} \times 100 \qquad (1)$$

where I_c and I_a are the scattering intensities of crystalline region and amorphous region, respectively.

2.5.3. Thermal Gravimetric Analysis (TGA)

The thermal stability of TSG particles was studied using a thermal gravimetric analysis (TGA) (Q-50, TA instruments Japan Inc., Tokyo, USA) at the heating rate of 10 °C/min from ambient temperature to 600 °C in a nitrogen atmosphere. The nitrogen flow rate was adjusted to 60 mL/min for the sample and 40 mL/min for the balance. The sample weight was more than 10 mg.

2.5.4. Thermal Mechanical Analysis (TMA)

The dimensional properties of TSG, LLDPE/CaCO$_3$, and TSG/LLDPE/CaCO$_3$ were measured using a TMA (Q-400, TA instruments Japan Inc., Tokyo, USA) operated in expansion mode of film/fiber. The samples were heated, cooled, and held under isothermal conditions by varying the loading applied to the samples. TMA tests were run from 10 to 70 °C using specimens punched out from hot-pressed 200–300 μm thick films. Test specimens were strips 20 mm long and 2 mm wide. TMA tests were run under the following conditions with two cycles: isothermal at 10 °C for 5 min, 10–70 °C at 5 °C/min, heating process force: 0.02 N, isothermal at 70 °C for 10 min, 70–10 °C at 2 °C/min, cooling process force: 0.00 N.

The coefficient of linear thermal expansion (α) over a temperature range was calculated according to the well-known formula (2):

$$\alpha = \frac{\Delta L}{L_0} \cdot \frac{1}{\Delta T} \qquad (2)$$

where ΔL represents the elongation (change in length), L_0 represents the initial specimen length, and ΔT is the change in temperature [30].

2.5.5. Pressure-Volume-Temperature (PVT) Test

The PVT tests were performed to measure the volume thermal expansion using a P-V-T test system (MPPS-1, Toyo Seiki Seisaku-sho, Ltd., Tokyo, Japan) under 1 MPa pressure. The cooling rate was 5 °C/min. In the case of TSG, the measurement was made in the 30–120 °C range, and with the LLDPE/CaCO$_3$ composite, in the 30–200 °C range [31,32].

2.5.6. Tensile Testing

The mechanical properties of the specimens were determined by tensile tests using a versatile tension-testing machine (STROGRAPH VGS-1-E, Toyo Seiki Seisaku-sho, Ltd., Tokyo, Japan). The specimens were dumbbell-shaped, punched out from hot-pressed 200–300 μm thick films, with dimensions of 35 mm × 6 mm × 2 mm (test specimen No. 7, JIS K6251 standard). The specimens were elongated at a constant rate of 300 mm/min until the specimens rupture at room temperature. The tensile strength at break, the Young's modulus, and elongation at break were evaluated from the tensile stress-strain curves. Testing was run on a minimum of five specimens. The reported data were the average of the results of five specimens.

2.5.7. Scanning Electron Microscope (SEM)

The morphology of TSG particles and LLDPE/CaCO$_3$ and TSG/LLDPE/CaCO$_3$ composite films was observed using a scanning electron microscope (SEM) (Miniscope TM-1000, Hitachi High-Technologies Corporation, Tokyo, Japan) at accelerating voltages of 10–15 kV. Samples were sputter-coated with platinum-palladium for SEM analyses and observed in a magnification range from 100× to 1000×. The sizes of TSG particles and the pores of microporous composite films were measured from the SEM micrographs. To obtain the mean size, at least 100–150 particles and pores were selected randomly.

2.5.8. Air Permeability Measurement

The air permeability of the biaxially-oriented microporous LLDPE/CaCO$_3$ and TSG/LLDPE/CaCO$_3$ films was characterized through a Gurley-type densometer (G-B3C, Toyo Seiki Seisaku-sho, Ltd., Tokyo, Japan), according to JIS P 8117 and ISO 5635. The time (t) for a settled volume (100 mL) of air to pass through the sample with a fixed area (6.45 cm^2) at 20–58 °C was measured. The temperature cycles were defined as follows: 20 °C, then 50 °C, and 58 °C to finally measure the air permeability at 20 °C. The temperature was progressively increased and controlled by an oven sensor placed on the flat module surface.

2.5.9. Temperature Control Experiment

To verify the controllable air permeability of TSG/LLDPE/CaCO$_3$ composite films, temperature control experiments were conducted using the handmade heating pouches by using the biaxially-oriented composite films as the outer layer. Inside the heating pouch, an iron mixture that was taken out of a commercial portable hand warmer was used. The main ingredients of the iron mixture are iron powder, a water absorbent material, and activated carbon [33]. The oxygen and water react with the iron powder located inside to form iron oxide and release heat. If the pouch admits more oxygen (air), the reaction occurs more quickly, eventually, the quick release of too much heat leads to the rapid rise in temperature. A temperature-measuring instrument (Kohjin Film and Chemicals Co., Ltd., Yatsushiro, Japan) was used to detect the temperature changes of the heating pouches with time.

3. Results and Discussion

3.1. Size, Morphology, and Thermal Properties of TSG Particles

The synthesized TSG sheet and TSG particles made from frozen, ground TSG sheets are shown in Figure 1. The SEM image of the TSG particles is shown in Figure 2. It was found that TSG particles have a size in the range of 10 to 400 μm, and their shape is angular, like crushed stones. The DSC melting curve for TSG particles, registered during the second heating at a rate of 10 °C/min, is shown in Figure 3a. A clear single endothermal peak at 55.3 °C was observed, which belongs to the melting point of TSG crystallites. Figure 3b is the intensity vs. 2θ plot for TSG particles measured by wide-angle X-ray scattering (WAXS). The degree of crystallinity (W_c) was estimated from the areas under the curves of the crystal region and amorphous region [29,34], and the value of W_c was about 41%, indicating TSG particles have high crystallinity, and the crystals melt when increases the particles' temperature to nearly 55.3 °C. Figure 4 shows the thermogravimetric analysis (TGA) of TSG particles. The TGA plot shows that the TSG particles undergo thermal degradation beginning at 206 °C and with a total mass loss of 99.3%. It also shows that the temperatures at which there was a mass loss of 10 wt % ($T_{10\%}$) and 50 wt % ($T_{50\%}$) are 305 °C and 381 °C, respectively. This suggests TSG particles have enough high thermal durability resistance to the general polymer processing temperature and are suitable for being used as functional fillers to develop polymer composites [3,35].

TSG sheet   TSG particles

(a)   (b)

Figure 1. TSG sheet synthesized via bulk photopolymerization (**a**) and TSG particles prepared by freeze-grinding technique (**b**).

Figure 2. SEM micrograph of TSG particles. TSG particles have a size in the range of from 10 to 400 μm, and their shape is angular, like crushed stones.

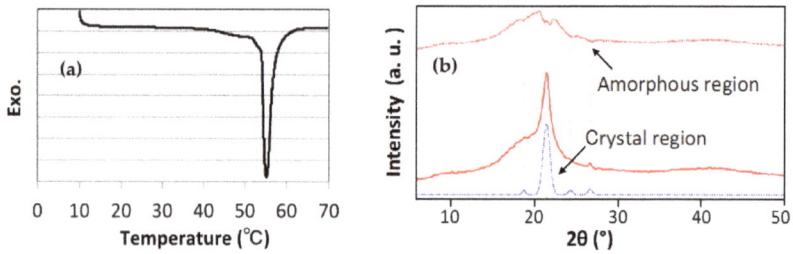

Figure 3. Differential scanning calorimetry (DSC) melting curve (**a**) and wide angle X-ray scattering (WAXS) (**b**) of TSG particles.

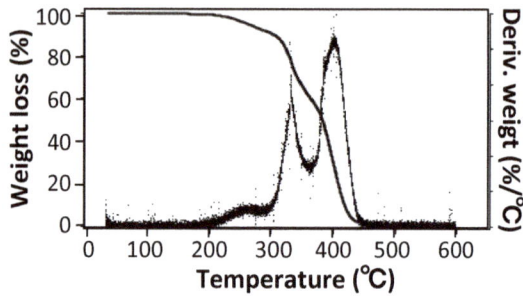

Figure 4. The weight loss vs. temperature plot and its derivative curve measured by thermogravimetric analysis (TGA) for TSG particles.

3.2. Linear and Volumue Thermal Expansion of TSG

In Figure 5, the TMA expansion curve of the TSG is presented, showing the linear dimensional changes versus temperature. A sharp dimension increase appears between 50 and 60 °C, which is near the melting temperature of the TSG. The maximum value of the dimension change reached 3784.9 μm. The values of the linear coefficient of thermal expansion for the temperature range 10–70 °C are given in Table 1. When the temperature approaches 55 °C, which is in the vicinity of the melting point of TSG, α begins to raise, signaling the beginning of the crystal melting. When the melting point is approached, the molecular mobility of TSG chains increases and the glassy solid becomes a viscous liquid, thus α increases rapidly from $322.2 \times 10^{-5}/°C$ at 55–65 °C to $2817.0 \times 10^{-5}/°C$ at 65–70 °C. After the first cycle, the curve of the second cycle overlapped well, suggesting that the TSG has a good shape memory property. The rapid increase of α ensures the pores decrease in size, or partially close. Hence, introducing TSG particles into the LLDPE/CaCO$_3$ composite having microporosity enables us to give the film pore-size adjustment function responding to the change in temperature. Furthermore, the shape memory property makes the reduced pores re-expand with decreasing temperature, which reveals that the pore-size adjustment function is reversible and hopeful to sustainable applications.

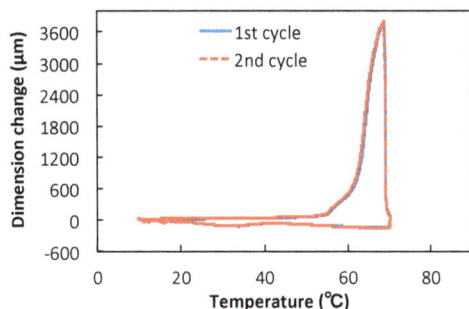

Figure 5. TMA curves of the dimensional change versus temperature obtained in the first cycle and second cycle for the TSG.

Table 1. The coefficient of linear thermal expansion (α) at various temperature intervals obtained by TMA measurement for TSG films.

Sample	Coefficient of Linear Thermal Expansion, α (10^{-5}/°C)		
	10–50 °C	55–65 °C	65–70 °C
TSG	4.5	322.2	2817.0

[1] The initial specimen length L_0 is 23308.3 μm at 10 °C.

The volume thermal expansion of TSG, and that of LLDPE/CaCO$_3$ as a comparison, were measured by PVT testing. The volume was calculated from specific volume obtained from PVT results. The volumes and the change rates of volume for TSG and LLDPE/CaCO$_3$ were summarized in Table 2. The change rate of the volume of the TSG is 3.5 times that of the LLDPE/CaCO$_3$ composite. The high volume thermal expansion of the TSG near the melting temperature is critical to make the pores smaller to control the air permeability for the microporous TSG/LLDPE/CaCO$_3$ composite films.

Table 2. The volumes and change rates of volume for TSG and LLDPE/CaCO$_3$ samples measured by the PVT test.

Samples	Volume (cm^3)		Change Rate of Volume (%)
	30 °C	60 °C	30–60 °C
TSG	0.443	0.459	3.530
LLDPE/CaCO$_3$	0.391	0.395	1.111

3.3. Effect of TSG on Mechanical Properties and Thermal Expansion

Tensile stress-strain curves of LLDPE/CaCO$_3$ and TSG/LLDPE/CaCO$_3$ are presented in Figure 6a. The average values of the mechanical properties are given in Table 3. The tensile strength at break, Young's modulus, and elongation at break of TSG/LLDPE/CaCO$_3$ composites and non-TSG-added LLDPE/CaCO$_3$ composites were compared. The Young's modulus was measured in the strain range of 0–0.04% in which the stress is proportional to the strain. Although a decrease in tensile strength at break was noted, the elongation of TSG/LLDPE/CaCO$_3$ composites increased from 571% to 699%, while the Young's modulus maintained almost the same value with the addition of 10 wt % TSG particles. The addition of TSG did not impair the balanced mechanical properties of the composites.

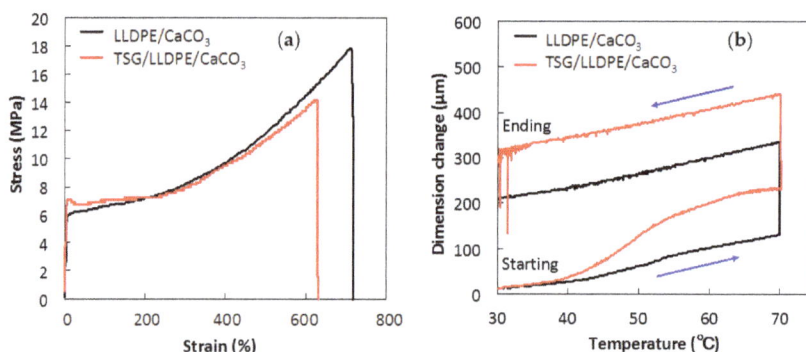

Figure 6. Tensile stress-strain curves (**a**) and dimension change as a function of temperature supplied by TMA testing (**b**) for LLDPE/CaCO$_3$ and TSG/LLDPE/CaCO$_3$ composites.

Table 3. Mechanical properties of LLDPE/CaCO$_3$ and TSG/LLDPE/CaCO$_3$ composites.

Samples	Tensile Strength at Break (MPa)	Young's Modulus (MPa)	Elongation (%)
LLDPE/CaCO$_3$	39 ± 2	147.7 ± 0.3	571 ± 5
TSG /LLDPE/CaCO$_3$	28.32 ± 2	145.5 ± 0.5	699 ± 5

In Figure 6b, the TMA curves of LLDPE/CaCO$_3$ and TSG/LLDPE/CaCO$_3$ composites are presented. The dimension change is recorded as a function of temperature. When temperature increases to 40 °C, close to the melting point of the TSG, a rapid and large increase in dimension change occurred for TSG/LLDPE/CaCO$_3$ composites, while for the non-TSG-added LLDPE/CaCO$_3$ composites, the dimension change was not so obvious. The dimension change of TSG/LLDPE/CaCO$_3$ composites at 70 °C reached 232.07 μm, which is 1.8 times that of LLDPE/CaCO$_3$ composites (129.71 μm). The total dimension change after one heating/cooling cycle increased 1.5 times from 210.94 to 314.47 μm by adding the TSG. Namely, the effect of adding the TSG on the thermal expansion rise of the composites, which would be crucial to adjust the pore size, was clearly demonstrated.

3.4. Effect of TSG on Air Permeability

The microporous structure of the prepared LLDPE/CaCO$_3$ and TSG/LLDPE/CaCO$_3$ composite films were observed by SEM. The film samples stretched 2.5 times biaxially were used and their surfaces were observed. The biaxial stretch forms pores, and the actually-used films in commercial products are the stretched films. In order to compare with the commercial LLDPE/CaCO$_3$ films, the actual application condition of 2.5 times the biaxial stretch was used to prepare the SEM samples. Figure 7 shows their SEM micrographs. Long and narrow pores along machine direction (MD) were observed for non-TSG-added LLDPE/CaCO$_3$ films. The length and width of pores along MD were 5–50 μm and 5–15 μm, respectively. Clearly, the pore enlargement in the transverse direction (TD) was limited. For TSG-added TSG/LLDPE/CaCO$_3$ films, it was found that the limitation of pore enlargement on TD was attenuated. Pores with a uniform size of 5–40 μm on both MD and TD were well-formed probably due to the contribution of soft molecular chains of the TSG, benefitting from its lower melting temperature. According to the SEM observation results, the TSG showed a significant effect in pore size and shape adjustment.

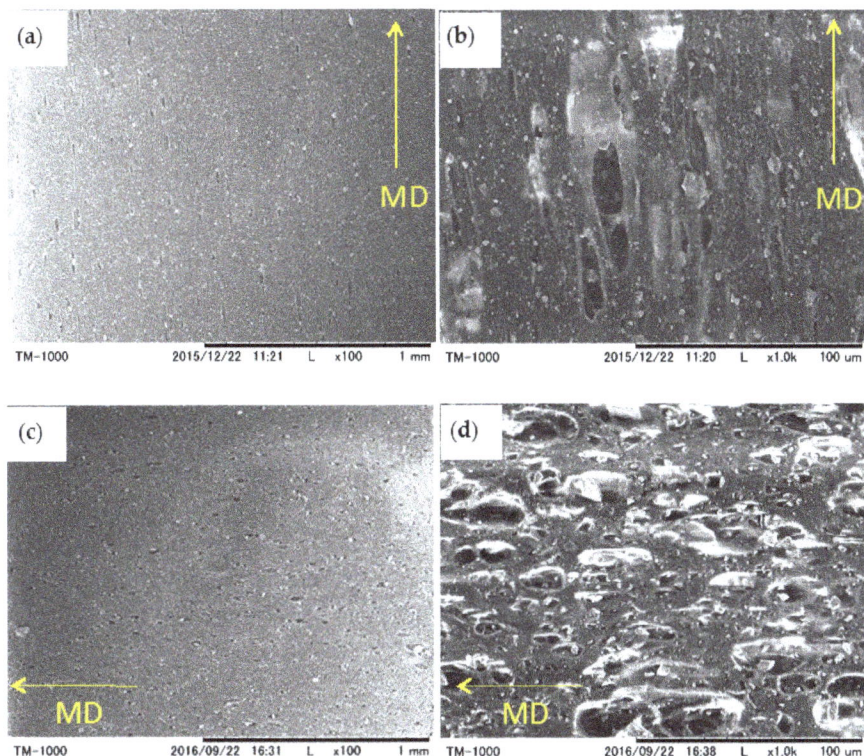

Figure 7. SEM micrographs of the surface for biaxially-oriented LLDPE/CaCO$_3$ (**a,b**) and TSG/LLDPE/CaCO$_3$ (**c,d**) films. MD represents machine direction; (**a,c**) are low-magnification pictures; and (**b,d**) are relatively high-magnification pictures.

To further explore the effect of TSG, the air permeability testing of LLDPE/CaCO$_3$ and TSG/LLDPE/CaCO$_3$ composite films was conducted. The temperature dependence of permeability for air transport through composite film samples with two heating/cooling cycles is presented in Figure 8. As Figure 8 shows, the testing temperature range is 40–58 °C. With an increase in temperature, permeation time per 100 mL air became longer for both LLDPE/CaCO$_3$ (Figure 8a) and TSG/LLDPE/CaCO$_3$ (Figure 8b) films. The difference in permeability of LLDPE/CaCO$_3$ films at 40 and 58 °C was 900–2100 s/mL, while, for TSG/LLDPE/CaCO$_3$ films, it was 2900–4600 s/mL, more than twice the value of the LLDPE/CaCO$_3$ films. That is, the addition of TSG lengthened the permeation time more than two times when the temperature increased close to the melting point of the TSG. The first and second testing cycles showed a similar trend. Generally, a longer time corresponds to low air permeability, implying a smaller size of pores. This means the air permeability (pore size) adjustment function of the TSG works very well in TSG-added TSG/LLDPE/CaCO$_3$ films. Figure 9 represents the working mechanism related to thermal expansion and pore size readjustment for the microporous TSG/LLDPE/CaCO3 composite films. During heating, the crystals of TSG near the pores melt and becomes a viscous liquid, thus, the volume increases to decrease the pore size. On the other hand, during cooling, the recrystallization of TSG crystals lowers the volume to increase the pore size once again. The pore size adjustment behavior is reversible in response to temperature. The controllable air permeability of microporous composite films was realized successfully by adding the TSG.

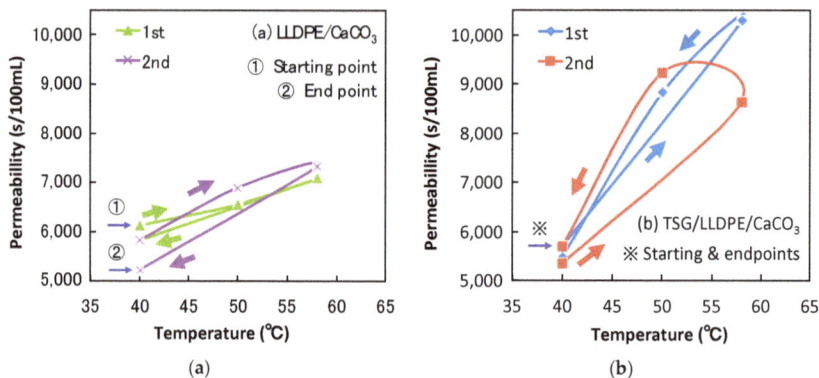

Figure 8. Temperature-permeability dependency of LLDPE/CaCO$_3$ (a) and TSG/LLDPE/CaCO$_3$ (b) composite films.

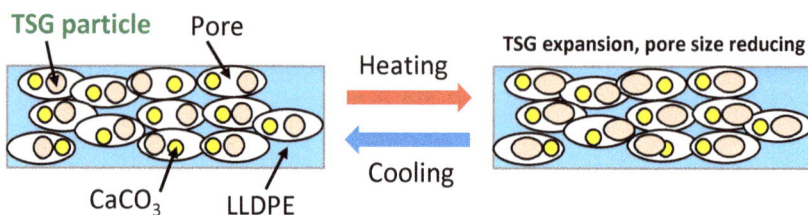

Figure 9. The working mechanism related to thermal expansion and pore size readjustment for the microporous TSG/LLDPE/CaCO$_3$ composite films. As the temperature increases, TSG expands and, as a result, the pore size reduces. The pore size adjustment behavior is reversible in response to temperature.

3.5. Effect of TSG on Temperature Control

The demonstration temperature control experiments were further conducted to prove the pore-size adjustment effect of the TSG. The heating pouches made from LLDPE/CaCO$_3$ and the TSG/LLDPE/CaCO$_3$ composite films and iron mixture are shown in Figure 10. The temperature-measuring instrument is shown in Figure 11. The experiments were conducted from the ambient temperature. The detected temperature change with time was plotted and given in Figure 12. The maximum temperature of using LLDPE/CaCO$_3$ was 59 °C, 7 °C higher than that of using TSG/LLDPE/CaCO$_3$. After 150 min, the temperature holding was also better for using TSG/LLDPE/CaCO$_3$. The temperature increase was mild from 40 °C, and the high temperature maintained a longer time when using TSG/LLDPE/CaCO$_3$ composite films. This suggests less oxygen entered the pouch to react with the iron powder to generate heat, that is, the pore size reduced with the temperature increasing for the TSG/LLDPE/CaCO$_3$ composite films. This is consistent with all the results mentioned above. The addition of TSG can give the composite films' pore-size adjustment (controllable gas-permeability) function, and this was further proved to be feasible for practical applications by the demonstration temperature control experiments.

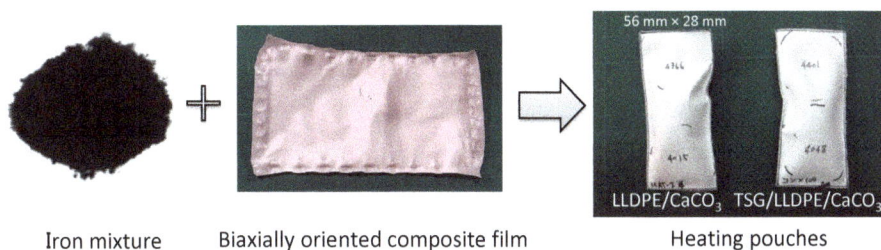

Iron mixture   Biaxially oriented composite film   Heating pouches

Figure 10. Hand-made heating pouches. The size of the pouches is 56 mm in length and 28 mm in width.

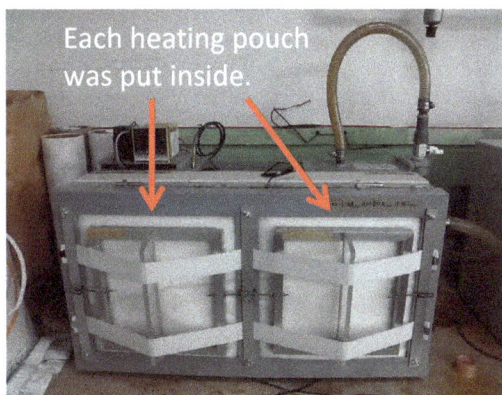

Figure 11. Temperature-measuring instrument. Each heating pouch was closely pressed against the temperature sensor plate inside the sealed box.

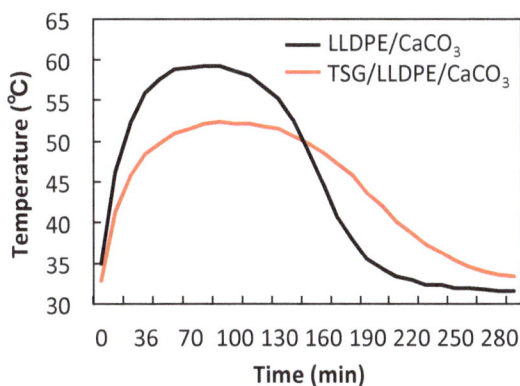

Figure 12. Temperature as a function of time for heating pouches of LLDPE/CaCO$_3$ and TSG/LLDPE/CaCO$_3$ composite films.

4. Conclusions

In this work, we synthesized a novel TSG with good thermal durability and a lower melting temperature below 60 °C. TSG particles with a size of 10–400 μm were also produced successfully.

Polymers **2018**, *10*, 486

The coefficient of linear thermal expansion (α) was as high as $2817.0 \times 10^{-5}/°C$ at 65–70 °C, and the excellent thermal expansion of TSG was revealed by TMA and PVT testing.

To effectively utilize the thermal expansion of the TSG, we proposed the usage of TSG particles to reduce the pore size to adjust the gas permeability for breathable films under temperature simulation. By introducing the TSG into $LLDPE/CaCO_3$, we prepared the microporous $TSG/LLDPE/CaCO_3$ composite films. The $TSG/LLDPE/CaCO_3$ composite films had well-balanced mechanical properties, their thermal expansion behavior was also clarified, according to the tensile testing and TMA results.

The SEM micrographs indicated the uniformly-porous structure and the pore size was in the range of 5–40 μm for the $TSG/LLDPE/CaCO_3$ composite films. The addition of TSG lengthened the permeation time almost two time at around 60 °C, showing the pore size reduced due to the thermal expansion effect of the TSG near and above the melting temperature. Furthermore, to verify the practical feasibility of the functional $TSG/LLDPE/CaCO_3$ composite films, we conducted the demonstration temperature control experiments. The experimental results clearly proved the addition of TSG offered the effect of pore-size adjustment (controllable gas-permeability), more promisingly, the great practical value and application prospects of this strategic effect for a temperature-sensitive breathable film was clarified.

The global functional porous membranes/films market is expected to grow during the forecast period [36]. The major drivers of this growth will be the increasing awareness and concern for hygiene and healthy life. With the increase in the food and beverage industry, the packaging industry will also be affected. The breathable films for the packaging industry to maintain the quality and freshness of the food items during the shelf period is also a quite promising application field of $TSG/LLDPE/CaCO_3$ films, because we are able to provide a humidity adjustment function readily at the same time with the microporous films.

Author Contributions: Jin Gong and Hiroshi Ito conceived and designed the experiments; Eiichi Hosaka, Kohei Sakai and Yoshikazu Shibata performed the experiments; Jin Gong, Hiroshi Ito, Eiichi Hosaka, Kohei Sakai, Yoshikazu Shibata, and Kohei Sakai analyzed the data; Yoshikazu Shibata, Kosei Sato, Dai Nakanishi, Shinichiro Ishihara and Kazuhiro Hamada contributed reagents/materials/analysis tools; and Jin Gong and Hiroshi Ito wrote the paper.

Conflicts of Interest: The authors declare no conflict of interest.

References

1. Gong, J.; Shitara, M.; Serizawa, R.; Makino, M.; Kabir, M.H.; Furukawa, H. 3D printing of meso-decorated gels and foods. *Mater. Sci. Forum* **2014**, *783–786*, 1250–1254. [CrossRef]
2. Gong, J.; Igarashi, S.; Sawamura, K.; Furukawa, H. Gel engineering materials meso-decorated with polymorphic crystals. *Adv. Mater. Res.* **2013**, *746*, 325–329. [CrossRef]
3. Gong, J.; Watanabe, Y.; Watanabe, Y.; Hidema, R.; Kabir, M.H.; Furukawa, H. Development of a novel standard type of gel engineering materials via simple bulk polymerization. *J. Solid Mech. Mater. Eng.* **2013**, *7*, 455–462. [CrossRef]
4. Kabir, M.H.; Gong, J.; Watanabe, Y.; Makino, M.; Furukawa, H. Hard-to-soft transition of transparent shape memory gels and the first observation of their critical temperature studied with scanning microscopic light scattering. *Mater. Lett.* **2013**, *108*, 239–242. [CrossRef]
5. Gong, J.; Furukawa, H. Smart optical device of varifocal lens developed with high transparent shape memory gels. *Expect. Mater. Future* **2013**, *13*, 5–8.
6. Amano, Y.; Hidema, R.; Gong, J.; Furukawa, H. Creation of shape memory gels with inter-crosslinking network structure. *Chem. Lett.* **2012**, *41*, 1029–1031. [CrossRef]
7. Harada, S.; Hidema, R.; Gong, J.; Furukawa, H. Intelligent button developed with smart soft and wet materials. *Chem. Lett.* **2012**, *41*, 1047–1049. [CrossRef]
8. Kwon, Y.M.; Kim, S.W. Thermosensitive biodegradable hydrogels for the delivery of therapeutic agents. In *Drugs and the Pharmaceutical Sciences, Polymer Drug Delivery System*, 1st ed.; Kwon, G.S., Ed.; Taylor & Francis Group: Boca Raton, FL, USA, 2005; Volume 148, pp. 251–274, ISBN 0-8247-2532-8.

9.   Chandra, A.; Meyer, W.H.; Best, A.; Hanewald, A.; Wegner, G. Modifying thermal expansion of polymer composites by blending with a negative thermal expansion material. *Macromol. Mater. Eng.* **2007**, *292*, 295–301. [CrossRef]

10.   Takenaka, K.; Ichigo, M. Thermal expansion adjustable polymer matrix composites with giant negative thermal expansion filler. *Compos. Sci. Technol.* **2014**, *104*, 47–51. [CrossRef]

11.   Poveda, R.L.; Gupta, N. Thermal expansion of CNF/polymer composites. In *Carbon Nanofiber Reinforced Polymer Composites*; Springer eBook; Springer: Berlin, Germany, 2016; pp. 53–62, ISBN 978-3-319-23787-9.

12.   Kiba, S.; Suzuki, N.; Okawauchi, Y.; Yamauchi, Y. Prototype of low thermal expansion materials: Fabrication of mesoporous silica/polymer composites with densely filled polymer inside mesopore space. *Chem. Asian J.* **2010**, *5*, 2100–2105. [CrossRef] [PubMed]

13.   Yamashina, N.; Isobe, T.; Ando, S. Low Thermal Expansion Composites Prepared from Polyimide and ZrW2O8 Particles with Negative Thermal Expansion. *J. Photopolym. Sci. Technol.* **2012**, *25*, 385–388. [CrossRef]

14.   Wei, C.; Srivastava, D.; Cho, K. Thermal Expansion and Diffusion Coefficients of Carbon Nanotube-Polymer Composites. *Nano Lett.* **2002**, *2*, 647–650. [CrossRef]

15.   Nishino, T.; Kotera, M.; Sugiura, Y. Residual stress of particulate polymer composites with reduced thermal expansion. In *Journal of Physics: Conference Series*; IOP Publishing: Bristol, UK, 2009; Volume 184.

16.   Sideridou, I.; Achilias, D.S.; Kyrikou, E. Thermal expansion characteristics of light-cured dental resins and resin composites. *Biomaterials* **2004**, *25*, 3087–3097. [CrossRef] [PubMed]

17.   Yang, B.; Yang, W. Thermo-sensitive switching membranes regulated by pore-covering polymer brushes. *J. Membr. Sci.* **2003**, *218*, 247–255. [CrossRef]

18.   Albo, J.; Wang, J.; Tsuru, T. Gas transport properties of interfacially polymerized polyamide composite membranes under different pre-treatments and temperatures. *J. Membr. Sci.* **2014**, *449*, 109–118. [CrossRef]

19.   Pilar Castejón, P.; Habibi, K.; Saffar, A.; Ajji, A.; Martínez, A.B.; Arencón, D. Polypropylene-Based Porous Membranes: Influence of Polymer Composition, Extrusion Draw Ratio and Uniaxial Strain. *Polymers* **2018**, *33*, 1–21. [CrossRef]

20.   Park, Y.; Gutierrez, M.P.; Lee, L.P. Reversible self-actuated thermo-responsive pore membrane. *Sci. Rep.* **2016**, *6*, 2045–2322. [CrossRef] [PubMed]

21.   Sershen, S.R.; Westcott, S.L.; Halas, N.J.; West, J.L. Temperature-sensitive polymer-nanoshell composites for photothermally modulated drug delivery. *J. Biomed. Mater. Res. Part A* **2000**, *51*, 293–298. [CrossRef]

22.   He, Y.; Moreira, E.; Overson, A.; Nakamura, S.H.; Bider, C.; Briscoe, J.F. Thermal characterization of an epoxy-based underfill material for flip chip packaging. *Thermochim. Acta* **2000**, *357–358*, 1–8. [CrossRef]

23.   Choi, Y.J.; Yamaguchi, T.; Nakao, S.I. A novel separation system using porous thermosensitive membranes. *Ind. Eng. Chem. Res.* **2000**, *39*, 2491–2495. [CrossRef]

24.   Park, Y.S.; Ito, Y.; Imanishi, Y. Permeation control through porous membranes immobilized with thermosensitive polymer. *Langmuir* **1998**, *14*, 910–914. [CrossRef]

25.   Shtanko, N.I.; Kabano, V.Y.; Apel, P.Y.; Yoshida, M. The use of radiation-induced graft polymerization for modification of polymer track membranes. *Nucl. Instrum. Methods Phys. Res. Sect. B* **1999**, *151*, 416–422. [CrossRef]

26.   Chung, D.J.; Ito, Y.; Imanishi, Y. Preparation of porous membranes grafted with poly (spiropyran-containing methacrylate) and photocontrol of permeability. *J. Appl. Polymer Sci.* **1994**, *51*, 2027–2033. [CrossRef]

27.   Iwata, H.; Oodate, M.; Uyama, Y.; Amemiya, H.; Ikada, Y. Preparation of temperature-sensitive membranes by graft polymerization onto a porous membrane. *J. Membr. Sci.* **1991**, *55*, 119–130. [CrossRef]

28.   Hotpress Gilfillan, W.N.; Moghaddam, L.; Bartley, J.; Doherty, W.O.S. Thermal extrusion of starch film with alcohol. *J. Food Eng.* **2015**, *170*, 92–99. [CrossRef]

29.   Ruland, W. X-ray determination of crystalinity and diffuse disorder scattering. *Acta Crystallogr.* **1961**, *14*, 1180–1185. [CrossRef]

30.   Roger, B. *Handbook of Polymer Testing—Physical Methods*, 1st ed.; Marcel Dekker, Inc.: New York, NY, USA, 1999; pp. 341–342, ISBN 0-8247-0171-2.

31.   Landsberg, M.I.; Winston, G. Relationship between measurements of air Permeability by two machines. *Text. Res. J.* **1947**, *17*, 214–221. [CrossRef]

32.   Chakravorty, S. PVT testing of polymers under industrial processing conditions. *Polym. Test.* **2002**, *21*, 313–317. [CrossRef]

33. Huang, D.H.; Tran, T.N.; Yang, B. Investigation on the reaction of iron powder mixture as a portable heat source for thermoelectric power generators. *J. Therm. Anal. Calorim.* **2014**, *116*, 1047–1053. [CrossRef]

34. Mo, Z.; Zhang, H. The degree of crystallinity in polymers by wide-angle x-ray diffraction (WAXD). *J. Macromol. Sci. Part C Polym. Rev.* **1995**, *35*, 55–580. [CrossRef]

35. Hosaka, E.; Gong, J.; Ito, H.; Shibata, Y.; Nakanishi, D.; Ishihara, S. Processing and Thermal Response Properties of Temperature Sensitive Gel/Polymer composites. Design, Manufacturing and Applications of Composites. In Proceedings of the Eleventh Joint Canada-Japan Workshop on Composites: First Joint Canada-Japan-Vietnam Workshop on Composites, Ho Chi Minh, Vietnam, 8–10 August 2016; Available online: dpi-proceedings.com/index.php/dmac/article/view/5246/0 (accessed on 15 May 2016).

36. Technavio. *Top 5 Vendors in the Breathable Films Market from 2017 to 2021*; Technavio: London, UK, August 2017; pp. 1–71. Available online: technavio.com/report/global-breathable-films-market (accessed on 14 September 2017).

polymers

MDPI

Article

Single-Point Incremental Forming of Two Biocompatible Polymers: An Insight into Their Thermal and Structural Properties

Luis Marcelo Lozano-Sánchez [1,*], Isabel Bagudanch [2], Alan Osiris Sustaita [1], Jackeline Iturbe-Ek [1], Luis Ernesto Elizalde [3], Maria Luisa Garcia-Romeu [2] and **Alex Elías-Zúñiga [1]**

[1] Tecnologico de Monterrey, Escuela de Ingeniería y Ciencias, Ave. Eugenio Garza Sada 2501, Monterrey 64849, México; alan.sustaita@itesm.mx (A.O.S.); jackeline.iturbe.ek@itesm.mx (J.I.-E.); aelias@itesm.mx (A.E.-Z.)
[2] Mechanical Engineering and Industrial Construction Department, Campus Montilivi, University of Girona, 17071 Girona, Spain; isabel.bagudanch@udg.edu (I.B.); mluisa.gromeu@udg.edu (M.L.G.-R.)
[3] Centro de Investigación en Química Aplicada, Blvd. Ing. Enrique Reyna No. 140, Saltillo 25253, México; luis.elizalde@ciqa.edu.mx
* Correspondence: lmarcelo.lozano@gmail.com or marcelo.lozano@itesm.mx; Tel.: +52-(81)-8358-2000 (ext. 5653)

Received: 6 March 2018; Accepted: 28 March 2018; Published: 1 April 2018

Abstract: Sheets of polycaprolactone (PCL) and ultra-high molecular weight polyethylene (UHMWPE) were fabricated and shaped by the Single-Point Incremental Forming process (SPIF). The performance of these biocompatible polymers in SPIF was assessed through the variation of four main parameters: the diameter of the forming tool, the spindle speed, the feed rate, and the step size based on a Box–Behnken design of experiments of four variables and three levels. The design of experiments allowed us to identify the parameters that most affect the forming of PCL and UHMWPE. The study was completed by means of a deep characterization of the thermal and structural properties of both polymers. These properties were correlated to the performance of the polymers observed in SPIF, and it was found that the polymer chains are oriented as a consequence of the SPIF processing. Moreover, by X-ray diffraction it was proved that polymer chains behave differently on each surface of the fabricated parts, since the chains on the surface in contact with the forming tool are oriented horizontally, while on the opposite surface they are oriented in the vertical direction. The unit cell of UHMWPE is distorted, passing from an orthorhombic cell to a monoclinic due to the slippage between crystallites. This slippage between crystallites was observed in both PCL and UHMWPE, and was identified as an alpha star thermal transition located in the rubbery region between the glass transition and the melting point of each polymer.

Keywords: polycaprolactone; ultra-high molecular weight polyethylene; incremental forming; SPIF; XRD; chain orientation

---

1. Introduction

In recent years, great attention has been paid to the low-cost manufacturing process known as Incremental Sheet Forming (ISF), which can be described in a general way as a process where a plastic deformation is applied locally and in a consecutive manner on a flat sheet until a final part with desired geometry is obtained [1]. One of the variations of ISF is the so-called Single-Point Incremental Forming process (SPIF), in which a simple-shaped tool moves horizontally and vertically by a toolpath program until the desired final part is formed [2] as depicted in Figure 1. The main motivation to further develop the SPIF process, and ISF in general, is the possibility to produce many different parts without the need to manufacture tooling, i.e., the toolpath defines the part geometry, so a new path can be programmed and used without incurring costs of tool development and switchover of setup [3]. For this reason, SPIF is considered as a manufacturing process with high economic payoff for rapid prototyping and small-batch production [4].

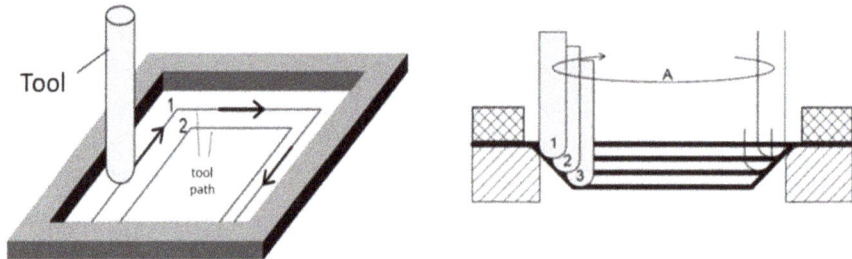

Figure 1. Schematic representation of the Single-Point Incremental Forming (SPIF) process.

At its inception, SPIF emerged as a process used for the fabrication of sheet metal components [2,5–7]. However, around a decade ago, the use of SPIF was extended to fabricate workpieces from polymers, specifically thermoplastics [8–11], because of two main reasons: the structural and thermal properties of thermoplastics make them particularly suitable to applications in which a high strength/mass ratio and good formability are required, e.g., in the medical, aerospace, and automotive sectors. The second important reason is that SPIF enables the deformation of polymers at room temperature. This represents a significant cost saving because besides not requiring energy nor special equipment to melt the polymer, a mold is not needed either.

The research on SPIF of polymers has been directed towards the development of strategies for selecting the most appropriate polymers for SPIF [12], taking into account the ductility of the polymer and the required level of geometrical part accuracy, and for selecting the optimal parameters for the SPIF process. Some of the early work generated in this context was reported by Martins et al. [11], who made the first attempt towards the development of criteria for the selection of polymers that are suitable for SPIF by testing five different polymers: polyoxymethylene (POM), polyethylene (PE), polyamide (PA), polyvinylchloride (PVC), and polycarbonate (PC). In this work, the authors concluded that the feed rate, the thickness of the sheet, the diameter of the forming tool, and the step size play a key role in SPIF of polymers. In another work developed with PVC, Bagudanch et al. [13] concluded that the spindle speed has a significant influence on the temperature variation during the forming process and in the maximum force achieved, in such a way that the force needed to form polymers in SPIF is reduced as the temperature is increased.

Nevertheless, a lack of research has been detected regarding an in-depth study of the properties of polymers, mainly the thermal and structural ones, and their correlation with their performance in SPIF. For instance, the work reported by Martins et al. [11] studied the formability of the five thermoplastics previously mentioned based on their mechanical properties. Similarly, Marques et al. [4] based a study of the formability in SPIF of some polymers previously used, such as PVC, PC, and PA, and adding polyethylene terephthalate (PET) to this study, on a merely mechanical characterization and a qualitative consideration of the crystallinity using a classification from high-crystalline to amorphous. Davarpanah et al. [14] quantitatively analyzed, by using Differential Scanning Calorimetry (DSC), how different parameters varied in SPIF affect the crystallinity of polylactic acid (PLA). Lately, the use of characterization techniques based on X-ray have allowed for the observation that polymer chains are oriented as a consequence of the SPIF process. In a previous work, Lozano-Sanchez et al. [15] reported the orientation of polypropylene (PP) chains and suspected a difference in the behavior of polymer chains between the inner and outer surfaces of the analyzed sheets. This difference was related to different forces exerted on each surface, which was later supported by the work of Jiménez et al. [16] who found that the residual stresses were changing from tensile to compressive along the inner and outer surfaces of aluminum sheets processed by SPIF. Although this last work used metal sheets, it has been assumed that the same effect occurs on any material formed by SPIF, since this effect is attributed to the processing rather than to the material itself. Moreover, although it is widely accepted that the friction between the forming tool and the sheet generates enough heat to soften the polymer, there is no study that considers the thermal properties of polymers, since in thermoplastics the temperature plays a key role in its processing.

On the other hand, only few works related to the performance of biocompatible polymers in SPIF are found in the literature [14,17]. This represents an interesting area of opportunity considering that most of the materials tested up to now are non-biocompatible thermoplastics, and that one of the major advantages of SPIF is the possibility to manufacture customized products that can be applied in the medical field. Among the biocompatible polymers, polycaprolactone (PCL) is of great interest due to some of its properties, such as a low melting point (59–64 °C), exceptional blend-compatibility, high flexibility, and a medium Young's modulus at room temperature, which has stimulated extensive research on its potential application in the biomedical field since 1980 [18]. Nowadays, the main commercial application of PCL is in the manufacture of biodegradable bottles and films, but according to Khan et al. [19], bio-medical applications, such as synthetic wound dressings, encapsulants for drug release systems, and contraceptive implants, are becoming increasingly common. Likewise, ultra-high molecular weight polyethylene (UHMWPE) has been widely used since the late 1960s in medical implants, such as total joint replacements in hips and knees [20,21], due to its biocompatibility coupled with its high strength and ductility resulting from the semi-crystalline structure of its long chains [22]. So, the use of these biocompatible polymers in SPIF have just gained interest in the last few years. For instance, PCL has lately been used for manufacturing cranial geometries [23–25], while both PCL and UHMWPE were included in a work in which was analyzed the influence of the main parameters, i.e., tool diameter, spindle speed, feed rate, and step down, in the maximum temperature reached during the SPIF process [26].

In this work, sheets of PCL and UHMWPE were fabricated by compression molding and then used in SPIF to fabricate pyramid-shaped parts with circular generatrix. The SPIF processing of both polymers was conducted using a Box–Behnken design of experiments varying four parameters: the diameter of the forming tool, the spindle speed, the feed rate, and the step down. The performance of these biocompatible polymers in SPIF was assessed through the formability, defined in terms of the maximum depth of the pyramid-shaped parts, the forming force, and the maximum temperature reached during the forming process. The PCL and UHMWPE sheets were characterized before and after being processed by SPIF in order to correlate their thermal and structural properties with the behavior displayed in SPIF. The results here presented seek to contribute to a better understanding of the behavior of polymers formed by SPIF in addition to encouraging the use of biocompatible polymers to achieve the development of this process aimed at one of its most interesting potential applications: the manufacture of custom medical implants.

2. Materials and Methods

2.1. PCL and UHMWPE Sheets Preparation

Fifty-two grams of PCL pellets (Sigma Aldrich, St. Louis, MO, USA, ≈3 mm, average Mn = 80,000) were placed into the square cavity of a stainless steel mold with dimensions of 150 mm × 150 mm × 2 mm. The PCL pellets were compressed at 100 °C by using a CARVER 4122 hydraulic heating press (Carver Inc., Wabash, IN, USA), initially applying a load of 2 metric tons for 5 min to ensure the melting of the material. After that, the load was increased to 30 metric tons and maintained for 2 min. Finally, the load was removed and the mold was cooled to room temperature in a cooling press under a load of 30 metric tons as well. The fabrication of the UHMWPE sheets was carried out using the same procedure described before, using 46 g of UHMWPE powder (Sigma Aldrich, average *Mw*: 3,000,000–6,000,000) compressed at 170 °C in the hydraulic heating press.

2.2. Materials Characterization

The PCL and UHMWPE sheets were studied before the SPIF processing by Vicat softening temperature (VST), differential scanning calorimetry (DSC), thermal gravimetric analysis (TGA), dynamic mechanical analysis (DMA), and tensile tests. The VST was determined by using a DTUL/Vicat tester model HDT-III (Custom Scientific Instruments, Easton, PA, USA). Tests were carried out with a total load of 50 N and a heating rate of 120 °C/h following the guidelines of the ASTM D-1525 standard. TGA and DSC analysis were carried out simultaneously in an SDT Q600 (TA Instruments, New Castle, DE, USA) with a precision of the thermobalance of 0.0001 mg. For this, small samples of about 20 mg were cut from the pre-fabricated sheets. Each sample was heated to 600 °C at 10 °C/min under an argon flow of 100 mL/min. DMA was carried out in a Dynamic Mechanical Thermal Analyzer model Q800 from TA Instruments used in the cantilever mode with a frequency of 1 Hz and amplitude of 20 μm. DMA was performed to determine the viscoelastic behavior of the materials through the measurement of the storage modulus (E'), loss modulus (E''), and loss factor (tan $\delta = E''/E'$) over a temperature range of 30–70 °C for PCL, and 30–150 °C for UHMWPE, with a heating rate of 5 °C/min in both cases. The tensile tests were performed in a United universal testing machine equipped with a load cell of 4450 N, according to the ASTM 638 standard, using specimens of type 1. An extensometer was employed to measure the elongation of the specimens. The used rate was 50 mm/min for the analysis of the tensile strength and 5 mm/min for the modulus calculation. The initial separation between gags was 114 mm. The mechanical tensile properties of PCL and UHMWPE were taken as the average values obtained from a total of five specimens tested of each polymer. Some selected final parts fabricated by SPIF were analyzed by X-ray diffraction (XRD). The XRD analyzes were performed on a Empyrean diffractometer (PANalytical, Almelo, Netherlands) equipped with a Bragg–Brentano module and by using a X-ray tube source of Cu Ka radiation (λ = 1.5406 Å) operated at 40 kV and 45 mA. The diffraction patterns were obtained over the 2θ range of 5°–60° with steps of 0.02° and 50 s per step.

2.3. SPIF of PCL and UHMWPE Sheets

SPIF of PCL and UHMWPE sheets was conducted in a Kondia®HS1000 three-axis milling machine (Kondia, Spain). A fixed system consisting of a hollow support die was bolted over a dynamometer. The polymer sheets were placed between the clamping and the top plates with an effective working area of 120 mm × 120 mm. The incremental forming was made with a hemispherical Vanadis 23 steel tool. Around 5 mL of vegetable oil was spilled on the testing sheet in order to decrease friction effects between the forming tool and the testing material. The performance of the polymer sheets was evaluated through a Box–Behnken (BB) design of experiments considering four variables. BB designs are three-level designs that allow for fitting second-order response surfaces efficiently. The considered parameters and levels are listed in Table 1.

Table 1. Parameters and levels considered in the Box–Behnken (BB) design for SPIF of PCL and UHMWPE sheets.

Parameter	−1	0	1
D_t (tool diameter, mm)	6	10	14
S (spindle speed, rpm)	0	1000	2000
F (feed rate, mm/min)	1500	2250	3000
Δz (Step down, mm)	0.2	0.35	0.5

The BB design with four factors consists of 27 experimental runs (Table 2) that can be split into three blocks with one center point at each block. During the SPIF processing, the force in the z-axis (F_z)

was measured by using a Kistler 9257B dynamometer and the temperature of the polymer sheets was measured with an Irbis ImageIR 3300 thermographic camera (InfraTec GmbH, Dresden, Germany). A summary of the experimental methodology used in this work is displayed in Figure 2.

Table 2. List of 27 experiments for SPIF of PCL and UHMWPE sheets.

Exp.	Tool Diameter, D_t (mm)	Spindle Speed, S (rpm)	Feed Rate, F (mm/min)	Step down, Δz (mm)
1	6	0	2250	0.35
2	6	2000	2250	0.35
3	14	0	2250	0.35
4	14	2000	2250	0.35
5	10	1000	1500	0.20
6	10	1000	1500	0.50
7	10	1000	3000	0.20
8	10	1000	3000	0.50
9	6	1000	2250	0.20
10	6	1000	2250	0.50
11	14	1000	2250	0.20
12	14	1000	2250	0.50
13	10	0	1500	0.35
14	10	0	3000	0.35
15	10	2000	1500	0.35
16	10	2000	3000	0.35
17	6	1000	1500	0.35
18	6	1000	3000	0.35
19	14	1000	1500	0.35
20	14	1000	3000	0.35
21	10	0	2250	0.20
22	10	0	2250	0.50
23	10	2000	2250	0.20
24	10	2000	2250	0.50
25	10	1000	2250	0.35
26	10	1000	2250	0.35
27	10	1000	2250	0.35

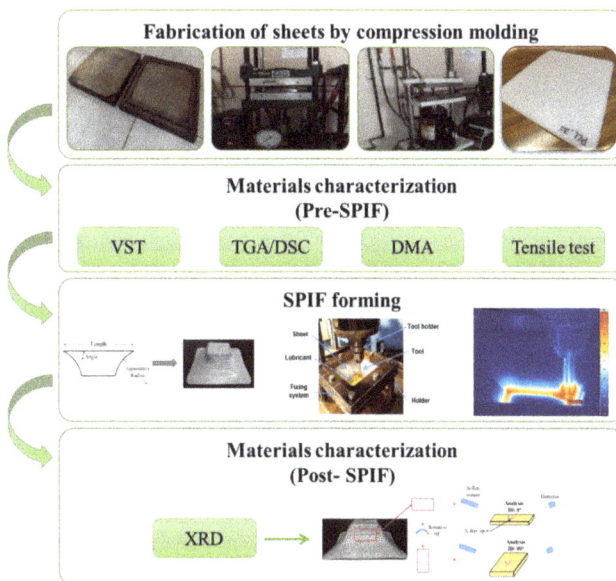

Figure 2. Summary of the experimental methodology.

3. Results and Discussion

3.1. Characterization of PCL and UHMWPE Sheets

The thermal properties of the PCL and UHMWPE sheets were studied by DSC, TGA, and Vicat test. From the Vicat test, the corresponding analysis determined that the Vicat softening temperature (VST) for PCL and UHMWPE is around 44 and 90 °C, respectively. The VST is considered as the temperature at which a specimen is penetrated to a depth of 1 mm by a flat-ended needle with a 1 mm^2 circular or square cross-section. Is worth mentioning that the VST is a specific temperature value, while the melt point of thermoplastic polymers is usually taken within a whole range of temperature. So, for applications in SPIF, it will be more useful to consider the VST to analyze the formability of polymers once the sheet reaches the VST due to the friction caused by the contact between the sheet surface and the forming tool. The DSC and TGA curves for PCL are shown in Figure 3. The melting behavior of PCL displayed in Figure 3a shows a melting peak around 71 °C. From the TGA curve (Figure 3b), it can be observed that PCL shows a degradation step in the range from 394 to 433 °C, with a temperature of maximum decomposition rate of 409 °C. These results agree with the values reported in the literature [19,20].

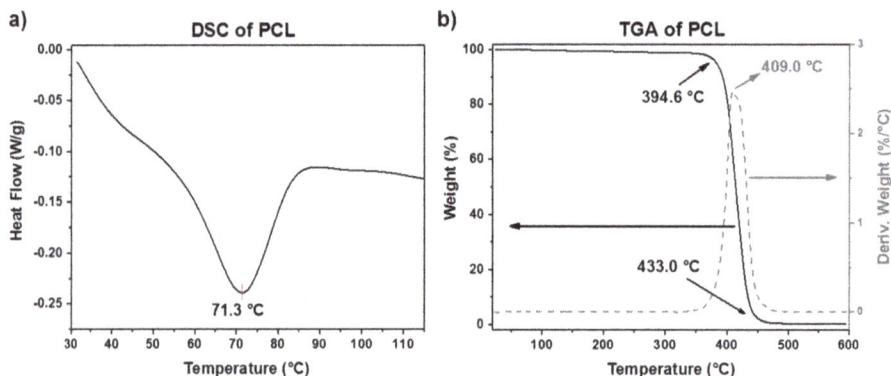

Figure 3. (a) DSC and (b) TGA plots of PCL.

The DSC and TGA curves of UHMWPE are shown in Figure 4a,b, respectively. The DSC curve shows a melting peak at around 139 °C. On the other hand, the TGA curve also shows a single-step degradation behavior in the range from 467 to 494 °C, with a temperature of maximum decomposition rate around 484 °C. The higher thermal stability of UHMWPE compared to PCL is attributed to its longer polymer chains. The thermal properties of the two polymers are summarized in Table 3.

Table 3. Thermal properties of PCL and UHMWPE.

Sample	VST (°C)	T_m (°C)	Temp. of Initial Decomposition (°C)	Temp. of Maximum Decomposition Rate (°C)
PCL	44.3	71.3	394.6	409.0
UHMWPE	89.7	138.9	467.3	483.5

DMA results of PCL in the temperature range of 30–70 °C are shown in Figure 5 for the temperature dependence of the storage (E') and loss (E") moduli and the loss factor (tan δ). Observed at the initial temperature of the analysis were values of E' and E" equal to 528 and 113.7 MPa, respectively. The curve of E" shows a peak that has been attributed to a transition occurring in the polymer associated with the slippage between crystallites. This type of transition has often been identified as an alpha star transition ($T_\alpha*$) and detected in semicrystalline polymers [27]. This peak has a maximum value at 43.6 °C, which is close to the VST previously determined for PCL. The alpha star transition was not detected by DSC because DMA is much more sensitive and can easily measure transitions not apparent in other thermal methods. This $T_\alpha*$ can be correlated with the VST if it is considered that at a temperature around 44 °C there is a crystal–crystal slippage within the PCL, which facilitates the penetration of the needle in the VST test. The curve of tan δ shows a peak at 68.8 °C, which corresponds to the melting temperature (T_m) and is similar to that previously measured by DSC.

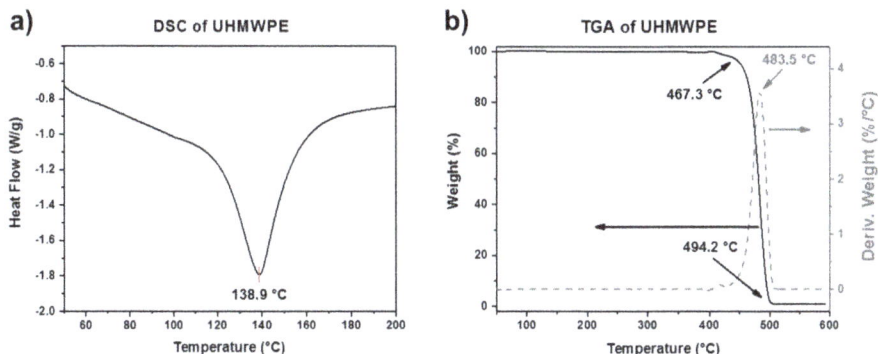

Figure 4. (**a**) DSC and (**b**) TGA plots of UHMWPE.

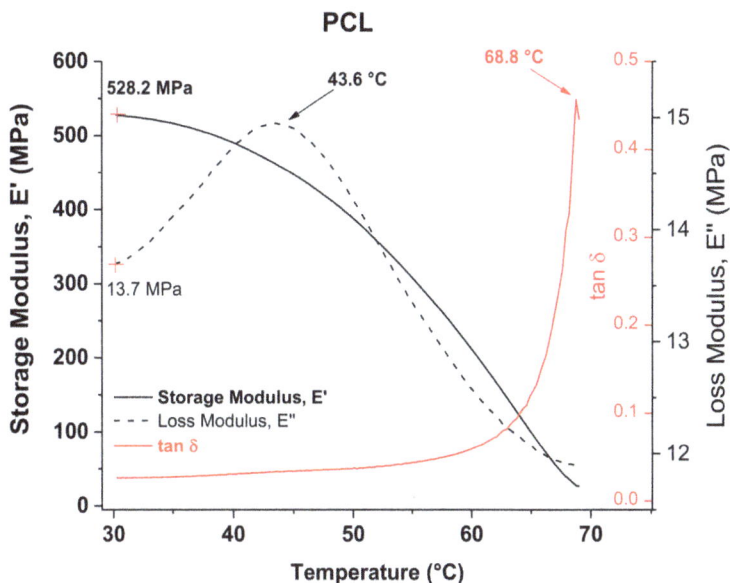

Figure 5. DMA graphs of PCL.

Figure 6 shows the DMA results of UHMWPE. Similar to what occurs with PCL, the *E''* curve of UHMWPE exhibits a peak at 55.6 °C, which can also be identified as an alpha star transition associated with the slippage between crystallites, considering that it is a semicrystalline polymer as well. Unlike PCL, the values of VST and $T_\alpha*$ of UHMWPE differ considerably from each other. This can be attributed to the presence of long molecular chains in UHMWPE, so that even when there is a crystal–crystal slippage at around 55 °C, it becomes more complicated to move or pass through a polymer composed of long chains that must also be highly entangled as occurs in UHMWPE. Therefore, it is necessary to continue heating to allow for sufficient movement of the long chains so that the VST test needle penetrates in the UHMWPE structure. The T_m obtained by DMA, taken as the peak of the tan δ curve (i.e., 131.2 °C), is similar to that obtained by DSC.

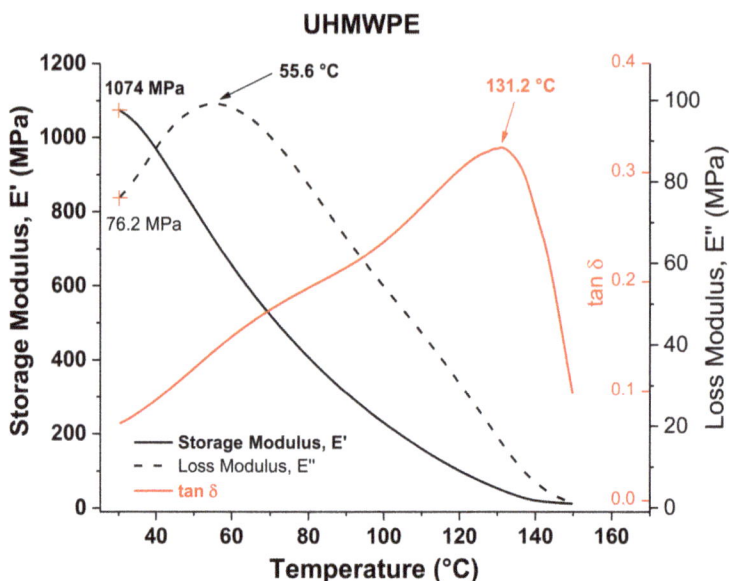

Figure 6. DMA graphs of UHMWPE.

The storage and loss moduli of PCL and UHMWPE have been plotted together in Figure 7, and the tan δ curves are shown in the inset graph. For both polymers, E' reduced gradually as temperature increased, which is typical in thermoplastics and represents that less force is required for deformation; however, it is observed that E' of PCL decreases more sharply and UHMWPE displays higher E' values meaning a more rigid structure. In other words, based on the DMA results it would be more difficult to plastically deform UHMWPE than PCL. In the inset figure, the tan δ curve of PCL shows a pronounced increase as it approaches the T_m, unlike the tan δ curve of UHMPE that exhibits a continuing increase throughout the whole temperature sweep. The temperature transitions in UHMWPE, specifically the melting transition, occur more gradually due to the presence of long molecular chains in a manner that before the melt is reached, i.e., where large-scale chain slippage occurs and the material flows, the coiled long chains must be disentangled first.

The results from tensile tests of PCL and UHMWPE are graphically shown in Figure 8. For this, the graph of only one specimen of each material was taken, but it is properly representative of the behavior observed in all the specimens tested. During the tensile test, PCL shows a decrease in strength after the yield point, although it is maintained in a stable value. This behavior is typically observed when the specimen undergoes a necking effect. On the other hand, UHMWPE shows a continuous increase of the supported tensile stress after the yield point, which is clear evidence that a strain hardening effect occurs in the specimen. The inset graph in Figure 8 represents the initial stage of the stress versus strain curve, where a higher Young's modulus is observed in UHMWPE indicating a more rigid material as was also observed by DMA. The mechanical properties from the tensile tests are summarized in Table 4. From these data, it was determined that the ultimate tensile strength of PCL and UHMWPE was 16.4 and 20.2 MPa, respectively. It is worth mentioning that the tensile strength of

Polymers **2018**, *10*, 391

UHMWPE was taken at a point on the stress versus strain curve near where the plastic deformation begins, so the strain hardening effect is not considered (see Figure 8). Both materials showed a highly ductile behavior with an average elongation at the end of the test of more than 450% and 340% for PCL and UHMWPE, respectively. Moreover, due to the strain hardening effect observed in UHMWPE, it is clear that it has higher toughness than PCL.

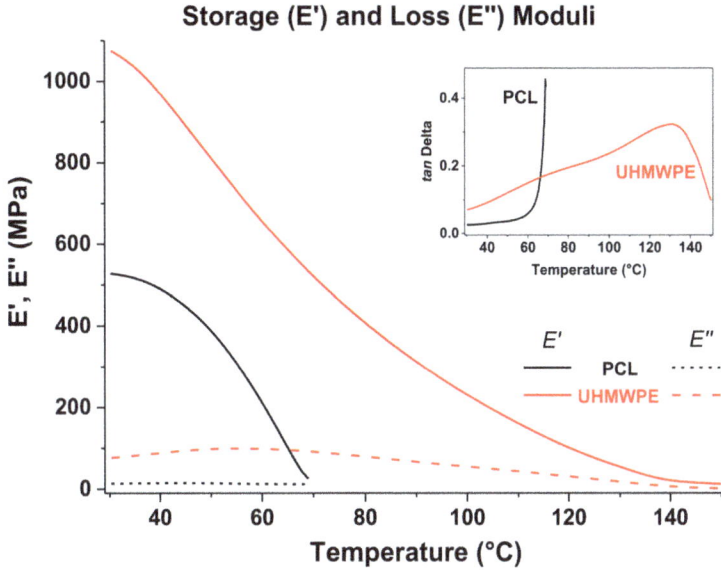

Figure 7. Storage and loss moduli from DMA for PCL and UHMWPE. The inset graph corresponds to the plots of tan δ.

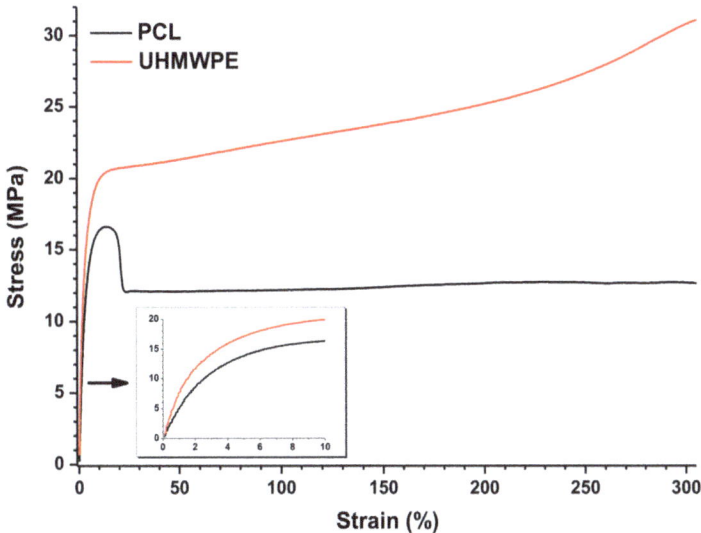

Figure 8. Tensile tests of PCL and UHMWPE.

Table 4. Mechanical properties of PCL and UHMWPE from uniaxial tensile tests.

Sample	Young's Modulus (MPa)	Yield Strength (MPa)	Elongation (%)
PCL	374.7	16.4	451.2
UHMWPE	445.3	20.2	343.0

3.2. SPIF of PCL and UHMWPE Sheets

The results obtained from the BB design of experiments were statistically analyzed by means of the response surface methodology; however, a behavior similar to that previously reported by Bagudanch et al. [26] was found even when the PCL and UHMWPE sheets corresponded to polymers with different properties to those used in this work. So, the detailed analysis of the response surface methodology can be reviewed in detail in [26]. In the present work, the effect that the parameters considered in the design of experiments have when processing PCL and UHMWPE by SPIF will be briefly analyzed by using box diagrams in order to make a correlation between the behavior shown in the forming process and the properties studied by the different characterization techniques used here.

The maximum formed depth (in percentage), forming force in the z-axis (F_z), and maximum temperature reached (T_{max}) in each test done from the BB design of experiments are summarized in Table 5 for the PCL sheets and in Table 6 for the UHMWPE sheets. For convenience, the conditions of each test (enlisted previously in Table 2) have been listed again. The results for PCL (see Table 5) show that most of the experiments reached a depth of 100%. The images in Figure 9 represent sheets of PCL completely formed and with failure. The pyramid-shaped part of Figure 9b corresponds to the sheet formed under parameters of experiment #15, which failed at a depth of 95%. This experiment is particularly interesting because it registered the maximum temperature (67.52 °C) among all of the tests made with PCL, which caused a very irregular surface as observed on the walls of the pyramid-like part. The temperature reached in this experiment is close to the T_m of PCL, so that at the end of the SPIF processing, the polymer likely behaved more as a viscous liquid than as a solid.

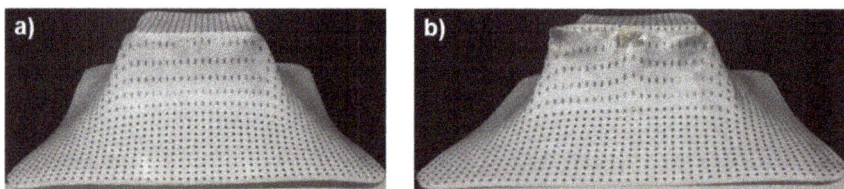

Figure 9. Pyramid-shaped parts made of PCL. (**a**) Sheet completely formed; (**b**) Sheet with failure.

From the design of experiments, it was found that D_t and S are the parameters that mostly affect F_z and T_{max}. The box diagrams displayed in Figures 10 and 11 show the tendency of T_{max} and F_z, respectively, in PCL as an effect of D_t and S. The effect of the feed speed and the step size (data not shown) was found to be negligible for T_{max} and F_z. It is noticed that T_{max} increases as D_t is increased (see Figure 10a). When a forming tool of larger diameter is used, the area in contact with the polymer sheet increases, generating a greater friction between both surfaces and consequently a higher temperature. A similar behavior associated with the friction occurs when S is varied (see Figure 10b), since there is more friction between surfaces when the tool rotates faster, i.e., the temperature increases when S is increased. In fact, it has been reported that the spindle speed has the most important role in the temperature variation in the SPIF of polymers [13].

Table 5. SPIF results of PCL sheets.

Exp.	D_t (mm)	S (rpm)	F (mm/min)	Δz (mm)	Depth (%)	F_z (N)	T_{max} (°C)
1	6	0	2250	0.35	100	213.31	41.50
2	6	2000	2250	0.35	100	431.23	53.23
3	14	0	2250	0.35	100	192.02	45.32
4	14	2000	2250	0.35	100	323.42	61.90
5	10	1000	1500	0.20	100	308.92	50.91
6	10	1000	1500	0.50	100	299.78	49.23
7	10	1000	3000	0.20	100	298.03	49.13
8	10	1000	3000	0.50	100	309.45	49.11
9	6	1000	2250	0.20	100	301.49	50.16
10	6	1000	2250	0.50	100	193.24	45.76
11	14	1000	2250	0.20	100	362.43	50.90
12	14	1000	2250	0.50	100	196.85	49.67
13	10	0	1500	0.35	100	362.68	54.84
14	10	0	3000	0.35	100	366.14	52.25
15	10	2000	1500	0.35	95.08	262.67	67.52
16	10	2000	3000	0.35	100	324.69	47.13
17	6	1000	1500	0.35	100	262.24	58.27
18	6	1000	3000	0.35	100	287.48	50.04
19	14	1000	1500	0.35	100	214.77	48.69
20	14	1000	3000	0.35	100	348.03	52.77
21	10	0	2250	0.20	100	203.74	41.69
22	10	0	2250	0.50	100	362.48	52.58
23	10	2000	2250	0.20	100	354.03	64.50
24	10	2000	2250	0.50	100	274.20	53.75
25	10	1000	2250	0.35	100	327.08	46.35
26	10	1000	2250	0.35	100	273.95	56.94
27	10	1000	2250	0.35	100	291.83	51.56

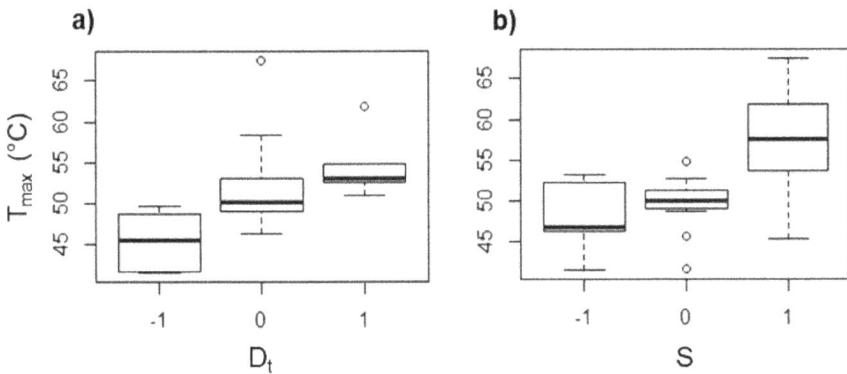

Figure 10. Box diagrams of T_{max} (in °C) for PCL as a function of (**a**) D_t and (**b**) S.

On the other hand, F_z increased with larger D_t (see Figure 11a). This could be a result of the higher contact area between the forming tool and the polymer sheet, and with this, there is a greater amount of material that must be pushed down. Meanwhile, the force is reduced as S is increased (see Figure 11b), which can be directly associated with the softening of the polymer as a consequence of the temperature increase due to the friction between the tool and the polymer sheet as was mentioned before. In general, when the temperature is increased, the forming force is reduced because the polymer undergoes a softening.

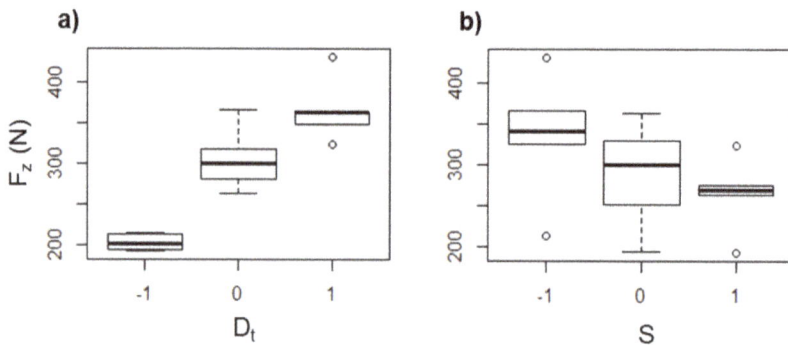

Figure 11. Box diagrams of F_z (in N) for PCL as a function of (**a**) D_t and (**b**) S.

For the case of UHMWPE, Table 6 shows that most of the sheets tested in SPIF formed to 100% of the final geometry; however, more sheets fractured compared to PCL. Figure 12 shows a pyramid-like part completely formed and and another with failure. At first sight, it is observed that the geometric precision of the pyramids is low since the shape of the manufactured parts has a curvature in the walls that is very different from that originally designed by the toolpath.

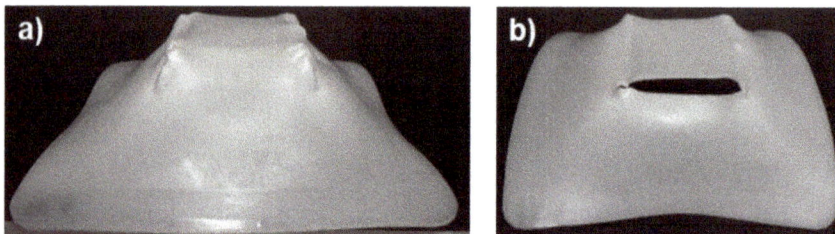

Figure 12. Pyramid-shaped parts made of UHMWPE (**a**) Sheet completely formed; (**b**) Sheet with failure.

An increase of S leads to a remarkable rise of the heat generated due to the tool–sheet friction as determined from the values of T_{max} measured with rotation of the tool. In this regard, in some experiments with UHMWPE sheets, T_{max} values of up to 120 °C were reached with $S = 2000$ rpm, while for some experiments without rotation, the T_{max} registered was below 50 °C. In general, the values of T_{max} measured for the UHMWPE sheets show more variation than that observed in the experiments with PCL, as it can be seen in the box diagrams of Figure 13, and additionally, there is not a clear tendency of T_{max} as a function of D_t (see Figure 13a), so it could be stated that the long chains of UHMWPE can generate a random behavior during the SPIF processing. Nonetheless, it can be still concluded that T_{max} increases as S increases (Figure 13b), which reaffirms that, regardless of the molecular structure of the polymer, S is the parameter in SPIF that most influences the temperature reached in the sheet.

Table 6. SPIF results of UHMWPE sheets.

Exp.	D_t (mm)	S (rpm)	F (mm/min)	Δz (mm)	Depth (%)	F_z (N)	T_{max} (°C)
1	6	0	2250	0.35	96.72	503.16	83.82
2	6	2000	2250	0.35	96.72	433.23	120.36
3	14	0	2250	0.35	100	1003.60	48.24
4	14	2000	2250	0.35	100	619.47	120.85
5	10	1000	1500	0.20	100	685.67	60.89
6	10	1000	1500	0.50	100	617.41	67.63
7	10	1000	3000	0.20	100	713.51	75.75
8	10	1000	3000	0.50	100	641.53	75.03
9	6	1000	2250	0.20	100	494.03	107.84
10	6	1000	2250	0.50	100	452.33	70.01
11	14	1000	2250	0.20	100	799.40	102.14
12	14	1000	2250	0.50	100	785.14	104.01
13	10	0	1500	0.35	100	831.28	48.95
14	10	0	3000	0.35	100	754.38	52.71
15	10	2000	1500	0.35	83.61	658.87	83.06
16	10	2000	3000	0.35	84.43	577.47	85.39
17	6	1000	1500	0.35	100	466.49	79.48
18	6	1000	3000	0.35	100	449.15	75.89
19	14	1000	1500	0.35	100	742.55	101.54
20	14	1000	3000	0.35	100	745.89	105.94
21	10	0	2250	0.20	100	782.94	55.55
22	10	0	2250	0.50	79.07	580.94	84.80
23	10	2000	2250	0.20	100	788.07	51.23
24	10	2000	2250	0.50	100	653.74	85.46
25	10	1000	2250	0.35	100	665.85	67.52
26	10	1000	2250	0.35	100	660.19	65.57
27	10	1000	2250	0.35	100	699.30	62.69

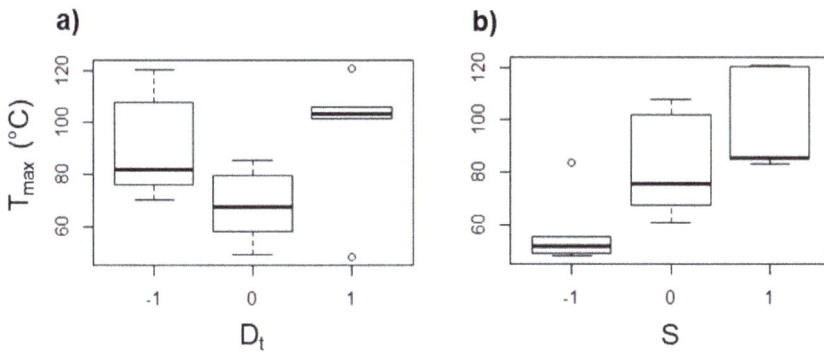

Figure 13. Box diagrams of T_{max} (in °C) for UHMWPE as a function of (**a**) D_t and (**b**) S.

The F_z in UHMWPE shows the same tendency previously observed in PCL as a function of D_t and S, i.e., F_z increases as D_t increases (Figure 14a) and decreases as S increases (Figure 14b), which is attributed to the same as discussed earlier for the case of PCL, i.e., a greater D_t represents a greater amount of material pushed down, and so the F_z is higher, while a higher S generates more heat due the friction between the tool and the sheet, which in turn softens the polymer, and consequently F_z is reduced.

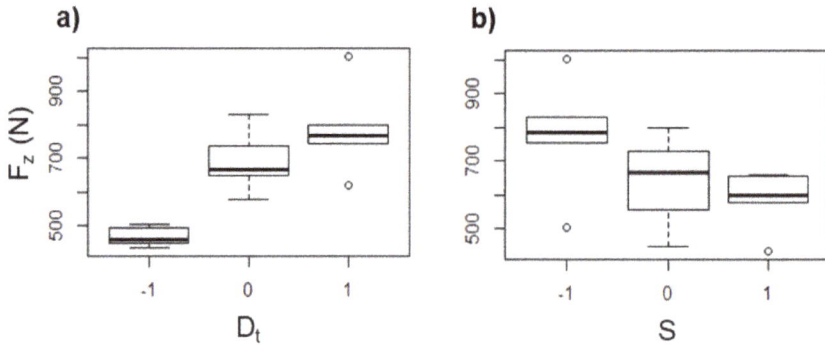

Figure 14. Box diagrams of T_{max} (in °C) for UHMWPE as a function of (**a**) D_t and (**b**) S.

3.3. Characterization by XRD after the SPIF Processing

Pyramid-shaped parts of PCL and UHMWPE formed by SPIF were analyzed by XRD in order to assess the molecular chain orientation as a result of the SPIF processing. In a previous work, Lozano-Sánchez et al. [15] observed by small- and wide-angle X-ray scattering (SWAXS) the orientation of Polypropylene chains in the vertical direction of cone-shaped parts as a result of SPIF processing. Nonetheless, it should be pointed out that the SWAXS analysis is completed through-thickness of the sample. Here, XRD was used to separately analyze the inner and outer surfaces of the wall of the pyramid-shaped parts based on the fact that compressive and tensile stresses are exerted on each side of sheets formed by SPIF as was previously concluded by Jiménez et al. [16]. Due to the large amount of samples formed from the design of experiments, the XRD analysis was completed in the pyramid-like parts obtained from only four different experiments: 9, 10, 11, and 12, which vary D_t and Δz from the lowest value to the highest (D_t = 6, 14 mm; Δz = 0.2, 0.5 mm), and the spindle speed and feed rate remained unchanged and in the middle value (see Table 2). Only variations of D_t and Δz were considered since these parameters directly influence the shaping of the polymer in the vertical direction, i.e., the direction of descent of the forming tool, during SPIF processing.

For the XRD analysis, a rectangular sample was taken from the wall of the pyramid-shaped parts. All of the samples analyzed correspond to the side opposite to the descent side of the forming tool. The samples were first analyzed in one direction with respect to the X-ray incident beam, and then a second analysis was completed after rotating the sample 90° as is schematically illustrated in Figure 15. This rotation of the samples was done in order to detect differences in the intensity of the diffraction peaks associated with the crystallographic planes of the polymer unit cell, mainly the {hk0} plane groups, that is, those that are parallel to the *c*-axis. If the molecular chains of the polymer are preferably oriented in one direction, a greater intensity could be seen in the diffraction peaks since the planes would also be elongated in the longitudinal axis of the oriented polymer chains, i.e., along the *c*-axis. Here, it should be emphasized that *x*, *y*, *z* coordinates are used as reference directions in the pyramid-shaped parts, where *z* corresponds to the vertical direction or the direction of descent of the forming tool, being *x* and *y* the directions of the horizontal plane, while the *a*, *b*, *c* coordinates are used as reference directions for the crystal unit cell for both PCL and UHMWPE crystalline structures.

First, the XRD patterns of the PCL and UHMWPE reference samples, i.e., unformed sheets, are shown in Figure 16a. These samples were analyzed in only one side considering that they are unformed, so the existence of inner and outer surfaces does not apply. In fact, it is expected that there is no difference in the order of the molecular chains on both sides of the unformed sheets. The XRD patterns show peaks of the diffraction planes (110) and (200) of the orthorhombic unit cell in around 21.4° and 23.7°, respectively, for both PCL and UHMWPE. The pattern of PCL shows a small peak at 22° associated with planes (111), which are also of the orthorhombic unit cell [28]. From these

XRD patterns, it is concluded that both polymers have a semi-crystalline structure, evidenced by the well-defined, high-intensity peaks observed, which correspond to the crystalline part, and to the wide, low-intensity signal that can be seen at 2θ angles below 21°, which corresponds to the amorphous part. As was expected, the XRD patterns of the unformed sheets of PCL and UHMWPE show no difference between the measurements made at 0° and 90°, indicating that there is no chain orientation in these samples. The diffraction planes (110) and (200) of an orthorhombic unit cell are schematically represented in Figure 16b. It should be noted that the polymer chains extend along the *c*-axis. The structures of the repeat unit of PCL and UHMWPE are shown in Figure 16c. A repeat unit corresponds to each of the gray spheres that make up the orthorhombic cell shown in Figure 16b.

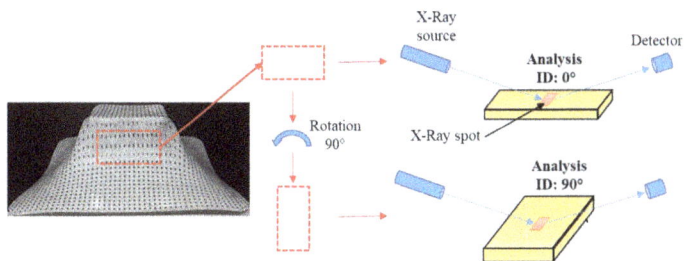

Figure 15. Configuration of the XRD analysis on a rectangular sample taken from the wall of the pyramid-shaped parts. The analysis were identified as "0°" and "90°" after rotation of the analyzed specimen as illustrated.

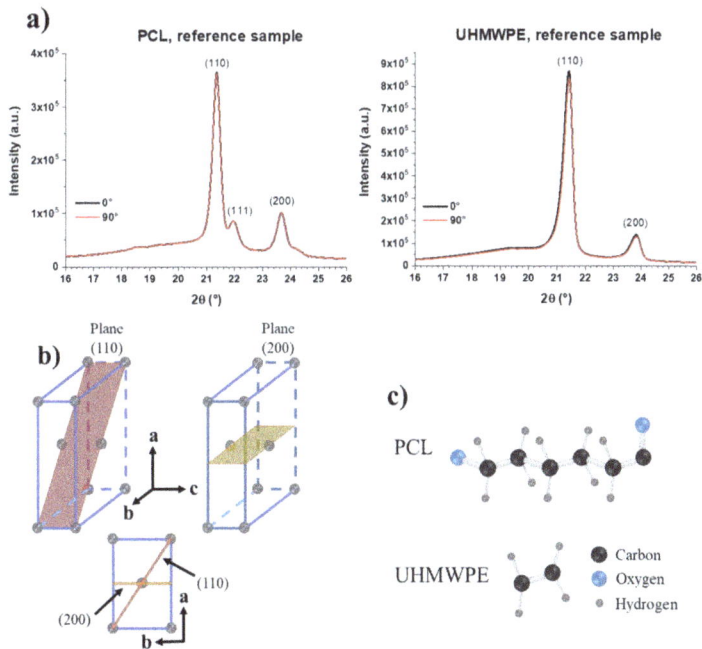

Figure 16. (**a**) XRD patterns of PCL and UHMWPE reference samples, i.e., unformed sheets; (**b**) Schematic representation of planes (110) and (200) of an orthorhombic crystal cell; (**c**) Repeat units of PCL and UHMPWE. A repetitive unit is located in each of the gray points that form the orthorhombic unit cell in (**b**).

The XRD patterns of the PCL final parts fabricated through the experiments 9, 10, 11, and 12 are shown in Figure 17. The dotted lines correspond to the measurements made on the inner surface and the solid lines to those made on the outer surface of the wall of the pyramid-shaped parts. The XRD patterns of formed parts still show the peaks of (110) and (200) planes at around 21.4° and 23.7°, respectively. However, the peak attributed to the (111) plane is rather observed as an overlapped shoulder with the (110) peak, which evidenced a reduced crystallinity in the formed samples compared to the unformed sheet (i.e., reference PCL, see Figure 16a). In fact, some peaks show a clear broadening that also proves a reduction in crystallinity. In addition, it is possible to observe a clear shift of the diffraction peaks in the XRD patterns of the PCL formed sheets, which is due to the inherent curvature in the final parts resulting from the forming process. For a structural comparison between the analyzed samples, this shift will be dismissed and attention will be paid only to the maximum intensities and width of the peaks.

Furthermore, the XRD patterns in Figure 17 demonstrate a different molecular behavior between the inner and outer surfaces, except in experiment 10 where diffraction patterns are practically equal. This different behavior is more evident in experiments 11 and 12, where patterns of the outer surface are wider than those of the inner surface. In fact, in "Exp. #12", the broad peaks obtained on the outer surface show a crystallinity that is well below the crystallinity observed on the inner surface. Interestingly, on the outer surface, the patterns obtained at 0° are more intense than those obtained at 90°, proving that PCL molecular chains are oriented in the vertical direction of the pyramid-shaped parts, that is, the direction of descent of the forming tool, according to the experimental setup used in this work for the XRD analysis (see Figure 15). On the other hand, the XRD patterns obtained on the inner surface show more intense peaks at 90° than at 0°, especially in experiments 9 and 11, indicating that in the inner surface, the PCL chains are preferentially oriented in the horizontal direction. These results suggest that PCL molecular chains are oriented horizontally in the inner surface due to the action of the forming tool, because it moves almost completely along the x and y directions (except for the small region where the tool steps down) and in this movement it could be "pulling" the molecular chains in the same direction. The difference between the patterns obtained at 0° and 90° on the inner surface is small, perhaps due to the fact that the movement of the forming tool is alternating, that is, a bidirectional contouring, which would be inhibiting a greater chain orientation.

In the outer surface, the polymer chains are not in contact with the forming tool, but these are incrementally stretched along the vertical direction at the same time and in the same way as the pyramid-shaped part is incrementally formed. As was mentioned before, the orientation of polymer chains in the vertical direction was observed by means of SWAXS in polypropylene sheets [15]; however, in those previous results, the horizontal orientation of chains in the inner surface could not be observed because the SWAXS analysis is performed through-thickness of the sheet. Nonetheless, in that previous work the authors obtained a hint of the difference in molecular behavior of the polymer between the inner and outer surfaces when they noticed a whitening on the polypropylene sheets after the SPIF processing and highlighted that this whitening only occurred on the outer surface of the formed parts. This whitening has also been observed in PVC [4], and is attributed to a crazing effect, which, according to McLeish et al. [29], is related to the disentanglement of chains.

Figure 18 shows the XRD patterns in the 2θ range from 16° to 26° of the pyramid-shaped parts of UHMWPE. The shift of peaks is more evident than in the case of the PCL sheets because the curvature of the UHMWPE parts is greater than in the parts of PCL, as can be observed in the pyramid parts shown in the images of Figures 9 and 12. All patterns of the UHMWPE formed sheets show the peaks of (110) and (200) planes, and it is evident that both surfaces, inner and outer, behave differently when they are processed by SPIF. Regarding the inner surface, in experiments 9 and 10, the patters obtained at 90° are more intense than those at 0°, suggesting that polymer chains are oriented horizontally, similar to what happens with PCL. However, in experiments 11 and 12, the opposite occurs, i.e., the peaks at 0° are more intense, indicating that in the formed parts fabricated through these experiments,

Polymers **2018**, *10*, 391

the polymer chains are preferentially oriented along the vertical direction. Experiments 11 and 12 were performed with the larger D_t, which should contribute to this difference in chain orientation.

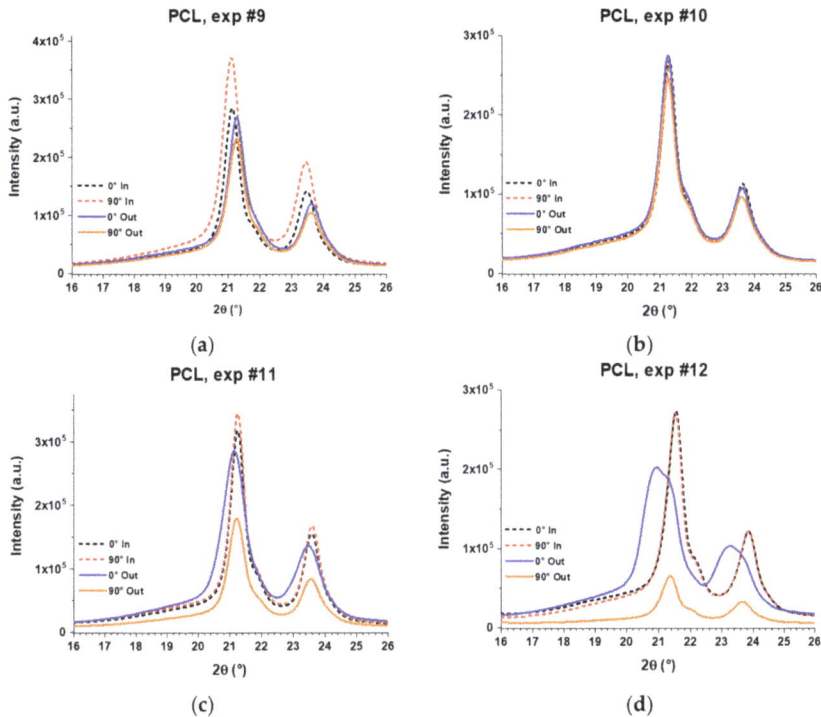

Figure 17. XRD patterns of PCL sheets formed by SPIF through experiments (**a**) #9; (**b**) #10; (**c**) #11; and (**d**) #12. The XRD analysis was carried out separately for the inner surface (in contact with the forming tool) and the outer surface (not in contact with the forming tool) according to the experimental setup described in Figure 15.

Regarding the outer surface, the XRD patterns of UHMPWE formed parts show a slight difference in intensities between the measurements obtained at $0°$ and $90°$ in experiments 9, 10, and 11, but in experiment 12, the difference in intensities is more marked, being greater for the measurement carried out at $0°$, which indicates that the polymer chains are oriented in the vertical direction. It is worth noting that in experiments 9 and 10, the patterns obtained on the outer surface show a peak in the 2θ range of $19.3°$–$19.7°$, which is associated to the (001) diffraction plane of the monoclinic structure of UHMWPE [30]. Thus, it is possible to infer that the movement of chains in a preferential direction generates a distortion in the initial orthorhombic unit cell. In fact, a monoclinic unit cell differs from an orthorhombic in only one of the angles between planes, specifically the angle β between the cell parameters *a* and *c*, being $\beta \neq 90°$ in the monoclinic unit cell, and $\beta = 90°$ in the orthorhombic unit cell. This distortion in the unit cell, which is schematically represented in Figure 19, can be related to the slippage between crystallites that was previously detected by DMA and identified as the alpha star transition. Such distortion of the unit cell was not observed in PCL, probably due to the fact that the crystalline region of the UHMWPE is more compact because it contains a linear structure with carbon-carbon bonds in the main chain while the structure of PCL contains carbonyl groups (see Figure 16c).

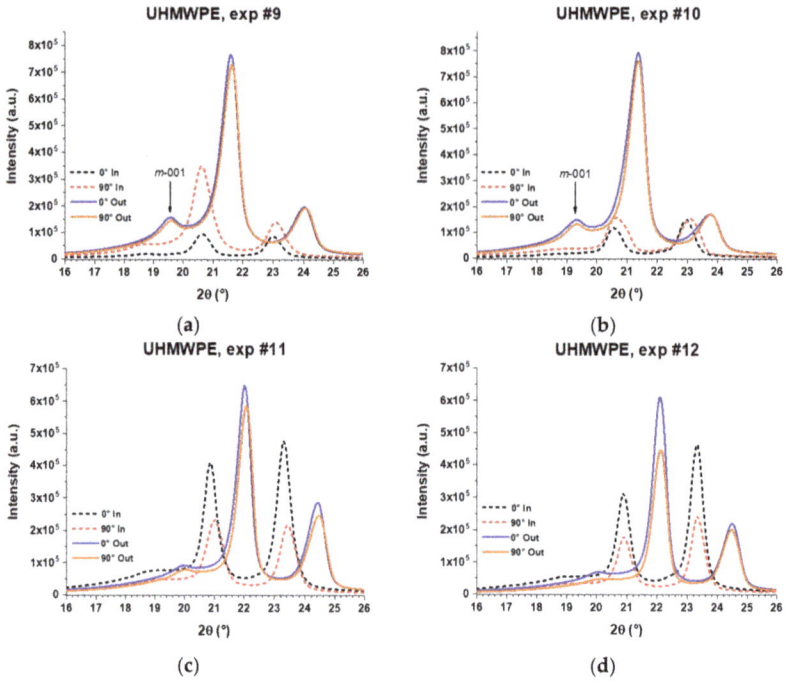

Figure 18. XRD patterns of UHMWPE sheets formed by SPIF through experiments (**a**) #9; (**b**) #10; (**c**) #11; and (**d**) #12. The XRD analysis was carried out separately for the inner surface (in contact with the forming tool) and the outer surface (not in contact with the forming tool) according to the experimental setup described in Figure 15.

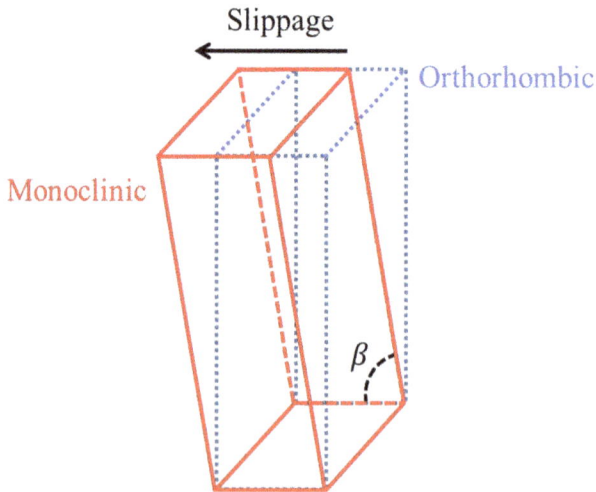

Figure 19. Schematic representation of the distortion from the orthorhombic unit cell to a monoclinic unit cell.

The chain orientation in the outer surface of the UHMWPE formed parts was less than that observed in the PCL ones because of the larger size of the UHMWPE chains. The forming tool generates the movement of chains on the surface of the sheet that is in direct contact with the tool (i.e., the inner surface). This produces a rearrangement of the polymer chains that continues along the thickness of the sheet, that is, towards the outer surface. However, this rearrangement requires the movement of the long, coiled chains of UHMWPE, which is hardly carried out. In fact, the DMA results showed that in UHMWPE, more energy is lost due to friction and internal movements compared to PCL (the loss modulus, E'', of UHMWPE is higher); however, the tan δ curve of UHMWPE showed a gradual increase as the temperature is also increased, while that of PCL increased more abruptly. Considering that tan δ is a measure of how efficiently a material loses energy due to molecular rearrangements, these results indicate that in the case of UHMWPE, the lost energy is scarcely used for molecular rearrangement, while in PCL the opposite occurs, mainly as it approaches the melting temperature where the energy is efficiently harnessed for molecular motion and the consequent flow of the material.

In order to analyze how the parameters D_t and Δz affect the chain orientation in the formed parts, Figure 20 presents the values of the ratio I_0/I_{90} for the diffraction peak associated to the (110) plane, where I_0 corresponds to the maximum intensity obtained in the measurement at 0° and I_{90} corresponds to the maximum intensity in the measurement at 90°. In this way, a value of $I_0/I_{90} = 1$ represents that the polymer chains are not oriented in a preferential direction, a value of $I_0/I_{90} > 1$ indicates that the chains are oriented vertically in the pyramid-shaped parts, and $I_0/I_{90} < 1$ indicates that the chains are oriented horizontally. If the value of I_0/I_{90} is farther from 1, it means that the chain orientation is greater. The graphs show the results for the inner surface (small internal square) and for the outer surface (large external square). The central red values represent the I_0/I_{90} ratio of the corresponding reference sample (unformed sheet) of each polymer. This value for PCL and UHMWPE is close to 1, since the polymer chains are not oriented in these unformed sheets as was mentioned before.

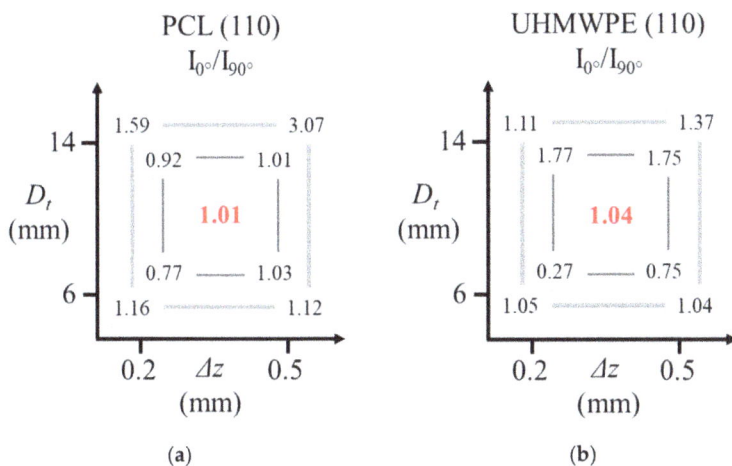

Figure 20. Values of the ratio I_0/I_{90} for the diffraction peak associated to the (110) plane (**a**) PCL; (**b**) UHMWPE; where I_0 is the maximum intensity in the measurement at 0° and I_{90} is the maximum intensity in the measurement at 90° according to the experimental setup described in Figure 15.

As has been discussed, in the formed parts of PCL, the molecular chains are preferentially oriented in the horizontal direction in the inner surface. The experiment performed with $D_t = 6$ mm and $\Delta z = 0.2$ mm generates the greater orientation of chains in the inner surface. One can think that a small tool concentrates the pulling of chains in a smaller region, and thus the crystalline parts of the polymer are oriented in the feed direction of the tool. A larger D_t implies a less-concentrated molecular movement, so that it could be causing the movement of a large amount of crystalline parts, but also of a large amount of the amorphous part of the polymer. To some extent, a small Δz also represents a way of concentrating the movement of chains in a small size region. If the step size is small, in a second step down, the chains of the previous step can still be pulled by the forming tool, something that would not happen if the step size were larger, where the chains that were pulled in a previous step would hardly be affected by the displacement of the tool in the subsequent steps. Based on the aforementioned, a smaller Δz causes a higher chain orientation, although this parameter is not as significant as D_t is for the orientation of chains in the inner surface.

Concerning the outer surface, D_t causes the most marked impact in the orientation of chains, which could be attributed to the same reason explained before regarding the large amount of material that is pulled by the forming tool when its diameter is larger. However, in the outer surface the orientation is preferentially towards the vertical direction of the pyramid-shaped parts because on this surface, the tool path in the *x*-*y* plane does not have an impact. Therefore, regardless of the direction, the orientation of polymer chains in the outer surface is higher than in the inner surface because in the outer surface the polymer is pulled only in one direction, that is, downwards.

In the case of UHMWPE, the parameter that most influences the orientation is D_t. For a polymer that contains very long chains, such as UHMWPE, a tool of greater diameter creates a greater contact area between the tool and the surface of the sheet, which can be a more significant parameter compared to a polymer whose chains are not that big (the case of PCL). On the inner surface, a small D_t generates the horizontal orientation, while the larger D_t gives rise to the vertical orientation. The orientation of the chains on the outer surface is less evident than in the case of the PCL due to the energy lost by the molecular rearrangement from the inner surface to the outer one. Only a combination of parameters consisting of the largest D_t and the largest Δz is able to generate a greater orientation on the outer surface because one could think that in this way a greater amount of material is pulled down and so a greater molecular rearrangement among the UHMWPE chains takes place.

The characterization of PCL and UHMWPE performed in this work before and after being processed by SPIF has revealed important thermal and structural properties that certainly should be considered when forming thermoplastics by SPIF. Even though the XRD analyzes were imprecise due to the curvature of the samples evaluated, this characterization technique allowed us to separately study the surface of the sheet that is in contact with the forming tool and the opposite surface that is not, and in this manner, it was clearly revealed that polymer chains behave differently on each surface. It is strongly believed that the use of techniques such as DMA and XRD may become a very useful tool for SPIF of polymers and that the results presented here can be extended to other semicrystalline thermoplastics.

4. Conclusions

In summary, sheets of the biocompatible polymers PCL and UHMWPE were fabricated by compression molding and shaped by SPIF with pyramid-like geometry through a BB design of experiments in which four parameters were varied. The performance of PCL and UHMWPE in SPIF was evaluated in terms of the maximum depth, F_z, and T_{max} reached during the SPIF processing. The results indicated that F_z, and T_{max} are mostly affected by the tool diameter and the spindle speed.

The thermal and structural properties of the polymers used were thoroughly studied. DMA revealed the existence of an alpha star transition ($T_\alpha{}^*$) occurring in both polymers. During SPIF processing, it is key to reach the temperature of this transition for the forming of PCL and UHMWPE since it is associated with slippage between crystallites. XRD results proved that the polymer chains are preferentially oriented in different directions in the inner and outer surfaces of the pyramid-shaped parts. In the inner surface, the tool path contributes to the horizontal orientation of chains, while in the outer surface, the polymer chains are mainly pulled downwards, generating the vertical orientation of chains. The slippage of crystallites as a consequence of the forming process induces the distortion of the unit cell of UHMWPE, passing from orthorhombic to monoclinic.

It has been demonstrated that the characterization techniques used in this work are very useful in SPIF of polymers based on the thermal and structural information that is presented here for the first time. For instance, the use of XRD allowed the structural study of inner and outer surfaces separately. On the other hand, DMA supplied information about thermal transitions not readily identifiable by other methods, which makes it a very powerful tool in SPIF of polymers since the temperature is critical when forming these materials, especially thermoplastics.

Acknowledgments: This work was funded by Tecnológico de Monterrey, Campus Monterrey, through the Research Chair in Nanotechnology and Devices Design, and by Consejo Nacional de Ciencia y Tecnología (CONACYT), Mexico (projects number 242269 and 255837) by the University of Girona (MPCUdG2016/036) and by the Spanish Ministry of Education (DPI2016-77156-R). The authors gratefully acknowledge the valuable help of Jorge Espinosa (Centro de Investigación en Química Aplicada, Mexico) and Tomás Martínez (Tecnológico de Monterrey, campus Monterrey, Mexico) in the use of facilities and equipment to carry out the preparation of materials. Luis Marcelo Lozano-Sánchez appreciates the PhD scholarship from CONACYT (grant number 369647). Isabel Bagudanch appreciates the Spanish grant FPU12/05402.

Author Contributions: Isabel Bagudanch, Maria Luisa Garcia-Romeu, and Luis Marcelo Lozano-Sánchez conceived and designed the experiments. Luis Marcelo Lozano-Sánchez and Alan Osiris Sustaita fabricated the sheets. Isabel Bagudanch performed the SPIF experiments. Jackeline Iturbe-Ek performed the XRD analysis. Luis Marcelo Lozano-Sánchez, Isabel Bagudanch, Maria Luisa Garcia-Romeu, Alan Osiris Sustaita, Jackeline Iturbe-Ek, and Luis Ernesto Elizalde analyzed the data. Luis Marcelo Lozano-Sánchez, Maria Luisa Garcia-Romeu, and Alex Elias-Zúñiga directed the research. Luis Marcelo Lozano-Sánchez wrote the manuscript. The manuscript was finalized through contributions from all authors, and all authors approve the final manuscript.

Conflicts of Interest: The authors declare no conflict of interest.

References

1. Jackson, K.; Allwood, J. The mechanics of incremental sheet forming. *J. Mater. Process. Technol.* **2009**, *209*, 1158–1174. [CrossRef]
2. Kim, Y.H.; Park, J.J. Effect of process parameters on formability in incremental forming of sheet metal. *J. Mater. Process. Technol.* **2002**, *130–131*, 42–46. [CrossRef]
3. Jackson, K.P.; Allwood, J.M.; Landert, M. Incremental forming of sandwich panels. *J. Mater. Process. Technol.* **2008**, *204*, 290–303. [CrossRef]
4. Marques, T.A.; Silva, M.B.; Martins, P.A.F. On the potential of single point incremental forming of sheet polymer parts. *Int. J. Adv. Manuf. Technol.* **2012**, *60*, 75–86. [CrossRef]
5. Fan, G.; Gao, L.; Hussain, G.; Wu, Z. Electric hot incremental forming: A novel technique. *Int. J. Mach. Tools Manuf.* **2008**, *48*, 1688–1692. [CrossRef]
6. Park, J.J.; Kim, Y.H. Fundamental studies on the incremental sheet metal forming technique. *J. Mater. Process. Technol.* **2003**, *140*, 447–453. [CrossRef]
7. Duflou, J.R.; Callebaut, B.; Verbert, J.; De Baerdemaeker, H. Improved SPIF performance through dynamic local heating. *Int. J. Mach. Tools Manuf.* **2008**, *48*, 543–549. [CrossRef]
8. Le, V.S.; Ghiotti, A.; Lucchetta, G. Preliminary studies on single point incremental forming for thermoplastic materials. *Int. J. Mater. Form.* **2008**, *1*, 1179–1182. [CrossRef]
9. Franzen, V.; Kwiatkowski, L.; Neves, J.; Martins, P.A.F.; Tekkaya, A.E. On the Capability of Single Point Incremental Forming for Manufacturing Polymer Sheet Parts. *Int. Conf. Technol. Plast.* **2008**, *2008*, 890–895.
10. Franzen, V.; Kwiatkowski, L.; Martins, P.A.F.; Tekkaya, A.E. Single point incremental forming of PVC. *J. Mater. Process. Technol.* **2009**, *209*, 462–469. [CrossRef]
11. Martins, P.A.F.; Kwiatkowski, L.; Franzen, V.; Tekkaya, A.E.; Kleiner, M. Single point incremental forming of polymers. *CIRP Ann. Manuf. Technol.* **2009**, *58*, 229–232. [CrossRef]
12. Brosius, A.; Hermes, M.; Khalifa, N.B.; Trompeter, M.; Tekkaya, A.E. Innovation by forming technology: Motivation for research. *Int. J. Mater. Form.* **2009**, *2*, 29–38. [CrossRef]
13. Bagudanch, I.; Garcia-Romeu, M.L.; Centeno, G.; Elías-Zúñiga, A.; Ciurana, J. Forming force and temperature effects on single point incremental forming of polyvinylchloride. *J. Mater. Process. Technol.* **2015**, *219*, 221–229. [CrossRef]

14. Davarpanah, M.A.; Mirkouei, A.; Yu, X.; Malhotra, R.; Pilla, S. Effects of incremental depth and tool rotation on failure modes and microstructural properties in Single Point Incremental Forming of polymers. *J. Mater. Process. Technol.* **2015**, *222*, 287–300. [CrossRef]

15. Lozano-Sánchez, L.M.; Sustaita, A.O.; Soto, M.; Biradar, S.; Ge, L.; Segura-Cárdenas, E.; Diabb, J.; Elizalde, L.E.; Barrera, E.V.; Elías-Zúñiga, A. Mechanical and structural studies on single point incremental forming of polypropylene-MWCNTs composite sheets. *J. Mater. Process. Technol.* **2017**, *242*, 218–227. [CrossRef]

16. Jiménez, I.; López, C.; Martinez-Romero, O.; Mares, P.; Siller, H.R.; Diabb, J.; Sandoval-Robles, J.A.; Elías-Zúñiga, A. Investigation of residual stress distribution in single point incremental forming of aluminum parts by X-ray diffraction technique. *Int. J. Adv. Manuf. Technol.* **2017**, *91*, 2571–2580. [CrossRef]

17. Clavijo-Chaparro, S.L.; Iturbe-Ek, J.; Lozano-Sánchez, L.M.; Sustaita, A.O.; Elías-Zúñiga, A. Plasticized and reinforced poly(methyl methacrylate) obtained by a dissolution-dispersion process for single point incremental forming: Enhanced formability towards the fabrication of cranial implants. *Polym. Test.* **2018**. [CrossRef]

18. Fiorentino, A.; Marzi, R.; Ceretti, E. Preliminary results on Ti incremental sheet forming (ISF) of biomedical devices: Biocompatibility, surface finishing and treatment. *Int. J. Mechatron. Manuf. Syst.* **2012**, *5*, 36–45. [CrossRef]

19. Khan, R.A.; Parsons, A.J.; Jones, I.A.; Walker, G.S.; Rudd, C.D. Preparation and Characterization of Phosphate Glass Fibres and Fabrication of Poly(Caprolactone) Matrix Resorbable Composites. *J. Reinf. Plast. Compos.* **2009**. [CrossRef]

20. López-Cervantes, A.; Domínguez-López, I.; Barceinas-Sánchez, J.D.O.; García-García, A.L. Effects of surface texturing on the performance of biocompatible UHMWPE as a bearing material during in vitro lubricated sliding/rolling motion. *J. Mech. Behav. Biomed. Mater.* **2013**, *20*, 45–53. [CrossRef] [PubMed]

21. Slouf, M.; Vackova, T.; Nevoralova, M.; Pokorny, D. Micromechanical properties of one-step and sequentially crosslinked UHMWPEs for total joint replacements. *Polym. Test.* **2015**, *41*, 191–197. [CrossRef]

22. Ansari, F.; Gludovatz, B.; Kozak, A.; Ritchie, R.O.; Pruitt, L.A. Notch fatigue of ultrahigh molecular weight polyethylene (UHMWPE) used in total joint replacements. *J. Mech. Behav. Biomed. Mater.* **2016**, *60*, 267–279. [CrossRef] [PubMed]

23. Bagudanch, I.; Lozano-Sánchez, L.M.; Puigpinós, L.; Sabater, M.; Elizalde, L.E.; Elías-Zúñiga, A.; Garcia-Romeu, M.L. Manufacturing of Polymeric Biocompatible Cranial Geometry by Single Point Incremental Forming. *Procedia Eng.* **2015**, *132*, 267–273. [CrossRef]

24. Centeno, G.; Bagudanch, I.; Morales-Palma, D.; García-Romeu, M.L.; Gonzalez-Perez-Somarriba, B.; Martinez-Donaire, A.J.; Gonzalez-Perez, L.M.; Vallellano, C. Recent Approaches for the Manufacturing of Polymeric Cranial Prostheses by Incremental Sheet Forming. *Procedia Eng.* **2017**, *183*, 180–187. [CrossRef]

25. Centeno, G.; Morales-Palma, D.; Gonzalez-Perez-Somarriba, B.; Bagudanch, I.; Egea-Guerrero, J.J.; Gonzalez-Perez, L.M.; García-Romeu, M.L.; Vallellano, C. A functional methodology on the manufacturing of customized polymeric cranial prostheses from CAT using SPIF. *Rapid Prototyp. J.* **2017**, *23*, 771–780. [CrossRef]

26. Bagudanch, I.; Vives-Mestres, M.; Sabater, M.; Garcia-Romeu, M.L. Polymer incremental sheet forming process: Temperature analysis using response surface methodology. *Mater. Manuf. Process.* **2017**, *32*, 44–53. [CrossRef]

27. Menard, H.P. *Dynamic Mechanical Analysis*; Taylor Fr: Boca Raton, FL, USA, 2008; pp. 98–112.

28. Chen, E.-C.; Wu, T.-M. Isothermal crystallization kinetics and thermal behavior of poly(ε-caprolactone)/multi-walled carbon nanotube composites. *Polym. Degrad. Stab.* **2007**, *92*, 1009–1015. [CrossRef]

29. McLeish, T.C.B.; Plummer, C.J.G.; Donald, A.M. Crazing by disentanglement: Non-diffusive reptation. *Polymer* **1989**, *30*, 1651–1655. [CrossRef]

30. Joo, Y.L.; Han, O.H.; Lee, H.K.; Song, J.K. Characterization of ultra high molecular weight polyethyelene nascent reactor powders by X-ray diffraction and solid state NMR. *Polymer* **2000**, *41*, 1355–1368. [CrossRef]

polymers

MDPI

Article

Orthogonal Templating Control of the Crystallisation of Poly(ε-Caprolactone)

Geoffrey R. Mitchell [1,*] and Robert H. Olley [2]

[1]   Centre for Rapid and Sustainable Product Development, Institute Polytechnic of Leiria,
     2430-028 Marinha Grande, Portugal
[2]   EMLAB, University of Reading, Reading RG6 6AF, UK; hinmeigeng@hotmail.com
*   Correspondence: geoffrey.mitchell@ipleiria.pt; Tel.: +351-244-569-441

Received: 8 January 2018; Accepted: 2 March 2018; Published: 11 March 2018

Abstract: The crystal growth of poly(ε-caprolactone) can be very effectively directed through the use of small amounts of dibenzylidene sorbitol in conjunction with modest flow fields to yield extremely high levels of the preferred polymer crystal orientation. We show that by introducing small quantities of a terpolymer, based on polyvinyl butyral we can switch the symmetry axis of the final lamellar orientation from parallel to perpendicular to the melt flow direction. During shear flow of the polymer melt, the dibenzylidene sorbitol forms highly extended nanoparticles which adopt a preferred alignment with respect to the flow field and on cooling, polymer crystallisation is directed by these particles. The presence of the terpolymer, based on polyvinyl butyral, limits the aspect ratio of the dibenzylidene sorbitol (DBS) particles, such that the preferred orientation of the particles in the polymer melt changes from parallel to normal to the flow direction. The alignment of lamellar crystals perpendicular to the flow direction has important implications for applications such as scaffolds for tissue engineering and for barrier film properties.

Keywords: crystallisation; morphology; nanoparticles; shear; flow; orientation; poly(ε-caprolactone); polyvinyl butyral

1. Introduction

Dibenzylidene sorbitol (DBS) (see Scheme 1) [I] is a low molar mass gelator [1–4]. In a variety of solvents including polymers at very low concentrations, <1% w/w, it self-assembles into extended crystalline nanofibrils. When DBS is dispersed in a polymer matrix, the application of modest shear flow leads to a macroscopic alignment of the fibrils [5–7]. When the matrix is a crystallisable polymer, subsequent crystallisation of the matrix is templated by the DBS nanofibrils [5,6]. In this work, we centre our attention on the control of the templating direction.

The properties of polymer based materials depend on the processing procedures used to shape the object as well as the ingenuity of the molecule maker in terms of the chemical configuration of the polymer chains. In the case of a crystallisable polymer such as poly(ε-caprolactone), the properties of the final product are especially dependent on the arrangement of the chain folded lamellar crystals [8]. This is termed the morphology, they may be aligned as a consequence of a common orientation of the row nuclei, the crystals may be extended in a fibrillary structure providing high modulus [9], or they may be arranged with a variety of orientations within spherulites which is the most common morphology [10] if no particular action is taken. In the case of spherulites, the distribution of crystals may lead to a reduction of the dielectric breakdown strength in the case of insulators [11]. As well as mechanical properties, the distribution of crystals will impact on the optical properties [12], on the degradation rate of the polymer [13] and the permeability of the polymer film to gases such as oxygen [14].

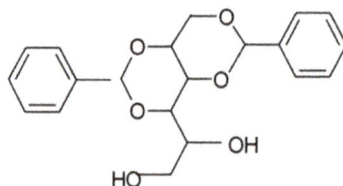

Scheme 1. Dibenzylidene Sorbitol (DBS).

The role of dibenzylidene sorbitol as a nucleating agent for isotactic polypropylene has been known for some time, although the mechanism is less clear. Smith et al., proposed that it stabilized the helical structure through the cleft in the V shaped nature of the molecular structure [15]. Mandelkern has reviewed the nucleation mechanisms and identified the importance of epitaxy but also states "it is not the only one that can be involved" [16]. Lotz and co-workers have identified the strong role of atomistic epitaxy in the nucleation in which there is matching of the crystal structures of the two components [17]. More recently we have explored the use of dibenzylidene sorbitols and some derivatives to direct the crystallization of different polymers. The ubiquity of the process and the similarity to the use of graphene nanoflakes, carbon nanotubes, and other preformed nanoparticles has lead us to the conclusion that the underlying mechanism is most likely to be graphoepitaxy which is a process in which the orienting influence on the growth of a layer on the substrate can be determined by various factors (forces) distinctive of the crystalline lattice [18].

Key factors with regard to dibenzylidene sorbitol enabling the direction of crystallisation are the formation of highly anisotropic nanoparticles which easily align in low shear flow field [7]. The solubility of dibenzylidene sorbitol in a polymer melt is strongly related to the polarity of the solvent as has been shown with low molar mass solvents [7]. Previous work has shown the nucleating effect of dibenzylidene sorbitol in poly(ε-caprolactone) saturates at 3% whereas in polyethylene the effect saturates at ~1% [6].The difference is related to the solubility of the dibenzylidene sorbitol in the polymeric solvent and therefore how many nanoparticles are formed to direct the crystallization.

The work reported here is based on the use of self-assembling nanoparticles rather than engineered nanoparticles which have significant environmental impact issues and will limit the recycling or disposal of objects prepared using this approach. As a compound approved for use in food contact applications DBS has none of the potential hazards of engineered nanoparticles such as carbon nanotubes or metallic particles.

DBS is widely employed as a clarifying agent with isotactic polypropylene where it yields films with a high level of optical transparency for packaging [12]. It is also widely used in cosmetics and small quantities are thus unlikely to have any adverse effect in the environment. In this work, we focus on poly(ε-caprolactone) (PCL) which is a biocompatible and biodegradable polymer widely used for the manufacture of biomedical devices including scaffolds for tissue engineering and drug delivery systems. It is sufficiently stable thermally to be considered for implants in the human body, its low melting point (55–60 °C) has allowed its use in "fantastic plastic" demonstrations in which it is melted in hot water and moulded by hand. Keith et al., reported that the introduction of small quantities of polyvinyl butyral (PVB) could profoundly affect the nucleation of PCL spherulites with major changes in the banding in the spherulites [19]. Use of PVB in conjunction with PCL to produce large spherulites has been exploited to allow detailed studies of the spherulite banding using microbeam wide-angle and small-angle X-ray scattering (WAXS and SAXS) techniques [20,21] and electron tomography [22]. Such studies have shown PCL and PVB to be compatible in the melt. PVB has similar suppressive, although less pronounced, effects on the nucleation in other polyesters [19]. Here we consider a PCL system which contains both the nucleating and directing additive DBS as well as the nucleating suppression agent PVB. We use in-situ time-resolving X-ray scattering techniques to explore the potential consequences of the interactions between these competing additives on the flow and crystallisation of these novel mixtures.

2. Materials and Methods

The polyvinyl butyral based terpolymer (commonly abbreviated in industrial practice as PVB) used in this work is a random terpolymer containing ~88% of polyvinyl butyral together with ~0.5% polyvinyl acetate and ~11.5% polyvinyl alcohol units and was obtained from Aldrich (St. Louis, MO, USA). The molecular weight is 90,000 to 120,000 Da. 3% w/w dibenzylidene sorbitol (Ciba, Basel, Switzerland) and 3% w/w polyvinyl butyral were dispersed alone and together in PCL (Aldrich, St. Louis, MO, USA) (MW = 80,000 Da) by solution blending using butanone as the co-solvent, cooling the solution to room temperature and allowing the butanone to evaporate under the draught of a fume cupboard. Moulded discs for the shear flow experiments were prepared from these mixtures by melt moulding at 80 °C, followed by cooling to 0 °C to allow the PCL to crystallise fully before warming to room temperature.

Samples of PCL, PCL/DBS, PCL/PVB, and PCL/DBS/PVB were subjected to controlled flow and thermal profiles using a specially designed parallel plate shear flow system [23,24] as shown in Figure 1. The sample was held between two thin mica discs supported on slotted metal plates mounted within a small oven equipped with both electrical heating and refrigerated gas cooling, providing a controlled temperature range from −20 to 300 °C with heating and cooling rates up to 20 °C/min. This shear flow cell system enabled X-ray scattering experiments to be performed during and following controlled shear flow. The slots in the metal plates allowed the X-ray beam to pass through the sample for ~85% of a revolution. Collimation and masks prior to the rotating plate in the beam line were used to minimize the parasitic scattering from the spokes of the rotating plate as they cut the incident beam. An intelligent motor control system ensured that when the rotation was stopped, this was at a rotation angle where the beam was not obscured by a metal spoke.

Time resolved X-ray scattering experiments were performed on the fixed wavelength (λ = 1.54 Å) beam-line 2.1 [25] at the Daresbury Synchrotron Radiation Source (UK). Small-angle X-ray scattering (SAXS) data were recorded using a 2-D RAPID detector which allowed scattering data to be accumulated in the range $|Q|$ ~0.01 to 0.2 Å^{-1} ($|Q|$ = $4\pi \sin \theta/\lambda$ where λ is the incident wavelength and 2θ is the scattering angle) with a data accumulation time of 10 s. The sample to detector distance and the detector pixel size were calibrated in terms of geometry using wet collagen.

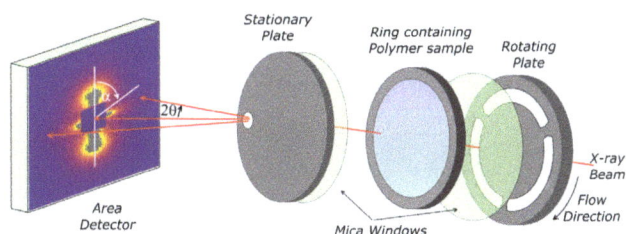

Figure 1. Schematic of X-ray shear cell system used in this work Adapted from [24] (Wangsoub, S.; Oiley, R.H.; Mitchell, G.R. Directed Crystallisation of Poly(ε-caprolactone) using a Low-Molar-Mass Self-Assembled Template. *Macromol. Chem. Phys.* **2005**, *206*, 1826–1839).

Wide-angle X-ray Scattering (WAXS) were obtained at room temperature on static samples using a symmetrical transmission diffractometer equipped with a graphite monochromator and a copper X-ray sealed tube source. The intensity was measured as a function of $|Q|$ and α, the angle between the symmetry axis of the sample and Q, in steps of $|\Delta Q|$ = 0.02 Å^{-1} and in steps $\Delta \alpha$ = 5°.

Specimens for optical microscopy were prepared on glass coverslips in a Metter FP hotstage by melting for 2 min at 150 °C, then cooling rapidly (~30 °C/min) to 60 °C followed by cooling at a rate of 1 °C/min to 30 °C before removing to room temperature. Optical microscopy was performed at room temperature using a Carl Zeiss GFL polarised light microscope.

3. Results

Figure 2a shows an optical micrograph with the typical spherulitic morphology for PCL crystallized under quiescent conditions while cooling from 60 to 30 °C at 1 °C/min, after melting at 150 °C. The picture is taken at the edge of a film approximately 0.1 mm thick—such films are used to ensure that the sample is representative of the material as a whole, instead of possibly being dominated by nucleation on the glass slide and cover slip surfaces. The bulk film (lower half of picture) is nucleated heavily so that individual spherulites appear to be of the order of 10 μm or less in diameter. At the top of the picture is the edge of the specimen, where a much thinner film is left as a result of smearing on one of the glass surfaces. Here individual objects are more easily discernible, generally up to 20 μm in diameter, but larger spherulites always tend to form in such situations in many kinds of polymers, so this cannot be taken as representative of the material as a whole. The addition of 3% PVB (Figure 2b) gives rise to a specimen containing much larger spherulites, typically 50 μm in diameter. This is consistent with the earlier observations by Keith et al. [19] that PVB greatly reduced the nucleation density in PCL.

For samples of PCL containing 3% DBS, in-situ SAXS experiments have shown that at 150 °C the DBS will be fully dissolved in the PCL [24]. The PCL/DBS sample (Figure 2c) prepared from a melt at 150 °C as described above gives rise to a much greater nucleation density, such that the individual objects are not at all properly discernible at this magnification. The texture has the appearance of large grainy spherulites. These are made up of small PCL crystals which have crystallised preferentially with respect to fibrils of DBS which crystallised at a higher temperature from solution in molten PCL [24].

The specimen containing 3% DBS and 3% PVB was prepared by melting at 170 °C before being given the same treatment as the previous samples. This resultant micrograph appears to be the same as shown in Figure 2b for the PCL/PVB sample, suggesting either that the PVB is preventing the DBS from crystallising during cooling, or it is completely inhibiting the nucleation effect of the DBS on the PCL. If we repeat the treatment but cooling from a lower melt temperature where the DBS will not be fully dissolved, the morphology is more mixed between that shown in Figure 2a,d suggesting that the PVB has the capacity to suppress or modify the crystallisation of the DBS as well as that of the PCL.

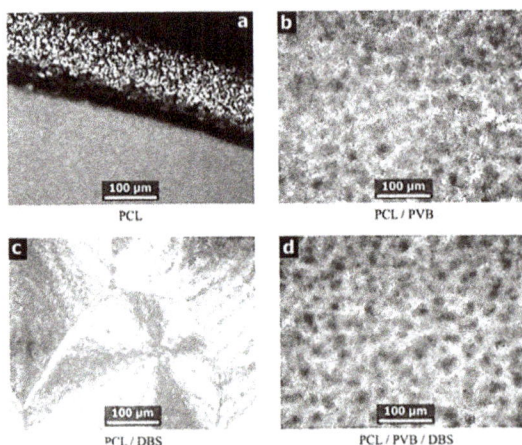

Figure 2. Optical Micrographs obtained in a polarised light microscope at room temperature of samples of (**a**) PCL; (**b**) PCL with 3%PVB; (**c**) PCL with DBS; and (**d**) PCL with 3% PVB and 3% DBS which were crystallised from the melt as described in the text.

Samples of PCL, PCL/DBS, PCL/PVB, and PCL/DBS/PVB were processed using a defined thermal and shear profile which consisted of heating to 80 °C at 20 °C/min, holding for 5 min and then subjected to a shear flow field of 10 s^{-1} for 100 s (shear strain = shear rate x time = 1000 su). After cessation of shearing the sample was cooled to room temperature at rate of 10 °C/min. Figure 3a shows the WAXS intensity map for the PCL sample, which, although it has been crystallised from a sheared melt, nevertheless exhibits an isotropic pattern typical of a semi-crystalline polymer. The pattern exhibits several crystalline rings (seen as arcs), of which by far the strongest is from the (110) planes at $|Q| \sim 1.5$ Å^{-1}. This is in-line with previous studies and reflects the zero memory of the shear field due to the rapid relaxation of the PCL chains after shearing before crystallisation take place [24]. The corresponding PCL/PVB specimen shows the same isotropic pattern (Figure 3b). The PCL/DBS specimen, on the other hand, shows a considerable level of anisotropy in the WAXS pattern (Figure 3c). The specimen is mounted so that the shear direction is vertical and the 110 reflection is most intense on the equatorial section normal to the flow direction. This indicates that lamellae are aligned normal to the flow direction although there is an ambiguity as the preferential alignment of the c-axis of the PCL crystals parallel to the melt flow. This distribution of intensity is consistent with the model in which the PCL crystals are directed by DBS fibrils which have been previously preferentially aligned parallel to the flow direction in the PCL melt phase [24].

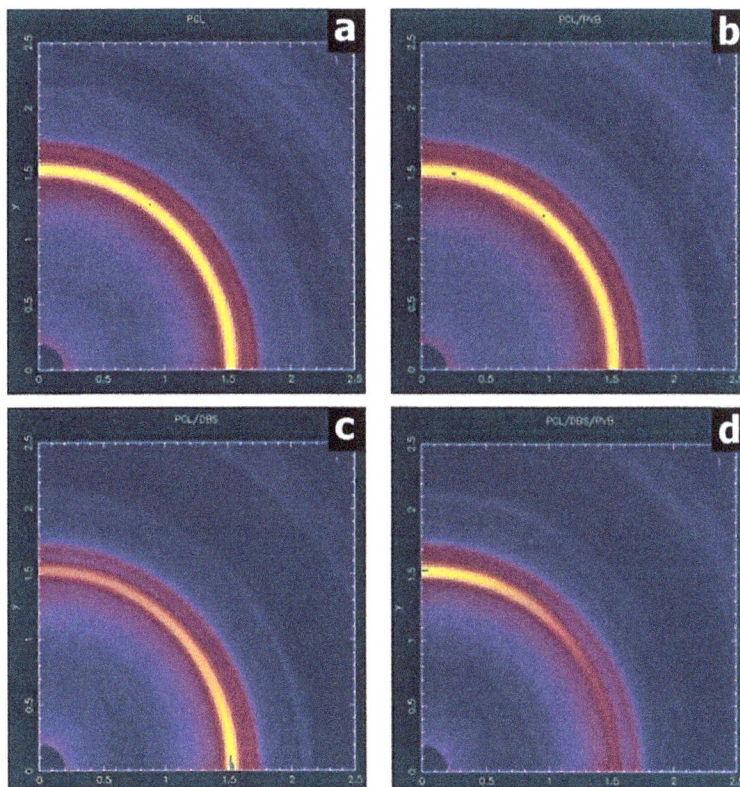

Figure 3. Maps of the wide-angle X-ray scattering (WAXS) intensity obtained for samples of (**a**) PCL; (**b**) PCL with 3%PVB; (**c**) PCL with 3% DBS; and (**d**) PCL with PVB and 3% DBS crystallised from a melt after a shear flow of I0 s^{-1} for 100 s at 80 °C. The melt flow direction is vertical.

The WAXS pattern for the PCL/PVB/DBS specimen (Figure 3d) is also substantially anisotropic. In contrast to the PCL/DBS sample, the 110 reflection is most intense on the meridional section, indicating an orientation of the PCL lamellae which is orthogonal to that displayed by the PCL/DBS sample.

Figure 4 shows the SAXS patterns for the four different materials considered here recorded at the point where the invariant (see below) reaches a maximum during the cooling stage of the thermal and shear profile described above.

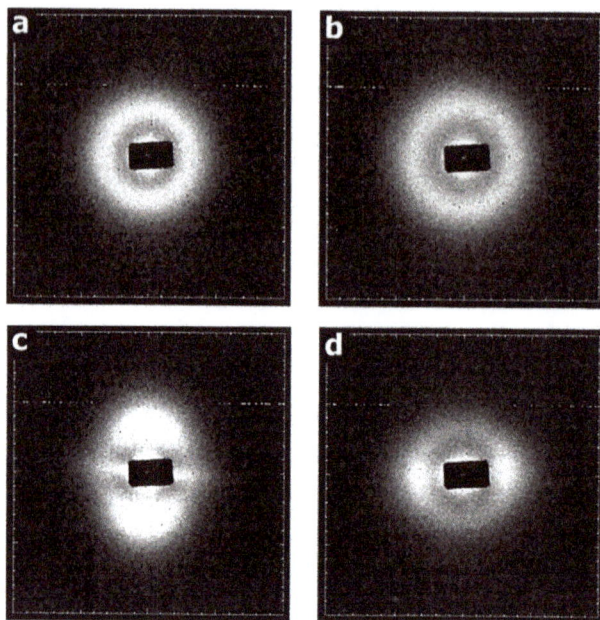

Figure 4. Small-angle X-ray scattering (SAXS) patterns obtained for samples of (**a**) PCL at 20 °C; (**b**) PCL with 3% PVB at 7 °C; (**c**) PCL with DBS at 22 °C and (**d**) PCL with 3% PVB and 3% DBS at 19 °C, all crystallised from a melt after a shear flow of 10 s^{-1} for 100 su at 80 °C. The melt flow direction is vertical. Each pattern is shown for the temperature where the invariant recorded during crystallisation was at its maximum.

The PCL sample exhibits a SAXS pattern which is typical for many semi-crystalline polymers [26]. The maximum in intensity at $|Q| \sim 0.032 \text{ Å}^{-1}$, corresponding to a long period of ~195 Å, arising from the lamellar structure. The constant intensity as a function of the azimuthal angle shows that the lamellae have no preferred orientation with respect to the melt flow direction. A similar pattern is displayed by the PCL/PVB sample (Figure 4b) with the lamellar peak at a $|Q|$ value of $\sim 0.0385 \text{ Å}^{-1}$, corresponding to a long period ~165 Å. The PCL/DBS sample exhibits a SAXS pattern (Figure 4c) which is strongly anisotropic. The intensity scattering on the meridional axis indicates a strong preferred orientation of the lamellae normal to what was the melt flow direction (vertical on the page); this is consistent with the WAXS result. The $|Q|$ value for the maximum is $\sim 0.0325 \text{ Å}^{-1}$, corresponding to a long period of ~195 Å. This confirms observations made previously that the DBS directs the lamellar growth direction but does not otherwise change the lamellar characteristics [7,26]. There is also a horizontal streak on the equatorial section which was not present in the patterns recorded for the PCL and PCL/PVB samples. This relates to the presence of highly extended DBS fibrils which are ~150 Å in width and 1000 Å or more in length. In striking contrast the PCL/PVB/DBS (Figure 4d) shows a similar pattern but one in which the lamellar peaks are now on the equator indicating that the lamellae are preferentially aligned parallel to the flow direction in accord with the crystal plane orientation information available from the WAXS patterns. Here the $|Q|$ value is ~0.0335 Å^{-1}, corresponding to a long period ~190 Å. Examination of the scattering around the beamstop reveals that there is no highly anisotropic equatorial streak as observed in the PCL/DBS sample. In fact, the portion of the scattering appears almost isotropic but there is a substantial level of scattering here as compared to that observed in the patterns for the PCL and PCL/PVB samples. Analysis of this scattering gives a particle radius of ~40 Å diameter which is smaller than that observed for the PCL/DBS system. The lamellar scattering is also noticeably weaker than the equivalent pattern observed for the PCL/DBS.

During the thermal and shear profile for each sample, we recorded SAXS patterns on a time resolving basis with a time cycle of 10 s throughout the processing cycle. This enables us to follow the formation of structure including the crystallisation. For each SAXS pattern we have calculated the so-called invariant Ω [27]:

$$\Omega = \int_{\alpha=0}^{\pi/2} \int_{Q=0}^{Q_{max}} |Q|^2 I(|Q|, \alpha) \sin \alpha \, dQ d\alpha \qquad (1)$$

The invariant is related to the volume fraction of crystals, at least in the initial stage of crystallisation, as all other things being equal, the invariant would also go through a maximum when the volume occupied by lamellae reaches 50%, and it will decrease slightly on cooling as the electron densities of crystalline and amorphous material approach each other. We have used the plots of invariant versus temperature to identify the temperature at which the maximum rate of crystallisation occurs. Figure 5 shows the results. The vertical axis plots of the differential of the invariant (which in practice corresponds very closely to a crystallization exotherm in differential scanning calorimetry), are derived from the data using an algorithm based on the method of Savitzky and Golay [28]. The four peaks indicate the maximum crystallisation rate for each of the materials. Compared to the PCL sample, the peak for PCL-DBS occurs sooner (higher temperature), as expected by increased nucleation [24], while that for PCL-PVB occurs much later (lower temperature), owing to the suppression of nucleation as also observed by optical microscopy (Figure 2b). The maximum value of the derivative is noticeably lower corresponding to a slower crystallisation rate and the peak is much broader indicating crystallisation taking place over a wider temperature range. This concurs with the observation that the invariant itself reaches a maximum about 11 °C below the point of greatest crystallisation rate, as compared with 8–9° below for all the other specimens.

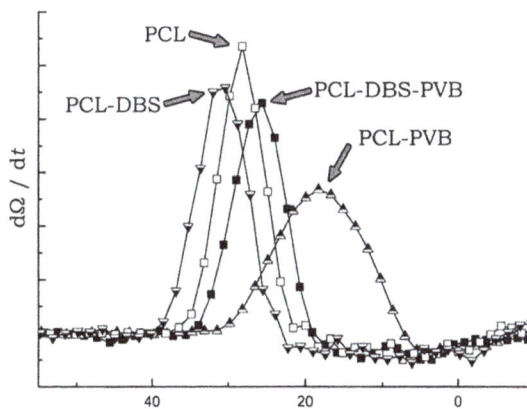

Figure 5. Plots of the derivative of the invariant against the temperature of measurements for each of the four materials considered in this work during the crystallisation stage of the processing cycle described in the text. The uncertainties associated with each data point are horizontally ±1 °C, vertically 2% of the vertical range.

4. Discussion

The maximum rate of crystallisation for the PCL/DBS/PVB sample is about 5 °C below that observed for the PCL sample. However this is considerably above the peak temperature for PCL/PVB. The addition of the PVB to the PCL/PVB appears to have significantly deactivated the DBS as a nucleant as measured by a shift in the maximum crystallisation temperature of about 10 °C, and this corresponds to the smaller lamellar long period indicated by the SAXS measurements. We emphasise that shear flow is imposed on the polymer melt ~40 °C above the crystallisation temperature of PCL. As the crystallisation of the sheared PCL melt reveals, at the point of crystallisation the melt itself has no memory of the shear flow. The behaviour reported above is the consequence of the distribution and nature of the nanoparticles and their influence on the nucleation of the PCL. There is the possibility that the presence of the slightly higher molecular weight PVB could alter the flow behaviour of the overall system. However, we can discount this possibility as the WAXS (Figure 2b) and SAXS (Figure 3b) patterns for the PCL/PVB sample indicate clearly an isotropic distribution of PCL crystals. Moreover, if PVB chains were significantly extended in the shear flow this would promote the development of the type of lamellar orientation observed in Figure 4c rather than that actually observed (Figure 4d).

It is more probable that the PVB acts to suppress the crystallisation of the DBS. The optical microscopy shows that this is indeed the case at least in part. The morphology shown in Figure 2d was obtained by crystallising from a melt in which the DBS was fully dissolved. From the optical specimen, it appears that once dissolved, the DBS is hindered from recrystallising by the PVB such that the morphology is similar to that of the PCL/PVB sample. However, from the presence of a significant level of small-angle scattering around the beam stop for the PCL/DBS/PVB sample (Figure 4d) we can deduce any suppression is only partial: we estimate the reduction ~50%. However, the fact that this scattering is not highly anisotropic is significant. We propose that the presence of the PVB during the preparation partially inhibits the growth of the DBS particles. This limits the length to breadth ratio of the DBS particles and modifies the behaviour of the particles in the shear flow stage of the processing cycle.

Jeffrey showed particles in a fluid during shear flow may exhibit a number of orientation states [29]. A number of studies have considered the effect of the anisotropy of the particle for example [30,31] and observations of both alignments parallel to the flow direction and to the vorticity direction have been reported for carbon nanotubes in shear polymer melts [32]. Gunes et al., have explored the effects of both the particle aspect ratio and the rheological parameters of the matrix polymer on the orientation behaviour of the particles in shear flow [33]. In materials with a low Deborah number (product of relaxation time and shear rate), the fluid elasticity can damp the particle rotations. In this study, we focus on fluid systems with higher Deborah numbers and Leal has proposed a critical shear rate for the transition from a vorticity orientation (normal to the flow) to flow axis orientation which is inversely related to the particle aspect ratio [33]. The detailed experimental study of Gunes et al., reveals a rich array of behaviour which matches theory in part.

We propose that the addition of PVB to the PCL/DBS system results in fewer and less anisotropic particles which when subjected to modest shear flow adopt an alignment preferentially parallel to the vorticity axis and normal to the flow direction. We anticipate this is principally due to the reduced aspect ratio in line with the prediction of Leal [34], but there are other consequences of the change of concentration for example, the elasticity of the system. Without the PVB, highly extended DBS particles are formed which align preferentially parallel to the flow direction in line with Leal's prediction under the same shear rate as used in the PCL/DBS/PVB system. On cooling, these particles act to nucleate and direct the crystallisation of the PCL so that the lamellae grow normal to the surface of the particle. As a consequence, for PCL/DBS/PVB and PCL/DBS, the subsequent lamellar growth directions are very different. In both cases, the lamellae grow out radially from the DBS particles but with symmetry axes which are orthogonal to each other. We anticipate that by adjusting the proportion of the PVB introduced in to the system, we can control the inhibiting extent of the PVB on the DBS and thereby prepare samples with a range of different crystalline texture.

Controlling the morphology of a semi-crystalline polymer can be achieved in a variety of ways yielding major changes to properties. The addition of nucleating agents to polymers may be used to control the crystal structure as in the case of β-nucleating agents for isotactic polypropylene [35], to control of the spherulite size in packaging films [12] and in general the addition of nucleating agents leads to stiffer products [12]. We have shown in a series of papers how nucleating agents can be exploited to provide a common orientation of row nuclei to generate structures with a common lamellar crystal orientation [5,7,24]. Figure 6 shows the geometry of this process. We need to bear in mind that this nucleating effect will be cylindrically symmetric and hence any compounds diffusing from the surface of the sample in to the interior will have facile access to the amorphous regions between the crystalline lamellar. As a consequence a number of approaches have been tried to provide polymer films with lower permeability for oxygen diffusion to provide higher performance barrier films for food and medical packaging [36]. The ideal arrangement of the crystal lamellae is an arrangement in which the lamellar normal is perpendicular to the polymer film as shown in Figure 6b This provides for such an arrangement so that the lamellar crystals provide the maximum obstruction to the diffusing oxygen pathways thereby increasing tortuosity increasing the diffusion pathways [37]. PCL is a biodegradable polymer and is biocompatible. It is widely used to manufacture implants such as scaffolds for tissue engineering. The degradation of the PCL involves the diffusion of bodily fluids in to the polymer initiating hydrolytic degradation [38,39]. The same arguments relevant to gas diffusion also apply to such degradation of the morphology shown in Figure 6b will be much slower than that observed for the morphology shown in Figure 6a [39]. Of course these different morphologies will also lead to changes in mechanical properties such as stiffness due to the anisotropy of the properties of the PCL crystals.

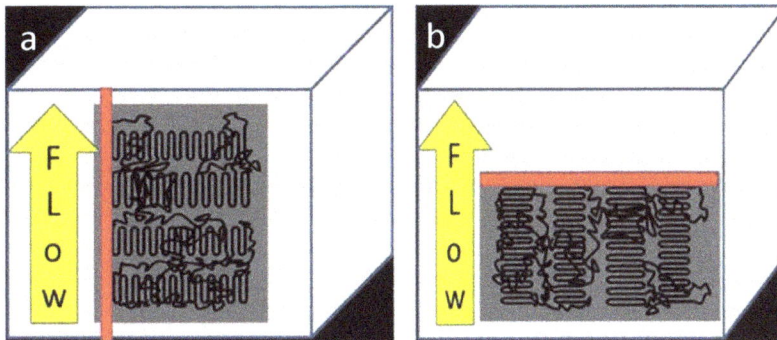

Figure 6. Schematic of the geometry of the templating mechanism for PCL/DBS. The red rod represents the DBS nanofibril aligned (**a**) (**left**) parallel with the flow direction and (**b**) (**right**) normal to the flow direction. This directs the growth direction of the lamellar crystals to be normal to the long axis of the DBS fibril as described in the text.

5. Conclusions

When poly(ε-caprolactone) is nucleated by dibenzylidene sorbitol in the presence of a terpolymer based on polyvinyl butyral there is an orthogonal switch in the symmetry axis of the resultant lamellar orientation as compared with PCL nucleated by DBS alone.

Previously, it has been shown that the aligned extended nanoparticles formed through the use of small amounts of DBS in conjunction with modest flow fields will nucleate lamellae with their c-axes parallel to the particles and to the flow direction. We confirm that small quantities of a terpolymer based on polyvinyl butyral inhibit the nucleation in PCL by itself and show that its presence can also modify the behaviour of the DBS. In particular, the presence of the PVB limits the aspect ratio of the DBS particles which form and this alters the preferred orientation of the DBS particles in the sheared

Polymers **2018**, *10*, 300

polymer melt from parallel to normal to the flow direction. This morphology is suitable for barrier property films and will influence the degradation rates in bioabsorbable biomedical products prepared from PCL.

Acknowledgments: The SAXS data were obtained at the STFC SRS at Daresbury UK and we thank Guenter Grossman for his help with the SAXS experiments. The microscopy was performed in the Centre for Advanced Microscopy at the University of Reading. The research work at IPL was supported by the Fundação para a Ciência e a Tecnologia (FCT) through the Project references: UID/Multi/04044/2013; PAMI-ROTEIRO/0328/2013 (No 022158), MATIS (CENTRO-01-0145-FEDER-000014-3362) and Project UC4EP.

Author Contributions: Robert H. Olley and Geoffrey R. Mitchell conceived and designed the experiments; Geoffrey R. Mitchell and Robert H. Olley performed the experiments; Geoffrey R. Mitchell analyzed the data; Robert H. Olley contributed to the discussion of the results; Geoffrey R. Mitchell wrote the paper. Both approved the final version.

Conflicts of Interest: The authors declare no conflict of interest. The funding sponsors had no role in the design of the study; in the collection, analyses, or interpretation of data; in the writing of the manuscript, and in the decision to publish the results.

References

1. Smith, D.K. Lost in translation? Chirality effects in the self-assembly of nanostructured gel-phase materials. *Chem. Soc. Rev.* **2009**, *38*, 684–694. [CrossRef] [PubMed]

2. Sangeetha, N.M.; Maitra, U. Supramolecular gels: Functions and uses. *Chem. Soc. Rev.* **2005**, *34*, 821–836. [CrossRef] [PubMed]

3. Weiss, R.G.; Terech, P. *Molecular Gels: Materials with Self-Assembled Fibrillar Networks*; Springer: New York, NY, USA, 2006; Chapter 8; pp. 233–244.

4. Terech, P.; Weiss, R.G. Low Molecular Mass Gelators of Organic Liquids and the Properties of Their Gels. *Chem. Rev.* **1997**, *97*, 3133–3159. [CrossRef] [PubMed]

5. Nogales, A.; Mitchell, G.R.; Vaughan, A.S. Anisotropic Crystallization in Polypropylene Induced by Deformation of a Nucleating Agent Network. *Macromolecules* **2003**, *36*, 4898–4906. [CrossRef]

6. Nogales, A.; Olley, R.H.; Mitchell, G.R. Directed Crystallisation of Synthetic Polymers by Low-Molar-Mass Self-Assembled Templates. *Macromol. Rapid Commun.* **2003**, *24*, 496–502. [CrossRef]

7. Mitchell, G.R.; Wangsoub, S.; Nogales, A.; Davis, F.J.; Olley, R.H. Controlling Morphology Using Low Molar Mass Nucleators. In *Controlling the Morphology of Polymers: Multiple Scales of Structure and Processing*; Mitchell, G.R., Tojeira, A., Eds.; Springer: Basel, Switzerland, 2016; Chapter 5; ISBN 978-3-319-39320-9.

8. Ping, Q. *Polymer Morphology: Principles, Characterization, and Processing*; Wiley: Hoboken, NJ, USA, 2016.

9. Kimata, S.; Sakurai, T.; Nozue, Y.; Kasahara, T.; Yamaguchi, N.; Karina, T.; Shibyama, M.; Kornfield, J.A. Report Molecular Basis of the Shish-Kebab Morphology in Polymer Crystallization. *Science* **2007**, *316*, 1014–1017. [CrossRef] [PubMed]

10. Bassett, D.C. Lamellar Organization in Polymer Spherulites. In *Integration of Fundamental Polymer Science and Technology*; Kleintjens, L.A., Lemstra, P.J., Eds.; Springer: Dordrecht, The Netherlands, 1986.

11. Vaughan, S.; Barré, L.L.; Zhao, Y.; Sutton, S.J.; Swingler, S.G. On additives, morphological evolution and dielectric breakdown in low density polyethylene. *Eur. Polym. J.* **2003**, *39*, 355–365.

12. Zweifel, H. (Ed.) *Plastics Additives Handbook*; Hanser: Munich, Germany, 2001; Chapter 18.

13. Planell, J.A.; Best, S.M.; Lacroix, D.; Merolli, A. *Bone Repair Biomaterials*; Woodhead: Oxford, UK, 2009.

14. Kim, D.; Kim, S.-W. Barrier property and morphology of polypropylene/polyamide blend film. *Korean J. Chem. Eng.* **2003**, *20*, 776–782. [CrossRef]

15. Smith, T.L.; Masilamani, D.; Bui, L.-G.; Khanna, Y.-P.; Bray, R.G.; Hammond, W.B.; Curran, S.; Belles, J.J.; Binder-Castelli, S. The Mechanism of Action of Sugar Acetals as Nucleating Agents for Polypropylene. *Macromolecules* **1994**, *27*, 3147–3155. [CrossRef]

16. Mandelkern, L. *Crystalization of Polymers*, 2nd ed.; CUP: Cambridge, UK, 2004; Volume 2.

17. Thierry, A.; Straupé, C.; Wittmann, J.-C.; Lotz, B. Organogelators and Polymer Crystallisation. *Macromol. Symp.* **2006**, *241*, 103–110. [CrossRef]

18. Sundararajan, P.R. *Physical Aspects of Polymer Self-Assembly*; Wiley: Hoboken, NJ, USA, 2016; Section 5.9.

19. Keith, H.D.; Padden, F.J.; Russell, T.P. Morphological changes in polyesters and polyamides induced by blending with small concentrations of polymer diluents. *Macromolecules* **1989**, *22*, 666–675. [CrossRef]

20. Nozue, Y.; Kurita, R.; Hirano, S.; Kawasaki, N.; Ueno, S.; Iida, A.; Nishi, T.; Amemiya, Y. Spatial distribution of lamella structure in PCL/PVB band spherulite investigated with microbeam small- and wide-angle X-ray scatterin. *Polymer* **2003**, *44*, 6397–6405. [CrossRef]

21. Nozue, Y.; Hirano, S.; Kurita, R.; Kawasaki, N.; Ueno, S.; Iida, A.; Nishi, T.; Amemiya, Y. Co-existing handednesses of lamella twisting in one spherulite observed with scanning microbeam wide-angle X-ray scattering. *Polymer* **2004**, *45*, 8299–8302. [CrossRef]

22. Ikehara, T.; Jinnai, H.; Kaneko, T.; Nishioka, H.; Nishi, T. Local lamellar structures in banded spherulites analyzed by three-dimensional electron tomography. *J. Polym. Sci. Part B Polym. Phys.* **2007**, *45*, 1122–1125. [CrossRef]

23. Nogales, A.; Thornley, S.A.; Mitchell, G.R. Shear Cell for In Situ WAXS, SAXS, and SANS Experiments on Polymer Melts Under Flow Fields. *J. Macrom. Sci. Phys.* **2004**, *B43*, 1161–1170. [CrossRef]

24. Wangsoub, S.; Oiley, R.H.; Mitchell, G.R. Directed Crystallisation of Poly(ε-caprolactone) using a Low-Molar-Mass Self-Assembled Template. *Macromol. Chem. Phys.* **2005**, *206*, 1826–1839. [CrossRef]

25. Towns-Andrews, E.; Berry, A.; Bordas, J.; Mant, G.R.; Murray, P.K.; Roberts, K.; Sumner, I.; Worgan, J.S.; Lewis, R.; Gabriel, A. Time-resolved x-ray diffraction station: X-ray optics, detectors, and data acquisition. *Rev. Sci. Instrum.* **1989**, *60*, 2346–2349.

26. Mohan, S.; Olley, R.H.; Vaughan, A.S.; Mitchell, G.R. Evaluating Scales of Structure in Polymers. In *Controlling Controlling the Morphology of Polymers: Multiple Scales of Structure and Processing*; Mitchell, G.R., Tojeira, A., Eds.; Springer: Basel, Switzerland, 2016; Chapter 2; ISBN 978-3-319-39320-9.

27. Roe, R.-J. *Methods of X-ray and Neutron Scattering in Polymer Science*; Oxford University Press: New York, NY, USA, 2000.

28. Savitzky, A.; Golay, M.J.E. Smoothing and Differentiation of Data by Simplified Least Squares Procedures. *Anal. Chem.* **1964**, *36*, 1627–1639. [CrossRef]

29. Jeffreys, G.B. The motion of ellipsoidal particles immersed in a viscous fluid. *Proc. R. Soc. Lond.* **1922**, *102*, 201–211.

30. Iso, Y.; Koch, D.L.; Cohen, C. Orientation in simple shear flow of semi-dilute fiber suspensions 1. Weakly elastic fluids. *J. Non-Newton. Fluid Mech.* **1996**, *62*, 115–134. [CrossRef]

31. Iso, Y.; Cohen, C.; Koch, D.L. Orientation in simple shear flow of semi-dilute fiber suspensions 2. Highly elastic fluids. *J. Non-Newton. Fluid Mech.* **1996**, *62*, 135–153. [CrossRef]

32. Hobbie, E.K.; Wang, H.; Kim, H.; Lin-Gibson, S.; Grulke, S. Orientation of carbon nanotubes in a sheared polymer melt. *Phys. Fluids* **2003**, *15*, 1196. [CrossRef]

33. Gubes, D.Z.; Scirocco, R.; Mewis, J.; Vermant, J. Flow-induced orientation of non-spherical particles: Effect of aspect ratio and medium rheology. *J. Non-Newton. Fluid Mech.* **2008**, *155*, 39–50.

34. Leal, L.G. The slow motion of slender rod-like particles in a second-order fluid. *J. Fluid Mech.* **1975**, *69*, 305–337. [CrossRef]

35. Papageorgiou, D.G.; Chrissafis, K.; Bikiaris, D.N. β-Nucleated Polypropylene: Processing, Properties and Nanocomposites. *Polym. Rev.* **2015**, *55*, 596–629. [CrossRef]

36. Mitchell, G.R. Characterisation of safe nanostructured polymers. In *Ecosustainable Polymer Nanomaterials for Food Packaging*; Silvestre, C., Cimmino, S., Eds.; Taylor and Francis: Boca Raton, FL, USA, 2013; ISBN 978-90-04-20737-0 eBook, ISBN 978-90-04-20738-7.

37. Bharadwaj, R.K. Modeling the Barrier Properties of Polymer-Layered Silicate Nanocomposites. *Macromolecules* **2001**, *34*, 9189–9192. [CrossRef]

38. Lam, C.X.; Savalani, M.M.; Teoh, S.H.; Hutmacher, D.W. Dynamics of in vitro polymer degradation of polycaprolactone-based scaffolds: Accelerated versus simulated physiological conditions. *Biomed. Mater.* **2008**, *3*, 034108. [CrossRef] [PubMed]

39. Mitchell, G.R.; Davis, F.J.; Olley, R.H. Scales of Structure in Polymers. In *Controlling the Morphology of Polymers: Multiple Scales of Structure and Processing*; Mitchell, G.R., Tojeira, A., Eds.; Springer: Basel, Switzerland, 2016; Chapter 1, ISBN 978-3-319-39320-9.

polymers

MDPI

Article

Development of Crystalline Morphology and Its Relationship with Mechanical Properties of PP/PET Microfibrillar Composites Containing POE and POE-*g*-MA

Maja Kuzmanović [1], Laurens Delva [1], Dashan Mi [1,2], Carla Isabel Martins [3], Ludwig Cardon [1] and Kim Ragaert [1,*

[1] Centre for Polymer and Material Technologies, Department of Materials, Textiles and Chemical Engineering, Faculty of Engineering and Architecture, Ghent University, Technologiepark 915, Zwijnaarde 9052, Belgium; maja.kuzmanovic@ugent.be (M.K.); laurens.delva@ugent.be (L.D.); dashan.mi@ugent.be (D.M.); ludwig.cardon@ugent.be (L.C.)

[2] College of Polymer Science and Engineering, State Key Laboratory of Polymer Materials Engineering, Sichuan University, Chengdu 610065, China

[3] Institute for Polymer and Composites/I3N, University of Minho, Campus de Azurém, 4800-058 Guimarães, Portugal; cmartins@dep.uminho.pt

* Correspondence: kim.ragaert@ugent.be; Tel.: +32-933-103-92

Received: 9 February 2018; Accepted: 5 March 2018; Published: 8 March 2018

Abstract: The main goal of this research is to study the development of crystalline morphology and compare it to various mechanical properties of microfibrillar composites (MFCs) based on polypropylene (PP) and poly(ethylene terephthalate) (PET), by adding a functional compatibilizer and a non-functional rubber in two different steps in the processing sequence. The MFCs were prepared at a weight ratio of 80/20 PP/PET by twin screw extrusion followed by cold drawing and injection moulding. The non-functionalized polyolefin-based elastomer (POE) and the functional compatibilizer (i.e., POE grafted with maleic anhydride (POE-*g*-MA)) were added in a fixed weight percentage at two stages: during extrusion or during injection moulding. The morphology observations showed differences in crystalline structure, and the PP spherulite size was reduced in all MFCs due to the presence of PET fibrils. Their relationship with the mechanical performances of the composite was studied by tensile and impact tests. Adding the functional compatibilizer during extrusions showed better mechanical properties compared to MFCs. Overall, a clear relationship was identified between processing, structure and properties.

Keywords: microfibrillar composites; crystalline morphology; crystallinity; mechanical properties

---

1. Introduction

Fibre reinforced composites have attracted great attention in the last decades. It is well known that incorporating different inorganic fibres into polymer matrices such as glass fibres or carbon fibres significantly improves mechanical performance, by enhancing strength and stiffness [1–3]. However, these types of fibre composites are not very environment-friendly, as they are difficult to mechanically recycle because of issues in the separation of the different components. This may be avoided with polymer-polymer reinforced composites, which can easily be mechanically recycled. In this respect, microfibrillar composites (MFCs) could be interesting as their improved mechanical properties allow them to be used in a wide range of applications. The MFC concept is a methodology developed in the early nineties by Fakirov and Evstatiev [4]. MFCs are a type of polymer-polymer composites in which a high-melting fibrillar thermoplastic polymer reinforces a lower-melting one [4].

In our recent study [5], the importance of the processing parameters during the production of MFCs has been pointed out, but besides these parameters, the composition of the starting mixture, the viscosity ratio and the compatibility of the components are equally important [6].

The research on polymer blends and polymer-polymer composites has also led to an increased interest in compatibilization. Numerous studies have been conducted on the compatibilization of blends, using different compatibilizers [7–11] and improving the dispersion of the second phase. The compatibilizer can be concentrated in the interface between two polymers during blending, thus preventing coalescence and resulting in smaller and finer dispersions as well as better adhesion between the phases [12–14].

On the other hand, it is also known that the MFC concept relies on the incompatibility between the matrix and dispersed phase, and that the final mechanical properties depend on the aspect ratio of the reinforcement and the interfacial adhesion between the matrix and reinforcement [4]. Various authors [4,15] have proposed that combining these two approaches, the MFC concept and compatibilization, could improve the properties of the final composites. According to Fakirov et al. [4], MFCs without compatibilizer can reach fibril lengths of up to 200 μm because the coalescence will take place during blending and drawing. In cases with a compatibilizer, Friedrich et al. [16] found a decrease in the tensile modulus and strength for the PP/PET in situ compatibilized microfibrillar blends. They attributed this reduction to the shorter microfibrils caused by the use of a compatibilizer, which covers the PET particles during melt blending and prevents their coalescence during drawing. This change in morphology can be seen in Figure 1. Fakirov et al. [15] reported in one of their studies that a compatibilizer affects the length of the fibrils depending on in which step of the processing sequence it is added. They have therefore suggested adding a compatibilizer to the drawn blend, in the final processing step, in injection or compression moulding. At this stage, the compatibilizer should facilitate distribution of the fibrils, improve the interfacial adhesion between matrix and fibrils, and enhance mechanical properties without reducing the aspect ratio of the fibrils. However, there are no experimental results to support this theory. To fill this gap, the present study investigates the effect of adding a compatibilizer during both extrusion and injection moulding. To this end, we have selected both a non-functionalized rubber and a functional compatibilizer, as a difference in migration to the interface and reactivity is expected.

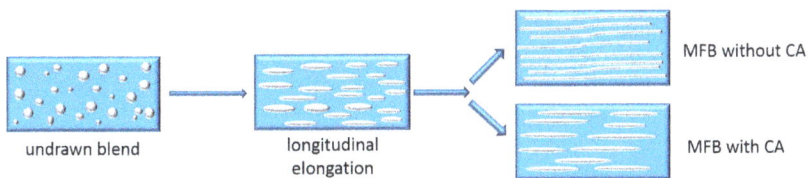

undrawn blend   longitudinal   MFB without CA
                    elongation   MFB with CA

Figure 1. Morphology of blend during cold drawing with and without compatibilizer added during extrusion [4].

To reveal the origin of ductile or brittle behaviour, researchers typically focus on the influence of crystallinity and crystalline structure of the composites. Embrittlement is known to be the result of the high crystallinity of semicrystalline polymers, but the size and perfection of the spherulites also play an important role in this behaviour [17,18]. In the case of semicrystalline polymers and their composites, different processing conditions can affect the crystalline structure, such as the perfection of crystallites, spherulite growth and orientation of the lamellae [19]. The process of making the MFCs may cause changes in the crystalline morphology of the matrix. The fibres could act as heterogeneous nucleating agents for the matrix, in which these nuclei can induce the crystal growth in the lateral direction [2,20,21].

To shed more light on this, the aim of this study is to examine the relationships between the development of the microstructure, the crystallinity and the mechanical properties of the PP/PET microfibrillar composites (MFCs). A polyolefin-based elastomer (POE) and POE grafted with maleic anhydride (POE-*g*-MA) will be added in a fixed weight percentage during extrusion and injection moulding, and we will investigate how this affects crystalline morphology and properties.

This research study will provide better insight into the morphological and crystallinity development of MFCs during processing. An alternative approach will be suggested to achieve good mechanical properties of MFCs.

2. Materials and Methods

2.1. Materials

Polypropylene (PP) was purchased from Sabic (Sabic 575P, Bergen op Zoom, The Netherlands) with a melt flow rate (MFR) of 11 g/10 min (2.16 kg, 230 °C), and the used PET was LIGHTER C93 from Equipolymers (Schkopau, Germany), which is a bottle-grade material with an intrinsic viscosity of 0.80 ± 0.02 dL/g. PET was dried in a vacuum oven for 15 h at 80 °C, and 2 h before processing at 120 °C, while PP was used as received. As additives, rubber and a compatibilizer were used in this research. POE (Vistamaxx^{TM} 6102), which is an ethylene-propylene elastomer with a MFR of 3 g/10 min (2.16 kg, 230 °C), was kindly provided by ExxonMobil (Machelen, Belgium). POE-*g*-MA used in this study was Acti-Tech 16MA13, which is a Vistamaxx-based compatibilizer, kindly donated by Nordic Grafting Company (NGC, Hellerup, Denmark). The grafting percentage of the MA group onto the backbone of the compatibilizer was 1.3 wt %, according to the data sheet. Both rubber and compatibilizer were dried at 60 °C for 15 h before processing.

2.2. Preparation of MFCs

In this study, PP was used as a matrix and PET as a reinforcing element. The samples were prepared in a weight ratio of 80/20 PP/PET and POE or POE-*g*-MA were added in 6 wt %, while the same PP/PET ratio was maintained. The PP/PET weight ratio was determined based on previous proprietary research. The MFCs were prepared by extrusion followed by cold drawing and injection moulding. Five different samples were prepared: non-compatibilized MFC, MFCs with POE and POE-*g*-MA added in the extrusion step (POE$_{EXT}$ and POE-*g*-MA$_{EXT}$), and MFCs with POE and POE-*g*-MA added in the injection moulding step (POE$_{IM}$ and POE-*g*-MA$_{IM}$) (Table 1).

Table 1. Formulations of the composites.

Material	PP/PET wt %	CA wt %
MFC	80/20	0
POE$_{EXT}$	75.2/18.8	6 $_{EXT}$
POE-*g*-MA$_{EXT}$	75.2/18.8	6 $_{EXT}$
POE$_{IM}$	75.2/18.8	6 $_{IM}$
POE-*g*-MA$_{IM}$	75.2/18.8	6 $_{IM}$

The melt blending of polymers with and without additive was conducted with a twin-screw extruder (Coperion ZSK18, Stuttgart, Germany) with two co-rotating screws of 18 mm diameter, $L/D = 40$ and a die opening of 19 mm × 2 mm. The screw speed was set at 120 rpm and the barrel temperatures were set between 205 and 260 °C. The extrudate was obtained as a sheet with dimensions of 25 mm × 1 mm, by passing through calender rolls, which were cooled down to ~15 °C. The received cooled extrudate entered directly into a hot oven (200 °C, 55.5 cm × 60 cm) and cold drawn by a pair of rolls above the glass transition temperature of PET. During drawing, the surface temperature of the extrudate was measured and amounted to approximately 95 °C. The speed of the rolls was adjusted to obtain a draw ratio of 8. Injection moulding was performed on an Engel 80T, with a temperature profile of 180, 190, 200 and 210 °C in a standard mould with a temperature of 30 °C, thus obtaining both tensile (114 × 6.45 × 4 mm^3, with a gauge length of 33 mm) and impact specimens (126 × 13 × 3 mm^3).

2.3. Characterization

2.3.1. Structural Characterization

Polarized optical microscopy (POM) Leica DM 2500 P (Wetzlar, Germany) was used to study the morphology of the specimens. Thin slices of 15 μm were cut from the injection moulded samples with a microtome Leica RM2245 (Wetzlar, Germany) in the direction parallel to the injection flow. They were subsequently inserted between two microscope cover glasses and glued with Canadian balm. Samples were analysed with a Leica Camera type DFC425 and DFC360FX (Wetzlar, Germany).

To determine the spherulites size, small angle light scattering (SALS) experiments were performed. For this, injection moulded samples were microtomed into 15 μm thick layers parallel to the flow direction on the FD-TD plane [22]. To suppress surface scattering, they were immersed in Canadian balm between two microscopic slides. Next, to obtain the SALS patterns, a 632.8 nm He-Ne laser with beam size of 1 mm was used as the source of polarized monochromatic light. SALS H_v patterns were captured using a Hamamatsu digital camera (Hamamatsu Photonics K.K., Hamamatsu City, Japan) and analysed with Hipic 6.3.0 software (Hamamatsu Photonics K.K., Hamamatsu City, Japan). The equivalent radius (R_0) of the spherulites was estimated with the following Equation (1) [23]:

$$R_0 = \frac{1025}{\pi} \frac{\lambda}{\sin(\theta_{max}/2)},\tag{1}$$

λ is the wavelength of light in the medium. The distance from the centre of the H_v pattern to the intensity maximum in one lobe, in conjugation with the known sample-to-film distance, is a measure of the polar angle θ_{max}. (θ_{max} = tan (distance from the centre of the H_v pattern to the intensity maximum in one lobe/sample-to-film distance)).

To study the phase and crystalline morphology of MFCs, we used scanning electron microscopy (SEM) FEG SEM JEOL JSM-7600F 202 (Tokyo, Japan). The samples were immersed in liquid nitrogen and subsequently fractured. For the observation of the phase morphology, the PP matrix together with POE was selectively dissolved in hot xylene for 1–3 h. For the crystalline morphology observation of the composites, the amorphous phase of the PP and PET was chemically etched in a solution H_2SO_4–H_3PO_4–$KMnO_4$ at 70 °C for 5–6 h. Furthermore, the samples surfaces were sputtered with gold by a Bal-Tec SCD005 sputter coater (Bal-Tec, Balzers, Liechtenstein). Micrographs were obtained with an accelerating voltage of 20 kV. The average diameter of the fibrils was calculated with Image J software (National Institute of Health, Bethesda, MD, USA). For the calculation, at least 50 measurements were used.

X-ray diffraction (XRD) measurements were carried out to confirm the crystal modification of PP. Tests were performed on a Bruker D8 Discover XRD system (Bruker Nederland BV, Leiden, The Netherlands) equipped with a Cu X-ray source (λ = 1.5406 Å) and a linear X-ray detector. The samples were put on a Si sample cup on the sample heating stage. θ–2θ measurements were carried out in air at atmospheric pressure at a temperature of 24 °C.

The structure-related thermal properties of the composites were determined via differential scanning calorimetry (DSC). Analysis was performed on a Netzsch DSC 214 Polyma device (Selb, Germany) in one cycle of heating-cooling in the temperature range between 20–200 °C. The tests ran under nitrogen atmosphere; the flow of nitrogen gas was 20 mL·min^{-1} and the heating/cooling rate was 10 °C·min^{-1}. Crystallinity (α_c) was calculated for the PP phase based on the theoretical enthalpy for 100% crystalline polymer and taking the mass percentage into account (Equation (2)):

$$\alpha_c = \frac{\Delta H^{exp}}{\Delta H^{\circ} w_f} \cdot 100\%,\tag{2}$$

where standard enthalpy (ΔH°) for PP is 207 J·g^{-1} [24], and w_f is the weight fraction of the relevant polymer in the PP/PET mixture. The mean thermal properties were averaged from three measurements and the differences were calculated by comparing population means by t-independent test via the software package SPSS Statistics 24 (Armonk, NY, USA).

2.3.2. Mechanical Characterization

Mechanical characterization was conducted under controlled conditions (23 °C and 50% relative humidity), after the samples had been conditioned for a minimum of 48 h within this controlled environment. The standard tensile bars were tested with an Instron 5565 tensile device (Norwood, MA, USA) according to standard ISO 527. During the tests, different test speeds were used before and after the Instron dynamic extensometer was removed (type catalogue 2620-603 with a gauge length of 12.5 mm), 1 mm/min and 5 mm/min, respectively. Analysis was performed with Bluehill software.

The notched Charpy impact test was used to evaluate the toughness of the samples by using a Tinius Olsen IT 503 Pendulum Impact Tester (Ulm, Germany) according to ISO 179. The specimens were notched in the middle of the sample with a depth of 2 mm, placed horizontally with the notch oriented away from the pendulum and broken by a hammer with an energy of 2 J. At least five specimens were tested for both tensile and impact tests. The differences between the samples are calculated by t-independent test preceded by a Levene's test for equality of variance via the software package SPSS Statistics 24 (Armonk, NY, USA) with a probability value of 0.05.

3. Results and Discussion

3.1. Morphology Development

Polarized optical microscopy was found to be a simple method to distinguish the changes in crystalline structure, such as the growth of crystals and their orientation [25]. Micrographs of neat PP sample are represented in Figure 2A,B. As can be seen, the micrographs show a clear spherulitic structure (Figure 2A) of PP, and due to the injection moulding process a typical "skin–core" structure (Figure 2B) can be discerned [19]. The average PP crystal size, measured quantitively by SALS, was found to be 22.3 µm (Table 2).

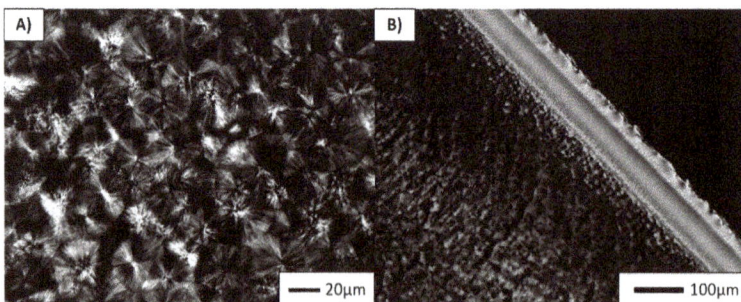

Figure 2. POM micrographs of neat PP sample in transferred direction to the injection flow.

Table 2. Average diameter of PP spherulites and PET fibrils in the composites.

Material	Spherulite Size (µm)	Fibril Diameter (µm)
PP	22.3 ± 0.8	-
MFC	8.9 ± 0.4	0.6 ± 0.2
POE$_{EXT}$	5.5 ± 0.7	0.8 ± 0.3
POE-g-MA$_{EXT}$	4.2 ± 0.3	0.5 ± 0.1
POE$_{IM}$	7.2 ± 0.4	0.7 ± 0.3
POE-g-MA$_{IM}$	7.6 ± 0.4	0.6 ± 0.2

The dispersion and distribution of the PET fibrils were examined on different scales of magnification using both POM (Figure 3, left column) and SEM (Figure 3, right column). Figure 3A represents a micrograph of a non-compatibilized MFC sample. Various dark regions can be found along the analysed sample, which are in fact clusters of PET fibrils. As the matching SEM picture (Figure 3A′) confirms, the dispersion and distribution of the PET phase are not very adequate. It is known that during drawing the coalescence effect causes the formation of very long microfibrils. However, as they have high aspect ratios, they may break during the injection moulding under high shear rate and may therefore stick together, thus forming fibril clusters [5]. Although it is difficult to determine the length of the PET microfibrils, they are assumed to be quite long, with an average diameter of 0.60 μm (Table 2).

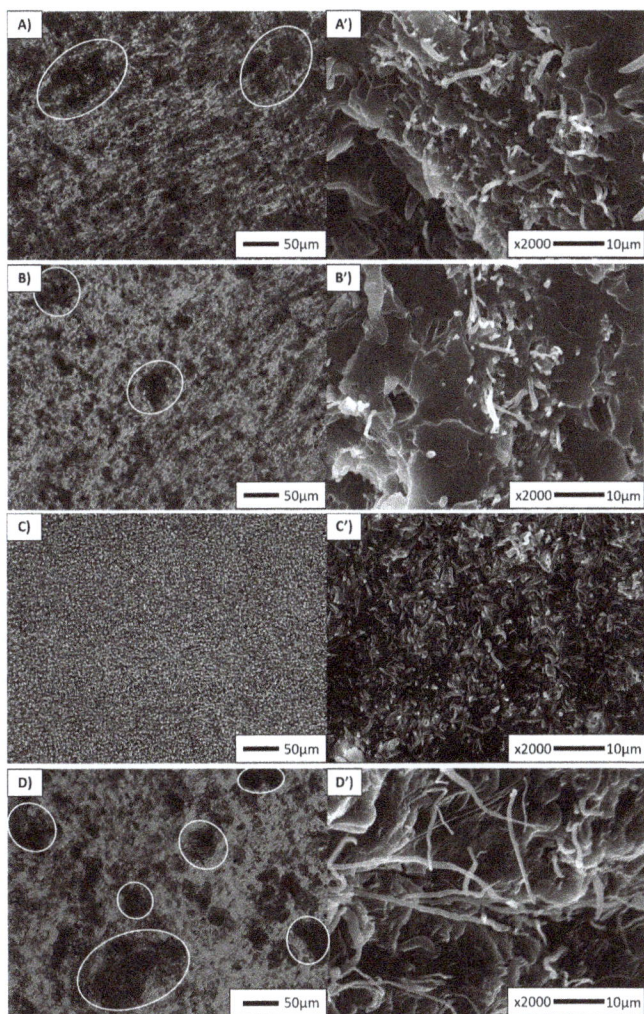

Figure 3. *Cont.*

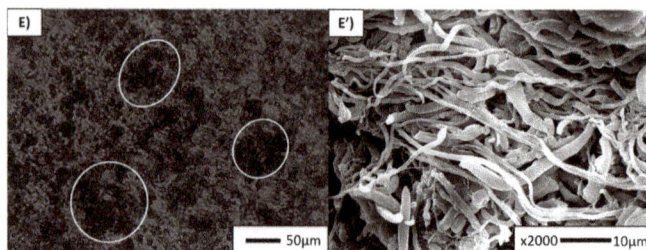

Figure 3. Microstructures of MFCs samples obtained via POM along the flow direction: (**A**) MFC; (**B**) POE$_{EXT}$; (**C**) POE-g-MA$_{EXT}$; (**D**) POE$_{IM}$; (**E**) POE-g-MA$_{IM}$. Microstructures of MFCs samples obtained via SEM in transverse direction: (**A′**) MFC (PP partially etched after 1.5 h in hot xylene); (**B′**) POE$_{EXT}$ (PP partially etched after 1.5 h in hot xylene); (**C′**) POE-g-MA$_{EXT}$ (PP partially etched after 3 h in hot xylene); (**D′**) POE$_{IM}$ (PP partially etched after 3 h in hot xylene); (**E′**) POE-g-MA$_{IM}$ (PP partially etched after 3 h in hot xylene).

Similarly, the POE$_{EXT}$ and POE$_{IM}$ samples are depicted in Figure 3B,D, respectively. Both POM and SEM micrographs again show a non-uniform distribution of the microfibrils and some fibril bundles along the analysed sample surfaces. Table 2 indicates that the fibril diameter of POE$_{EXT}$ was higher than in other samples (i.e., 0.80 μm, compared to 0.70 μm in POE$_{IM}$). In POE$_{IM}$, the long fibrils made during the cold drawing were preserved during injection moulding. It is quite clear that non-functionalized rubber will not have a significant effect, whether it is added during extrusion or injection moulding, because it will only act as a third phase due to non-existent functional group.

On the other hand, POE-g-MA$_{EXT}$ displays morphologies with both good dispersion and distribution of the PET microfibrils (Figure 3C,C′). The microfibrils appear much shorter compared to the other samples and the average fibril diameter was found to be lower (0.50 μm). In this case, the addition of compatibilizer first prevents coalescence during blending, thus reducing the starting diameter of the PET spheres and therefore also the length and diameter of PET fibrils in the MFC [15].

The POE-g-MA$_{IM}$ sample is represented in Figure 3E. Although this sample shows morphology with poorly distributed fibrils, the fibrils appear to be quite long, with an average diameter of 0.60 μm. It was expected to preserve the long microfibrils made during cold drawing, as the compatibilizer was added in the injection moulding step. Fakirov et al. [15] stated that, if that was the case, the high aspect ratio of the fibrils would not be reduced. However, their distribution is not as effective as expected. This could be due to the PET fibrils being in a solid state during injection moulding, which hinders both the migration of POE-g-MA to the interface and the reactivity of the MA group towards the end hydroxyl groups.

3.2. Development of Crystalline Morphologies

Additional high-magnification SEM experiments were carried out to investigate the location of the additives and the influence of the PET fibrils rubber and compatibilizer on the PP crystalline structure. Although SEM is not the preferred method to visualize the spherulitic structure, we could observe some crystalline structures under high magnifications. The average spherulite size was measured with SALS (Supporting Information, Figure S1), and the resulting diameters are listed in Table 2. It can be noted that PP spherulite size is drastically lowered in all MFCs compared to the neat PP sample, making the detection via SEM more difficult. However, in Figure 4, showing high magnification micrographs, the spherulite structure can be detected around the hole of the etched PET fibril in the MFC sample (Figure 4A). The average spherulite diameter in the MFC was found to be 9 μm, which is roughly 60% lower than the crystal size in neat PP. This would indicate that α-crystals are present in the composite. In addition to this, XRD measurements have confirmed the presence of PP α-spherulites in all samples, as the planes (110), (040), (130), (111) and (041) were observed at 2θ = 14.1°, 17.1°, 18.6°,

21.3° and 22°, respectively (Supporting information, Figure S2). These are the typical reflections of the α-crystals [26,27].

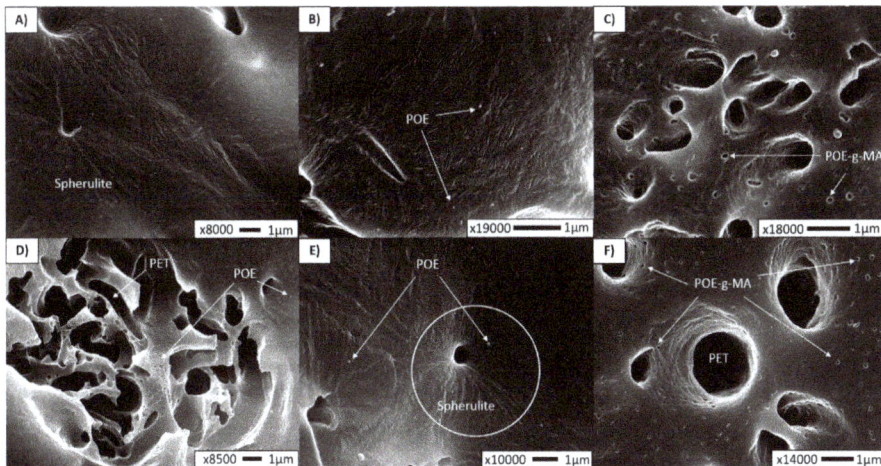

Figure 4. SEM micrographs of cryogenically fractured surface under liquid nitrogen of the injection moulded samples: (**A**) MFC (spherulite around fibril hole); (**B**) POE$_{EXT}$ (randomly oriented lamellae); (**C**) POE-g-MA$_{EXT}$ (non-reacted compatibilizer particles dispersed into matrix); (**D**) POE$_{IM}$ (rubber located at the interface); (**E**) POE$_{IM}$ (spherulite orientation around the fibril hole); and (**F**) POE-g-MA$_{IM}$ (compatibilizer particles located at the interface).

Furthermore, POE$_{EXT}$ and POE-g-MA$_{EXT}$ show the lowest spherulitic radius of 5.5 and 4.2 μm, while POE$_{IM}$ and POE-g-MA$_{IM}$ exhibit diameters of 7.2 and 7.6 μm, respectively. POE-g-MA$_{EXT}$ (Figure 4C) possesses well dispersed and distributed PET fibres which will have a strong nucleating effect on the PP matrix, regardless of whether they are covered by compatibilizer. The functionalized compatibilizer will be more prone to migrate towards the interface than the rubber. This is evidenced in Figure 4D (POE$_{IM}$) and 4F (POE-g-MA$_{IM}$), which indicate the difference in location of the rubber versus the compatibilizer.

Besides POE-g-MA at the interface, non-reacted compatibilizer particles were found in the PP matrix as well (Figure 4C,F). As there is always some amount of the compatibilizer that will not react with the PET during melt blending, this amount is dispersed through the matrix and between the microfibrils. These isolated POE-g-MA particles may also act as nucleation sites for PP [28], which explains why the nucleating effect is the most pronounced and the lowest PP crystal size is achieved. Additionally, in POE$_{IM}$, a spherulitic orientation is observed around the fibril hole (Figure 4E), which confirms the nucleating effect of the PET.

As far as the SEM observations are concerned, it is challenging to discuss the orientation of the lamellae. It seems that the random orientation of lamellae exists in the sample POE$_{EXT}$ represented in micrograph 4B. Moreover, as Friedrich et al. [29] explained, the organization of lamellae depends on how close the crystallites are to the surface of the microfibril. Far away from the fibril, in the bulk polymer, the lamellae are randomly dispersed with no preferred direction of orientation, which could confirm our previous statement. However, various research studies [21,29,30] conducted within the same or similar compositions (PP/PET, LDPE/PET) have shown the lamellae orientation in a normal direction to the fibril.

3.3. Crystallinity Development

To study the melting and crystallization behaviour of the composites, the samples were analysed via differential scanning calorimetry (DSC). Table 3 lists the melting temperature (T_m), peak crystallization temperature (T_c), the temperature at the beginning and end of crystallization during cooling (T_c^{onset} and T_c^{endset}), and the calculated percentage of crystallinity (α_c). These results are considered along those already presented in Table 2 (i.e., the PP spherulitic sizes and the PET fibril diameters). Statistically, there is a significant difference between the T_m of neat PP and all MFC-based materials, which in turn do not significantly differ from one another. Due to relatively high variations, there are no significant differences in α_c between all reported materials. Apparent differences in means will be discussed, however.

Table 3. Thermal properties of PP, MFC, POE$_{EXT}$, POE-*g*-MA$_{EXT}$, POE$_{IM}$, POE-*g*-MA$_{IM}$ during heating and cooling.

Material	T_m^{PP} (°C)	T_c^{PP} (°C)	T_c^{onset} (°C)	T_c^{endset} (°C)	α_c^{PP} (%)
PP	171.5 ± 0.2	118.8 ± 0.1	122.5 ± 1.2	108.1 ± 1.4	47.05 ± 0.8
MFC	170.2 ± 0.8	122.8 ± 0.7	127.2 ± 0.2	112.8 ± 0.4	47.35 ± 2.2
POE$_{EXT}$	170.1 ± 0.4	123.1 ± 0.6	127.1 ± 0.2	113.2 ± 0.1	48.40 ± 0.5
POE-*g*-MA$_{EXT}$	170.1 ± 0.4	118.1 ± 0.6	121.5 ± 0.2	109.3 ± 0.9	50.01 ± 1.9
POE$_{IM}$	169.3 ± 0.9	122.9 ± 0.4	127.1 ± 0.2	113.2 ± 0.3	51.65 ± 2.9
POE-*g*-MA$_{IM}$	170.1 ± 1.2	116.4 ± 0.6	123.0 ± 0.1	108.3 ± 0.5	49.44 ± 2.8

All materials show a single melting peak of PP, thus confirming the continued presence of α-crystalline modification, as detected via XRD.

The average T_c of neat PP is found at 118.8 °C but in pure MFC, POE$_{EXT}$ and POE$_{IM}$ is increased and amounts to approximately 123 °C.

As has already been observed in the study of crystalline morphology, long PET microfibrils will act as nucleating agents for the PP matrix. This potentially results in imperfect growth of the PP crystals, which become smaller and more numerous [21,31,32]. In this case, the presence of long microfibrils in MFC enables the crystallization to start roughly 5 °C earlier than in neat PP. The total crystalline fraction is unaffected here, but indeed spherulite sizes are severely reduced, from around 22 to around 9 μm.

A similar trend can be observed in POE$_{EXT}$ and POE$_{IM}$ concerning the onset of crystallization, indicating that the nucleating function of the PET fibres remains uninhibited. As POE contains no structural elements that could interact with PET, it is considered to be dispersed within the PP matrix, thus not affecting the PET fibre shape or the PP-PET interface. However, there is a noticeable effect on the crystallinity of the PP matrix. It is well known that POE-type polymers will act as a nucleating agent for the a-crystals of PP [28,33,34]. Moreover, Danesi et al. [35] demonstrated many years ago that a secondary POE phase will be finely and uniformly dispersed if the viscosity of the POE is significantly lower than that of the PP matrix, as is the case here. This was confirmed in the SEM images above, which show fine droplets of POE (Figure 4D). Since POE$_{EXT}$ benefits from already having POE present in a twin screw compounding step, it stands to reason that the dispersion of the rubber will be markedly better for POE$_{EXT}$ than for POE$_{IM}$. It is this increased dispersion of the rubber throughout the matrix that is responsible for the smaller crystallite sizes of PP for POE$_{EXT}$. Average crystallinity appears to be higher for both POE materials, compared to binary MFC. This is not only due to the nucleating effect of the dispersed rubber but, as postulated by Martuscelli et al. [34], POE might also selectively extract from the PP more defective polymer chains into its amorphous phase, thus leaving a more stereoregular PP behind and increasing crystallinity.

Next, the composites with compatibilizer will be considered, for POE-*g*-MA$_{EXT}$ crystallization is once more shifted to the level of neat PP. T_c^{onset} and T_c^{endset} were observed for all samples. For POE-*g*-MA$_{EXT}$, PP crystallization started later at 121.5 and finished at 109.3 °C, which implies that it also crystalizes faster than other samples. Crystallite sizes are the lowest for this material, while overall crystallinity remains in the higher levels. It is our understanding that, given the affinity of the MA group to PET groups, the compatibilizer at least partially migrates towards the PP-PET interface during the compounding (EXT) step and there reacts with the PET. This has several effects: (i) during MFC production, coalescence of PET is inhibited, leading to shorter fibres (as was shown in SEM image 3C); and (ii) the compatibilizer will to some degree cover the PET fibres with regard to the PP matrix, thus inhibiting the nucleating effect of PET for the PP matrix. A nucleation resulting in a high α_c still occurs via the compatibilizer phase, but this will not affect the crystallization onset of PP as the PET fibres did: PET fibres are already in solid phase at the moment of potential PP nucleation, whereas the POE backbone is mostly amorphous and the small amount of crystallization that could occur, does so at much lower temperatures. This was confirmed in DSC analysis of the POE and POE-*g*-MA component (Supporting information, Figure S3). Some of the POE-*g*-MA is assumed to be dispersed throughout the matrix as well, given the high α_c and the very homogeneous structures observed in POM (Figure 4C).

In the case of POE-*g*-MA$_{IM}$, the same significant decrease in T_c (compared to the MFC and POE materials) was noticed, indicating that some of the compatibilizer does migrate to the PP-PET interface, even when added during injection moulding. It is remarkable that injection moulding temperatures are much lower than compounding temperatures. The PET is in solid state, which hinders both the migration of POE-*g*-MA to the interface and lowers the reactivity of the MA groups towards the end hydroxyl groups. PET fibres remain relatively long, as there is no compatibilizer yet to inhibit their coalescence during drawing.

It was observed in the morphology study (Figure 4F) that the compatibilizer was located in both the matrix and at the interface. As a result, here as well, the POE backbones can provide matrix-wide nucleation of the PP. However, as with the difference between POE$_{EXT}$ and POE$_{IM}$, the dispersion of POE-*g*-MA is less efficient when the compounding step is missing, leading to spherulite sizes of the same order as POE$_{IM}$. Logically, the effect of a seemingly faster crystallization as with POE-*g*-MA$_{EXT}$ is not noted here.

4. Mechanical Properties

4.1. Tensile Behaviour

The tensile modulus and yield strength of the composites are given in Figure 5. Strain-at-break ε_b and strain-at-yield ε_y are reported in Table 4.

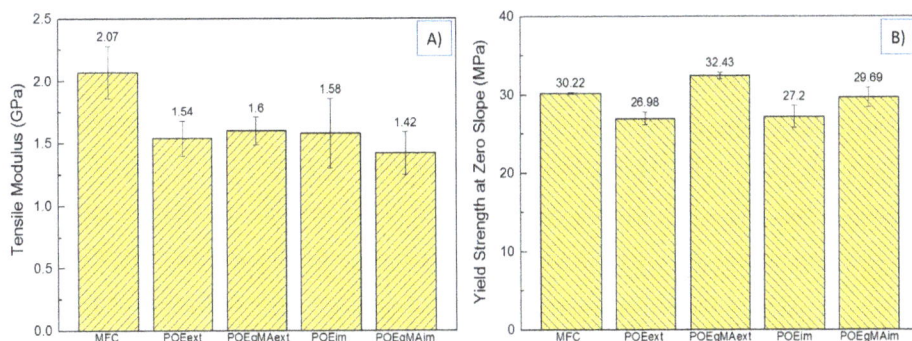

Figure 5. Tensile properties of MFC composites (**A**) Tensile modulus; (**B**) Yield strength at zero slope.

Table 4. Strain at break of the composites.

Material	Strain at Yield (%)	Strain at Break (%)
MFC	3.74 ± 0.9	5.29 ± 0.9
POE$_{EXT}$	6.84 ± 0.2	10.69 ± 1.7
POE-g-MA$_{EXT}$	11.90 ± 0.3	190.37 ± 162.8
POE$_{IM}$	6.52 ± 1.1	9.70 ± 2.3
POE-g-MA$_{IM}$	8.91 ± 1.4	11.42 ± 2.4

The bar charts indicate that MFC obtains a relatively high modulus. This could be explained by the extensive load-bearing capacity of the long PET fibrils. The interfacial area between microfibrils and matrix is large enough and some interfacial contact is assumed to exist, even without the presence of a compatibilizer. During the elongation, the matrix is expected to exert pressure on the fibrils, thus producing a high frictional force and preventing the composite from deforming. This results in constrained cavitation formation and very small ultimate elongation (Figure 6A) [36], as confirmed by the low ε_b.

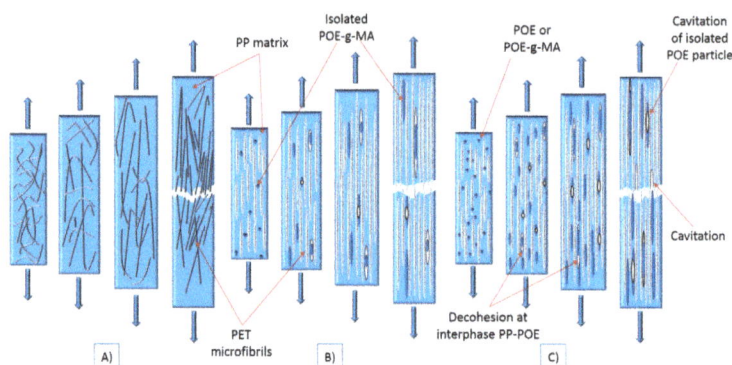

Figure 6. Tensile fracture model: (**A**) Case MFC; (**B**) POE-g-MA$_{EXT}$; (**C**) Case POE$_{EXT}$, POE$_{IM}$ or POE-g-MA$_{IM}$.

Among all other composites containing POE and POE-g-MA added during extrusion or injection moulding, no significant differences are found for the modulus. These composites all obtain a lower stiffness due to the presence of the rubber and the compatibilizer with a soft backbone.

A significant increase was found in the yield strength of POE-g-MA$_{EXT}$ compared to non-compatibilized MFC, POE$_{EXT}$, POE$_{IM}$ and POE-g-MA$_{IM}$ ($p < 0.001$). The reason for this lies in the lower interfacial tension between the POE-g-MA and PET achieved during extrusion, which enhances the interfacial adhesion of PP and PET in the final composite. Figure 6B represents a tensile fracture model which can be applied to a compatibilized composite such as POE-g-MA$_{EXT}$. In this model, stress transfer between matrix and fibre is excellent: both strain together while the PET fibres do most of the load bearing, until they finally fail together. Such deformation behaviour is only possible with great adhesion between the phases, which in this case is demonstrated by the high ε_y and ε_b values.

As a result, while neat MFC—being the only composite not containing some rubber fraction—obtains the highest stiffness, POE-g-MA$_{EXT}$ surpasses it in terms of strength and strain behaviour due to the presence of the compatibilizer, even if the aspect ratio of the PET fibrils is reduced and the composite contains a rubber phase. This corroborates previous experimental results for in situ compatibilized MFC by Yi et al. [6].

It can also be noted that the yield strength for POE$_{EXT}$, POE$_{IM}$ and POE-g-MA$_{IM}$ is lower due to elastomeric chains in the polymer backbone, as mentioned earlier. No differences were found between POE_{EXT} and POE_{IM} ($p = 0.772$), while there were significant differences when POE-g-MA$_{IM}$ was compared to POE_{EXT} ($p = 0.003$) and POE_{IM} ($p = 0.017$). The somewhat higher yield stress for POE-g-MA$_{IM}$ can be attributed to a limited interaction between the compatibilizer and PET, as proposed in Section 3.3.

G'Sell et al. [37] explained that several mechanisms could contribute to the general deformation of an MFC sample including elastomers under tension: interface decohesion, cavitation at the PP/PET interface, and the cavitation in isolated POE-g-MA particles. These are summarized in the fracture model for the MFCs made with POE_{EXT}, POE_{IM} or POE-g-MA$_{IM}$, as shown in Figure 6C.

In these composites, mostly isolated POE or POE-g-MA particles were found during morphology study. As observed earlier, non-functionalized POE cannot react with PET and thus a three-phase microstructure was created. The same effect was largely present in POE-g-MA$_{IM}$. Under tensile test, the strain rate is low and when critical stress is reached, the rubber takes the shape of fibril. With further stretching, the rubber fibrils continuously transfer stress to the matrix and they become elongated when their yield strength is exceeded. Rubber fibril structures will be preserved during the whole fracture process [38]. In all three cases, during continuous stretching there is a substantial risk of decohesion to appear at the interphases between PET and PP, as well as POE particles and PP matrix and the cavitation in isolated POE particle [37]. Strain levels are improved by the addition of the elastomer, but effective decohesion between the PP matrix and the PET fibres is likely, given the much larger ε_y and ε_b demonstrated by POE-g-MA$_{EXT}$.

G'Sell et al. [37] reported that in the case of decohesion of the matrix from a nodule or cavitation in the rubber particle (Figure 6C) by stretching the matrix, large voids can occur on poles of the rubber. These voids play a significant role at the end of deformation because their presence make a composite fracture easier, causing a decrease in modulus and strength.

4.2. Impact Behaviour

Numerous studies have explored how a high fracture toughness may be achieved, for example by using intrinsically tough matrices or rubber modified or by incorporating different fibres as reinforcements [1–4,37,39,40]. Similarly, the impact behaviour of semicrystalline polymers has been studied extensively, as well as how this may be improved [17,18,39,41–43].

Comparing the toughness of the composites (Figure 7A), we observed that the POE-g-MA$_{EXT}$ gave the highest value for the impact energy. This could be explained by the fact that there is high interfacial adhesion between the fibrils and matrix in POE-g-MA$_{EXT}$, as polar carboxyl groups of MA grafted onto POE backbone improve the adhesion with the PET phase. In addition to the interfacial adhesion between the oriented PET fibrils and the PP matrix, as has been mentioned earlier, both the size and amount of PP crystallites play an important role in determining impact strength. The PP spherulites become more imperfect and smaller due to the presence of PET fibrils, which may increase toughness [5,18]. As mentioned earlier, the spherulite size in POE-g-MA$_{EXT}$ was the lowest, which obviously contributed to an increase in toughness.

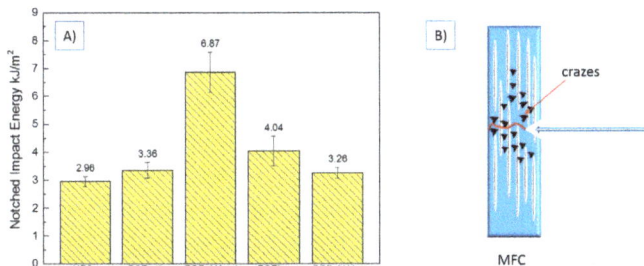

Figure 7. (**A**) Comparison of impact strength of MFC composites; (**B**) Impact fracture model of MFC.

Although the fibrils formed in this composite are shorter, they have a higher resistance by better dissipation of the impact energy. It is known that the rubber phase could initiate crazes thus contributing to the absorption of the impact energy to block the crack propagation [44]. As was reported by Perkins et al. [18], polymer toughness can be improved by optimizing the crystalline morphology, incorporating a discrete rubbery phase or by adding fibres as reinforcements to polymer matrices.

Figure 7B represents the impact fracture model of the MFC and shows that the crazing mechanism is in fact present. Crazing is one of the preferential deformation mechanisms which may prevent the further development of the craze into crack along the impact direction [44]. In the composite with good fibrils dispersion and distribution, the crack will propagate along the impact direction, but it will also deflect for an angle from the impact direction. The fibrils may induce a crack deflection perpendicular to the impact direction and transmit the stress to the matrix. This will make the matrix participate more actively in the stress transfer, which in turn will increase absorbed energy.

Compared to pure MFC, composites POE$_{EXT}$, POE$_{IM}$ and POE-g-MA$_{IM}$ show a slight but still significant increase in impact energy. As was explained earlier, all these composites have obtained three-phase morphology, where besides the PET fibrils in the matrix, POE or POE-g-MA particles are also dispersed. The highest increase in impact strength was achieved in POE$_{IM}$, and compared with POE$_{EXT}$ ($p = 0.035$) and POE-g-MA$_{IM}$ ($p = 0.015$), significant differences were found. The reason for this could be the existence of both spherical POE particles and long PET microfibrils, which both have acted as nucleating sites for PP matrix and could more effectively include the matrix in absorbing the energy. As Wang et al. [44] have explained, if the composite contains spherical rubber reinforcements, the material could be toughened only when the stress field around the rubber particles overlap and go through the matrix. No significant differences could be observed between POE$_{EXT}$ and POE-g-MA$_{IM}$ ($p = 0.533$), but compared to pure MFC ($p = 0.030$), significant differences can in fact be found. These differences may be the result from the third rubber phase, as stress transfer is not continuous and the PET microfibrils cannot reach higher levels of the energy absorption due to their poor dispersion and distribution.

4.3. Comparison to Non-Fibrillated Blends

In the end, by adding the compatibilizer, we have seen that POE-g-MA$_{EXT}$ improved in terms of both yield and impact strength, as well as in strain at break compared to all MFCs. Not unexpectedly, this comes at a cost in terms of stiffness. Given that an increased stiffness is one of the largest gains achieved by producing MFCs rather than just using non-fibrillated blends, we found it necessary to hold our results against similar experiments for neat PP, blend or additivated undrawn blends (here referred to as IMB, Injection Moulded Blends), with the subscript denominating the additive (added in the compounding step, EXT). These results are summarized in Table 5, together with the results of MFC and POE-g-MA$_{EXT}$, which have been added for clarity's sake. The highest value achieved is marked in bold.

Table 5. Mechanical properties of injection moulding blends.

Material	Tensile Modulus GPa	Yield Strength MPa	Strain at Break %	Impact Strength kJ/m^2
PP	1.47 ± 0.1	17.3 ± 1.2	>500	1.45 ± 0.2
IMB	1.72 ± 0.1	24.8 ± 0.4	4.78 ± 0.5	2.06 ± 0.4
IMB$_{POEext}$	1.58 ± 0.1	27.59 ± 0.7	10.43 ± 2.6	3.04 ± 0.5
IMB$_{POE-g-MAext}$	1.21 ± 0.1	29.54 ± 0.1	10.48 ± 4.8	6.55 ± 1.9
POE-g-MA$_{EXT}$	1.60 ± 0.1	32.43 ± 0.4	190.32 ± 162.8	6.87 ± 0.7
MFC	2.07 ± 0.2	30.22 ± 0.1	5.29 ± 0.9	2.96 ± 0.2

None of the composites, including POE-based additives, achieve the high modulus of neat MFC. IMB$_{POEext}$ and POE-g-MA$_{EXT}$ both manage to maintain a modulus of around 1600 MPa, which is close to that of the neat IMB. However, only POE-g-MA$_{EXT}$ shows an additional large increase in impact strength and strain levels, as well as a small improvement in yield strength. IMB$_{POE-g-MAext}$ does manage to match its impact strength, but at an unacceptable cost in stiffness.

The formulation of POE-g-MA$_{EXT}$ shows that also short fibrils may act as good reinforcement for the PP matrix and that at least equal importance should be given to the stress transfer possibilities between matrix and fibre, which in this case is effectively facilitated by the compatibilizer POE-g-MA.

POE-g-MA$_{EXT}$ is considered the best formulation for the manufacture of a PP/PET MFC with all-round good mechanical properties.

5. Conclusions

This paper presented a comprehensive study on the relationships between the microstructure and mechanical properties of PP/PET microfibrillar composites (MFCs). MFCs were prepared in a weight ratio 80/20 PP/PET via three-step processing with addition of 6 wt % of POE or POE-g-MA during extrusion or injection moulding.

Under POM microscopy, the presence of PP spherulites was observed in neat PP, and the skin-core structure was established. Both POM and SEM microscopies confirmed the non-uniform dispersion and distribution of the fibrils in the samples MFC, POE$_{EXT}$, POE$_{IM}$ and POE-g-MA$_{IM}$, while in POE-g-MA$_{EXT}$ the dispersion and distribution were found to be very good. Furthermore, we were able to examine the effect of adding POE or POE-g-MA in different steps of processing by means of the SEM micrographs. In the composites POE$_{EXT}$, POE$_{IM}$ and POE-g-MA$_{IM}$ the three-phase microstructure was developed, but the long microfibrils were preserved. Shorter fibrils were obtained in POE-g-MA$_{EXT}$, due to the addition of the compatibilizer during extrusion, which prevented a coalescence of the PET particles. In addition to the development of phase morphology, crystalline morphology was investigated. The MFCs without an amorphous phase were analysed via SEM under high magnification. The spherulite structure developed in most of the samples, and a random orientation of the lamellae was noted in bulk polymer, as well as some spherulite orientation around the fibril holes. PP spherulite size was drastically lower in all MFC samples, supporting the theory that PET fibrils may act as heterogeneous nucleating points for the PP matrix.

XRD measurements and melting behaviour have confirmed the formation of α-spherulites in all composites. DSC analysis, along with SALS measurements of spherulite sizes, showed that both POE and PET fibrils are good nucleators for PP, in which only PET also effects a shift in T_c. Furthermore, these combined results suggest that POE will be dispersed within the matrix exclusively, while POE-g-MA will migrate to the PP-PET interface. It does so, even when added only during injection moulding, but the strong compatibilizing effect will only occur if it is already added during the compounding of the blend.

The mechanical results have confirmed that the microstructure and properties are significantly affected by adding the rubber and the compatibilizer in different processing steps. The impact strength of POE-g-MA$_{EXT}$ was found to be superior compared to all other composites. All composites have shown a brittle breakage during the tensile tests, except POE-g-MA$_{EXT}$, where necking was observed. In POE-g-MA$_{EXT}$, a significant increase in yield strength and at strain at break was noted, as the compatibilizer added during extrusion caused a better interfacial adhesion between PP and PET in the final composite.

It has been demonstrated that POE-g-MA$_{EXT}$ shows better all-round mechanical properties compared to both IMBs and MFCs. The PET short fibrils can act as excellent reinforcement for the PP matrix, when they are produced with an addition of rubber-based compatibilizer during extrusion. The main objective of this study was to evaluate the potential of adding the compatibilizer only in the injection moulding step, as suggested by Fakirov et al. Considering all experimental results, we can conclude that, while postponing the compatibilizer addition does conserve the long fibrils, it does not create the best mechanical properties for a compatibilized PP/PET MFC. Mixing in the compatibilizer prior to drawing does achieve this, despite the reduction in fibril length.

Supplementary Materials: The following are available online at www.mdpi.com/xxx/s1.

Acknowledgments: We gratefully acknowledge to Ghent University, Belgium for financial support.

Author Contributions: Kim Ragaert, Laurens Delva, Ludwig Cardon and Maja Kuzmanović conceived and designed the experiments; Maja Kuzmanović performed the experiments; Dashan Mi and Carla Isabel Martins supported experiments and analysis related to crystalline morphology. Maja Kuzmanović, Laurens Delva and Kim Ragaert analysed the data and wrote the paper.

Conflicts of Interest: The authors declare no conflict of interest. The founding sponsors had no role in the design of the study; in the collection, analyses, or interpretation of data; in the writing of the manuscript, and in the decision to publish the results.

References

1. Fu, S.-Y.; Lauke, B.; Mäder, E.; Yue, C.-Y.; Hu, X. Tensile properties of short-glass-fiber-and short-carbon-fiber-reinforced polypropylene composites. *Compos. Part A Appl. Sci. Manuf.* **2000**, *31*, 1117–1125. [CrossRef]

2. Bao, S.; Liang, G.; Tjong, S. Part II: Polymer-Polymer Composites with Premade Fibrous Reinforcement: Fracture Behavior of Short Carbon Fiber Reinforced Polymer Composites. In *Synthetic Polymer–Polymer Composites*; Elsevier: Amsterdam, The Netherlands, 2012; pp. 117–143.

3. McCardle, R.; Bhattacharyya, D.; Fakirov, S. Effect of Reinforcement Orientation on the Mechanical Properties of Microfibrillar PP/PET and PET Single-Polymer Composites. *Macromol. Mater. Eng.* **2012**, *297*, 711–723. [CrossRef]

4. Fakirov, S. *The Concept of Micro-or Nanofibrils Reinforced Polymer-Polymer Composites*; Carl Hanser Verlag: Munich, Germany, 2012.

5. Kuzmanović, M.; Delva, L.; Cardon, L.; Ragaert, K. The Effect of Injection Molding Temperature on the Morphology and Mechanical Properties of PP/PET Blends and Microfibrillar Composites. *Polymers* **2016**, *8*, 355. [CrossRef]

6. Yi, X.; Xu, L.; Wang, Y.-L.; Zhong, G.-J.; Ji, X.; Li, Z.-M. Morphology and properties of isotactic polypropylene/poly(ethylene terephthalate) in situ microfibrillar reinforced blends: Influence of viscosity ratio. *Eur. Polym. J.* **2010**, *46*, 719–730. [CrossRef]

7. Chiu, H.-T.; Hsiao, Y.-K. Compatibilization of poly(ethylene terephthalate)/polypropylene blends with maleic anhydride grafted polyethylene-octene elastomer. *J. Polym. Res.* **2006**, *13*, 153–160. [CrossRef]

8. Moini Jazani, O.; Khoramabadi, M.A.; Salehi, M.M.; Riazi, H.; Soltanokottabi, F. Effective parameters on the phase morphology and mechanical properties of PP/PET/SEBS ternary polymer blends. *J. Part. Sci. Technol.* **2015**, *1*, 39–48.

9. Jayanarayanan, K.; Thomas, S.; Joseph, K. Effect of compatibilizer on the morphology development, static and dynamic mechanical properties of polymer-polymer composites from LDPE and PET. *Int. J. Plast. Technol.* **2015**, *19*, 84–105. [CrossRef]

10. Heino, M.; Kirjava, J.; Hietaoja, P. Compatibilization of polyethylene terephthalate/polypropylene blends with styrene–ethylene/butylene–styrene (SEBS) block copolymers. *J. Appl. Polym. Sci.* **1997**, *65*, 241–249. [CrossRef]

11. Asgari, M.; Masoomi, M. Thermal and impact study of PP/PET fibre composites compatibilized with glycidyl methacrylate and maleic anhydride. *Compos. Part B Eng.* **2012**, *43*, 1164–1170. [CrossRef]

12. Zhang, X.; Li, B.; Wang, K.; Zhang, Q.; Fu, Q. The effect of interfacial adhesion on the impact strength of immiscible PP/PETG blends compatibilized with triblock copolymers. *Polymer* **2009**, *50*, 4737–4744. [CrossRef]

13. Pang, Y.; Jia, D.; Hu, H.; Hourston, D.; Song, M. Effects of a compatibilizing agent on the morphology, interface and mechanical behaviour of polypropylene/poly(ethylene terephthalate) blends. *Polymer* **2000**, *41*, 357–365. [CrossRef]

14. Entezam, M.; Khonakdar, H.A.; Yousefi, A.A.; Jafari, S.H.; Wagenknecht, U.; Heinrich, G.; Kretzschmar, B. Influence of interfacial activity and micelle formation on rheological behavior and microstructure of reactively compatibilized PP/PET blends. *Macromol. Mater. Eng.* **2012**, *297*, 312–328. [CrossRef]

15. Fakirov, S.; Bhattacharyya, D.; Lin, R.; Fuchs, C.; Friedrich, K. Contribution of coalescence to microfibril formation in polymer blends during cold drawing. *J. Macromol. Sci. Part B Phys.* **2007**, *46*, 183–194. [CrossRef]

16. Friedrich, K.; Evstatiev, M.; Fakirov, S.; Evstatiev, O.; Ishii, M.; Harrass, M. Microfibrillar reinforced composites from PET/PP blends: Processing, morphology and mechanical properties. *Compos. Sci. Technol.* **2005**, *65*, 107–116. [CrossRef]

17. Schrauwen, B.B. Deformation and Failure of Semi-Crystalline Polymers: Influence of Micro and Molecular Structure. Ph.D. Thesis, Technische Universiteit Eindhoven, Eindhoven, The Netherlands, January 2003.

18. Perkins, W.G. Polymer toughness and impact resistance. *Polym. Eng. Sci.* **1999**, *39*, 2445–2460. [CrossRef]

19. Zhao, Z.; Yang, Q.; Kong, M.; Tang, D.; Chen, Q.; Liu, Y.; Lou, F.; Huang, Y.; Liao, X. Unusual hierarchical structures of micro-injection molded isotactic polypropylene in presence of an in situ microfibrillar network and a β-nucleating agent. *RSC Adv.* **2015**, *5*, 43571–43580. [CrossRef]

20. Harel, H.; Marom, G. On crystalline interfaces in composite materials. *Acta Polym.* **1998**, *49*, 583–587. [CrossRef]

21. Quan, H.; Li, Z.-M.; Yang, M.-B.; Huang, R. On transcrystallinity in semi-crystalline polymer composites. *Compos. Sci. Technol.* **2005**, *65*, 999–1021. [CrossRef]

22. Yalcin, B.; Cakmak, M. Superstructural hierarchy developed in coupled high shear/high thermal gradient conditions of injection molding in nylon 6 nanocomposites. *Polymer* **2004**, *45*, 2691–2710. [CrossRef]

23. Meeten, G. Refraction and extinction of polymers. In *Optical Properties of Polymers*; Elsevier Applied Science Publishers Ltd.: Amsterdam, The Netherlands, 1986; pp. 1–62.

24. Jayanarayanan, K.; Bhagawan, S.; Thomas, S.; Joseph, K. Morphology development and non isothermal crystallization behaviour of drawn blends and microfibrillar composites from PP and PET. *Polym. Bull.* **2008**, *60*, 525–532. [CrossRef]

25. Xu, J.; Ye, H.; Zhang, S.; Guo, B. Organization of Twisting Lamellar Crystals in Birefringent Banded Polymer Spherulites: A Mini-Review. *Crystals* **2017**, *7*, 241. [CrossRef]

26. Zhang, S.; Lin, W.; Zhu, L.; Wong, C.P.; Bucknall, D.G. γ-Form Transcrystals of Poly propylene) Induced by Individual Carbon Nanotubes. *Macromol. Chem. Phys.* **2010**, *211*, 1348–1354. [CrossRef]

27. Jayanarayanan, K.; Thomas, S.; Joseph, K. In situ microfibrillar blends and composites of polypropylene and poly(ethylene terephthalate): Morphology and thermal properties. *J. Polym. Res.* **2011**, *18*, 1–11. [CrossRef]

28. Marcinčin, A. Modification of fiber-forming polymers by additives. *Prog. Polym. Sci.* **2002**, *27*, 853–913. [CrossRef]

29. Friedrich, K.; Ueda, E.; Kamo, H.; Evstatiev, M.; Krasteva, B.; Fakirov, S. Direct electron microscopic observation of transcrystalline layers in microfibrillar reinforced polymer-polymer composites. *J. Mater. Sci.* **2002**, *37*, 4299–4305. [CrossRef]

30. Li, Z.M.; Li, L.B.; Shen, K.Z.; Yang, W.; Huang, R.; Yang, M.B. Transcrystalline morphology of an in situ microfibrillar poly(ethylene terephthalate)/poly(propylene) blend fabricated through a slit extrusion hot stretching-quenching process. *Macromol. Commun.* **2004**, *25*, 553–558. [CrossRef]

31. Denchev, Z.; Dencheva, N. Preparation, mechanical properties and structural characterization of microfibrillar composites based on polyethylene/polyamide blends. In *Synthetic Polymer-Polymer Composites*; Hanser: Munich, Germany, 2012; pp. 465–524.

32. Xu, L.; Zhong, G.-J.; Ji, X.; Li, Z.-M. Crystallization behavior and morphology of one-step reaction compatibilized microfibrillar reinforced isotactic polypropylene/poly(ethylene terephthalate)(iPP/PET) blends. *Chin. J. Polym. Sci.* **2011**, *29*, 540–551. [CrossRef]

33. Karger-Kocsis, J.; Kallo, A.; Szafner, A.; Bodor, G.; Senyei, Z. Morphological study on the effect of elastomeric impact modifiers in polypropylene systems. *Polymer* **1979**, *20*, 37–43. [CrossRef]

34. Martuscelli, E.; Silvestre, C.; Abate, G. Morphology, crystallization and melting behaviour of films of isotactic polypropylene blended with ethylene-propylene copolymers and polyisobutylene. *Polymer* **1982**, *23*, 229–237. [CrossRef]

35. Danesi, S.; Porter, R.S. Blends of isotactic polypropylene and ethylene-propylene rubbers: Rheology, morphology and mechanics. *Polymer* **1978**, *19*, 448–457. [CrossRef]

36. Chen, Y.; Zhong, G.; Li, Z. *Microfibril Reinforced Polymer-Polymer Composites via Hot Stretching: Preparation, Structure and Properties*; Hanser: Munich, Germany, 2012.

37. G'Sell, C.; Bai, S.-L.; Hiver, J.-M. Polypropylene/polyamide 6/polyethylene–octene elastomer blends. Part 2: Volume dilatation during plastic deformation under uniaxial tension. *Polymer* **2004**, *45*, 5785–5792. [CrossRef]

38. Shi, D.; Liu, E.; Tan, T.; Shi, H.; Jiang, T.; Yang, Y.; Luan, S.; Yin, J.; Mai, Y.-W.; Li, R.K. Core/shell rubber toughened polyamide 6: An effective way to get good balance between toughness and yield strength. *RSC Adv.* **2013**, *3*, 21563–21569. [CrossRef]

39. Argon, A.; Cohen, R. Toughenability of polymers. *Polymer* **2003**, *44*, 6013–6032. [CrossRef]

40. Evstatiev, M.; Fakirov, S.; Bechtold, G.; Friedrich, K. Structure-property relationships of injection-and compression-molded microfibrillar-reinforced PET/PA-6 composites. *Adv. Polym. Technol.* **2000**, *19*, 249–259. [CrossRef]

41. Redakcji, O. Plastic deformation and cavitation in semicrystalline polymers studied by X-ray methods. *Polimery* **2014**, *59*, 533. [CrossRef]

42. Meijer, H.E.; Govaert, L.E. Mechanical performance of polymer systems: The relation between structure and properties. *Prog. Polym. Sci.* **2005**, *30*, 915–938. [CrossRef]

43. Galeski, A. Strength and toughness of crystalline polymer systems. *Prog. Polym. Sci.* **2003**, *28*, 1643–1699. [CrossRef]

44. Wang, J.; Wu, H.; Guo, S. Realizing simultaneous reinforcement and toughening in polypropylene based on polypropylene/elastomer via control of the crystalline structure and dispersed phase morphology. *RSC Adv.* **2016**, *6*, 1313–1323. [CrossRef]

polymers

MDPI

Article

PLA Melt Stabilization by High-Surface-Area Graphite and Carbon Black

Luciana D'Urso [1], Maria Rosaria Acocella [1], Gaetano Guerra [1], Valentina Iozzino [2], Felice De Santis [2] and Roberto Pantani [2,*]

[1] Department of Chemistry and Biology, University of Salerno, Via Giovanni Paolo II 132, 84084 Fisciano (SA), Italy; ldurso@unisa.it (L.D.); macocella@unisa.it (M.R.A.); gguerra@unisa.it (G.G.)
[2] Department of Industrial Engineering, University of Salerno, Via Giovanni Paolo II 132, 84084 Fisciano (SA), Italy; viozzino@unisa.it (V.I.); fedesant@unisa.it (F.D.S.)
* Correspondence: rpantani@unisa.it; Tel.: +39-089-964-141

Received: 31 December 2017; Accepted: 30 January 2018; Published: 1 February 2018

Abstract: Small amounts of carbon nanofillers, specifically high-surface-area graphite (HSAG) and more effectively carbon black (CB), are able to solve the well-known problem of degradation (molecular weight reduction) during melt processing, for the most relevant biodegradable polymer, namely poly(lactic acid), PLA. This behavior is shown by rheological measurements (melt viscosity during extrusion experiments and time sweep-complex viscosity) combined with gel permeation chromatography (GPC) experiments. PLA's molecular weight, which is heavily reduced during melt extrusion of the neat polymer, can remain essentially unaltered by simple compounding with only 0.1 wt % of CB. At temperatures close to polymer melting by compounding with graphitic fillers, the observed stabilization of PLA melt could be rationalized by scavenging traces of water, which reduces hydrolysis of polyester bonds. Thermogravimetric analyses (TGA) indicate that the same carbon fillers, on the contrary, slightly destabilize PLA toward decomposition reactions, leading to the loss of volatile byproducts, which occur at temperatures higher than 300 °C, i.e., far from melt processing conditions.

Keywords: poly(lactic acid); carbon black; graphite

---

1. Introduction

It is well known that melt processing of poly(lactic acid) (PLA) (typically conducted at temperatures close to 200 °C) generally leads to degradation, i.e., high molecular weight reduction [1–4], even in nitrogen atmospheres [2–5].

Many different approaches have been proposed to achieve PLA melt stabilization [6–14]. In particular, stabilization by compounding with commercial antioxidants and water scavengers [6,7], or with chain extenders (i.e., molecules that reconnect polymer chains broken due to moisture at elevated temperatures), such as organic phosphites [8–10] or functional polysilsesquioxane microspheres [11], has been described. Additional thermal stabilization approaches involve polymer crosslinking by suitable agents [12–14] and polymer-end protection by acetyl groups [15].

Many reports show, on the basis of thermogravimetric analyses (TGA), that several nanofillers, such as clays and organoclays [16–21], silica [22,23], lignin [24], and silk [25] and cellulose [26] nanocrystals, stabilize PLA with respect to decomposition reactions, leading to loss of low-molecular-weight byproducts, which occur at temperatures higher than 300 °C (i.e., very far from melt processing conditions). In recent years, many papers have been published on PLA composites with graphite-based nanofillers [27–46] and carbon black [47–50]. These papers report significant improvements in PLA's physical (mainly electrical) properties with only a small amount of filler. Some of these papers report that some carbon fillers, like other nanofillers [16–26], can stabilize

PLA with respect to decomposition reactions, leading to volatile byproducts at a temperature higher than 300 °C [27,29,32,34,43–46,48,49]. For instance, the addition of exfoliated graphite increases of 5–15 °C decomposition temperatures corresponding to a 5% weight loss ($T_{d,5\%}$) [27,29,32].

Stabilization of the polymer toward decomposition reactions leading to volatile byproducts at high temperatures (for PLA for T > 300 °C) does not assure maintenance of polymer molecular weight during melt processing (for PLA at T ≈ 200 °C). In fact, some fillers (e.g., clays) that stabilize PLA toward high-temperature decomposition reactions destabilize PLA toward degradation reactions, leading to more pronounced polymer molecular weight reductions, in melt processing conditions [51–56]. In this respect, it is also worth adding that the temperatures corresponding to a 10% weight loss ($T_{d,10\%}$) for PLA and isotactic polypropylene (PP) were evaluated as 320 °C and 270 °C, while PP was incomparably more stable than PLA during melt processing [51].

In this paper, the influence of different graphite-based fillers—A low-surface-area graphite (LSAG), a high-surface-area graphite (HSAG), and a carbon black (CB)—On the stability of PLA melt at different temperatures is reported. In particular, PLA stability during melt processing (at 200 °C) was studied via melt viscosity measurements during extrusion, via evaluations of molecular weight distributions of the extruded unfilled and filled samples with gel permeation chromatography (GPC), and via rheological measurements (time sweep-complex viscosity). PLA stability at higher temperatures (above 300 °C), where decomposition reactions lead to loss of volatile byproducts, was studied via TGA.

It was found that the considered graphite-based fillers do not improve (or even slightly reduce) PLA thermal stability above 300 °C, i.e., they marginally affect high temperature decomposition reactions, leading to low-molecular-mass byproducts. However, very small amounts (as low as 0.1 wt %) of high-surface-area graphite and of carbon black, are sufficient to inhibit degradation reactions, leading to molecular weight reductions without a loss of volatile byproducts, which occur during processing at 200 °C.

2. Experimental

In this work, a commercial grade of PLA produced by NatureWorks (Minnetonka, MN, USA) with the trade name of 4032D was adopted. This PLA grade has a D-enantiomer content of approximately 2% and a maximum degree of crystallinity of about 45%. A thermal and rheological characterization of the material can be found in the literature [57–60].

The material was dried at 60 °C under vacuum overnight before any processing and testing operation.

Primary Synthetic Graphite TIMREX® SFG6 with a low surface area (LSAG, of about 17 m²/g), an average particle size of 6 μm, and a carbon amount of 99.6% was provided by Timcal Graphite & Carbon (Bodio, Switzerland). Synthetic Graphite TC 307 with a high surface area (HSAG, of about 352 m²/g), a primary particle size less than 1 μm, a carbon amount of 99.92%, and a high shape anisotropy of the crystallites [61] was purchased from Asbury Graphite Mills, Inc (Asbury, NJ, USA). The used carbon black sample (CB) of grade N660, with a surface area of 33 m²/g and a particle size around 49–60 nm, was purchased from Cabot Company (Boston, MA, USA).

Wide-angle X-ray diffraction (WAXD) patterns were obtained by an automatic Bruker D8 Advance diffractometer (Bruker Corp, Billerica, MA, USA), in reflection, at 35 kV and 40 mA, using nickel-filtered Cu Kα radiation (0.15418 nm). The *d* spacings were calculated using Bragg's law, and the observed integral breadths (β_{obs}) were determined by a fit with a Lorentzian function of the diffraction patterns. The instrumental broadening (β_{inst}) was also determined by fitting of a Lorentzian function to line profiles of a standard silicon powder 325 mesh (99%). The corrected integral breadths of the 002 peak were determined by subtracting the instrumental broadening of the closest silicon reflection from the observed integral breadths, β = β_{obs} − β_{inst}. The correlation lengths (*D*) were determined using Scherrer's equation:

$$D = \frac{K\lambda}{\beta \cos \theta} \tag{1}$$

where λ is the wavelength of the incident X-rays and θ the diffraction angle, assuming the Scherrer constant $K = 1$.

Melt compounding was carried out by a microcompounder (twin screws, counter-rotating, Haake Mini-lab II, Thermo Fisher Scientific, Schwerte, Germany). Thanks to a backflow channel and a bypass valve, it is possible to define the residence time in the microcompounder. Furthermore, this device allows one to estimate the viscosity as the material is compounded in the back flow channel that is a rectangular slit with two pressure transducers [62,63]. The tests were conducted at a temperature of 200 °C and a screw rotation speed of 100 rpm. Under these conditions, the estimated shear rate at which the viscosity was calculated is about 350 s^{-1}. After 15 min, the bypass valve is opened and the compound is extruded.

The zero-shear rate viscosity of a polymer can be related to the molecular weight of the polymer by the following equation:

$$\eta = c \, Mw^a \tag{2}$$

where a is an exponent whose value is generally accepted to be 3.4 [64], and c is a parameter that depends on temperature. According to Equation (2), due to the exponent a, the viscosity is extremely sensitive to changes in molecular weight, so rheological measurements are an extremely powerful means of assessing the degradation in the molten state.

The melt compounding was carried out for 15 min (900 s) at 200 °C. The materials were then taken from the microcompounder and used for the subsequent analysis: GPC and rheology.

GPC measurements were conducted by a Waters Breeze GPC system (Waters, Milford, MA, USA), equipped with a refractive index (RI) detector, by using a set consisting of four Styragel HT columns with (102, 103, 104, and 105 Å pore size) and 10 μm (particle size). Tetrahydrofuran, THF, was used as eluent at 35 °C at a flow rate of 1.0 mL·min^{-1}. The calibration curve was established with polystyrene standards.

Time sweep experiments were performed by means of a Haake Mars II (Thermo Scientific) rotational rheometer in a plate–plate configuration ($D = 20$ mm) under a dry nitrogen atmosphere. A constant stress of 100 Pa and a frequency of 1 rad/s were applied during the tests. In this condition, all measurements were carried out within the linear response domain and within the Newtonian plateau for all materials. The tests were carried out at a temperature of 200 °C for about 3 h.

TGA analyses were conducted with a TG 209 F1, manufactured by Netzsch Geraetebau (Selb, Germany), with a heating rate of 10 K/min under an N$_2$ flow.

DSC scans were conducted at a heating rate of 10 K/min. and the results are reported in the supplementary material (Figure S1).

3. Results and Discussion

3.1. WAXD Characterization of Carbon Fillers

WAXD patterns, as collected by an automatic powder diffractometer, of the used low-surface-area (blue curve) and high-surface-area (red curve) graphites are compared in Figure 1. It is immediately apparent that their crystalline structures are largely different. In particular, LSAG exhibits a much more ordered structure, with a large number of narrow diffraction peaks. These peaks can be easily indexed by assuming the presence of both hexagonal or rhombohedral phases [65,66]. Particularly informative is the pattern region with $42° < 2\theta_{Cu \, K\alpha} < 46°$, where four well defined peaks are present, with two peaks at $d = 0.213$ nm and $d = 0.205$ nm indexed as (100) and (101) reflections of the hexagonal phase and two peaks at $d = 0.209$ nm and $d = 0.197$ nm indexed as (101) and (102) reflections of the rhombohedral phase. The fraction of rhombohedral modification is approximately 30%, as derived by comparing the integrated intensities for the above cited hexagonal and rhombohedral peaks.

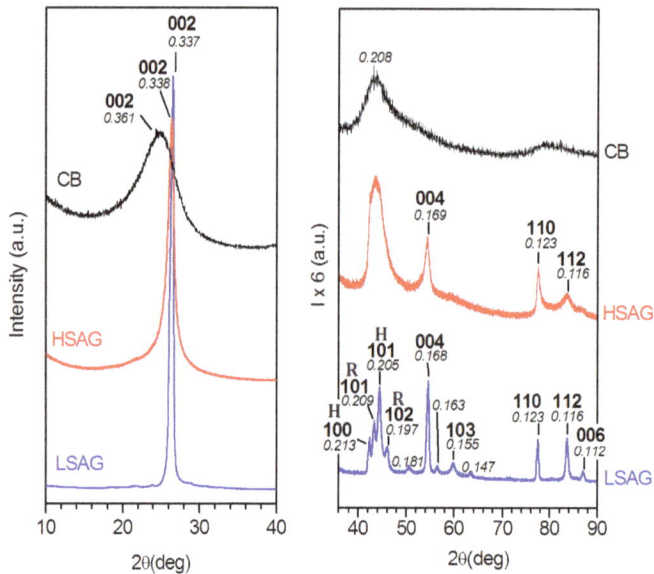

Figure 1. WAXD patterns (Cu Kα), as collected by an automatic powder diffractometer, of used carbon fillers: LSAG (lower blue curve); HSAG (intermediate red curve); CB (upper black curve). H and R labels refer to reflections being specific of hexagonal and rhombohedral phases, respectively.

HSAG exhibits a much more disordered structure, with a strongly reduced number of diffraction peaks. In particular, besides (00*l*) reflections and the in-plane 110 reflection at *d* = 0.123 nm, only a broad diffraction halo is present that is roughly centered at *d* = 0.208 nm. This clearly indicates the occurrence of a turbostratic graphite with a nearly complete disorder in the relative position of parallel graphitic layers [61,65,67].

For both LSAG and HSAG, the distance between parallel graphitic layers is equal to *d* = 0.337 nm, while the corresponding correlation length perpendicular to the graphitic planes (as evaluated by breadths of the 002 peak) is much lower for HSAG, with $D_{\perp,HSAG}$ = 12 nm and $D_{\perp,LSAG}$ = 26 nm.

For the sake of comparison, the WAXD pattern of the used carbon black is also shown in Figure 1. As discussed in detail in a recent paper [68], WAXD patterns of CB (as well as of oxidized CB, oCB) samples suggest that they are prevailingly constituted by a disordered spatial arrangement of highly defective structural layers with short in-plane correlation lengths (2–3 nm). This was confirmed by the ability of oCB to form ordered intercalation compounds [68].

3.2. Melt Compounding of PLA in the Presence of Different Kinds and Amounts of Carbon Fillers

Viscosity values measured during compounding are reported in Figure 2. The data are normalized with respect to the initial values measured for each material, so that all values start from 1. The curve which refers to the neat PLA is reported for comparison in all the plots of Figure 2. It can be seen that, in agreement with that reported in the literature [5,69], the viscosity of pure PLA during compounding immediately starts to reduce significantly, such that viscosity becomes about one half of the starting value after about 15 min. According to Equation (2), this reduction suggests a reduction of about 20% in the molecular weight of the material.

All the fillers adopted, with the exception of graphite LSAG at the lowest used percentage of 0.1%, introduce a significant stabilizing effect, such that the reduction of viscosity with time is limited to about 10%. For graphite HSAG and CB, this effect is reached already for filler contents of 0.1%, which is an extremely significant result for PLA processing.

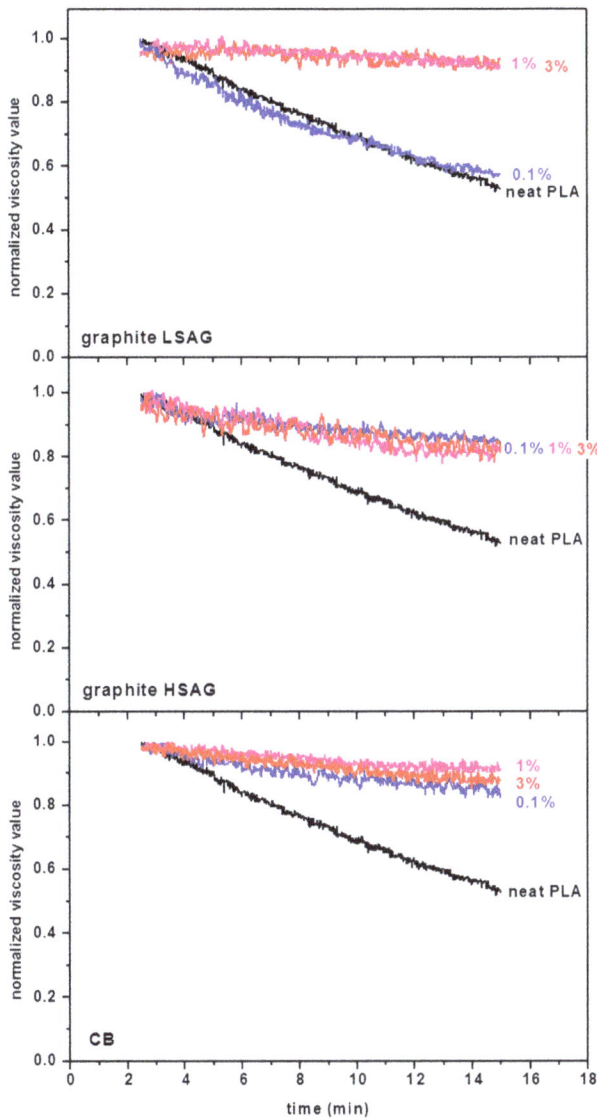

Figure 2. Time evolution of normalized viscosity (with respect to the initial value) in the microcompounder at $T = 200\ ^{\circ}$C and 100 rpm. The estimated value of shear rate is about 350 s^{-1}.

GPC curves of PLA pellet and of extruded PLA compounds, with different kinds and amounts of carbon fillers, are shown in Figure 3.

Elution times of GPC curves of PLA are definitely lower before extrusion (green lowest curve in Figure 3) than after extrusion (black curve in Figure 3). This confirms that, as generally observed for PLA, extrusion processes lead to a substantial polymer degradation. As shown in Table 1,

GPC curves indicate a reduction of about 25% of the initial number-average molecular weights (*Mn* and *Mw*). This is consistent with a reduction in viscosity of about 40%, in agreement with the results shown in Figure 2.

GPC curves of the extruded PLA compounds indicate that, for all the considered carbon fillers, a concentration of 3 wt % (continuous lines in Figure 3) is able to eliminate the adverse effect of the considered PLA processing on molecular mass. Moreover, for HSAG and CB, a concentration as low as 0.1 wt % is sufficient to stabilize PLA to the molecular mass of the virgin pellet (dotted curves in Figure 3, 7th and 10th columns in Table 1). Again, these results are consistent with rheological data reported in Figure 2.

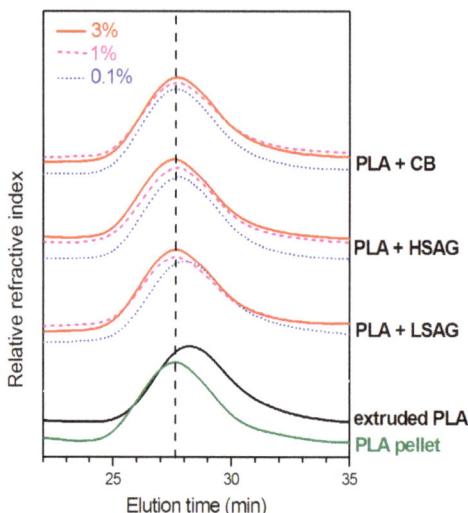

Figure 3. GPC curves in THF at 35 °C of the PLA pellet (green) of extruded PLA (black) as well as of extruded PLA compounds, as obtained by processes whose viscosity reduction is shown in Figure 2. Compounds contain 0.1 wt % (dotted lines), 1 wt % (dashed lines), and 3 wt % (continuous lines) of different graphitic fillers: LSAG, HSAG, and CB.

Table 1. Number average molecular weight (*Mn*), weight average molecular weight (*Mw*), and polydispersity index (PDI) as evaluated by GPC curves, for PLA pellet and extruded compounds with LSAG, HSAG, and CB. The evaluated variance is of ±3 kDa.

			Extruded PLA Samples								
			LSAG			HSAG			CB		
	PLA Pellet	PLA Neat	0.1%	1%	3%	0.1%	1%	3%	0.1%	1%	3%
Mn [kDa]	120 ± 3	88	95	120	120	120	121	120	121	121	120
Mw [kDa]	195 ± 3	146	165	195	196	190	194	196	192	193	193

The PLA pellet and the extruded PLA compounds were analyzed by time sweep rheological tests. The results are reported in Figure 4 and show even more clearly the thermal stabilization effect of carbon fillers. These measurements, which were carried out for a very long time (about 3 h) indicate a degradation at high temperature after the microcompounding step. Under these conditions, it is possible to discriminate between stabilization effects of HSAG and CB, which would appear to be very similar on the basis of the viscosity measurements of Figure 2 as well as on the basis of the GPC data of Figure 3 and Table 1, which indicate the effect of the microcompounding step.

The viscosity evolution reported in Figure 4 can be interpreted in terms of molecular weight according to Equation (2). The results are reported in Figure 5.

The initial molecular weight, $Mw(t = 0)$, indicates the value after the microcompounding step for all the samples except the PLA in pellet. The stabilizing effect of the fillers is clearly evidenced: the virgin polymer (both neat and extruded) presents a reduction of about 30% in molecular weight during the 3 h of the test at 200 °C. All considered carbon fillers indicate a slower decrease in Mw, with an effect that generally depends on concentration: an increase from 0.1% to 1% induces a slower degradation. For CB, 0.1% and 1% induce the same effect and allow for a reduction of just 10% of Mw during the test.

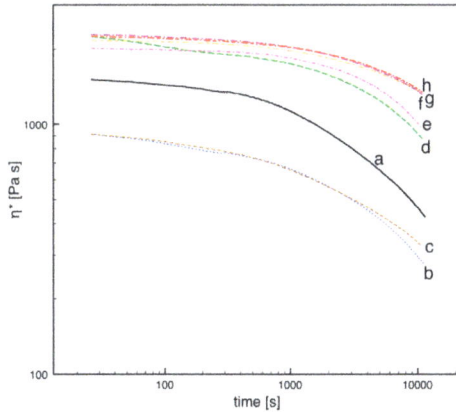

Figure 4. Time sweep-complex viscosity for the virgin pellet (**a**) and for extruded PLA samples after microcompounding: (**b**) neat and with (**c**) 0.1 wt % of LSAG; (**d**) 1 wt % of LSAG; (**e**) 0.1 wt % of HSAG; (**f**) 1 wt % of HSAG; (**g**) 0.1 wt % of CB; (**h**) 1 wt % of CB. Experimental conditions: $T = 200$ °C, $f = 1$ rad/s, plate–plate, gap = 200 μm.

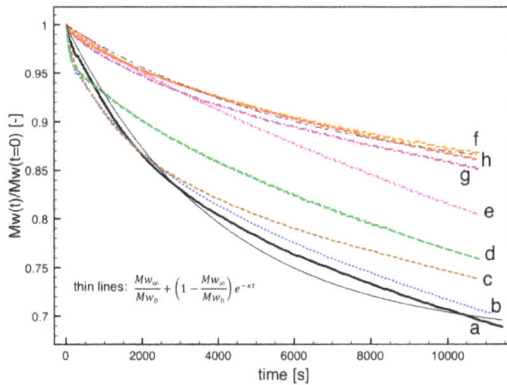

Figure 5. Time sweep-complex viscosity for the virgin PLA pellet (**a**) and for extruded samples after microcompounding: (**b**) neat and with (**c**) 0.1 wt % of LSAG; (**d**) 1 wt % of LSAG; (**e**) 0.1 wt % of HSAG; (**f**) 1 wt % of HSAG; (**g**) 0.1 wt % of CB; (**h**) 1 wt % of CB. Experimental conditions: $T = 200$ °C, $f = 1$ rad/s, plate–plate, gap = 200 μm.

The obtained results appear to be nearly independent of the surface area of carbon black. In fact, similar PLA melt stabilization was obtained using a different carbon black sample, exhibiting a definitely higher surface area (N110 with surface area of 150 m^2/g).

The observed polyester melt stabilization by HSAG and mainly by carbon black could be explained by the removal of traces of water from the melt, thus strongly reducing the hydrolysis of polyester bonds.

Indeed, on assuming that water is the only reason for molecular chain scission, namely assuming that hydrolysis is the only degradation mechanism (which is surely a simplification of more complex mechanisms taking place at high temperatures), one can relate the amount of water to the amount of carboxylic end groups according to the following equations:

$$\frac{dCa}{dt} = -\frac{dCc}{dt} \Rightarrow Ca = (Cc_0 + Ca_0) - Cc = Cc_\infty - Cc \tag{3}$$

in which Ca is the concentration of water inside the samples and Cc is the concentration of carboxylic end groups. The subscripts 0 and ∞ indicate the initial concentrations (at $t = 0$) and the final situation in which water molecules completely disappeared, respectively.

Considering that the concentration of carboxylic end groups is related to the concentration of polymeric chains

$$Cc = \frac{\rho}{Mn} \tag{4}$$

one simply obtains

$$Ca = \frac{\rho}{Mn_\infty}\left(1 - \frac{Mn_\infty}{Mn}\right) \tag{5}$$

and finally

$$Ca_0 = \frac{\rho}{Mn_\infty}\left(1 - \frac{Mn_\infty}{Mn_0}\right). \tag{6}$$

Assuming that, at least during early stages of degradation, the polydispersity index (Mw/Mn) can be considered to be constant, as also confirmed by the GPC data reported in Table 1, the data reported in Figure 5 can be considered to refer to Mn/Mn_0 during melt degradation. Those time evolutions can be fitted by a simple exponential curve, which can provide an estimate of the molecular weight at long times. For the samples a and h in Figure 5, the exponential fitting is reported. On knowing the density (1.2 g/cm^3) and the initial molecular weight from GPC data reported in Table 1, the initial water concentration is easily calculated. For the sample a (neat, virgin material), Ca_0 is 4.7 mol/m^3, corresponding to about 70 ppm. For the sample h (1 wt % of CB), Ca_0 is 2 mol/m^3, corresponding to about 30 ppm. The indication provided by the simplified model described above is that the amount of water taking part in the hydrolysis can be substantially reduced (of a factor 2 or more) by adding the fillers analyzed in this work.

This water absorption possibly occurs by specific interactions between water and oxidized groups on carbon surfaces [70].

3.3. Weight Loss in TGA Experiments of PLA in the Presence of Carbon Fillers

TGA scans of the extruded compounds indicate that PLA thermal stability, expressed in terms of weight loss, is slightly reduced by compounding with all the considered carbon fillers. Just as an example, TGA scans with a heating rate of 10 K/min for the PLA pellet, for the neat PLA-extruded sample, and for the extruded nanocomposites with 0.1 wt % of filler concentration are compared in Figure 6.

An increase in the decomposition temperature, e.g., corresponding to a 5% of weight loss ($T_{D,5\%}$), is observed going from the virgin ($T_{D,5\%} = 327$ °C) to the extruded neat ($T_{D,5\%} = 332$ °C) sample, although the latter has undergone a remarkable molecular weight reduction (GPC data of Table 1).

Moreover, compounding with all considered carbon fillers leads to small decreases of $T_{D,5\%}$, mainly for CB ($T_{D,5\%}$ = 328 °C).

The results of Section 3.2 clearly indicate that carbon fillers stabilize PLA toward degradation reactions occurring at temperatures close to 200 °C, not far from PLA's melting temperature, which leads to a reduction in the molecular masses, without any loss of volatile degradation products. The TGA results of this section indicate that the same carbon fillers, on the contrary, slightly destabilize PLA toward decomposition reactions occurring at temperatures higher than 300 °C, which leads to a loss of volatile byproducts.

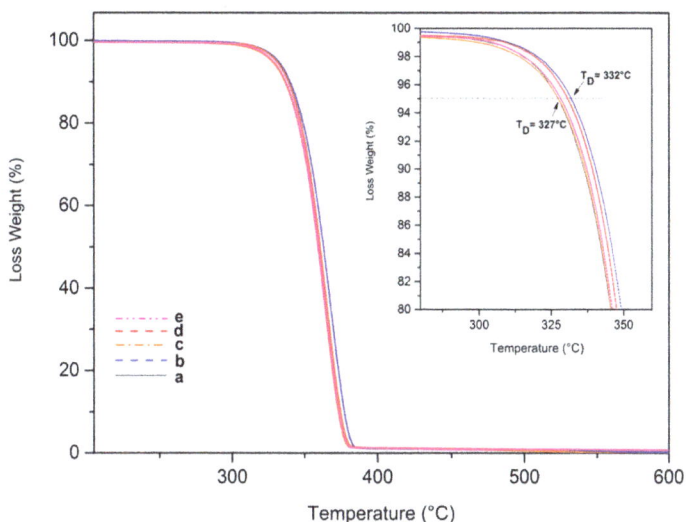

Figure 6. TGA scans for the virgin PLA pellet (**a**) and for extruded samples after microcompounding: (**b**) neat and with (**c**) 0.1 wt % of LSAG; (**d**) 0.1 wt % of HSAG; (**e**) 0.1 wt % of CB. Experimental conditions: a heating rate of 10 K/min under an N_2 atmosphere.

4. Conclusions

Different carbon fillers, specifically low- and high-surface-area graphite as well as carbon black, have been tested as possible stabilizers of PLA melt.

Melt viscosity measurements during extrusion processes and GPC experiments on the corresponding extruded samples show a remarkable PLA melt stabilization by all the considered nanofillers, at a temperature just above melting. In particular, melt stabilization leads to degradation reactions, leading to molecular weight reduction without any weight loss. For instance, for extrusions conducted at 200 °C, neat PLA exhibits a molecular weight reduction of about 25% while, by compounding with only 0.1 wt % of HSAG or of CB, PLA's molecular weight remains unaltered.

Time sweep-complex viscosity measurements, as carried out at 200 °C for about 3 h, confirm the ability of carbon nanofillers to stabilize PLA toward degradation reactions. In fact, PLA's molecular weight was reduced by about 30% in the virgin polymer (both neat and extruded), while it was reduced by amounts in the range 25–10% in the considered carbon composites. These measurements show that CB most effectively slowed down PLA degradation, for which a content of only 0.1 wt % leads to a molecular weight reduction close to 10%, much lower than that of the neat PLA (30%).

TGA analyses indicate that the considered carbon fillers, on the contrary, slightly destabilize PLA toward decomposition reactions, leading to a loss of volatile byproducts, which occur at temperatures higher than 300 °C, i.e., far from melt processing conditions.

The observed PLA stabilization by carbon fillers, at temperatures suitable for melt processing, can be explained by scavenging traces of water from the melt, which reduces the hydrolysis of polyester bonds.

In summary, PLA compounding with very small amounts (even 0.1 wt %) of HSAG and CB lead to remarkable PLA stabilization toward reactions, leading to molecular weight reduction. This contributes to the solution of the well-known problem of PLA degradation during processing. The same carbon nanofillers can also be effective in reducing degradation during processing for other polyesters that are sensitive to hydrolysis in the melt state.

Supplementary Materials: The following are available online at http://www.mdpi.com/2073-4360/10/2/139/s1, Figure S1: DSC heating scans of the extruded PLA and PLA compounded with 0.1wt % of HSAG, LSAG and CB.

Acknowledgments: We thank Mario Maggio of University of Salerno for useful discussions. Financial support of "Ministero dell' Istruzione, dell'Università e della Ricerca" is gratefully acknowledged.

Author Contributions: Gaetano Guerra and Roberto Pantani conceived and designed the experiments; Luciana D'Urso and Maria Rosaria Acocella prepared the fillers; Valentina Iozzino prepared the compounds; Luciana D'Urso and Valentina Iozzino performed the experiments; Maria Rosaria Acocella and Felice De Santis analyzed the data; Luciana D'Urso wrote the paper.

Conflicts of Interest: The authors declare no conflict of interest.

References

1. Degee, P.; Dubois, P.; Jerome, R. Bulk polymerization of lactides initiated by aluminium isopropoxide, 3. Thermal stability and viscoelastic properties. *Macromol. Chem. Phys.* **1997**, *198*, 1985–1995. [CrossRef]
2. Wachsen, O.; Platkowski, K.; Reichert, K.H. Thermal degradation of poly-L-lactide-studies on kinetics, modelling and melt stabilisation. *Polym. Degrad. Stab.* **1997**, *57*, 87–94. [CrossRef]
3. Signori, F.; Coltelli, M.B.; Bronco, S. Thermal degradation of poly(lactic acid) (PLA) and poly(butylene adipate-*co*-terephthalate) (PBAT) and their blends upon melt processing. *Polym. Degrad. Stab.* **2009**, *94*, 74–82. [CrossRef]
4. Carrasco, F.; Pagès, P.; Gámez-Pérez, J.; Santana, O.O.; Maspoch, M.L. Processing of poly(lactic acid): Characterization of chemical structure, thermal stability and mechanical properties. *Polym. Degrad. Stab.* **2010**, *95*, 116–125. [CrossRef]
5. Speranza, V.; De Meo, A.; Pantani, R. Thermal and hydrolytic degradation kinetics of PLA in the molten state. *Polym. Degrad. Stab.* **2014**, *100*, 37–41. [CrossRef]
6. Gruber, P.R.; Kolstad, J.J.; Ryan, C.M.; Hall, E.S.; Eichen, C.R.S. Melt-Stable Amorphous Lactide Polymer Film and Process for Manufacturing Thereof. U.S. Patent 5,484,881 A, 16 January 1996.
7. Amorin, N.S.Q.S.; Rosa, G.; Fernandes Alves, J.; Goncalves, S.P.C.; Franchetti, S.M.M.; Fechine, G.J.M. Study of thermodegradation and thermostabilization of poly(lactide acid) using subsequent extrusion cycles. *J. Appl. Polym. Sci.* **2014**, *131*, 40023. [CrossRef]
8. Cheung, M.F.; Carduner, K.R.; Golovoy, A.; Van Oene, H. Studies on the role of organophosphites in polyester blends: II. The inhibition of ester-exchange reactions. *J. Appl. Polym. Sci.* **1990**, *40*, 977–987. [CrossRef]
9. Cicero, J.A.; Dorgan, J.R.; Dec, S.F.; Knauss, D.M. Phosphite stabilization effects on two-step melt-spun fibers of polylactide. *Polym. Degrad. Stab.* **2002**, *78*, 95–105. [CrossRef]
10. Meng, X.; Shi, G.; Wu, C.; Chen, W.; Xin, Z.; Shi, Y.; Sheng, Y. Chain extension and oxidation stabilization of Triphenyl Phosphite (TPP) in PLA. *Polym. Degrad. Stab.* **2016**, *124*, 112–118. [CrossRef]
11. Han, T.; Xin, Z.; Shi, Y.; Zhao, S.; Meng, X.; Xu, H.; Zhou, S. Control of thermal degradation of poly(lactic acid) using functional polysilsesquioxane microspheres as chain extenders. *J. Appl. Polym. Sci.* **2015**, *132*, 41977. [CrossRef]
12. Yang, S.; Wu, Z.H.; Yang, W.; Yang, M.B. Thermal and mechanical properties of chemical crosslinked polylactide (PLA). *Polym. Test.* **2008**, *27*, 957–963. [CrossRef]
13. Yang, L.; Chen, X.; Jing, X. Stabilization of poly(lactic acid) by polycarbodiimide. *Polym. Degrad. Stab.* **2008**, *93*, 1923–1929. [CrossRef]

14. Bai, H.; Liu, H.; Bai, D.; Zhang, Q.; Wang, K.; Deng, H.; Chen, F.; Fu, Q. Enhancing the melt stability of polylactide stereocomplexes using a solid-state cross-linking strategy during a melt-blending process. *Polym. Chem.* **2014**, *5*, 5985–5993. [CrossRef]

15. Fan, Y.; Nishida, H.; Shirai, Y.; Endo, T. Thermal stability of poly(L-lactide): Influence of end protection by acetyl group. *Polym. Degrad. Stab.* **2004**, *84*, 143–149. [CrossRef]

16. Pluta, M.; Galeski, A.; Alexandre, M.; Paul, M.-A.; Dubois, P. Polylactide/montmorillonite nanocomposites and microcomposites prepared by melt blending: Structure and some physical properties. *J. Appl. Polym. Sci.* **2002**, *86*, 1497–1506. [CrossRef]

17. Paul, M.-A.; Alexandre, M.; Degee, P.; Henrist, C.; Rulmont, A.; Dubois, P. New nanocomposite materials based on plasticized poly(L-lactide) and organo-modified montmorillonites: Thermal and morphological study. *Polymer* **2003**, *44*, 443–450. [CrossRef]

18. Pluta, M. Melt compounding of polylactide/organoclay: Structure and properties of nanocomposites. *J. Polym. Sci. B* **2006**, *44*, 3392–3405. [CrossRef]

19. Wu, T.-M.; Wu, C.-Y. Biodegradable poly(lactic acid)/chitosan-modified montmorillonite nanocomposites: Preparation and characterization. *Polym. Degrad. Stab.* **2006**, *91*, 2198–2204. [CrossRef]

20. Iturrondobeitia, M.; Okariz, A.; Guraya, T.; Zaldua, A.M.; Ibarretxe, J. Influence of the processing parameters and composition on the thermal stability of PLA/nanoclay bio-nanocomposites. *J. Appl. Polym. Sci.* **2014**, *131*, 40747. [CrossRef]

21. Krishnaiah, P.; Ratnam, C.T.; Manickam, S. Development of silane grafted halloysite nanotube reinforced polylactide nanocomposites for the enhancement of mechanical, thermal and dynamic-mechanical properties. *Appl. Clay Sci.* **2017**, *135*, 583–595. [CrossRef]

22. Huang, J.-W.; Hung, Y.C.; Wen, Y.-L.; Kang, C.-C.; Yeh, M.-Y. Polylactide/nano and microscale silica composite films. I. Preparation and characterization. *J. Appl. Polym. Sci.* **2009**, *112*, 1688–1694. [CrossRef]

23. Wen, X.; Zhang, K.; Wang, Y.; Han, L.; Han, C.; Zhang, H.; Chen, S.; Dong, L. Study of the thermal stabilization mechanism of biodegradable poly(L-lactide)/silica nanocomposites. *Polym. Int.* **2011**, *60*, 202–210. [CrossRef]

24. Gordobi, O.; Delucis, R.; Egues, I.; Labidi, J. Kraft lignin as filler in PLA to improve ductility and thermal properties. *Ind. Crops Prod.* **2015**, *72*, 46–53. [CrossRef]

25. Tesfaye, M.; Patwa, R.; Gupta, A.; Kashyap, M.J.; Katiyar, V. Recycling of poly(lactic acid)/silk based bionanocomposites films and its influence on thermal stability, crystallization kinetics, solution and melt rheology. *Int. J. Biol. Macromol.* **2017**, *101*, 580–594. [CrossRef] [PubMed]

26. Lizundia, E.; Vilas, J.L.; León, L.M. Crystallization, structural relaxation and thermal degradation in Poly(L-lactide)/cellulose nanocrystal renewable nanocomposites. *Carbohydr. Polym.* **2015**, *123*, 256–265. [CrossRef] [PubMed]

27. Kim, I.H.; Jeong, Y.G. Polylactide/exfoliated graphite nanocomposites with enhanced thermal stability, mechanical modulus, and electrical conductivity. *J. Polym. Sci. B* **2010**, *48*, 850–858. [CrossRef]

28. Fukushima, K.; Murariu, M.; Camino, G.; Dubois, P. Effect of expanded graphite/layered-silicate clay on thermal, mechanical and fire retardant properties of poly(lactic acid). *Polym. Degrad. Stab.* **2010**, *95*, 1063–1076. [CrossRef]

29. Chen, Y.; Yao, X.; Zhou, X.; Pan, Z.; Gu, Q. Poly(lactic acid)/graphene nanocomposites prepared via solution blending using chloroform as a mutual solvent. *J. Nanosci. Nanotechnol.* **2011**, *11*, 7813–7819. [CrossRef] [PubMed]

30. Chieng, B.W.; Ibrahim, N.A.; Wan Yunus, W.M.Z. Optimization of Tensile Strength of Poly(Lactic Acid)/Graphene Nanocomposites Using Response Surface Methodology. *Polym.-Plast. Technol. Eng.* **2012**, *51*, 791–799. [CrossRef]

31. Mortazavi, B.; Hassouna, F.; Laachachi, A.; Rajabpour, A.; Ahzi, S.; Chapron, D.; Toniazzo, V.; Ruch, D. Experimental and multiscale modeling of thermal conductivity and elastic properties of PLA/expanded graphite polymer nanocomposites. *Thermochim. Acta* **2013**, *552*, 106–113. [CrossRef]

32. Li, X.; Xiao, Y.; Bergeret, A.; Longerey, M.; Che, J. Preparation of polylactide/graphene composites from liquid-phase exfoliated graphite sheets. *Polym. Compos.* **2014**, *35*, 396–403. [CrossRef]

33. Fu, Y.; Liu, L.; Zhang, J. Manipulating Dispersion and Distribution of Graphene in PLA through Novel Interface Engineering for Improved Conductive Properties. *ACS Appl. Mater. Interfaces* **2014**, *6*, 14069–14075. [CrossRef] [PubMed]

34. Fu, Y.; Liu, L.; Zhang, J.; Hiscox, W.C. Functionalized graphenes with polymer toughener as novel interface modifier for property-tailored polylactic acid/graphene nanocomposites. *Polymer* **2014**, *55*, 6381–6389. [CrossRef]

35. Mittal, V.; Chaudhry, A.U.; Luckachan, G.E. Biopolymer-Thermally reduced graphene nanocomposites: Structural characterization and properties. *Mater. Chem. Phys.* **2014**, *147*, 319–332. [CrossRef]

36. Norazlina, H.; Kamal, Y. Graphene modifications in polylactic acid nanocomposites: A review. *Polym. Bull.* **2015**, *72*, 931–961. [CrossRef]

37. Ibarra-Gomez, R.; Muller, R.; Bouquey, M.; Rondin, J.; Serra, C.A.; Hassouna, F.; El Mouedden, Y.; Toniazzo, V.; Ruch, D. Processing of nanocomposites PLA/graphite using a novel elongational mixing device. *Polym. Eng. Sci.* **2015**, *55*, 214–222. [CrossRef]

38. Kashi, S.; Gupta, R.K.; Baum, T.; Kao, N.; Bhattacharya, S.; Sati, N. Morphology, electromagnetic properties and electromagnetic interference shielding performance of poly lactide/graphene nanoplatelet nanocomposites. *Mater. Des.* **2016**, *95*, 119–126. [CrossRef]

39. Liu, C.; Ye, S.; Feng, J. Promoting the dispersion of graphene and crystallization of poly(lactic acid) with a freezing-dried graphene/PEG masterbatch. *Compos. Sci. Technol.* **2017**, *144*, 215–222. [CrossRef]

40. Scaffaro, R.; Botta, L.; Maio, A.; Gallo, G. PLA graphene nanoplatelets nanocomposites: Physical properties and release kinetics of an antimicrobial agent. *Compos. B* **2017**, *109*, 138–146. [CrossRef]

41. Prashantha, K.; Roger, F. Multifunctional properties of 3D printed poly(lactic acid)/graphene nanocomposites by fused deposition modeling. *J. Macromol. Sci. A* **2017**, *54*, 24–29. [CrossRef]

42. Yu, W.W.; Zhang, J.; Wu, J.R.; Wang, X.Z.; Deng, Y.H. Incorporation of graphitic nano-filler and poly(lactic acid) in fused deposition modeling. *J. Appl. Polym. Sci.* **2017**, *134*, 44703. [CrossRef]

43. Mngomezulu, M.E.; Luyt, A.S.; Chapple, S.A.; John, M.J. Effect of expandable graphite on thermal and flammability properties of poly(lactic acid)-starch/poly(ε-caprolactone) blend systems. *Polym. Eng. Sci.* **2017**. [CrossRef]

44. Botlhoko, O.J.; Ramontja, J.; Ray, S.S. Thermal, mechanical, and rheological properties of graphite- and graphene oxide-filled biodegradable polylactide/poly(ε-caprolactone) blend composites. *J. Appl. Polym. Sci.* **2017**, *134*. [CrossRef]

45. Wu, W.; Wu, C.-K.; Peng, H.; Sun, Q.; Zhou, L.; Zhuang, J.; Cao, X.; Roy, V.A.L.; Li, R.K.Y. Effect of nitrogen-doped graphene on morphology and properties of immiscible poly(butylene succinate)/polylactide blends. *Compos. B.* **2017**, *113*, 300–307. [CrossRef]

46. Goncalves, C.; Goncalves, I.C.; Magalhaes, F.D.; Pinto, A.M. Poly(lactic acid) composites containing carbon-based nanomaterials: A review. *Polymers* **2017**, *9*, 269. [CrossRef]

47. Yu, J.; Wang, N.; Ma, X. Fabrication and Characterization of Poly(lactic acid)/Acetyl Tributyl Citrate/Carbon Black as Conductive Polymer Composites. *Biomacromolecules* **2008**, *9*, 1050–1057. [CrossRef] [PubMed]

48. Wang, N.; Zhang, X.; Ma, X.; Fang, J. Influence of carbon black on the properties of plasticized poly(lactic acid) composites. *Polym. Degrad. Stab.* **2008**, *93*, 1044–1052. [CrossRef]

49. Wang, X.; Zhuang, Y.; Dong, L. Study of carbon black-filled poly(butylene succinate)/polylactide blend. *J. Appl. Polym. Sci.* **2012**, *126*, 1876–1884. [CrossRef]

50. Frackowiak, S.; Ludwiczak, J.; Leluk, K.; Orzechowski, K.; Kozlowski, M. Foamed poly(lactic acid) composites with carbonaceous fillers for electromagnetic shielding. *Mater. Des.* **2015**, *65*, 749–756. [CrossRef]

51. Lee, T.-W.; Jeong, Y.G. Enhanced electrical conductivity, mechanical modulus, and thermal stability of immiscible polylactide/polypropylene blends by the selective localization of multi-walled carbon nanotubes. *Compos. Sci. Technol.* **2014**, *103*, 78–84. [CrossRef]

52. Pluta, M.; Paul, M.-A.; Alexandre, M.; Dubois, P. Plasticized polylactide/clay nanocomposites. II. The effect of aging on structure and properties in relation to the filler content and the nature of its organo-modification. *Polym. Sci. Polym. Phys.* **2006**, *44*, 312–325. [CrossRef]

53. Zhou, Q.; Xanthos, M. Nanosize and microsize clay effects on the kinetics of the thermal degradation of polylactides. *Polym. Degrad. Stab.* **2009**, *94*, 327–338. [CrossRef]

54. Katiyar, V.; Gerds, N.; Koch, C.B.; Risbo, J.; Hansen, H.C.B.; Plackett, D. Melt processing of poly(L-lactic acid) in the presence of organomodified anionic or cationic clays. *J. Appl. Polym. Sci.* **2011**, *122*, 112–125. [CrossRef]

55. Carrasco, F.; Gamez-Perez, J.; Santana, O.O.; Maspoch, M.L. Processing of poly(lactic acid)/organo-montmorillonite nanocomposites: Microstructure, thermal stability and kinetics of the thermal decomposition. *Chem. Eng. J.* **2011**, *178*, 451–460. [CrossRef]

56. Gerds, N.; Katiyar, V.; Koch, C.B.; Hansen, H.C.B.; Plackett, D.; Larsen, E.H.; Risbo, J. Degradation of l-polylactide during melt processing with layered double hydroxides. *Polym. Degrad. Stab.* **2012**, *97*, 2002–2009. [CrossRef]

57. Androsch, R.; Di Lorenzo, M.L. Crystal nucleation in glassy poly(L-lactic acid). *Macromolecules* **2013**, *46*, 6048–6056. [CrossRef]

58. Jalali, A.; Huneault, M.A.; Elkoun, S. Effect of thermal history on nucleation and crystallization of poly(lactic acid). *J. Mater. Sci.* **2016**, *51*, 7768–7779. [CrossRef]

59. Volpe, V.; De Filitto, M.; Klofacova, V.; De Santis, F.; Pantani, R. Effect of mold opening on the properties of PLA samples obtained by foam injection molding. *Polym. Eng. Sci.* **2017**. [CrossRef]

60. Volpe, V.; Pantani, R. Effect of processing condition on properties of polylactic acid parts obtained by foam injection molding. *J. Cell. Plast.* **2016**, *53*, 491–502. [CrossRef]

61. Mauro, M.; Cipolletti, V.; Galimberti, M.; Longo, P.; Guerra, G. Chemically Reduced Graphite Oxide with Improved Shape Anisotropy. *J. Phys. Chem. C* **2012**, *116*, 24809–24813. [CrossRef]

62. Yous, M.; Alix, S.; Lebeau, M.; Soulestin, J.; Lacrampe, M.F.; Krawczak, P. Evaluation of rheological properties of non-Newtonian fluids in micro rheology compounder: Experimental procedures for a reliable polymer melt viscosity measurement. *Polym. Test.* **2014**, *40*, 207–217.

63. De Santis, F.; Pantani, R. Melt compounding of poly(lactic Acid) and talc: Assessment of material behavior during processing and resulting crystallization. *J. Polym. Res.* **2015**, *22*, 242. [CrossRef]

64. Moad, G.; Dagley, I.J.; Habsuda, J.; Garvey, C.J.; Li, G.; Nichols, L.; Simon, G.P.; Nobile, M.R. Aqueous hydrogen peroxide-induced degradation of polyolefins: A greener process for controlled-rheology polypropylene. *Polym. Degrad. Stab.* **2015**, *117*, 97–108. [CrossRef]

65. Lin, Q.; Li, T.; Liu, Z.; Song, Y.; He, L.; Hu, Z.; Guo, Q.; Ye, H. High-resolution TEM observations of isolated rhombohedral crystallites in graphite blocks. *Carbon* **2012**, *50*, 2347–2374. [CrossRef]

66. Freise, E.J.; Kelly, A. The deformation of graphite crystals and the production of the rhombohedral form. *Philos. Mag.* **1963**, *8*, 1519–1533. [CrossRef]

67. Li, Z.Q.; Lu, C.J.; Xia, Z.P.; Zhou, Y.; Luo, Z. X-ray diffraction patterns of graphite and turbostratic carbon. *Carbon* **2007**, *45*, 1686–1695. [CrossRef]

68. Maggio, M.; Acocella, M.R.; Guerra, G. Intercalation compounds of oxidized carbon black. *RSC Adv.* **2016**, *6*, 105565–105572. [CrossRef]

69. Gorrasi, G.; Pantani, R. Hydrolysis and biodegradation of Poly(lactic acid). *Adv. Polym. Sci.* **2017**, 1–33. [CrossRef]

70. Lopez-Ramon, M.V.; Stoeckli, F.; Moreno-Castilla, C.; Carrasco-Marın, F. Specific and non-specific interactions of water molecules with carbon surfaces from immersion calorimetry. *Carbon* **2000**, *38*, 825–829. [CrossRef]

polymers

MDPI

Article

Modelling of Rod-Like Fillers' Rotation and Translation near Two Growing Cells in Conductive Polymer Composite Foam Processing

Sai Wang [1], Amir Ameli [2], Vahid Shaayegan [1], Yasamin Kazemi [1], Yifeng Huang [3], Hani E. Naguib [4] and Chul B. Park [1,*]

[1] Microcellular Plastics Manufacturing Laboratory, Department of Mechanical and Industrial Engineering, University of Toronto, Toronto, ON M5S 3G8, Canada; swang@mie.utoronto.ca (S.W.); vahidsh@mie.utoronto.ca (V.S.); yasamin@mie.utoronto.ca (Y.K.)
[2] Advanced Composites Laboratory, School of Mechanical and Materials Engineering, Washington State University Tri-Cities, Richland, WA 99354, USA; a.ameli@wsu.edu
[3] Drinking Water Research Group, Department of Civil Engineering, University of Toronto, Toronto, ON M5B 1A4, Canada; yf.huang@utoronto.ca
[4] Smart and Adaptive Polymer Laboratory, Department of Mechanical and Industrial Engineering, University of Toronto, Toronto, ON M5S 3G8, Canada; naguib@mie.utoronto.ca
* Correspondence: park@mie.utoronto.ca; Tel.: +1-416-978-3053

Received: 11 January 2018; Accepted: 27 February 2018; Published: 2 March 2018

Abstract: We developed a simple analytical model to describe the instantaneous location and angle of rod-like conductive fillers as a function of cell growth during the foaming of conductive polymer composites (CPCs). First, we modelled the motion of the fillers that resulted from the growth of one cell. Then, by taking into account the fillers located at the line that connected the centres of the two growing cells, we found the final filler's angle and location. We identified this as a function of the corresponding cell size, filler size, and the filler's initial angle and location. We based the model's development on the assumption that a polymer melt is incompressible during cell growth. The two-cell growth model is better than the one-cell growth model because it describes the filler's movement in the cell wall between the two growing cells. The results revealed that the fillers near the cell were the ones most affected by the cell growth, while those at the midpoint between the two cells were the least affected. As a cell grows, its affected polymer area also increases. A dimensionless factor η was introduced to demonstrate the effects of the cell size and the filler length on the filler's interconnectivity in the CPC foams. It is vital to keep the filler length comparable to the cell size when preparing CPC foams with the desired electrical conductivity. Our research provides a deeper understanding of the mechanism through which foaming influences the filler connections in CPC foams.

Keywords: conductive filler; orientation; conductive polymer composites; foam; model

1. Introduction

In past decades, conductive polymer composites (CPCs) have been of great interest to those developing new generations of materials with unique properties and functions [1–4] and showed great potential for use in various applications such as high-dielectric materials for charge storage [5–8], electromagnetic interference (EMI) shielding [9–12], and bipolar plates of fuel cells [13–15]. The electrical conductivity can be easily established by adding micro- or nano-size conductive additives (that is, stainless-steel fiber [16], graphene [17–19], carbon fiber [20], carbon nanofiber [21–23], carbon black [4,24], and carbon nanotubes [25–27]) to the polymer matrix.

Efforts have been made to obtain a high electrical conductivity at low filler loadings [28–31]. Recent foaming technology, such as foam injection molding [32] and batch-type foaming [33], has shown promise in decreasing the electrical percolation threshold and increasing the CPC's conductivity. Foam injection molding can enhance the electrical conductivity in the through-plane direction because the preferentially aligned fillers, which have a machine-direction and/or in-plane orientations due to the shear force during the injection process, can be rotated toward the through-plane direction by the biaxial stretching in the polymer matrix caused by cell growth [34–36].

Unlike the foam injection molding, the initial filler distribution of the solid precursors in batch foaming can be controlled so as to be random. Previous research shows that cell growth may then slightly align the fillers, and thus may increase the possibility of filler connections, which ultimately will help to establish conductive networks [37,38]. However, Rizvi et al. [39] and Sun et al. [40] both found that the electrical conductivity of CPC foams was lower than their solid precursors. It is unknown whether these contradictory experimental results occurred because the optimal foaming and material conditions were only applied in some cases or because other factors, such as the polymer matrix's crystallinity, also influenced the conductivity [41]. Nevertheless, it remains clear that cellular growth, either in injection molding or in batch foaming, significantly affects the filler alignment in CPC foams. It is thus important to comprehend the correlations between the cellular growth and the filler movement. This will eventually make it possible to estimate the filler orientation's final state based on its initial orientation status and on the foaming process. This will result not only in a fundamental understanding of the foaming action effect on the electrical conductivity, but it will also help with the systematic optimization of the foaming processes, which will yield products with enhanced electrical properties.

Shaayegan et al. [20] developed a visualization system to observe the instantaneous fiber movement (i.e., rotation and translation) as a function of the corresponding cell size and of the fiber's initial orientation and location. However, their work was limited to the growth of one cell. The effect of multi-cell growth on the filler's motion has more importance in terms of practical applications.

Therefore, in our study, we introduced an analytical model to describe the filler's rotation and translation during cellular growth in polymer composite foams. Using this model, we have attempted to elucidate the effects of multiple factors, including the cell size, the filler length, and the filler's relative position with respect to the cell nucleus, on the displacement and re-orientation of conductive fillers. By analyzing the rotation and translation of the rod-like fillers in the cell walls during cell growth, we were able to better understand how cellular growth affects the filler's angle and location in real CPC foam. This would enable us to demonstrate the feasibility of foaming as a potential strategy to decrease the percolation threshold in CPCs.

2. Theoretical Estimation

This paper is about cell growth and how its deterministic modelling affects the movements of various rod-like fillers. First, we focused on how any filler can be moved by any one cell located close to it. We described the filler's movements as a combination of (i) the filler's rotation and (ii) the translation of the filler's centre point by the growing cell. Then, we extended this singular filler/cell case to a slightly more complicated case of one filler and two cells. We asked how one filler could be moved by two growing cells located close to it. This example showed us how the filler would move if there were multiple cells present. This approach provided us a clear insight into the filler's movement in relation to the cells growing near it.

Then, we extended the example of a single filler's movement to the movement of multiple fillers. Since the filler movement is independent, the one-filler's movement can be extended to the multiple-fillers' movement independently with respect to each other. Using this approach, we could accurately trace how all of the fillers were moving in the polymer matrix while the cells were growing.

2.1. One-Cell Model Approach

To model the movement of rod-like fillers during foaming, we first considered a simple case of one-cell growth. We made several assumptions prior to developing the model. In the case of one-cell growth, the following conditions apply: (1) The radial growth of the cell is symmetric; (2) The fillers are rigid and do not slip over the adjacent polymer matrix; (3) Upon cell growth, the filler only changes with respect to the line that passes through the cell centre and the filler midpoint; (4) The growth of a cell forces the filler to translate in the radial direction of the cell centre to the filler midpoint, and the filler rotates around its midpoint; (5) The filler does not interfere with the adjacent filler; (6) The fillers cannot penetrate into the cells and are always located in the polymer matrix; (7) The cell structure is a closed cell and no open cell is considered; (8) The polymer matrix is incompressible.

As Figure 1 shows, a set of two parameters was used to describe the fillers' locations and orientations as well as their changes during cellular growth. One parameter was the "filler location", which was defined as the radial distance of the filler midpoint, with respect to the corresponding cell centre (R_0 for the initial location in Figure 1a,c and R for instantaneous locations in Figure 1b,d). The second parameter was the "filler angle", which was defined as the angle of the rod-like filler axis with respect to the line passing through its midpoint and through the cell centre (α_0 for initial angle in Figure 1a,c and α for the instantaneous angle in Figure 1b,d).

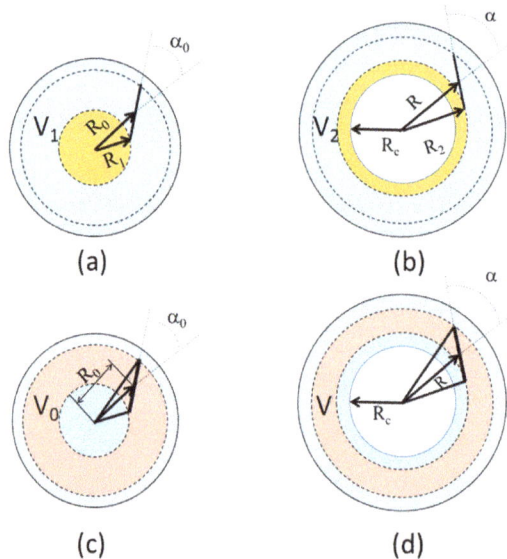

Figure 1. Two-dimensional (2-D) schematic illustration of filler alignment: (**a,c**) the initial state with an initial filler angle of α_0 and an initial filler location of R_0; (**b,d**) an instantaneous state with a filler angle α and a filler location R after expansion with a cell radius R_c. White area in the sphere represents the cell. Initial and instantaneous volume of incompressible polymer sphere is used to calculate the changes in "filler angle" and "filler location".

Due to the polymer's incompressibility, the initial volume, V_1, of a sphere of a radius R_1 (Figure 1a) remains unchanged during cellular growth. As shown in Figure 1a, the R_1 can be determined as follows:

$$V_1 = \frac{4}{3}\pi R_1{}^3 \tag{1}$$

and from trigonometry,

$$R_1^2 = R_0^2 + (\frac{l}{2})^2 - 2 \cdot R_0 \cdot \frac{l}{2} \cos \alpha_0 \tag{2}$$

where l is the length of the filler. The V_1 can then be written as follows:

$$V_1 = \frac{4}{3} \pi \left(R_0^2 + \frac{l^2}{4} - R_0 \cdot l \cdot \cos \alpha_0 \right)^{\frac{3}{2}} \tag{3}$$

After expansion, the polymer volume of the V_1 is transferred to a spherical ring with inner and outer radii of R_c and R_2, having a volume of V_2, as shown in Figure 1b, where R_c is the cell radius at any time. The V_2 can then be written as follows:

$$V_2 = \frac{4}{3} \pi R_2{}^3 - \frac{4}{3} \pi R_c{}^3 \tag{4}$$

and from trigonometry,

$$R_2^2 = R^2 + (\frac{l}{2})^2 - 2 \cdot R \cdot \frac{l}{2} \cos \alpha \tag{5}$$

Similar to the V_1, the V_2 can be written as follows:

$$V_2 = \frac{4}{3} \pi \left[\left(R^2 + \frac{l^2}{4} - R \cdot l \cdot \cos \alpha \right)^{\frac{3}{2}} - R_c{}^3 \right] \tag{6}$$

Furthermore, the volume of any spherical envelope surrounding the filler is invariable during the expansion phase because the expansion is symmetrical. This is schematically shown in Figure 1c,d, the volume of the initial and instantaneous spherical rings (envelopes) were V_0 and V, respectively. Therefore, we can write $V_0 = V$. Using trigonometric relations, they can be calculated as a function of the radii R_0 and R, respectively, as follows:

$$V_0 = \frac{4}{3} \pi \left\{ \left[R_0{}^2 + \left(\frac{l}{2} \right)^2 - 2R_0 \cdot \frac{l}{2} \cos(\pi - \alpha_0) \right]^{\frac{3}{2}} - \left[R_0{}^2 + \left(\frac{l}{2} \right)^2 - 2R_0 \cdot \frac{l}{2} \cos \alpha_0 \right]^{\frac{3}{2}} \right\} \tag{7}$$

$$V = \frac{4}{3} \pi \left\{ \left[R^2 + \left(\frac{l}{2} \right)^2 - 2R \cdot \frac{l}{2} \cos(\pi - \alpha) \right]^{\frac{3}{2}} - \left[R^2 + \left(\frac{l}{2} \right)^2 - 2R \cdot \frac{l}{2} \cos \alpha \right]^{\frac{3}{2}} \right\} \tag{8}$$

An equation set, which correlates the filler's location and its angle, before and after cellular growth, can then be written as follows:

$$V_1 = V_2 \tag{9}$$

$$V_0 = V \tag{10}$$

By solving this equation set, the instantaneous angle and the location of any filler can be expressed as a function of its initial location, initial angle, cell size, and filler length.

2.2. Two-Cell model Approach

In a real situation, the rod-like filler will not be affected by only one cell. Two or even more growing cells may co-influence its location and angle. Therefore, we attempted to model the filler's motion under the growth of two cells. To simplify the model, the fillers were assumed to be located at the centre line of the two cells. Even though this was a specialized general filler arrangement, it could still provide a straightforward way to analyze the co-influences of the multiple growing cells on the filler's movement. In addition to the assumptions made in our model of one-cell growth, additional assumptions were made in the two-cell case: (1) The face-centred cubic (FCC) close pack model (Figure 2a) was used to describe the cell distribution in the polymer matrix; (2) In this model, the shortest centre-to-centre distance for two adjacent cells corresponded to the distance between the cell in the cube's centre and the cell on the cube's vertex that is, the $2L_t$ in Figure 2a, where the L_t was half of the distance between two cell centres; (3) Only fillers located on the centreline of these two closest cells were considered; (4) The filler movement was assumed to be affected by only two bubbles; (5) All of the cells grew simultaneously and were the same size at any time; (6) Once the cells had nucleated, all of the cell nucleation sites were fixed, and the cells grew without any relative displacement during the entire foaming process.

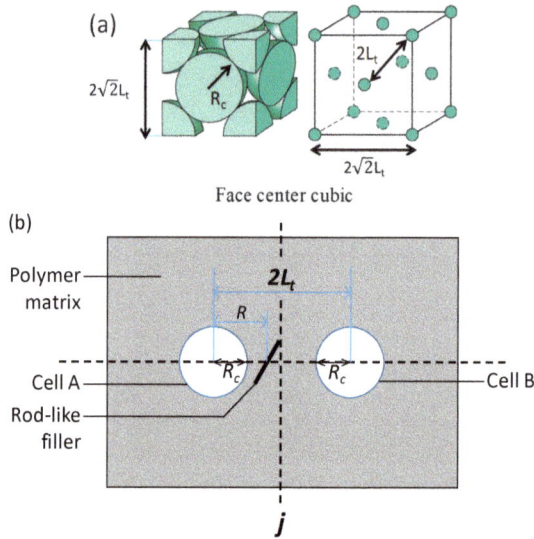

Face center cubic

Figure 2. Illustrations of (**a**) the face centred cubic close pack model and (**b**) the rod-like fillers located on the centreline of two closest cells.

In the FCC close pack model, the cell density can be expressed as follows:

$$\frac{4}{\left(2\sqrt{2}\cdot L_0\right)^3} = D \tag{11}$$

where $2L_0$ is the initial distance between the two closest cell nucleation sites at the beginning of the cellular growth, and D is the cell density. In one cube, the void fraction, β, can be expressed by Equation (12) as follows:

$$\frac{4}{3} \times \pi \times R_c^3 \times D = \beta \tag{12}$$

From Equations (11) and (12), we can express R_c as a function of L_0 and β:

$$R_c = L_0 \times \left(\frac{3\sqrt{2}\beta}{\pi}\right)^{\frac{1}{3}} \tag{13}$$

Since a two-cell growth symmetrical effect was expected, we used a different coordinate. This is shown in Figure 2b, where the origin point is the middle point of the two cell centres. Thus, the negative and positive values of the filler's location R represent the fillers close to Cell A and the fillers close to Cell B, respectively. At a specific cell density, the initial distance of the two neighboring cells can be calculated using Equation (11). Substituting the solved R_c with the known L_0 into Equations (9) and (10), we can then obtain the instantaneous angle α and the location R of any filler from the growth of Cell A. To define the mutual and related effects of the growth of the two cells, we had to consider the second Cell B's influence. We used a two-step cellular growth procedure to simulate the growth of two cells. As shown in Figure 3, in the polymer matrix around Cells A and B, at first Cell A started to grow to a radius of R_c. Then, Cell B began to grow based on Cell A's growth. Finally, we obtain the effects of the growth of two cells on filler movement. If we divide each step of cell growth into thousand sub-steps (that is, by dividing the growth of the R_c into thousand parts), we would confidently neglect the error found in the two-step method and obtain the pseudo-simultaneous growth of two cells. Using this strategy, we were able to estimate the functions between the final parameters and the initial status for two-cell growth. We used MATLAB (The MathWorks, Inc., Natick, MA, USA) to conduct all the calculations in this study.

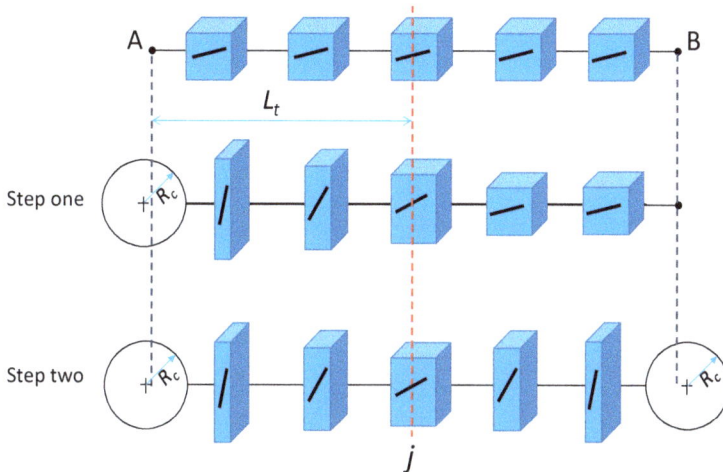

Figure 3. Illustration of two-step consecutive growth of Cells A and B used in modelling the growth of two cells.

3. Results and Discussion

3.1. Single Cell-Filler Interaction

Cellular growth in the foaming process can make the rod-like filler rotate and translate. The closer a filler is to the centre of a growing cell, the greater an impact it experiences. To quantitatively characterize the filler motion during cellular growth, we modelled the rotation and translation of one filler located around a single growing cell. Figure 4a shows the variation of the final radial angle, the α of a filler, with respect to its final location at various cell radii R_c; that is, assuming a cell is growing near a rod-like filler with a length of 260 nm and an initial angle of $10°$. R_c ranges from 0 to 260 nm. For a filler at a given distance from the cell nucleus centre, the changes in its angle and location are increased by an increased cell radius. For example, the filler initially located 50 nm away from the cell centre will move to 57.6 nm and will rotate from $10°$ to $33.3°$ when the cell radius is 52 nm. If the cell continues to grow up to 260 nm, the filler will then move 261 nm away from the cell centre and will rotate to $84°$. In addition, as cell radius increases, the area affected by cell growth, as well as the fillers located in it, will be increased. It is obvious that, with a 52 nm radius, a cell can, at best, affect an area of around 300 nm, while a cell with a 260 nm radius can affect fillers located as far as 1000 nm away.

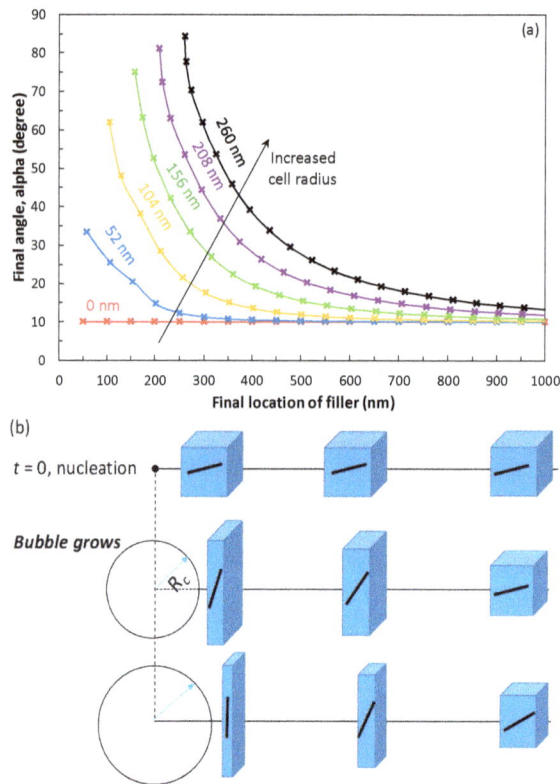

Figure 4. The effect of single cell growth on the filler orientation, expressed as (**a**) final angle, α vs. final location of filler; (**b**) schematic illustration of the filler's rotational and translational displacements. Initial conditions: $l = 260$ nm, $\alpha_0 = 10°$.

At a given cell radius, the impact of cellular growth on fiber motion deteriorates as the filler locates itself further away from the cell centre. The change in the final angle of the filler is decreased as the radial distance of the filler from the cell nucleus is increased. For example, in a 156 nm cell, the angle of the filler that was initially 0.4 nm away from the cell nucleus changed from 10° to 75°, but the filler that was initially 194 nm away from the cell only rotated to 20°. The same trend occurred in the translation of the filler. Both the rotation and translation trends of the filler indicated that the polymer domain closer to the cell had been more radially squeezed and bi-axially stretched. In Figure 4b, the schematic diagram shows the cell-growth-induced stretch or squeeze effect on the polymer matrix and how the filler rotation and translation was affected as a consequence of a single cell's growth.

3.2. Two Cell-Filler Interactions

In a specific CPC foam with a fixed cell size, the impact of one single cell on the filler displacement is restricted to a limited area. The cell has little influence on the fillers that are far from it. In fact, cells will grow simultaneously in the polymer matrix at the locations of the cell nuclei. Thus, the fillers are also simultaneously affected by the growth of multiple cells. This makes it necessary to model the displacement of the fillers driven by the growth of multiple cells. Therefore, we also modelled the growth of two cells and then attempted to use the result to interpret the filler's behaviour in the CPC foams during the cellular growth.

The motion of the rod-like fillers became far more complicated in relation to the growth of the two cells. To simplify the model, we studied only the fillers located on the line that connected the centres of the two cells. We also examined the fixed initial cell nucleation density from which the cell started to grow in each case. This was discussed later in this paper. We also studied the effects of several other factors; namely, the void fraction, the filler's length, and the initial filler location and angle on its own movement during the cell's growth.

3.2.1. Effect of Initial Filler Location and Void Fraction

First, we modelled the final angle of the filler as a function of the final filler's location at different void fractions, when the cell density was 2.54×10^{14} cells/cm^3. We chose this number because a previous work [42] had shown that the optimal percolation threshold of the CPC foams' conductivity occurred when the cell size was ~100 nm, cell density was ~2.54×10^{14} cells/cm^3, and the void fraction was ~30%. Thus, it was particularly important to investigate the filler motion at that cell density. At a fixed cell density, the cell radius is fixed once the void fraction is determined. Therefore, like the one-cell growth case, the two-cell-filler interaction of the cell radii at 0, 47, 62, 74, 86, and 98 nm, which corresponded to the void fractions at 0, 0.1, 0.2, 0.3, 0.4, and 0.5, respectively, were simulated using the proposed two-cell growth model. The filler length *l* and the initial filler angle α_0 were set at 260 nm and 10°, respectively, so as to be comparable with the one-cell growth case.

The results in Figure 5 show the effects of the two-cell growth on the filler's rotation and translation. At the beginning of the cellular growth, along the line connecting the two cell centres, the final filler angle decreased as the filler distance increased from one cell centre to the other. This reached its lowest value at the midpoint of the line connecting the two cell centres. Then, it increased, as the filler location approached the second cell's centre, and it presented a symmetrical pattern. This was logical because the two cells on either side of the filler had equally influenced the filler's motion. The filler at the axis of the symmetry had the lowest final angle because it was located the farthest away from both cells. The translation of the filler showed the same symmetrically changing trend as the filler rotation with respect to the distance of the filler from the cell centres. At the midpoint, the filler did not move at all. This was because it experienced squeezing pressures that were of the same magnitude, but which came from opposite directions; thus, the net force was zero. As the filler distance decreased in relation to either cell centre, it exhibited a greater degree of displacement.

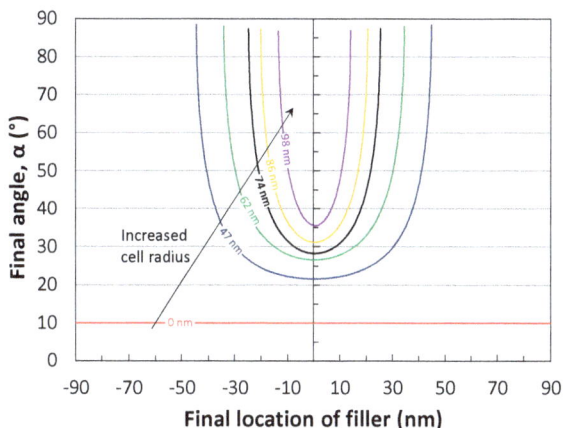

Figure 5. Effect of two-cell growth on the final angle of filler located in the cell centreline. Cell growth is simulated using the two-step consecutive growth of cells (Figure 4). Final angle vs. final location of filler at five different expansion ratios (i.e., five cell radii) was illustrated. *l* = 260 nm, α_0 = 10°, and *D* = 2.54×10^{14} cells/cm^3, cell size increases from 0 to 98 nm (void fraction increases from 0 to 0.5).

As the void fraction increases (via an increase in the cell radius R_c), the polymer between the two cells is radially squeezed and bi-axially stretched. Consequently, the forces exerted on the fillers in between will increase, and this will increase both the filler's rotation and its translation. For example, at the midpoint, the filler's angle increased from the initial 10° to about 35° when the cell grew to 98 nm, but only to 21° when the cell grew to 47 nm. With cell growth, the fillers that were initially evenly distributed between the cell nucleation centres were then squeezed into the narrowed areas in between two cells.

3.2.2. Effect of the Initial Filler Angle

To better understand the filler's rotation during cell growth, its rotation at a different initial filler angle α_0 at a given void fraction was also modelled and is shown in Figure 6. The α_0 varies from 10° to 90°. The fillers became harder to rotate as their initial angles were increased. In other words, the fillers with smaller angles were more sensitive to the cellular growth. If the filler is initially perpendicular to the cell centre-centre line (i.e., $\alpha_0 = 90°$), it cannot be rotated at all. This indicates that the fillers randomly distributed in the polymers will be aligned so as to be tangential to the cellular radii. For practical purposes, the alignment of the fillers must be controlled to a certain extent, so as to keep the fillers in contact with each other as much as possible.

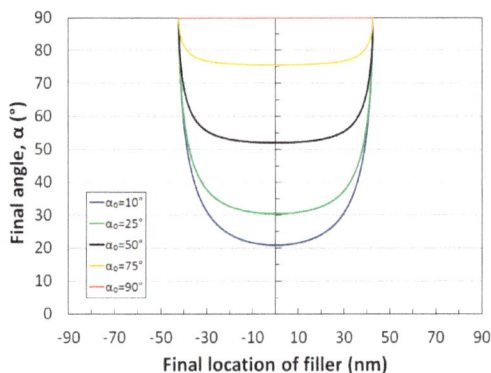

Figure 6. The rotation of filler at different initial angle during the two-cell growth. l = 260 nm, β = 0.1, $D = 2.54 \times 10^{14}$ cells/cm³, and R_c = 47 nm.

3.2.3. Effect of the Relative Size of the Filler and the Cell

The cell size could be different under the same void fraction (that is, the cell density can be different). This means that the electrical conductivity or the filler's interconnectivity might not be increased, even with a proper void fraction if the cell size, relative to the filler's length, was not also optimal. On the other hand, the filler's length also plays a critical role in its re-orientation during cellular growth. A shorter filler relative to the cell size is easier to rotate. Therefore, it was also important to investigate the effect of the relative size of the filler and the cell on the filler's motion.

To quantify the effect of the relative size of the filler and the cell, we simulated the filler rotation under different filler lengths in two representative cell structures (that is, nano- and micro-cellular foams). As shown in Figure 7a, in the nanocellular foam, when the filler length was smaller than the cell size or just comparable to it (that is, equal to or 2-fold larger), the filler located around the midpoint, where there was the least rotation, had rotated from 10° to about 35°. This showed that a short filler was easier to rotate. The filler rotation was more difficult when the filler length was increased. When the filler length was about 60 times larger than the cell radius, the fillers in between barely rotated, except those located very close to the cells. A similar effect of the filler's length on its rotation in microcellular foams is shown in Figure 7b. These results indicate that an appropriate filler length relative to the cell size is vitally important in preparing CPC foams with the desired electrical conductivity.

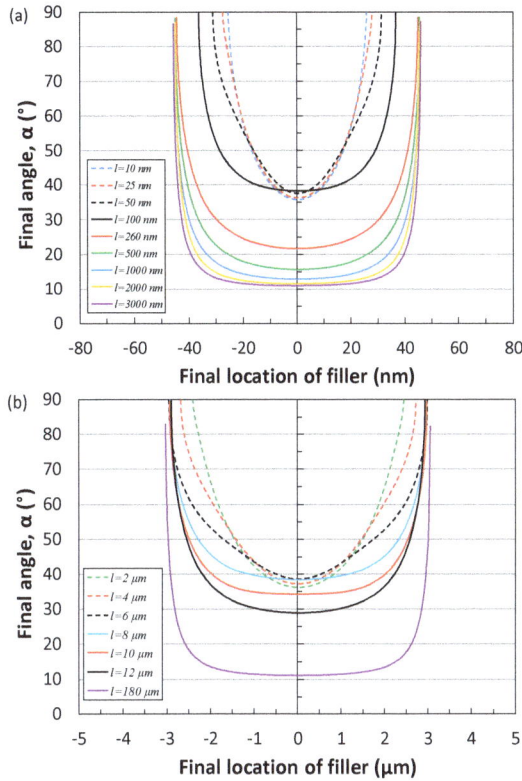

Figure 7. Effect of filler length and cell size on rotation of filler when the initial angle is 10°. (**a**) Nanocellular, $\beta = 0.1$, $D = 2.54 \times 10^{14}$ cells/cm^3, $R_c = 47$ nm; (**b**) microcellular, $\beta = 0.1$, $D = 9 \times 10^9$ cells/cm^3, $R_c = 3$ μm.

3.3. Illustration of the Two-Cell Growth

The effect of the two cells' growth on the filler's orientation and location is shown in a schematic diagram in Figure 8. At $t = 0$, when the bubbles were under the nuclei, all the fillers at different locations on the line connecting the two-cell centres were assumed to have the same initial angle. Once the bubbles started to grow, the polymer near the cells was efficiently bi-axially stretched. As a result, the filler rotated toward the same direction, which, in this case, was counter-clockwise. As the void fraction β increased, the filler continuously rotated, and the filler to filler distance was squeezed. Additionally, the fillers near the cells tended to become perpendicular to the cell-to-cell line. Moreover, at a fixed void fraction, the polymer close to the cells had more deformation than the polymers around the midpoint. Hence, the fillers in the polymer matrix next to the cells had more movement (both rotation and translation) than the fillers in the polymers close to the midpoint.

We only considered the fillers located in between the two cells. This was because each cell would only affect the fillers located close to it, but would not affect the fillers located further away. Hence, Cell A, as shown in Figure 8, would not affect the movement of the fillers located on the right side of Cell B. The fillers located on the right side of Cell B, rather than being located in between the two cells, were too far away from Cell A to receive any significant impact from it. In Figure 9, the one-cell case was compared with the two-cell case, where the fillers were on the right side of Cell B. The filler movement curve in the one-cell case almost completely overlapped the same curve in the two-cell case. This indicated that Cell A had not affected the filler rotation and translation at all. This observation theoretically supported the idea that the current two-cell growth model could also be used to describe the multi-cell growth system in all CPC foams.

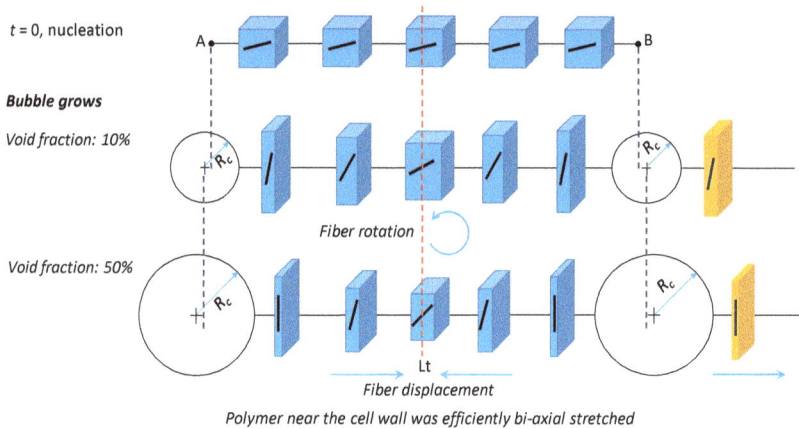

Figure 8. Schematic illustration of how the growth of two cells influence the fiber orientation and location. Fillers located in between the two cells and on the other side of one cell are both shown.

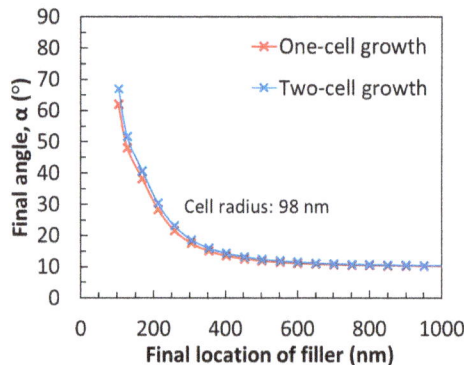

Figure 9. Filler movement under one-cell growth and two-cell growth with fillers located on the other side of cells rather than in between the two cells (illustrated in Figure 8).

3.4. Experimental Verification

We compared our theoretically predicted results with our earlier experimentally observed PP/MWCNT data [42]. Our previous experiments showed that, at a relative density of around 0.7 (that is, a void fraction of 0.3), the PP/MWCNT nanocomposite foam's electrical conductivity was enhanced and was even higher than in the solid sample (Figure 10) for all of the CNT content. Du et al. [43] and Gong et al. [44] both proposed that this conductivity enhancement (that is, a decreased percolation threshold) resulted from the slight CNT alignment. Chang et al. [45] also obtained similar results using a Monte Carlo simulation to estimate the percolation threshold of fillers with different filler orientations. It was found that the percolation threshold was minimized when the fillers were slightly oriented (both uniaxially and biaxially) rather than being completely isotropic. This corresponded with the modelling results in our study, which had shown that the conductive fillers were re-oriented during cell growth and that a relatively low void fraction caused a slight filler orientation and relocation. We thus hypothesized that the slight orientation of the conductive fillers at a proper void fraction in CPC foams can increase their conductivity and decrease their percolation threshold. More research is required to relate the filler interconnectivity (that is, the electrical conductivity) to foaming actions.

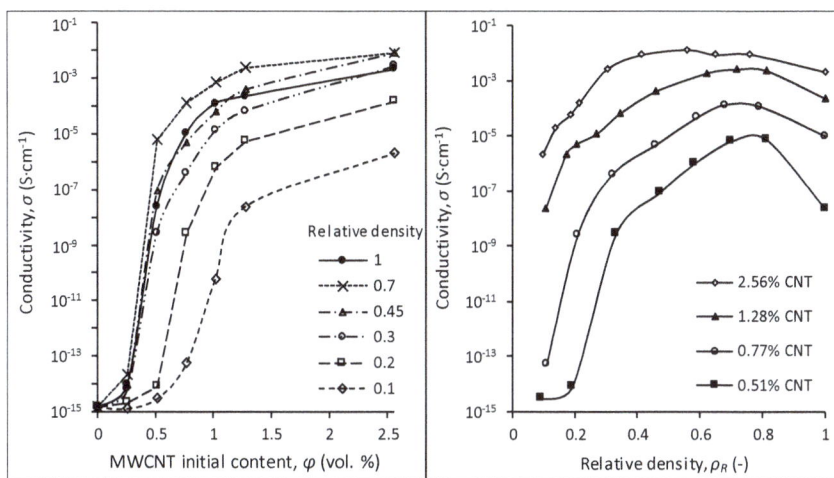

Figure 10. Electrical conductivity of PP/MWCNT nano/microcomposite foams as a function of (**a**) different MWCNT initial content and (**b**) relative density [42].

4. Conclusions

There is an absence of any quantitative analysis of the rotation and translation of conductive fillers due to cell growth in CPC foams. This includes those factors that would influence the filler's orientation and location. Although our modelling is simple and many assumptions are made, this study is a first attempt to initiate a discussion about this knowledge gap. The analytical model we developed successfully describes the instantaneous rod-like filler rotation and translation during the cellular growth in CPC foams. Our modelling results show that, as a cell grows, its affected polymer area also grows. The filler located in the polymer thus has more movement. Besides the void fraction, the relative size of the filler and cell, as well as the initial filler location and angle, has a substantial impact on the filler's orientation and translation.

Author Contributions: Sai Wang and Amir Ameli conceived and designed the experiments; Sai Wang performed the experiments and wrote the paper; Vahid Shaayegan, Yasamin Kazemi, and Yifeng Huang analyzed the data; Hani E. Naguib and Chul B. Park supervised the research project and gave input on the content and structure of the manscript.

Conflicts of Interest: The authors declare no conflict of interest.

References

1. Pang, H.; Xu, L.; Yan, D.-X.; Li, Z.-M. Conductive polymer composites with segregated structures. *Prog. Polym. Sci.* **2014**, *39*, 1908–1933. [CrossRef]
2. Stankovich, S.; Dikin, D.A.; Dommett, G.H.B.; Kohlhaas, K.M.; Zimney, E.J.; Stach, E.A.; Piner, R.D.; Nguyen, S.B.T.; Ruoff, R.S. Graphene-based composite materials. *Nature* **2006**, *442*, 282–286. [CrossRef] [PubMed]

3. Katsura, T.; Kamal, M.R.; Utracki, L.A. Electrical and thermal properties of polypropylene filled with steel fibers. *Adv. Polym. Technol.* **1985**, *5*, 193–202. [CrossRef]

4. Zhang, W.; Dehghani-Sanij, A.A.; Blackburn, R.S. Carbon based conductive polymer composites. *J. Mater. Sci.* **2007**, *42*, 3408–3418. [CrossRef]

5. Liu, H.; Shen, Y.; Song, Y.; Nan, C.-W.; Lin, Y.; Yang, X. Carbon nanotube array/polymer core/shell structured composites with high dielectric permittivity, low dielectric loss, and large energy density. *Adv. Mater.* **2011**, *23*, 5104–5108. [CrossRef] [PubMed]

6. Dang, Z.-M.; Zheng, M.-S.; Zha, J.-W. 1d/2d carbon nanomaterial-polymer dielectric composites with high permittivity for power energy storage applications. *Small* **2016**, *12*, 1688–1701. [CrossRef] [PubMed]

7. Song, S.; Zhai, Y.; Zhang, Y. Bioinspired graphene oxide/polymer nanocomposite paper with high strength, toughness, and dielectric constant. *ACS Appl. Mater. Interfaces* **2016**, *8*, 31264–31272. [CrossRef] [PubMed]

8. Tang, H.; Wang, P.; Zheng, P.; Liu, X. Core-shell structured batio3@polymer hybrid nanofiller for poly(arylene ether nitrile) nanocomposites with enhanced dielectric properties and high thermal stability. *Compos. Sci. Technol.* **2016**, *123*, 134–142. [CrossRef]

9. Zhao, B.; Zhao, C.; Li, R.; Hamidinejad, S.M.; Park, C.B. Flexible, ultrathin, and high-efficiency electromagnetic shielding properties of poly(vinylidene fluoride)/carbon composite films. *ACS Appl. Mater. Interfaces* **2017**, *9*, 20873–20884. [CrossRef] [PubMed]

10. Jia, L.-C.; Yan, D.-X.; Yang, Y.; Zhou, D.; Cui, C.-H.; Bianco, E.; Lou, J.; Vajtai, R.; Li, B.; Ajayan, P.M.; et al. High strain tolerant EMI shielding using carbon nanotube network stabilized rubber composite. *Adv. Mater. Technol.* **2017**, *2*, 1700078. [CrossRef]

11. Ren, F.; Shi, Y.; Ren, P.; Si, X.; Wang, H. Cyanate ester resin filled with graphene nanosheets and NiFe$_2$O$_4$–reduced graphene oxide nanohybrids for efficient electromagnetic interference shielding. *Nano* **2017**, *12*, 1750066. [CrossRef]

12. Lyu, J.; Zhao, X.; Hou, X.; Zhang, Y.; Li, T.; Yan, Y. Electromagnetic interference shielding based on a high strength polyaniline-aramid nanocomposite. *Compos. Sci. Technol.* **2017**, *149*, 159–165. [CrossRef]

13. Adloo, A.; Sadeghi, M.; Masoomi, M.; Pazhooh, H.N. High performance polymeric bipolar plate based on polypropylene/graphite/graphene/nano-carbon black composites for pem fuel cells. *Renew. Energy* **2016**, *99*, 867–874. [CrossRef]

14. Iwan, A.; Malinowski, M.; Pasciak, G. Polymer fuel cell components modified by graphene: Electrodes, electrolytes and bipolar plates. *Renew. Sust. Energy Rev.* **2015**, *49*, 954–967. [CrossRef]

15. Lopes De Oliveira, M.C.; Sayeg, I.J.; Ett, G.; Antunes, R.A. Corrosion behavior of polyphenylene sulfide-carbon black-graphite composites for bipolar plates of polymer electrolyte membrane fuel cells. *Int. J. Hydrogen Energy* **2014**, *39*, 16405–16418. [CrossRef]

16. Ameli, A.; Nofar, M.; Wang, S.; Park, C.B. Lightweight polypropylene/stainless-steel fiber composite foams with low percolation for efficient electromagnetic interference shielding. *ACS Appl. Mater. Interfaces* **2014**, *6*, 11091–11100. [CrossRef] [PubMed]

17. Ansari, S.; Giannelis, E.P. Functionalized graphene sheet-poly(vinylidene fluoride) conductive nanocomposites. *J. Polym. Sci. Pol. Phys.* **2009**, *47*, 888–897. [CrossRef]

18. Chen, M.; Duan, S.; Zhang, L.; Wang, Z.; Li, C. Three-dimensional porous stretchable and conductive polymer composites based on graphene networks grown by chemical vapour deposition and pedot: Pss coating. *Chem. Commun.* **2015**, *51*, 3169–3172. [CrossRef] [PubMed]

19. Noël, A.; Faucheu, J.; Rieu, M.; Viricelle, J.-P.; Bourgeat-Lami, E. Tunable architecture for flexible and highly conductive graphene-polymer composites. *Comp. Sci. Technol.* **2014**, *95*, 82–88. [CrossRef]

20. Shaayegan, V.; Ameli, A.; Wang, S.; Park, C.B. Experimental observation and modeling of fiber rotation and translation during foam injection molding of polymer composites. *Compos. Part A* **2016**, *88*, 67–74. [CrossRef]

21. Yang, Y.; Gupta, M.C.; Dudley, K.L.; Lawrence, R.W. Conductive carbon nanofiber polymer faom structures. *Adv. Mater.* **2005**, *17*, 1999–2003. [CrossRef]

22. Harikrishnan, G.; Singh, S.N.; Kiesel, E.; Macosko, C.W. Nanodispersions of carbon nanofiber for polyurethane foaming. *Polymer* **2010**, *51*, 3349–3353. [CrossRef]

23. Al-Saleh, M.H.; Sundararaj, U. A review of vapor grown carbon nanofiber/polymer conductive composites. *Carbon* **2009**, *47*, 2–22. [CrossRef]

24. Buys, Y.F.; Lokman, N.A.S. Conductive polymer composites from polylactic acid/natural rubber filled with carbon black. *Adv. Mat. Res.* **2015**, *1115*, 253–257. [CrossRef]

25. Wang, J.; Dai, J.; Yarlagadda, T. Carbon nanotube-conducting-polymer composite nanowires. *Langmuir* **2005**, *21*, 9–12. [CrossRef] [PubMed]

26. Lee, S.H.; Cho, E.; Jeon, S.H.; Youn, J.R. Rheological and electrical properties of polypropylene composites containing functionalized multi-walled carbon nanotubes and compatibilizers. *Carbon* **2007**, *45*, 2810–2822. [CrossRef]

27. Yang, Y.; Gupta, M.C. Novel carbon nanotube PS foam composites for electromagnetic interference shielding. *Nano Lett.* **2005**, *5*, 2131–2134. [CrossRef] [PubMed]

28. Pötschke, P.; Bhattacharyya, A.R.; Janke, A. Carbon nanotube-filled polycarbonate composites produced by melt mixing and their use in blends with polyethylene. *Carbon* **2004**, *42*, 965–969. [CrossRef]

29. Xu, X.B.; Li, Z.M.; Shi, L.; Bian, X.C.; Xiang, Z.D. Ultralight conductive carbon-nanotube-polymer composite. *Small* **2007**, *3*, 408–411. [CrossRef] [PubMed]

30. Al-Saleh, M.H.; Sundararaj, U. Electrically conductive carbon nanofiber/polyethylene composite: Effect of melt mixing conditions. *Polym. Adv. Technol.* **2011**, *22*, 246–253. [CrossRef]

31. Villmow, T.; Pötschke, P.; Pegel, S.; Häussler, L.; Kretzschmar, B. Influence of twin-screw extrusion conditions on the dispersion of multi-walled carbon nanotubes in a poly(lactic acid) matrix. *Polymer* **2008**, *49*, 3500–3509. [CrossRef]

32. Ameli, A.; Kazemi, Y.; Wang, S.; Park, C.B.; Pötschke, P. Process-microstructure-electrical conductivity relationships in injection-molded polypropylene/carbon nanotube nanocomposite foams. *Compos. Part A* **2017**, *96*, 28–36. [CrossRef]

33. Antunes, M.; Mudarra, M.; Velasco, J.G. Broad-band electrical conductivity of carbon nanofibre-reinforced polypropylene foams. *Carbon* **2011**, *49*, 708–717. [CrossRef]

34. Ameli, A.; Wang, S.; Kazemi, Y.; Park, C.B.; Pötschke, P. A facile method to increase the charge storage capability of polymer nanocomposites. *Nano Energy* **2015**, *15*, 54–65. [CrossRef]

35. Motlagh, G.H.; Hrymak, A.N.; Thompson, M.R. Improved through-plane electrical conductivity in a carbon-filled thermoplastic via foaming. *Polym. Eng. Sci.* **2008**, *48*, 687–696. [CrossRef]

36. Ameli, A.; Jung, P.U.; Park, C.B. Through-plane electrical conductivity of injection-molded polypropylene/carbon-fiber composite foams. *Comp. Sci. Technol.* **2013**, *76*, 37–44. [CrossRef]

37. Okamoto, M.; Nam, P.H.; Maiti, P.; Kotaka, T.; Nakayama, T.; Takada, M.; Ohshima, M.; Usuki, A.; Hasegawa, N.; Okamoto, H. Biaxial flow-induced alignment of silicate layers in polypropylene/clay nanocomposite foam. *Nano Lett.* **2001**, *1*, 503–505. [CrossRef]

38. Tran, M.-P.; Detrembleur, C.; Alexandre, M.; Jerome, C.; Thomassin, J.-M. The influence of foam morphology of multi-walled carbon nanotubes/poly(methyl methacrylate) nanocomposites on electrical conductivity. *Polymer* **2013**, *54*, 3261–3270. [CrossRef]

39. Rizvi, R.; Naguib, H. Porosity and composition dependence on electrical and piezoresistive properties of thermoplastic polyurethane nanocomposites. *J. Mater. Res.* **2013**, *28*, 2415–2425. [CrossRef]

40. Sun, Y.-C.; Terakita, D.; Tseng, A.C.; Naguib, H.E. Study on the thermoelectric properties of pvdf/mwcnt and pvdf/gnp composite foam. *Smart Mater. Struct.* **2015**, *24*, 085034. [CrossRef]

41. Kazemi, Y.; Kakroodi, A.R.; Wang, S.; Ameli, A.; Filleter, T.; Pötschke, P.; Park, C.B. Conductive network formation and destruction in polypropylene/carbon nanotube composites via crystal control using supercritical carbon dioxide. *Polymer* **2017**, *129*, 179–188. [CrossRef]

42. Ameli, A.; Nofar, M.; Park, C.B.; Pötschke, P.; Rizvi, G. Polypropylene/carbon nanotube nano/microcellular structures with high dielectric permittivity, low dielectric loss, and low percolation threshold. *Carbon* **2014**, *71*, 206–217. [CrossRef]

43. Du, F.; Fischer, J.; Winey, K. Effect of nanotube alignment on percolation conductivity in carbon nanotube/polymer composites. *Phys. Rev. B* **2005**, *72*, 121404. [CrossRef]

44. Gong, S.; Zhu, Z.H.; Meguid, S.A. Anisotropic electrical conductivity of polymer composites with aligned carbon nanotubes. *Polymer* **2015**, *56*, 498–506. [CrossRef]

45. Chang, E.; Ameli, A.; Mark, L.H.; Park, C.B. Effects of uniaxial and biaxial orientation on fiber percolation in conductive polymer composites. *AIP Conf. Proc.* **2015**, *1693*, 020027.

polymers

MDPI

Article

Effect of Polyhedral Oligomeric Silsesquioxane on the Melting, Structure, and Mechanical Behavior of Polyoxymethylene

Dorota Czarnecka-Komorowska * and Tomasz Sterzynski

Poznan University of Technology, Institute of Materials Technology, Polymer Processing Division, Piotrowo 3 Street, PL-61138 Poznan, Poland; Tomasz.Sterzynski@put.poznan.pl
* Correspondence: Dorota.Czarnecka-Komorowska@put.poznan.pl; Tel.: +48-61665-2732

Received: 8 December 2017; Accepted: 14 February 2018; Published: 17 February 2018

Abstract: The effects of octakis[(3-glycidoxypropyl)dimethylsiloxy]octasilsesquioxane (GPOSS) on the crystallinity, crystal structure, morphology, and mechanical properties of polyoxymethylene (POM) and POM/GPOSS composites were investigated. The POM/GPOSS composites with varying concentrations of GPOSS nanoparticles (0.05–0.25 wt %) were prepared via melt blending. The structure of POM/GPOSS composites was characterized by differential scanning calorimetry (DSC), wide angle X-ray diffraction (WAXD), and polarized light microscopy (PLM). The mechanical properties were determined by standardized tensile tests. The morphology and dispersion of GPOSS nanoparticles in the POM matrix were investigated with scanning electron microscopy (SEM) and energy dispersive X-ray (EDX) analysis. It was observed that the dispersion of the GPOSS nanoparticles was uniform. Based on DSC studies, it was found that the melting temperature, lamellar thickness, and the degree of crystallinity of the POM/GPOSS composites increased. The POM/GPOSS composites showed an increased Young's modulus and tensile strength. Finally, compared with the pure POM, the addition of GPOSS reduced the spherulites' size and improved the crystallinity of the POM, which demonstrates that the nucleation effect of GPOSS is favorable for the mechanical properties of POM.

Keywords: polyoxymethylene (POM); octakis[(3-glycidoxypropyl)dimethylsiloxy]octasilsesquioxane (GPOSS); composites; morphology; mechanical properties

1. Introduction

Composites based on polyoxymethylene (POM), or polyacetal with silsesesquioxanes, are an interesting group of engineering materials due to the significant modification capabilities of these polymers and their wide application in industry. POM, with ($-CH_2-O-$) as the main chain, is a major engineering plastic, with a high degree of mechanical strength, dimensional stability, and abrasion resistance. Two types of acetal are commercially available: homopolymers and copolymers [1].

Engineering applications require the creation of materials with significantly high strength properties and high thermal resistance. Thus, there is a need for research directed towards the modification of the structure and properties of materials, and such modification is the main aim of implementing various nanofillers. The nanoparticles incorporated into the polymer matrix usually significantly influence the polymer's crystallization as well as the crystal phase type [2]. Particle size, their quantity, the degree of dispersion in the polymer matrix, and the processing conditions are the most relevant parameters in terms of composite properties [3,4].

Polyhedral oligomeric silsesquioxanes (POSS) are nanostructures with the empirical formula $RSiO_{1.5}$, where R is a hydrogen atom or an organic group such as alkyl, alkylene, acrylate, hydroxyl, or epoxide units [4–8]. POSS may be referred to as a silica nanoparticle, consisting of a silica cage core and other organic functional groups attached to the corners of the cage [8]. The advantage of POSS as a nanofiller is due to the magnitude of its core; the core of octasilsesquioxanes can reach as high as approximately 0.5 nm. The whole molecule, depending on the substituent type, can range from 1 to 3 nm [9,10]. Comparing the average size of the macromolecular chains of these silsesquioxane molecule sizes, together with the possibility of implementation of the appropriate functional group, facilitates the deposition of POSS in the polymer matrix. Poly(methyl methacrylate) [10], polyethylene [11], polypropylene [12,13], poly(ethylene terephthalate) [14], polylactide [15], epoxide resins [16], polyurethanes [17], and polyoxymethylene [18,19] were all found to have an impact on certain physical properties such as glass transition, decomposition temperature, viscosity, fracture toughness, and barrier properties [20–24].

Few papers have been published concerning the effects of POSS molecules on the melting and crystallization, the microstructure, and the thermo-mechanical properties of polyoxymethylene [25–27]. Durmus et al. investigated the microstructure and isothermal melt-crystallization behavior of polyoxymethylene (POM) modified by methyl-polyhedral oligomeric silsesquioxanes (methyl-POSS) [25]. They pointed out that the addition of a PP/POSS nanocomposite dramatically enhances the crystallization rate of POM. It was concluded that the rate acceleration effect of the PP/POSS nanocomposite in POM crystallization is probably due to the tremendous increase in the number of nuclei due to the nano-POSS particles [25]. Lllescas et al. studied hybrid nanocomposites containing polyoxymethylene copolymer (POM) and four types of polyhedral oligomeric silsesquioxane (POSS) nanoparticles. They found that the added POSS nanoparticles can improve the thermo-mechanical properties of some polymers [26]. They reported that the formation of hydrogen bonding interactions between the POM and Si–OH groups of msib-POSS increased their mutual compatibility and led to nanometer-size dispersion of some msib-POSS molecules [26]. Pielichowski et al. studied the influence of the heating rate on the shape of the melting peak of polyoxymethylene (POM) using both high-speed differential scanning calorimetry (DSC) and StepScan DSC. They pointed out that a low heating rate facilitates the recrystallization of polyoxymethylene due to molecular nucleation [27].

The aim of this study was to determine the influence of a small amount of GPOSS on the crystallinity, crystal structure, morphology, and mechanical properties of POM, where the introduction of the nanoparticles is achieved via melt processing.

2. Experimental Section

2.1. Materials

The commercial form of POM (Tarnoform 300), with a melt flow index of 0.9 g/10 min (190 °C, 2.16 kg), was supplied by Grupa Azoty S.A (Zakłady Azotowe, Tarnów-Moscice, Poland) for use in this study. GPOSS was synthesized by the hydrosilylation process [28], using a methodology developed at the Faculty of Organic Chemistry of Adam Mickiewicz University (Poznan, Poland). The synthesis of GPOSS was conducted in accordance with the procedure described in [29,30].

The chemical structure of GPOSS was confirmed by spectroscopic methods (NMR, FT-IR), and the results of analysis are as follows:

^{1}**H NMR** (CDCl$_3$, 298 K, 300 MHz) δ (ppm) = 0.05 (OSiCH$_3$); 0.51 (SiCH$_2$); 1.51(CH$_2$); 2.47, 2.65 (CH$_2$O); 3.00 (CHO); 3.25 (CH$_2$O); 3.33, 3.56(OCH$_2$).

^{13}**C NMR** (CDCl$_3$, 298 K, 75.5 MHz) δ (ppm) = $-$0.66 (SiCH$_3$); 13.39 (SiCH$_2$); 22.89 (CH$_2$); 43.98 (CH$_2$O); 50.29 (CHO); 71.75 (OCH$_2$); 73.61 (CH$_2$O).

^{29}**Si NMR** (CDCl$_3$, 298 K, 59.6 MHz) δ (ppm) = 12.87 (OSi(CH$_3$)$_2$); $-$109.13 (SiOSi).

FT-IR (ATR): 2998, 2955, 2931, 2869, 1253, 1070, 902, 838, 547 cm^{-1}.

The chemical formula of GPOSS is presented in Figure 1.

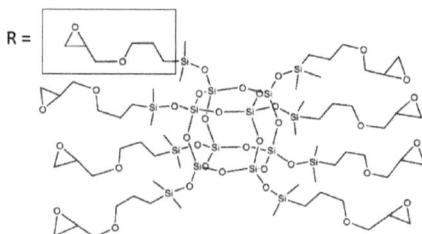

Figure 1. Chemical structure of the octakis[(3-glycidoxypropyl)dimethylsiloxy]octasilsesquioxane (GPOSS).

2.2. Preparation of the POM/GPOSS Nanocomposites

POM pellets were ground into a powder using a Tria grinder and dried at 80 °C for 4 h. The POM powder was premixed with various amounts of GPOSS (0.05, 0.1, and 0.25 wt %) using the rotary mixer Retsch GM 200 (Retsch GmbH, Haan, Germany) (*t* = 4 min, *n* = 2000 rpm). The melt-mixing of the blends was carried out in a Brabender single screw extruder, equipped with an intense mixing zone, operating at 190 °C with a screw rotational speed of 15 rpm. The extruded rod was cooled in a water bath and subsequently pelletized. Standard test samples were produced by injection molding using an Engel machine (ES 80/20HLS, Schwertberg, Austria) with a 22 mm screw and an L/D ratio of 18. The processing parameters were as follows: injection temperature along the barrel: 180/190/200 °C; nozzle temperature: 210 °C; mold temperature: 60 °C.

2.3. Characterization

2.3.1. Differential Scanning Calorimetry (DSC)

The melting and crystallization behavior of the composites were studied using a differential scanning calorimeter (Netzsch, DSC 204 F1 Phoenix, Selb, Germany) operating under nitrogen flow (150 mL/min). Samples of about 8 mg were heated up to 220 °C and held there in order to eliminate the thermal and mechanical prehistory. Next, the samples were cooled to 120 °C at a cooling rate of 20 °C/min and held there for 5 min and finally reheated again to 220 °C at the same rate.

The degree of crystallinity (X_c) of the samples was determined from the values of enthalpy melting registered during the second heating, using the following equation:

$$X_c(\%) = \frac{\Delta H_m}{(1 - \varphi)\Delta H_m^0} \tag{1}$$

where ΔH_m is the melting enthalpy of the samples, φ is the weight fraction of GPOSS nanofiller, and ΔH_m^0 is the melting enthalpy for a 100% crystalline of polymer—this latter measurement was taken as 186 J/g for the POM copolymer [31].

2.3.2. Optical Polarized Light Microscopy (PLM)

An optical polarized light microscope (PLM) (Nikon Eclipse E400, Kanagawa, Japan), equipped with a Linkam THMS 600 hot stage (Linkam Scientific Instruments Ltd., Tadworth, UK) was used to study the crystallization process via cooling. The samples were heated at 200 °C for 2 min to obtain a full melting of the POM, then quickly cooled (20 °C/min) to the isothermal crystallization temperature of approximately 148 °C. The growth of the spherulites was observed during crystallization using an Opta-Tech camera (Opta-Tech, Warsaw, Poland) at 200× magnification.

2.3.3. Wide-Angle X-ray Diffraction (WAXD)

The crystal structures of the POM and POM/GPOSS composites were evaluated with wide-angle X-ray diffraction (WAXD) measurements taken at room temperature, using a Brücker D2 diffractometer (Madison, Japan), equipped with Cu $K\alpha$ radiation (λ = 0.1542 nm), operating at 40 kV and 40 mA. Diffraction patterns were recorded for 2θ values ranging from 10 to 50°, at a scanning rate of 0.01°/s. The *d*-spacing of the POM/GPOSS was evaluated with the Bragg equation:

$$2d \sin \theta = n\lambda \qquad (2)$$

where θ is the Bragg's angle for the corresponding crystallographic plane (hkl), n is the order of diffraction, and λ is the incident wave length.

2.3.4. Mechanical Properties

The mechanical properties of the samples were determined by tensile tests performed using a universal testing machine: a Zwick/Roell Z005 (Ulm, Germany). The samples were dumbbell-shaped, with dimensions of 75 mm × 5 mm × 2 mm (test specimens type 1BA), according to the PN-EN ISO 527-2 standard. The tensile tests were run at room temperature. Tensile modulus was determined at a 1 mm/min cross head speed, whereas other tensile characteristics were measured at 50 mm/min speed. Young's modulus (E), ultimate tensile strength (σ_M), and tensile stress at break (σ_B) were evaluated from the tensile stress-strain curves. The reported data were the average of the results of 10 specimens.

2.3.5. Scanning Electron Microscope (SEM)

The morphology of the fractured surfaces of the POM/GPOSS composite, coated with gold, was observed using a scanning electron microscope (SEM)—TESCAN TS 5135 (Brno, Czech Republic). The dispersion of GPOSS nanoparticles in the POM matrix was investigated using the back-scattered electron (BSE) signal and secondary electron (SE) signal, with an accelerating voltage of 15 kV, and the Si distribution was subsequently mapped.

3. Results and Discussion

3.1. Effect of GPOSS on Melting and Crystallinity

The DSC melting curves for pure POM and POM with GPOSS nanoparticles, registered during the first heating at a rate of 20 °C/min, are shown in Figure 2.

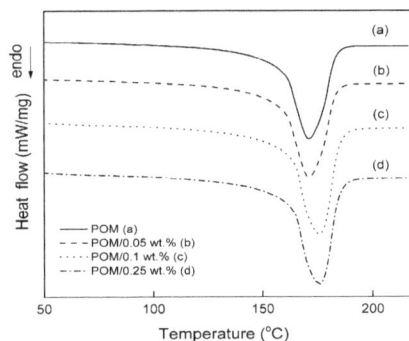

Figure 2. Differential scanning calorimetry (DSC) melting curves of pure POM (**a**) and its composites with different GPOSS concentrations at the first heating rate of 20 °C/min: (**b**) 0.05 wt %; (**c**) 0.1 wt %; (**d**) 0.25 wt %.

The onset of the melting temperature ($T_{m1\ onset}$), the maximum melting temperature (T_{m1}), the melting enthalpy (ΔH_{m1}), the lamellar thickness (l_c), the melting enthalpy (ΔH_{m2}), and the crystallinity (X_c) are listed in Table 1.

Table 1. DSC data of pure POM and POM/GPOSS composites (cooling and heating rate: 20°/min).

Samples	T_{m1} (°C)	$T_{m1\ onset}$ (°C)	ΔH_{m1} (J/g)	l_c (nm)	ΔH_{m2} (J/g)	X_c (%)
pure POM	169.2	158.0	−138.5	9.8	−134.3	72
POM/0.05 wt % GPOSS	171.2	159.0	−142.0	11.4	−135.8	73
POM/0.1 wt % GPOSS	173.7	160.0	−143.5	14.2	−137.4	74
POM/0.25 wt % GPOSS	175.7	163.2	−154.1	17.2	−149.9	81

A clear single endothermal peak within the temperature range of 169–180 °C for the pure POM and its composites was observed, indicating that, as the result of modification, compatible blends of POM with GPOSS were achieved. This suggests a strong interfacial interaction between POM and the glycid groups of POSS. As seen in Figure 2, it follows that the melting temperature of POM/GPOSS composites becomes higher as GPOSS content increases. For example, an increase in the T_m from 169 °C in pure POM to 176 °C (an increase of approximately 7 °C) in POM with 0.25 wt % GPOSS was observed, an effect consistent with our primary observation for polyoxymethylene modified by vinyl groups of POSS [19].

The melting temperature (T_{m1}) observed during the first DSC heating run may be related to the lamellar thickness, where a greater lamellar thickness corresponds to a higher polymer melting temperature. This relationship is described by the Gibbs-Thompson equation (Equation (3) [32]. Based on this equation, the lamellar thickness (l_c) of the POM/GPOSS composites was evaluated and compared with the corresponding values of the pure POM.

$$l_c = \frac{2\sigma_e \cdot T_m^0}{\Delta h_0(T_m^0 - T_m)} \tag{3}$$

where T_m is the observed melting temperature for a crystalline lamellar of thickness (l_c), T_m^0 is the equilibrium melting temperature of the crystalline lamella of an infinite thickness, σ_e is the lamellar basal surface free energy (12.5×10^6 J·cm^{-2} for POM) [33], and Δh_0 is the enthalpy of fusion for the crystalline phase (315×10^6 J·cm^{-3} for POM) [34].

Figure 3 illustrates the relationship between the amount of GPOSS and lamellar thickness (l_c) of the POM polymer single crystals.

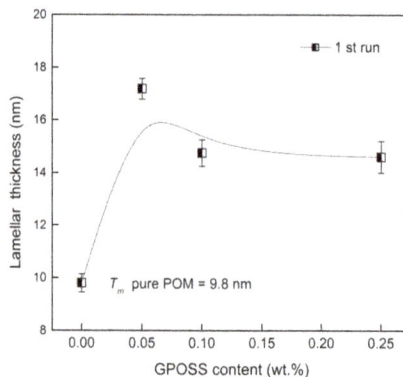

Figure 3. The variation in the lamellar thickness of pure POM and its composites, calculated from DSC (first run) experiments as a function of GPOSS content.

An increase in the lamellar thickness of POM with 0.25 wt % GPOSS to about 17 nm, compared with the pure POM where the lamellar thickness was equal to 9.8 nm, was noted. It is known that lamellar thickness increases with an increase in crystallization temperature [35]. A similar phenomenon has been observed in POM with carbon nanotubes (CNTs) [36].

As presented in Table 1, an increase in the X_c value from 72% in pure POM to 81% (i.e., an increase of about 13%) in POM modified with 0.25 wt % GPOSS was observed. A significant increase in enthalpy in the second melting (ΔH_{m2}) from 134.3 J/g in pure POM to 150 J/g in the highest amount of POM/0.25 wt % GPOSS may indicate that GPOSS additives play a heterogenic nucleation role in POM.

Similar effects have been observed by Hu and Ye through the analysis of the POM crystallization process with the addition of polyamide 6 (PA6). It has been shown that PA6 leads to an increase in the degree of crystallinity (X_c), improving the crystallization growth rate and resulting in a reduction in the size of spherulites in POM [37].

3.2. Effect of GPOSS on the Spherulitic Morphology

The spherulitic morphology of pure POM and POM/GPOSS composites was observed using optical polarized light microscopy (PLM), as shown in Figure 4.

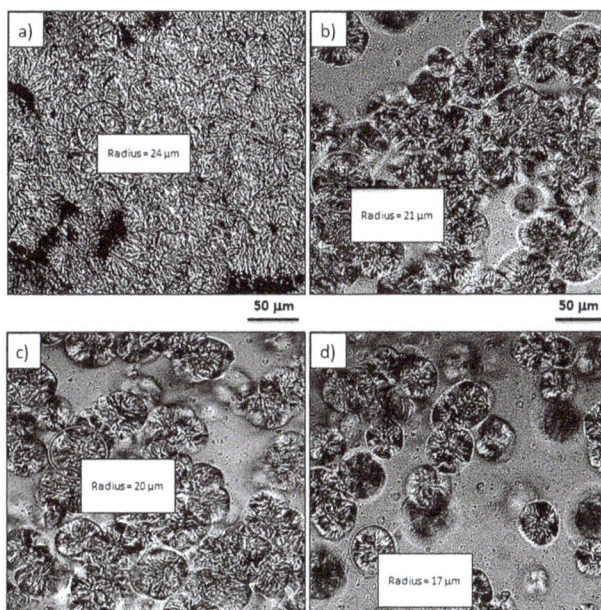

Figure 4. PLM micrographs of the crystallized pure POM (**a**) and its composites with different GPOSS concentrations: (**b**) 0.05 wt %; (**c**) 0.1 wt %; (**d**) 0.25 wt %. T = 148 °C. The cooling rate was set to 20 °C/min. (magnification 200×).

Significant differences between the structure of pure POM (Figure 4a) and POM modified with GPOSS can be seen in Figure 4b–d. For all cases, spherulite forms differing in magnitude, shape, spatial distribution, and well-developed spherulitic morphology can clearly be seen. It is obvious that the POM spherulites become smaller as GPOSS content increases (Figure 4b–d). Figure 5 illustrates the dependence of the size of spherulites in the POM on GPOSS content; the addition of 0.25 wt % GPOSS leads to a significant decrease in POM spherulite size—specifically, from 48×10^{-6} to about 35×10^{-6} m. Such an effect is due to the heterogeneous nucleation role of GPOSS.

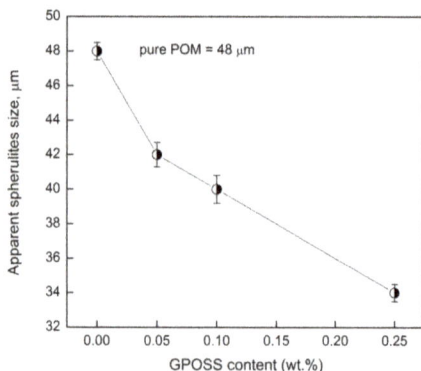

Figure 5. The dependence of the size of spherulites in pure POM and its composites with different GPOSS concentrations. The cooling rate was set to 20 °C/min.

3.3. Effect of GPOSS on the Crystal Structure

The crystal structure of POM was studied using the WAXD technique. The analysis of the WAXD allows us to determine the crystal atom structure, including the position/symmetry of the atoms in the unit cell, the unit cell size, and the shape/size of the nanocrystalline domain [38]. The X-ray diffraction patterns for the POM and POM/GPOSS composites with different GPOSS concentrations, as a function of Bragg's angle (2θ) and measured at room temperature, are presented in Figure 6.

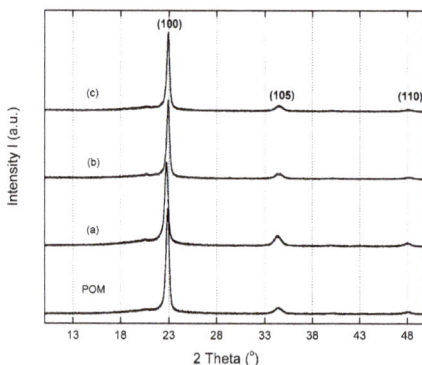

Figure 6. WAXD curves of pure POM and its composites with different GPOSS concentrations at room temperature: (**a**) 0.05 wt %; (**b**) 0.1 wt %; (**c**) 0.25 wt %.

As can be seen in Figure 6, two intense Bragg diffraction peaks can clearly be observed at the Bragg angles 2θ = 22.7° and 2θ = 34.14°, where the peak at 22.7° corresponds to the basal reflection (100), and the peak at 34.14° corresponds to the basal reflection (105). POM crystallizes in a hexagonal crystallographic form with unit cell dimensions of a = b = 4.46 Å and c = 17.3 Å [39]. In general, it was observed that the X-ray diffraction patterns of POM/GPOSS composites are similar to the diffraction pattern of the hexagonal form observed for the crystalline phase of pure POM. The crystallite size (L_{hkl}), according to the Scherrer formula (Equation (3)), was evaluated for the crystallographic plane (100), using the following equation:

$$L_{hkl} = \frac{K\lambda}{\beta_{hkl} \times \cos \theta_{hkl}} \tag{4}$$

where L_{hkl} is the apparent crystallite size of the normal direction of the {hkl} crystal plane (in Angstrom, 1 Å = 10^{-10} m), K is a Scherrer constant, normally taken as 0.9, β_{hkl} is the full-width at half-maximum (in radian) of the crystalline peak, nm, θ is the Bragg diffraction angle, and λ is the wavelength of entrance X-ray (0.154 nm for Cu), as presented in Table 2.

Table 2. Crystal size (L_{hkl}) evaluated according to the Scherrer formula.

Samples	L_{100} (Å)
pure POM	228
POM/0.05 wt % GPOSS	217
POM/0.1 wt % GPOSS	186
POM/0.25 wt % GPOSS	143

The highest WAXD diffractogram intensity was observed for the POM/GPOSS composites, indicating the higher crystalline structure order of the POM.

Based on WAXD studies, it was also shown that, with an increase in GPOSS content in the POM, an increase in the diffraction peak at half height width occurs, corresponding to a reduction in crystal size (from 228 to 143 Å).

In 0.05 and 0.25 wt % POSS, a clear decrease in the L_{100} crystallite size of composites, in comparison with pure POM, was found, suggesting that GPOSS has a nucleating effect on POM.

3.4. Effect of GPOSS on Mechanical Behaviour

In Figure 7, typical tensile stress-strain curves of pure POM and POM/GPOSS composites with different GPOSS nanoparticle concentrations are presented.

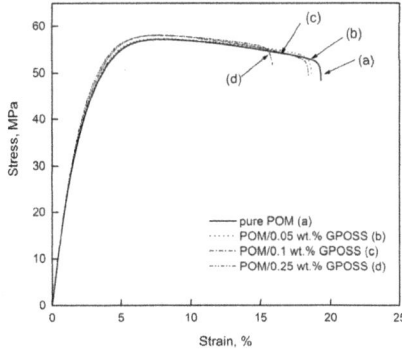

Figure 7. Tensile stress-strain curves of pure POM (**a**) and its composite with different GPOSS concentrations: (**b**) 0.05 wt %; (**c**) 0.1 wt %; (**d**) 0.25 wt %.

The average values and standard deviations of mechanical properties are given in Table 3.

Table 3. Mechanical properties, with standard deviations, of pure POM and POM/GPOSS composites.

Samples	Ultimate tensile strength (MPa)	Young's modulus (GPa)	Tensile stress at break (MPa)
pure POM	55.00 ± 0.52	1.83 ± 0.13	45.97 ± 0.65
POM/0.05 wt % GPOSS	57.50 ± 0.43	2.44 ± 0.23	49.78 ± 0.78
POM/0.1 wt % GPOSS	57.64 ± 0.90	2.48 ± 0.15	49.90 ± 0.69
POM/0.25 wt % GPOSS	57.80 ± 0.95	2.60 ± 0.25	50.25 ± 0.73

The Young's modulus (E), ultimate tensile strength (σ_M), and tensile stress at break (σ_B) values of POM/GPOSS composites and pure POM were compared, and an increase in tensile properties, such as strength and modulus, with increasing GPOSS concentrations was noted. The Young's modulus of the POM was increased from 1.8 to 2.6 GPa with the addition of 0.25 wt % GPOSS, which corresponds to an approximate 45% increase. Similarly, an increase in tensile strength was observed for POM/GPOSS composites compared to pure POM. The maximum value of the tensile strength (σ_M) was observed for the POM/GPOSS composite with the addition of 0.25 wt % GPOSS. Compared with the pure POM, the addition of GPOSS reduced the size of the spherulites and improved the crystallinity (X_c) of the POM, which demonstrates the nucleation effect of GPOSS, as the nucleus is favorable for the mechanical properties. These results are in agreement with the DSC results. The changes in mechanical properties may also be a result of a good dispersion of GPOSS nanoparticles in the POM matrix and a good adhesion of these two components.

3.5. Effect of GPOSS on Morphology

SEM microscopy was used to analyze the morphology and dispersion of particles of GPOSS in the POM matrix. Figure 8 presents SEM images that show the morphologies of pure POM (a) and POM/GPOSS composites with the addition of 0.05 wt % (b) and 0.25 wt % GPOSS (c).

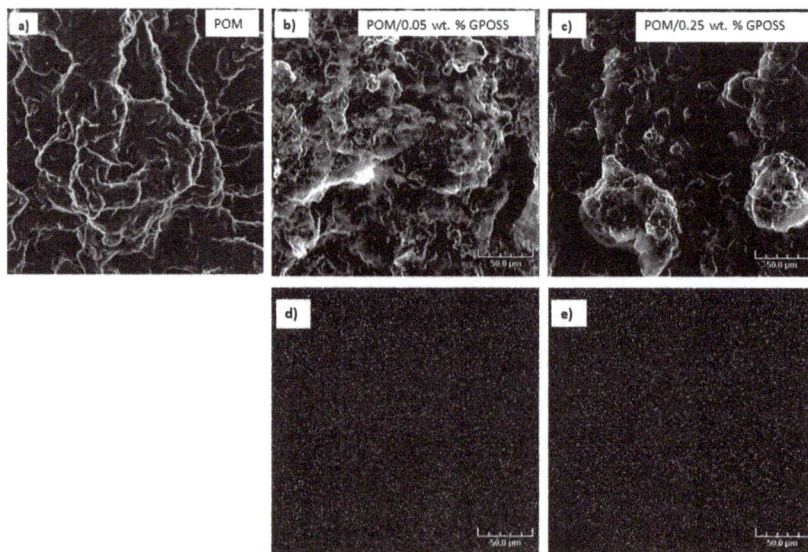

Figure 8. SEM images of the fractured surface (**a–c**) and Si mapping (**d–e**) of pure POM (**a**) and its composites with different GPOSS concentrations: (**b**) 0.05 wt %; (**c**) 0.25 wt %.

A relatively homogenous distribution of GPOSS particles in the POM matrix can be seen in Figure 8a–c. The dispersion of the GPOSS particles in the POM matrix was homogeneous, and a similar phenomenon was described in the literature for silsequioxanes with polyamide and polyethylene [30]. Figure 8d,e shows the energy dispersive X-ray spectrometer (EDS) composition distribution maps of the Si element on the surface of the POM/0.05 wt % GPOSS (d) and POM/0.25 wt % GPOSS composites (e). The white dots in this figure are the X radial signals radiating from the Si element. A random dispersion of 0.25 wt % GPOSS nanoparticles was found on the entire fractured surface of the POM. A similar effect was observed for polyoxymethylene and POM/HAp nanocomposites [40].

Polymers **2018**, *10*, 203

4. Conclusions

In the study reported herein, the effect of GPOSS on the melting behavior, structure, and mechanical properties of POM was investigated. The properties of POM/GPOSS composites were characterized as a function of the GPOSS content, which varied from 0 to 0.25 wt %. DSC measurements indicated a significant change in the melting temperature and crystallinity of the nanocomposites, relative to the matrix. POM/GPOSS composites of an evidently higher degree of crystallinity were achieved, indicating a heterogenic nucleation role of the GPOSS additive for pure POM.

Polarized microscopy results indicate that the POM spherulites become smaller as GPOSS content increases. During POM crystallization, small spherulites about 35 μm in diameter were produced. The incorporation of GPOSS reduced the mean size of the spherulites, suggesting a nucleation effect. X-ray diffraction results indicate no significant changes in the hexagonal structure of the POM in the presence of GPOSS.

The enhancement in the Young's modulus of the composites containing 0.25 wt % GPOSS, relative to the pure POM, was about 45%. This suggests that GPOSS should work as a reinforcement in the POM matrix. The reason for this is due to the good dispersion of GPOSS in the POM matrix. SEM results proved the existence of a uniform structure in the heterogeneous compatible blends, in which intensive interfacial interactions may be present.

Author Contributions: Dorota Czarnecka-Komorowska performed the experiments; Dorota Czarnecka-Komorowska and Tomasz Sterzynski analyzed the data and wrote the paper.

Conflicts of Interest: The authors declare no conflict of interest.

References

1. Sigrid, L.; Visakhn, P.M.; Chandran, S. *Polyoxymethylene Handbook*; Scrivener Publishing: Beverly, MA, USA, 2014.
2. Paula, D.R.; Robeson, L.M. Polymer nanotechnology: Nanocomposites. *Polymer* **2008**, *49*, 3187–3204. [CrossRef]
3. Piesowicz, E.; Irska, I.; Bryll, K.; Gawdzińska, K.; Bratychak, M. Poly(butylene terephthalate/carbon nanotubes nanocomposites. Part II. Structure and properties. *Polimery* **2016**, *24*, 61. [CrossRef]
4. Cordes, D.B.; Lickiss, P.D.; Rataboul, F. Recent Developments in the Chemistry of Cubic Polyhedral Oligosilsesquioxanes. *Chem. Rev.* **2010**, *110*, 2081–2173. [CrossRef] [PubMed]
5. Barczewski, M.; Czarnecka-Komorowska, D.; Andrzejewski, J.; Sterzyński, T.; Dutkiewicz, M.; Dudziec, B. Właściwości przetwórcze termoplastycznych tworzyw polimerowych modyfikowanych silseskwioksanami (POSS). *Polimery* **2013**, *58*, 805–815. [CrossRef]
6. Maciejewski, H.; Dutkiewicz, M.; Byczyński, Ł.; Marciniec, B. Silseskwioksany jako nanonapełniacze. Cz. I. Nanokompozyty z osnową silikonową. *Polimery* **2012**, *57*, 535–544. [CrossRef]
7. Przybylak, M.; Maciejewski, H.; Marciniec, B. Synteza i charakterystyka silseskwioksanów o strukturze niecałkowicie skondensowanych klatek. *Polimery* **2013**, *10*, 741–747. [CrossRef]
8. Hartmann-Thompson, C. *Applications of Polyhedral Oligomeric Silsesquioxanes*; Springer: Dordrecht, The Netherlands, 2011.
9. Mohamed, G.M.; Kuo, S.W. Polybenzoxazine/Polyhedral Oligomeric Silsesquioxane (POSS) Nanocomposites. *Polymers* **2016**, *8*, 225. [CrossRef]
10. Xue, Y.; Wang, H.; Yu, D.; Feng, L.; Dai, L.; Wang, X.; Lin, T. Superhydrophobic electrospun POSS-PMMA copolymer fibres with highly ordered nanofibrillar and surface structures. *Chem. Commun.* **2009**, *42*, 6418–6420. [CrossRef] [PubMed]
11. Furuheim, K.M.; Axelson, D.E.; Antonsen, H.W.; Helle, T. Phase structural analyses of polyethylene extrusion coatings on high-density papers. I. Monoclinic crystallinity. *J. Appl. Polym. Sci.* **2004**, *91*, 218–225. [CrossRef]
12. Barczewski, M.; Dudziec, B.; Dobrzyńska-Mizera, M.; Sterzyński, T. Synthesis and Influence of Sodium Benzoate Silsesquioxane Based Nucleating Agent on Thermal and Mechanical Properties of Isotactic Polypropylene. *J. Macromol. Sci. A* **2014**, *51*, 907–913. [CrossRef]

13. Barczewski, M.; Dudziec, B.; Dobrzyńska-Mizera, M.; Sterzyński, T. Influence of the cooling rate on the non-isothermal crystallization of isotactic polypropylene modified with sorbitol derivative and silsesquioxane. *Polimery* **2013**, *58*, 920–923. [CrossRef]

14. Didane, N.; Giraud, S.; Devaux, É.; Lemort, G. Development of fire resistant PET fibrous structures based on phosphinate-POSS blends. *Polym. Degrad. Stab.* **2012**, *97*, 879–885. [CrossRef]

15. Lee, J.H.; Jeong, Y.G. Preparation and crystallization behavior of polylactide nanocomposites reinforced with POSS-modified montmorillonite. *Fibers Polym.* **2011**, *12*, 180–189. [CrossRef]

16. Dutkiewicz, M.; Szołyga, M.; Maciejewski, H.J. Thiirane functional spherosilicate as epoxy resin modifier. *J. Therm. Anal. Calorim.* **2014**, *117*, 259–264. [CrossRef]

17. Pielichowski, K.; Jancia, M.; Hebda, E.; Pagacz, J.; Pielichowski, J.; Marciniec, B.; Franczyk, A. Poliuretany modyfikowane funkcjonalizowanym silseskwioksanem-synteza i właściwości. *Polimery* **2013**, *58*, 783–793. [CrossRef]

18. Illescas, S.; Arostegui, A.; Schiraldi, D.A.; Sánchez-Soto, M.; Velasco, J.I. The Role of Polyhedral Oligomeric Silsesquioxane on the Thermo-Mechanical Properties of Polyoxymethylene Copolymer Based Nanocomposites. *J. Nanosci. Nanotechnol.* **2010**, *10*, 1349–1360. [CrossRef] [PubMed]

19. Czarnecka-Komorowska, D.; Sterzynski, T.; Dutkiewicz, M. Polyhedral oligomeric silsesquioxanes as modifiers of polyoxymethylene structure. *AIP Conf. Proc.* **2015**, *1695*, 020013. [CrossRef]

20. Farmahini-Farahani, M.; Xiao, H.; Zhao, Y. Poly lactic acid nanocomposites containing modified nanoclay with synergistic barrier to water vapor for coated paper. *J. Appl. Polym. Sci.* **2014**, *131*, 40952. [CrossRef]

21. Colomines, G.; Ducruet, V.; Courgneau, C.; Guinault, A.; Domenek, S. Barrier properties of poly(lactic acid) and its morphological changes induced by aroma compound sorption. *Polym. Int.* **2010**, *59*, 818–826. [CrossRef]

22. Philips, S.H.; Haddad, T.S.; Tomczak, S.J. Developments in nanoscience: Polyhedral oligomeric silsesquioxane (POSS)-polymers. *Curr. Opin. Solid State Mater. Sci.* **2004**, *8*, 21–29. [CrossRef]

23. Hossain, M.D.; Yoo, Y.; Lim, K.T. Synthesis of poly(ε-caprolactone)/clay nanocomposites using polyhedral oligomeric silsesquioxane surfactants as organic modifier and initiator. *J. Appl. Polym. Sci.* **2011**, *119*, 936–943. [CrossRef]

24. Shao, Y.; Aizhao, P.; Lin, H. POSS end-capped diblock copolymers: Synthesis, micelle self-assembly and properties. *J. Colloid Interface Sci.* **2014**, *425*, 5–11. [CrossRef] [PubMed]

25. Durmus, A.; Kasgoz, A.; Erca, N.; Akn, D.; Şanl, S. Effect of Polyhedral Oligomeric Silsesquioxane (POSS) Reinforced Polypropylene (PP) Nanocomposite on the Microstructure and Isothermal Crystallization Kinetics of Polyoxymethylene (POM). *Polymer* **2012**, *53*, 5347–5357. [CrossRef]

26. Illescas, S.; Sanchez-Soto, M.; Milliman, H.; Schiraldi, D.A.; Arostegui, A. The morphology and properties of melt-mixed polyoxymethylene/monosilanolisobutyl-POSS composites. *High Perform. Polym.* **2011**, *23*, 457–467. [CrossRef]

27. Pielichowski, K.; Flejtuch, F. Some comments on the melting and recrystallization of polyoxymethylene by high-speed and StepScan differential scanning calorimetry. *Polimery* **2004**, *49*, 558–560.

28. Szubert, K.B.; Marciniec, M.; Dutkiewicz, M.; Potrzebowski, M.; Maciejewski, H. Functionalization of spherosilicates via hydrosilylation catalyzed by well-defined rhodium siloxide complexes immobilized on silica. *J. Mol. Catal. A: Chem.* **2014**, *391*, 150–157. [CrossRef]

29. Szwarc-Rzepka, K.; Marciniec, B.; Jesionowski, T. Immobilization of multifunctional silsesquioxane cage on precipitated silica supports. *Adsorption* **2013**, *19*, 483. [CrossRef]

30. Jeziórska, R.; Swierz-Motysia, B.; Szadkowska, A.; Marciniec, B.; Maciejewski, H.; Dutkiewicz, M.; Leszczyńska, I. Effect of POSS on Morphology, Thermal and Mechanical Properties of Polyamide 6. *Polimery* **2011**, *56*, 809–816.

31. Siengchin, S.; Psarras, G.C.; Karger-Kocsis, J. POM/PU/carbon nanofiber composites produced by water-mediated melt compounding: Structure, thermo-mechanical and dielectric properties. *J. Appl. Polym. Sci.* **2010**, *117*, 1804–1812. [CrossRef]

32. Wunderlich, B. *Macromolecular Physics: Crystal Nucleation, Growth, Annealing*; Academic Press: New York, NY, USA, 1976.

33. Plummer, C.J.G.; Menu, P.; Cudré-Mauroux, N.; Kausch, H.-H. The effect of crystallization conditions on the properties of polyoxymethylene. *J. Appl. Polym. Sci.* **1995**, *55*, 489–500. [CrossRef]

34. Mück, K.-F. Physical Constants of Poly(oxymethylene). In *Wiley Database of Polymer Properties*; Wiley: Hoboken, NJ, USA, 2003.
35. Pelzbauer, Z.; Galeski, A. Growth rate and morphology of polyoxymethylene supermolecular structures. *J. Polym. Sci. Polym. Symp.* **1972**, *8*, 23–32. [CrossRef]
36. Yousef, S.; Visco, A.; Galtieri, G.; Njuguna, J. Flexural, impact, rheological and physical characterizations of POM reinforced by carbon nanotubes and paraffin oil. *Polym. Adv. Technol.* **2016**, *27*, 1338–1344. [CrossRef]
37. Hu, Y.; Ye, L. Nucleation effect of polyamide on polyoxymethylene. *Polym. Eng. Sci.* **2005**, *45*, 1174–1179. [CrossRef]
38. Nobile, M.R.; Bove, L.; Somma, E.; Kruszelnicka, I.; Sterzynski, T. Rheological and structure investigation of shear-induced crystallization of isotactic polypropylene. *Polym. Eng. Sci.* **2005**, *45*, 153–162. [CrossRef]
39. Everaert, V.; Groeninck, G.; Koch, M.H.J.; Reynaers, H. Influence of fractionated crystallization on the semicrystalline structure of (POM/(PS/PPE)) blends. Static and time-resolved SAXS, WAXD and DSC studies. *Polymer* **2003**, *44*, 3491–3508. [CrossRef]
40. Pielichowska, K.; Szczygielska, A.; Spasówka, E. Preparation and characterization of polyoxymethylene-copolymer/hydroxyapatite nanocomposites for long-term bone implants. *Polym. Adv. Technol.* **2012**, *23*, 1141–1150. [CrossRef]

![polymers logo] *polymers*

MDPI

Article

Time-Resolving Study of Stress-Induced Transformations of Isotactic Polypropylene through Wide Angle X-ray Scattering Measurements

Finizia Auriemma [1,*], Claudio De Rosa [1], Rocco Di Girolamo [1], Anna Malafronte [1], Miriam Scoti [1], Geoffrey Robert Mitchell [2] and Simona Esposito [1]

[1] Dipartimento di Scienze Chimiche, Università di Napoli "Federico II", Complesso Monte Sant' Angelo, via Cintia, 80126 Napoli, Italy; claudio.derosa@unina.it (C.D.R.); rocco.digirolamo@unina.it (R.D.G.); anna.malafronte@unina.it (A.M.); miriam.scoti@unina.it (M.S.); Simona.Esposito@lamberti.com (S.E.)

[2] CDRSP—Centre for Rapid and Sustainable Product Development, Polytechnic Institute of Leiria, Centro Empresarial da Marinha Grande, 2430-028 Marinha Grande, Portugal; geoffrey.mitchell@ipleiria.pt

* Correspondence: finizia.auriemma@unina.it; Tel.: +81-081-674341

Received: 3 January 2018; Accepted: 6 February 2018; Published: 8 February 2018

Abstract: The development of a highly oriented fiber morphology by effect of tensile deformation of stereodefective isotactic polypropylene (iPP) samples, starting from the unoriented γ form, is studied by following the transformation in real time during stretching through wide angle X-ray scattering (WAXS) measurements. In the stretching process, after yielding, the initial γ form transforms into the mesomorphic form of iPP through mechanical melting and re-crystallization. The analysis of the scattering invariant measured in the WAXS region highlights that the size of the mesomorphic domains included in the well oriented fiber morphology obtained at high deformations increases through a process which involves the coalescence of the small fragments formed by effect of tensile stress during lamellar destruction with the domain of higher dimensions.

Keywords: isotactic polypropylene; stress-induced phase transitions; structural analysis; X-ray diffraction

---

1. Introduction

The structural and morphological transformations of isotactic polypropylene (iPP) by effect of stretching have been widely studied [1–20]. These studies are mainly concerned with iPP samples synthesized by Ziegler-Natta (ZN) heterogeneous catalyst systems [1–15]. In general, ZN-iPP crystallizes from the melt in the α form, which shows high rigidity, and breaks at low deformations before yielding [1–15,21]. Therefore, the key to probing the mechanical response of iPP, in the α form up to high deformations, generally consists in crystallizing iPP from the melt at high cooling rates, to obtain more easily deformable samples, due to formation of thin lamellar crystals or of the mesophase [1–10,14].

More recently, the deformation behavior of metallocene-made iPP samples with a uniform distribution of stereodefects has been also studied [16–20]. These samples contain only one kind of stereo-defects, namely isolated *rr* triads, at concentrations in the range 0.5–11 mol % [16]. They crystallize from the melt as mixtures of the α and γ forms [22]. In particular, the relative amount of γ form increases with increasing the concentration of *rr* defects. Simultaneously, the values of the melting temperature and the degree of crystallinity decrease from 160 to 84 °C, and from 70% to 40%, respectively, and also the values of Young modulus decrease, from 200 to 19 MPa [16,18,22]. However, in spite of the relatively high values of melting temperature, crystallinity and Young modulus, these samples show high ductility, with values of deformation at break increasing with increasing the concentration of defects [16,18,22]. They define a class of materials with properties which can be finely tuned by introducing the tailored amount of *rr* defects in the reactor step [17,22]. Typically, properties of stiff-plastic materials with high melting temperatures are achieved at low defect concentrations, whereas at intermediate *rr* concentrations, the samples show properties of flexible-materials melting in the temperature range 115–120 °C, and achieve properties of elastomeric materials with high strength due to the high crystallinity (higher than 40%), melting at 80–110 °C, at high concentration of *rr* defects [22]. This behavior is clearly different from that of ZN-iPP, for which the γ form may be obtained only at high pressures and in low molecular mass fractions [23].

The study of the structural and morphological transitions occurring by effect of stretching in metallocene-made iPP samples was studied both at unit cell and lamellar length scales, by performing wide (WAXS) and small (SAXS) angle X-ray scattering measurements [16–20]. In particular, WAXS analysis show that during the transformation of the initially isotropic spherulitic morphology into a well oriented fiber morphology by effect of stretching, the initial crystalline α and/or γ forms transform into a well-oriented mesophase [16–20], as in Ziegler-Natta iPP [1–15]. The transformation into mesophase is gradual and starts occurring at deformations higher than the yielding point, that is concomitant with lamellar fragmentation (mechanical melting) and successive recrystallization into fibrils [17–20,24,25]. The critical strain at which the phase transition starts occurring depends on the stability of the crystal blocks initially present in the sample and the entanglement density of the amorphous phase [17–20,24,25]. SAXS analysis, instead, reveals that the main events occurring at low deformations are interlamellar separation and lamellar reorientation [19,20,26–29]. With increasing the deformation, chevron-like textures develop, followed by fragmentation of lamellae, and cavitation [19,20,29–34]. The fibrillary morphology obtained upon lamellar fragmentation is characterized by rod-like entities consisting of mesomorphic elongated aggregates separated by amorphous regions, placed at uncorrelated longitudinal and lateral distances [19].

In this paper, we focus on an aspect often neglected in performing the analysis of the WAXS data collected in real time during tensile deformation, that is the behavior of the scattering invariant defined in the WAXS region as a function of deformation. The analysis is illustrated in the case of two stereodefective metallocene iPP samples and has allowed highlighting a process of coalescence of mesomorphic fragments included in the well oriented fibers obtained at high deformations with consequent formation of larger domains.

2. Materials and Methods

Samples, synthesized using single site organometallic catalysts activated with methylalumoxane as described in ref. [35], are analyzed. The samples are characterized by molecular mass of ≈120 and ≈210 kg/mol, a polydispersity index close to two and a concentration of stereo-defects consisting solely of isolated *rr* triads equal to 5.92 and 11.01 mol % [22]. The samples show high level of crystallinity (≈40–55%), relatively high melting temperature, glass transition temperature around 0 °C, and crystallize from the melt almost completely in the γ form [22]. The main characteristics of these samples are shown in Table 1.

Table 1. Molecular mass (M_v), melting temperature (T_m) and content of *rr* triads of iPP samples prepared by metallocene catalysts as described in ref. [22] [a].

Sample	M_v (kg/mol) [b]	T_m (°C) [c]	[*rr*] (mol %) [d]	x_c [f] (%)	f_γ [f] (%)
iPP-5.9	211	114	5.92	55	96
iPP-11.0	123	84	11.01	42	100

[a] The polydispersity index of molecular masses is close to 2. [b] From the intrinsic viscosities [15,22]. [c] The melting temperatures were obtained with a differential scanning calorimeter Perkin Elmer DSC-7 performing scans in a flowing N_2 atmosphere and heating rate of 10 °C/min [22]. [d] Determined from solution ^{13}C NMR analysis [22]. [f] Crystallinity index x_c and relative amount of γ form with respect to the α form f_γ evaluated from X-ray diffraction profiles of compression-moulded samples [16–20].

Specimens of rectangular shape and initial gauge length l_0 of 3.0 mm, width w_0 of 1.6 mm, and thickness t_0 of 0.25 mm were cut from films obtained by melting the samples in a hot press up to reach a temperature 20–30 °C higher than the melting temperature determined in the DSC scans, using only a small pressure, and by cooling the melt to room temperature while fluxing cold water in the refrigerating system of the press (average cooling rate $\approx$ 10 °C/min).

The stretching experiments are performed at 25 °C while collecting in situ bidimensional WAXS patterns using the high flux available at the Synchrotron Radiation Source in Daresbury (Cheshire, UK, beam-line 16.1). The applied deformation rate is 2.36 mm/min, the wavelength of incident radiation is λ = 1.4 Å, and the WAXS images are collected at a rate of 1 frame/5 s. The size of the primary X-ray beam at the sample position is 0.1×0.1 mm². The strong reflections of the unoriented film specimens are used to determine the correct beam center coordinates and the sample to detector distance. Roughly, the transmission factor is determined measuring the intensity of the direct beam (using an attenuator) in absence of the sample, and in presence of the sample, before stretching, and at the end of stretching experiment. In no case the sample was brought to the rupture. Raw WAXS data are reduced and analyzed using the home made software XESA [36]. In the experiments, the X-ray beam is incident on the central region of the specimens. In this way, the structural changes occurring by effect of mere uniaxial deformation with negligible shear components may be probed. The equatorial profiles are obtained by radial integration of WAXS intensity along arcs in slices spanning an angle of $\pm 20°$ with respect to the horizontal axis (equator) of the bidimensional WAXS images. We checked that in the stretching conditions adopted for the in situ WAXS measurements, the samples experience uniform deformation. We found that the thickness t decreases according to a power law $t = t_0 \, (l_0/l)^\nu$, with ν comprised between 0.4 and 0.5, depending on the deformation, l being the gauge length of the deformed specimens [19,20]. In the calculation we set, for the sake of simplicity, $\nu \approx 0.5$, in the whole deformation range, as expected for ideal rubbery materials [37].

3. Results

The stress-strain curves and representative WAXS images collected in situ during stretching of compression-moulded specimens of the samples iPP-5.9 and iPP-11.0 are shown in Figure 1. The equatorial profiles extracted from the bidimensional WAXS images collected at relevant deformations are reported in Figure 2. The initial compression-moulded samples are crystallized from the melt in the γ form almost completely (patterns a of Figure 1) [19,20].

The WAXS images indicate that the γ form gradually transforms into the mesomorphic form of iPP by effect of deformation [16–20]. The transformation into mesophase starts at deformation $\varepsilon_{start} \approx 200\%$ and is complete at $\varepsilon_{end} \approx 400$–500% (Figures 1 and 2). In all cases a highly oriented fibrillary morphology is obtained already at the deformations marking the end of transition [19,20].

The textural transformations of the initially isotropically oriented morphology into a well-oriented fiber morphology is indicated by the change in distribution of intensity of the reflections along the azimuthal angle χ defined in the image a of Figure 1A. This is particularly evident for the changes in the azimuthal spreading of intensity at $q \approx 10$ nm^{-1}, corresponding to the reflection $(111)_\gamma$ of γ form at low deformations (images a–c of Figure 1 and curves a of Figure 2), and to the tail of the mesomorphic halo centered at $q \approx 11$ nm^{-1} at high deformations (images d–e of Figure 1A, d of Figure 1B, curves d–e of Figure 2A, and d of Figure 2B). In fact, as shown in Figure 3, the azimuthal intensity distribution at $q \approx 10$ nm^{-1} is uniform for the undeformed samples (curves a of Figure 3) and becomes polarized on the equator with increasing the deformation (curves b–e of Figure 3A and b–d of Figure 3B).

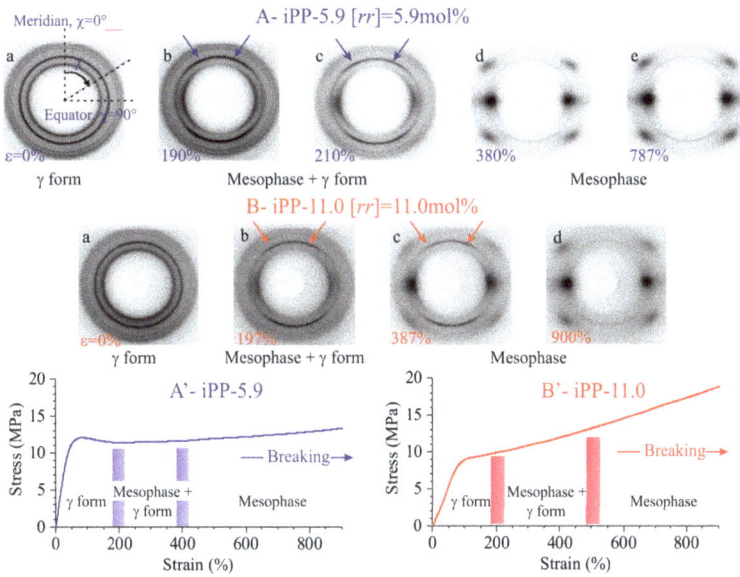

Figure 1. Stress strain curves (**A′**,**B′**) and bidimensional WAXS patterns of the sample iPP-5.9 (**A**) and iPP-11.0 (**B**) at the indicated deformations. The critical values of deformation marking the beginning (ε_{start}), and the complete (ε_{end}) transformation of the initial γ form into mesophase are indicated by the vertical bold lines in (**A′**,**B′**). The stretching direction is vertical. Arrows in b and c indicate the polarization of the (008)$_\gamma$ reflection of γ form, off the equator (horizontal direction) and off the meridian (vertical direction), defined in the image a of part A of the figure, occurring at low deformations, due to orientation of the crystals of the γ form with chain axes nearly perpendicular to the stretching direction (vide infra). The azimuthal angle χ is defined in the image a of part A.

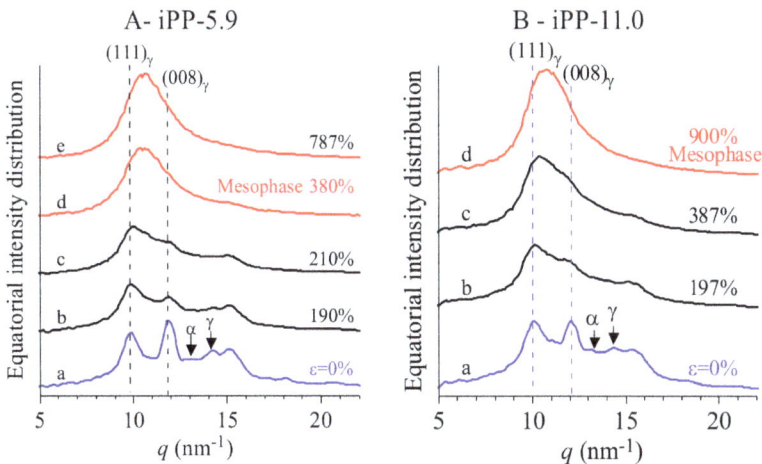

Figure 2. Equatorial profiles extracted from the bidimensional WAXS images of Figure 1 for the samples iPP-5.9 (**A**) and iPP-11.0 (**B**). The relevant reflections of α and γ forms are indicated.

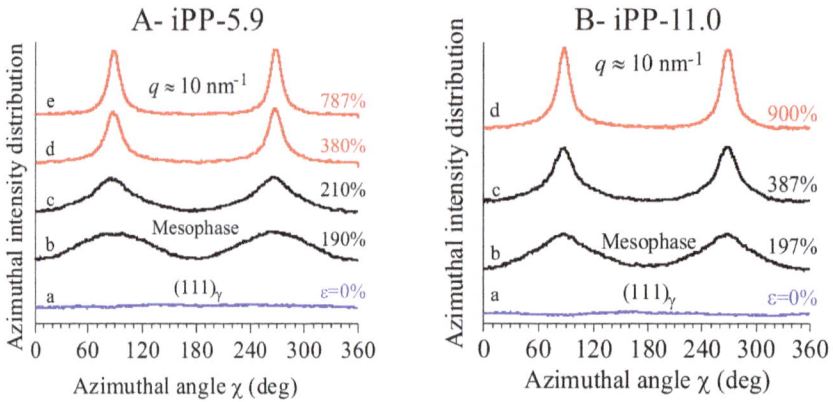

Figure 3. Azimuthal profiles of intensity centered in the *q* region around 10 nm^{-1}, extracted from the bidimensional WAXS images of Figure 1 for the samples iPP-5.9 (**A**) and iPP-11.0 (**B**).

The intensity at $q \approx 12$ nm^{-1} relative to the (008)$_\gamma$ reflections of γ form (curves a of Figure 2), instead, at low deformations is polarized on a layer line off the equator (horizontal direction) and off the meridian (vertical direction), as indicated by the arrows in the images b, c of Figure 1, and only when the transformation into the mesophase is complete, that is, at deformations higher than ε_{end}, the intensity at $q \approx 12$ nm^{-1} becomes polarized on the equator [19,20]. This indicates that part of the crystals of the initial γ form tend to become oriented at low deformations with the chain axes nearly perpendicular to the stretching direction, as shown in Figure 4, instead than parallel as in the standard fiber morphology [16,38]. It is worth noting that for the perpendicular chain axis orientation of γ form, the polarization of the (111)$_\gamma$ (at $q \approx 10$ nm^{-1}) reflection occurs on the equator [39].

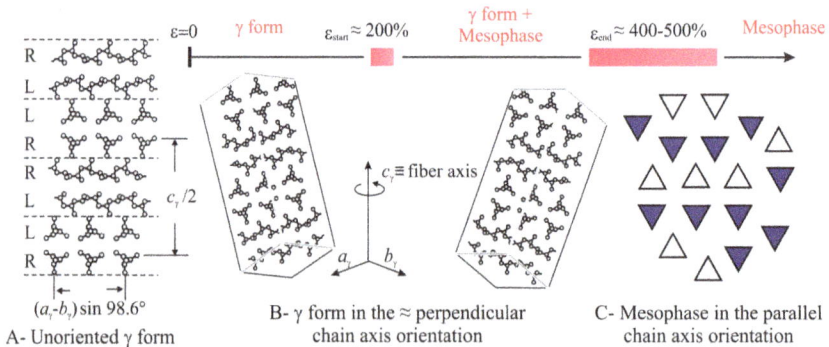

Figure 4. Structural model of γ form (**A**), scheme of the perpendicular chain axis orientation of γ crystals (**B**) and of mesomorphic domains of iPP (**C**), as they develop by effect of stretching, starting from unoriented specimens.

4. Data Analysis and Discussion

The change of orientation of crystals by stretching is quantitatively analyzed using the azimuthal intensity at $q \approx 10$ nm^{-1} of Figure 3. A numerical descriptor *O* that plays the role of an orientation order parameter is introduced from the calculation of the orientation function <P_2> [40]. In general, for uniaxially symmetric materials (such as fibers) the orientation function for a direction normal (pole) to any given {*hkl*} family of planes <P_2(cosχ)> with respect to a preferred direction (fiber axis) is defined as:

$$\langle P_2(\cos\chi)\rangle = \frac{1}{2}\left(3\langle\cos^2\chi\rangle - 1\right) \tag{1}$$

where <cosχ> is the average cosine of the angle χ that the poles make with the preferred direction and denotes the second order Legendre polynomial of argument <cosχ>. The values of <cosχ>, in turn, are calculated from the azimuthal intensity distribution of the *hkl* reflection as [40]:

$$\left\langle\cos^2\chi\right\rangle = \frac{\int\limits_0^{\pi/2} I(q)\cos^2\chi\sin\chi d\chi}{\int\limits_0^{\pi/2} I(q)\sin\chi d\chi} \tag{2}$$

According to Equation (1), <P_2> = 1 corresponds to an ideal case of perfect alignment of the poles in the preferential direction, <P_2> = 0 corresponds to isotropic case and <P_2> = −0.5 corresponds to an ideal case of perfect perpendicular orientation.

In our case, the order parameter is calculated considering the degree of polarization of the intensity at $q \approx 10$ nm^{-1} on the equator, with respect to the fiber axis. This would correspond to values of the orientation function <P_2>$_{exp}$ ranging from zero at zero deformation (completely isotropic case) to values less than zero with increasing the deformation. The parameter O is calculated by comparing the experimental values of <P_2>$_{exp}$ at $q \approx 10$ nm^{-1} with that of an ideal model characterized by an extremely narrow equatorial polarization of intensity at this q (that is <P_2>$_{id}$ = −0.5) as:

$$O = \frac{\langle P_2\rangle_{exp}}{\langle P_2\rangle_{id}} \tag{3}$$

According to Equation (3), the values of O are positive and range from zero (fully isotropic case) to 1 (fully oriented case).

Since the polarization intensity on the equator at $q \approx 10$ nm^{-1} originates from the perpendicular chain axis orientation of the residual γ form crystals at low and moderate deformations and from the parallel chain axis orientation of the already transformed mesomorphic domains at moderate and high deformations, the value of the parameter O represents a measure of the average degree of orientation of the chain axes with respect to the fiber axis only at high deformations, that is, after complete transformation of the original crystals of γ form into the mesophase, whereas at lower deformation, the values of O represents a measure of the average degree of orientation of the residual crystals of γ form in the perpendicular chain axis orientation, and of the transformed mesomorphic domains in the parallel chain axis orientation. The so obtained values of the parameter O derived for the iPP samples are reported in Figure 5A.

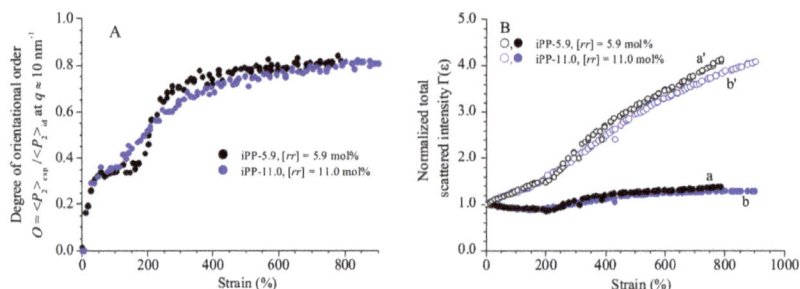

Figure 5. Orientation parameter O (**A**) and total WAXS scattered intensity (invariant) normalized for the integrated intensity of the unoriented sample $\Gamma(\varepsilon)$ (**B**) as a function of strain, for the samples iPP-5.9 (○, ●) and iPP-11.0 (○, ●). The parameter O is calculated from the azimuthal intensity distribution at $q \approx 10$ nm^{-1} in the bi-dimensional WAXS images of Figure 1, corresponding to the reflection $(111)_\gamma$ of γ form at low deformations, to the tail of the mesomorphic halo centered at $q \approx 11$ nm^{-1} at high deformations. It represents a measure of the average degree of orientation of the chain axes with respect to the fiber axis at high deformations ($>\varepsilon_{end}$), and to the average degree of orientation of the residual crystals of γ form in the perpendicular chain axis orientation, and of the transformed mesomorphic domains in the parallel chain axis orientation at deformations lower than ε_{end}. In (**B**), the curves a' and b' (open symbols) are obtained by multiplying the curves a and b (full symbols) by the factor $t_0/t = (l/l_0)^{1/2}$ to account for the thickness contraction (see Experimental section).

The evolution of orientational order O with deformation (Figure 5A) of the two samples follows a common trend involving different steps. For deformations ε lower than 100%, a steep increment of the orientational order occurs, due to orientation of the crystals in the γ form with chains axes perpendicular to stretching direction. In the deformation range between 100 and $\approx$150–200%, the values of O reach a quasi-plateau of 0.30–0.4. This indicates that at low deformations, changes in azimuthal intensity distribution of $(111)_\gamma$ reflections of γ forms essentially probe small reorientation of the lamellar crystals along the pathway toward the fibrillary morphology, due to lamellar rotations [26,27,40]. The leading deformation mechanisms in the first plateau region, instead, correspond to interlamellar shear, that is slip of the crystalline lamellae parallel to each other, and/or interlamellar separation [26–28,41,42]. At this stage, no phase transitions occur and all movements are assisted by the shear deformation of the compliant interlamellar amorphous phase. Successively, starting from deformations close to the critical strain at which transformation into mesophase starts ($\varepsilon_{start} \approx 200\%$, Figures 1 and 4), the orientational order O increases again up to reach a final plateau value of $\approx$0.8 at high deformations (Figure 5A). This indicates that after lamellar breaking, further deformation induces collective shear process, up to reach destruction of the preexisting crystals (mechanical melting) and successive recrystallization with formation of fibrils including mesomorphic aggregates [24,25].

A further quantitative descriptor of the changes at WAXS length scale is obtained using the ratio $\Gamma(\varepsilon) = Q(\varepsilon)/Q(0)$ between the total WAXS intensity $Q(\varepsilon)$ at deformation ε and the total scattered intensity $Q(0)$ at $\varepsilon = 0$. For samples exhibiting uniaxial symmetry (fibers), the total scattered intensity in the WAXS region $Q(\varepsilon)$ can be calculated from the WAXS intensity collected on a bidimensional detector (Figure 1) by integrating all over the sampled reciprocal space, that is between $q_{min} = 3$ nm^{-1} and $q_{max} = 22$ nm^{-1}, using Equation (4):

$$Q(\varepsilon) = \frac{1}{\pi} \int_{q_{min}}^{q_{max}} q^2 \left(\int_0^\pi I_\varepsilon(q, \chi) \sin\chi d\chi \right) - Bk(q)dq \qquad (4)$$

In Equation (4), it is implicitly assumed a uniform distribution of intensity along circles of radius q $\sin\chi$ because of the cylindrical symmetry of the fibers. Moreover, the intensity, at any q, was subtracted for the background contribution ($Bk(q)$) estimated from the equation of the straight line subtending the radially averaged profiles extracted from the bi-dimensional diffraction images. The ratio of the total scattered intensity and the integrated intensity of the unoriented sample gives the normalized value of the total scattered intensity:

$$\Gamma(\varepsilon) = \frac{Q(\varepsilon)}{Q(\varepsilon = 0)} \qquad (5)$$

Equation (4) corresponds to a reduced scattering invariant [43]. The values of the parameter $\Gamma(\varepsilon)$ are reported in Figure 5B as a function of deformation before (curves a, b) and after (curves a', b') correction for the thickness contraction.

Also in this case, we observe that the change of the parameter $\Gamma(\varepsilon)$ with deformation follows a common trend for the two samples (Figure 5B). In particular, before correction for the thickness contraction, $\Gamma(\varepsilon)$ decreases with increasing deformation up to $\varepsilon \approx 200\%$, then increases for further increment of deformation, up to reach a quasi-plateau at values close to 1.3–1.4 (curves a, b of Figure 5B). The decrease in $\Gamma(\varepsilon)$ values of Figure 5B at deformations lower than 200% can be attributed to a decrease in the sample thickness. In fact, after correction for this effect only an increase of $\Gamma(\varepsilon)$ with the deformation is observed (curves a', b' of Figure 5B). In particular, we use a correction factor equal to t_0/t to account for the thickness contraction, by setting $t_0/t = (l/l_0)^\nu$ with $\nu = 0.5$ (see Section 2). Therefore, curves a, b and a', b' of Figure 5B illustrate the change of the parameter $\Gamma(\varepsilon)$ with increasing the deformation, in the limiting cases of no thickness contraction (Poisson's ratio = 0, curves a, b) and in the rubbery limit of thickness contraction, corresponding to Poisson's ratio $\nu = 0.5$ (curves a', b'). The invariant is expected to change with deformation according to a behavior in between these two limiting cases.

After correction for the thickness contraction, the values of the WAXS reduced scattering invariant experience a remarkable increase with increasing the deformation, reaching values up to ≈ 4 times higher than those of the undeformed state. The remarkable increase of invariant with increasing deformation holds also after taking into account the absorption, since the correction of the scattered intensity for the transmission factor would decrease the values of the parameter $\Gamma(\varepsilon)$ at most by a factor of 1.1.

The increase of the total scattering intensity in the WAXS region by a factor 1.3–1.4 in the hypothesis of no thickness contraction, and a factor ≈ 4 in the hypothesis of thickness contraction according to the rubbery limit is unexpected. In fact, for a multi-phases system (in our case three phases, γ form, mesophase and amorphous component), the measured WAXS intensity $I_\varepsilon(q,\chi)$ can be written as a linear combination of the scattering intensities from each phase:

$$I_\varepsilon(q,\chi) = \sum_{i=1}^{3} x_i(\varepsilon) I_{\varepsilon i}(q,\chi) \qquad (6)$$

where $I_{\varepsilon i}(q,\chi)$ are the scattering intensities of the γ form (phase i = 1), mesophase (phase i = 2) and amorphous component (phase i = 3) and $x_i(\varepsilon)$ are the corresponding mass fractions. Therefore, also the total WAXS reduced invariant $Q(\varepsilon)$ (Equation (4)) is a linear combination of the contribution from components $Q_i(\varepsilon)$.

$$Q(\varepsilon) = \sum_{i=1}^{3} Q_i(\varepsilon) \qquad (7)$$

Each term of the invariant $Q_i(\varepsilon)$, in turn, is proportional to the following quantities [44,45]:

$$Q_i(\varepsilon) = \rho_{i\varepsilon}^2 N_{i\varepsilon} V_{i\varepsilon} \qquad (8)$$

where ρ_i and $N_{i\varepsilon}$, are the electron density and the number of domains of phase i with average volume of coherence equal to $V_{i\varepsilon}$. The $N_{i\varepsilon}$ domains contribute additively to the scattering, whereas the coherent (elastic) scattering occurs only within each domain. The density of γ form, mesophase and amorphous phase in isotactic polypropylene are 0.939, 0.91 and 0.854 g/cm^3 [46] and the corresponding values of electron density are 323, 313 and 294 electrons/nm^3 respectively. Based on previous analyses [16–18], the total degree of crystallinity, due to the γ form and mesophase contributions, apparently, does not greatly change during deformation ($x_1(\varepsilon) + x_2(\varepsilon) \approx 0.50$, see Table 1). Therefore, in the hypothesis that also the total volume of the coherently scattering crystalline domains does not change by effect of deformation ($N_{1\varepsilon} V_{1\varepsilon}, + N_{2\varepsilon} V_{2\varepsilon},$ = cost.), the scattering invariant is expected to remain constant or to slightly decrease with increasing the deformation, since the γ form transforms into the mesophase with slightly lower electron density.

The tendency of the scattering invariant to increase at deformations lower than ε_{start} marking the beginning of transformation of the initial γ form into mesophase, may be due to local "densification" of the amorphous intralamellar phase by effect of stretching, since the intralamellar amorphous phase experiences compressive forces in the direction parallel or perpendicular to the layers, depending on the orientation, perpendicular or parallel to the stretching direction, respectively, of the lamellar stacks in which the amorphous layers are embedded [2,20]. At deformations higher than ε_{start}, marking also the beginning of lamellar destruction, we hypothesize that the increase of invariant may be due to a slight increase of the size of coherent domains located in the mesomorphic aggregates, leading to a remarkable increase of their average volumes $V_{2\varepsilon}$, even though the total number of these domains $N_{2\varepsilon}$ does not necessarily change. This increase may involve the incorporation of isolated chains, or groups of 2–3 chains located at the boundaries of adjacent domains, into close neighboring larger domains through small movements and/or rotations facilitated by deformation, as shown in the model of Figure 6.

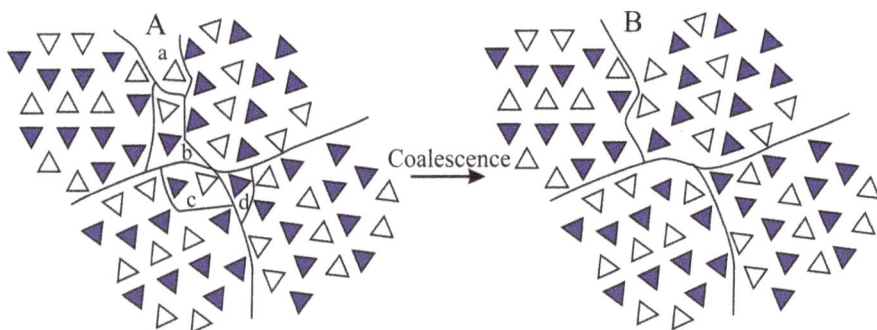

Figure 6. Model of coalescence of mesomorphic domains separated at the boundaries by isolated groups of 1–2 chains (sub-domains a–d) in a completely different arrangement (**A**) with formation of larger mesomorphic domains (**B**). By effect of stretching, the sub-domains a–d coalesce with close-neighbors mesomorphic domains.

Moreover, considering that the mesophase originates from the breaking of crystals of γ form through a process of mechanical melting and re-crystallization, and that the lamellar crystals of γ form are not characterized by chain folding because of the non-parallel arrangement of the chains (Figure 4A) [47], the size of the coherent domains in a mesomorphic aggregate may easily increase by effect of stretching also in the direction parallel to the chain axes, through incorporation of monomeric units located at the boundaries with the amorphous phase. We calculate that, on average, an increase of the coherent length by a factor of 1.1–1.6 along the three orthogonal directions x, y and z of a domain, leads to an increase of volume by a factor 1.1^3–$1.6^3 \approx 1.3$–4. As shown in the model of Figure 6, the increase of the coherent length of the mesomorphic domains by incorporation (coalescence) of the isolated sub-domains a-d within the mesomorphic domains, does not necessarily lead to a decrease of the number of domains $N_{2\varepsilon}$. As a consequence, the total scattering intensity increases by effect of deformation, because the product ($N_{2\varepsilon} V_{2\varepsilon}$) increases due to coalescence phenomena of the kind depicted in Figure 6.

On the other hand, the fact that the crystallinity index measured from the radially averaged WAXS profiles as the ratio between the area of the crystalline phase (A_c) and the total area of the diffraction profile (A_{tot}) does not apparently change (see Figure 7A), is due to the fact that A_c is often overestimated in this procedure, as it is obtained by the difference of A_{tot} and the diffraction area from the amorphous component (approximated in Figure 7A by the diffraction profile of an atactic polypropylene, curve b). In other terms, the crystallinity calculated from the radially averaged WAXS curves according to the procedure of Figure 7A, includes not only the Bragg contribution, but also diffuse scattering located in the diffraction region between the amorphous halo and the Bragg peaks (Figure 7A). Instead, in the experimental value of the scattering invariant, the diffuse scattering terms makes a contribution, which is at least one order of magnitude lower than the Bragg component. Diffuse scattering arises from the presence of disorder and its contribution to the scattering intensity is proportional to the fluctuations of the structure factor f (per monomeric unit), given by the term $<f^2> - <f>^2$, which represent the difference between the average square ($<f^2>$) and the square of average ($<f>^2$) of f [48]. Therefore, the higher the local structure factor deviates from the average value $<f>$ because of the presence of disorder, the higher the fluctuations, and the higher the diffuse scattering. By effect of coalescence of the small sub-domains into the large domains, the diffuse scattering decreases and the interference term (Bragg-like contribution) increases. The WAXS invariant increases, because the diffuse scattering term is always lower than the Bragg term.

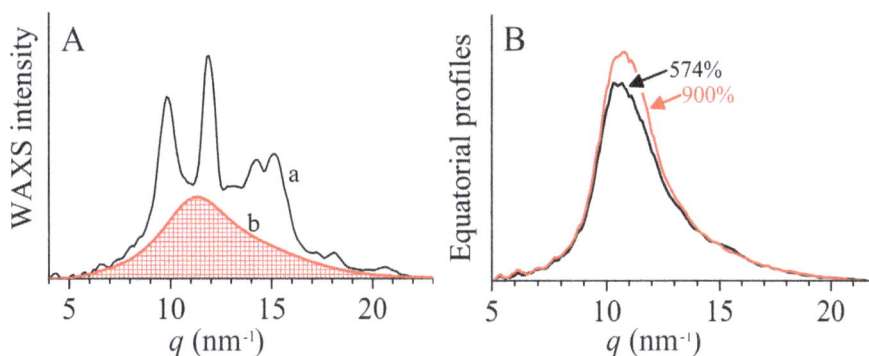

Figure 7. (**A**) Radially averaged WAXS profile of the undeformed sample iPP-5.9, extracted from the bidimensional WAXS image a of Figure 1A (a) and underlying amorphous contribution, approximated by the diffraction profile of an atactic polypropylene (b). (**B**) Equatorial profiles extracted from the bidimensional WAXS images of Figure 1B, of the sample iPP-11.0 stretched at 574 and 900% deformations. The diffraction curves are subtracted for the background contribution, approximated as a straight line subtending the whole profiles.

In order to probe the effective increase of the size of the mesomorphic domains by effect of stretching, we compare in Figure 7B the equatorial profiles of the sample iPP-11.0 stretched at 574 and 900% deformations. The curves are not corrected for the thickness contraction, but only for the transmission factor. Yet, it is evident that at 900% deformation the equatorial peak of the mesophase becomes narrower than that at 574% deformation, confirming the increase of the lateral size of the mesomorphic domains. It is worth noting that the maximum is shifted toward slightly higher q values at 900% deformation, indicating that the average correlation distance between the chain axes within the mesomorphic domains decreases, producing increase of electron density. Therefore, the increase of the scattering invariant is not only due to coarsening, but also to increase of electron density (densification).

Polymers **2018**, *10*, 162

5. Conclusions

The structural transformations occurring by effect of stretching in stereo-irregular isotactic polypropylene samples are studied in situ during deformation by collecting WAXS patterns. The metallocene-made iPP samples selected in this study show high ductility, allowing achieving high deformations before breaking.

WAXS analysis reveals that the crystals of γ form tend to assume a nearly perpendicular chain axis orientation instead than parallel to the stretching direction at low deformations. At $\approx 200\%$ deformation (ε_{start}) the γ form starts transforming into the mesomorphic form of iPP, and at 400–500% deformation (ε_{end}), the transformation is almost complete. The stress induced phase transition of the γ form into the mesophase occurs after the beginning of lamellar fragmentation, concomitant with destruction of the crystalline blocks (mechanical melting) and re-crystallization in fibrillary entities containing mesomorphic aggregates. At the end of transformation into mesophase, a fibrillary morphology develops with a high degree of orientational order of the chain axes parallel to the stretching direction. We show that the chains extracted from the original crystals re-crystallize forming aggregates of mesomorphic domains responsible for the coherent elastic scattering (Bragg-like contribution) which include at the boundaries isolated chains or groups of 2–3 chains, which contribute to diffuse scattering. The coalescence process of these chains with the larger mesomorphic domains, increases the volume of coherence of the mesomorphic domains. Since the number of coherent domains does not decrease, the WAXS scattering invariant increases by effect of coalescence during deformation, even if the sampled q region is narrow. Coalescence starts concomitant with the process of lamellar destruction, and beginning of transformation of the initial γ form into the mesophase and, after complete transformation into the mesomorphic form, continues facilitated by the mechanical stress field.

Acknowledgments: Paul Sotta, from CNRS/Solvay-France, is acknowledged for the useful discussions.

Author Contributions: Finizia Auriemma and Claudio De Rosa conceived the experiments. Finizia Auriemma wrote the manuscript. Simona Esposito, Geoffrey Robert Mitchell, Rocco Di Girolamo and Miriam Scoti performed the experiments. Geoffrey Robert Mitchell developed the program for data analysis. Finizia Auriemma, Simona Esposito and Anna Malafronte performed the data analysis.

Conflicts of Interest: The authors declare no conflict of interest.

References

1. Peterlin, A. Molecular model of drawing polyethylene and polypropylene. *J. Mater. Sci.* **1971**, *6*, 490–508. [CrossRef]
2. Aboulfaraj, M.; G'Sell, C.; Ulrich, B.; Dahoun, A. In situ observation of the plastic deformation of polypropylene spherulites under uniaxial tension and simple shear in the scanning electron microscope. *Polymer* **1995**, *36*, 731–742. [CrossRef]
3. Ran, S.; Zong, X.; Fang, D.; Hsiao, B.S.; Chu, B.; Phillips, R.A. Structural and morphological studies of isotactic polypropylene fibers during heat/draw deformation by in situ synchrotron SAXS/WAXS. *Macromolecules* **2001**, *34*, 2569–2578. [CrossRef]
4. Zuo, F.; Keum, J.K.; Chen, X.M.; Hsiao, B.S.; Chen, H.Y.; Lai, S.Y.; Wevers, R.; Li, J. The Role of Interlamellar Chain Entanglement in Deformation-Induced Structure Changes during Uniaxial Stretching of Isotactic Polypropylene. *Polymer* **2007**, *48*, 6867–6880. [CrossRef]
5. Pawlak, A.; Galeski, A. Cavitation during Tensile Deformation of Polypropylene. *Macromolecules* **2008**, *41*, 2839–2851. [CrossRef]
6. Ma, Z.; Shao, C.; Wang, X.; Zhao, B.; Li, X.; An, H.; Yan, T.; Li, Z.; Li, L. Critical stress for drawing-induced α-crystal-mesophase transition in isotactic polypropylene. *Polymer* **2009**, *50*, 2706–2715. [CrossRef]
7. Liu, Y.P.; Hong, Z.H.; Bai, L.G.; Tian, N.; Ma, Z.; Li, X.Y.; Chen, L.; Hsiao, B.S.; Li, L. A Novel Way to Monitor the Sequential Destruction of Parent-Daughter Crystals in Isotactic Polypropylene under Uniaxial Tension. *J. Mater. Sci.* **2014**, *49*, 3016–3024. [CrossRef]

8. Lu, Y.; Wang, Y.T.; Chen, R.; Zhao, J.Y.; Jiang, Z.Y.; Men, Y.F. Cavitation in Isotactic Polypropylene at Large Strains during Tensile Deformation at Elevated Temperatures. *Macromolecules* **2015**, *48*, 5799–5806. [CrossRef]

9. Krajenta, A.; Rozanski, A. The Influence of Cavitation Phenomenon on Selected Properties and Mechanisms Activated During Tensile Deformation of Polypropylene. *J. Polym. Sci. Part B* **2016**, *54*, 1853–1868. [CrossRef]

10. Kang, J.; Yuan, S.; Hong, Y.-L.; Chen, W.; Kamimura, A.; Otsubo, A.; Miyoshi, M. Unfolding of Isotactic Polypropylene under Uniaxial Stretching. *ACS Macro Lett.* **2016**, *5*, 65–68. [CrossRef]

11. Lezak, E.; Bartczak, Z.; Galeski, A. Plastic Deformation of the γ Phase in Isotactic Polypropylene in Plane-Strain Compression. *Macromolecules* **2006**, *39*, 4811–4819. [CrossRef]

12. Caelers, H.; Govaert, L.; Peters, G. The prediction of mechanical performance of isotactic polypropylene on the basis of processing conditions. *Polymer* **2016**, *83*, 116–128. [CrossRef]

13. Lu, Y.; Chen, R.; Zhao, J.; Jiang, Z.; Men, Y. Stretching Temperature Dependency of Fibrillation Process in Isotactic Polypropylene. *J. Phys. Chem. B* **2017**, *121*, 6969–6978. [CrossRef] [PubMed]

14. Caelers, H.J.M.; Troisi, E.M.; Govaert, L.E.; Peters, G.W.M. Deformation-Induced Phase Transitions in iPP Polymorphs. *Polymers* **2017**, *9*, 547. [CrossRef]

15. Xu, W.; Martin, D.; Arruda, E. Finite strain response, microstructural evolution and β→α phase transformation of crystalline isotactic polypropylene. *Polymer* **2005**, *46*, 455–470. [CrossRef]

16. De Rosa, C.; Auriemma, F.; De Lucia, G.; Resconi, L. From stiff plastic to elastic polypropylene: Polymorphic transformations during plastic deformation of metallocene-made isotactic polypropylene. *Polymer* **2005**, *46*, 9461–9475. [CrossRef]

17. De Rosa, C.; Auriemma, F. Structural-mechanical phase diagram of isotactic polypropylene. *J. Am. Chem. Soc.* **2006**, *128*, 11024–11025. [CrossRef] [PubMed]

18. De Rosa, C.; Auriemma, F. Stress-induced phase transitions in metallocene-made isotactic polypropylene. *Lect. Notes Phys.* **2007**, *714*, 345–371.

19. Auriemma, F.; De Rosa, C.; Di Girolamo, R.; Malafronte, A.; Scoti, M.; Mitchell, G.R.; Esposito, S. Deformation of Stereoirregular Isotactic Polypropylene across Length Scales: Influence of Temperature. *Macromolecules* **2017**, *50*, 2856–2870. [CrossRef]

20. Auriemma, F.; De Rosa, C.; Di Girolamo, R.; Malafronte, A.; Scoti, M.; Mitchell, G.R.; Esposito, S. Relationship Between Molecular Configuration and Stress-Induced Phase Transitions. In *Controlling the Morphology of Polymers*; Mitchell, G.R., Tojeira, A., Eds.; Springer: Berlin, Germany, 2016; pp. 287–327.

21. De Rosa, C.; Auriemma, F. The deformability of polymers: The role of disordered mesomorphic crystals and stress-induced phase transformations. *Angew. Chem. Int. Ed.* **2012**, *51*, 1207–1211. [CrossRef] [PubMed]

22. De Rosa, C.; Auriemma, F.; Di Capua, A.; Resconi, L.; Guidotti, S.; Camurati, I.; Nifant'ev, I.E.; Laishevtsev, I.P. Structure-property correlations in polypropylene from metallocene catalysts: Stereodefective, regioregular isotactic polypropylene. *J. Am. Chem. Soc.* **2004**, *126*, 17040–17049. [CrossRef] [PubMed]

23. Mezghani, K.; Phillips, P. The γ-phase of high molecular weight isotactic polypropylene: III. The equilibrium melting point and the phase diagram. *Polymer* **1998**, *39*, 3735–3744. [CrossRef]

24. Hiss, R.; Hobeika, S.; Lynn, C.; Strobl, G. Network Stretching, Slip Processes, and Fragmentation of Crystallites during Uniaxial Drawing of Polyethylene and Related Copolymers. A Comparative Study. *Macromolecules* **1999**, *32*, 4390–4403. [CrossRef]

25. Jiang, Z.Y.; Tang, Y.J.; Rieger, J.; Enderle, H.F.; Lilge, D.; Roth, S.V.; Gehrke, R.; Wu, Z.H.; Li, Z.H.; Men, Y.F. Structural Evolution of Tensile Deformed High-Density Polyethylene at Elevated Temperatures: Scanning Synchrotron Small- and Wide-Angle X-Ray Scattering Studies. *Polymer* **2009**, *50*, 4101–4111. [CrossRef]

26. Bowden, P.B.; Young, R.J. Deformation mechanisms in crystalline polymers. *J. Mater. Sci.* **1974**, *9*, 2034–2051. [CrossRef]

27. Lin, L.; Argon, A.S. Structure and plastic deformation of polyethylene. *J. Mater. Sci.* **1994**, *29*, 294–323. [CrossRef]

28. Schultz, J.M. *Polymer Materials Science*; Prentice Hall: Englewood Cliffs, NJ, USA, 1974.

29. Galeski, A. Strength and toughness of crystalline polymer systems. *Prog. Polym. Sci.* **2003**, *28*, 1643–1699. [CrossRef]

30. Pawlak, A.; Galeski, A. Plastic deformation of crystalline polymers: The role of cavitation and crystal plasticity. *Macromolecules* **2005**, *38*, 9688–9697. [CrossRef]

31. Pawlak, A.; Galeski, A. Cavitation and morphological changes in polypropylene deformed at elevated temperatures. *J. Polym. Sci. Part B* **2010**, *48*, 1271–1280. [CrossRef]

32. Pawlak, A. Cavitation during tensile deformation of isothermally crystallized polypropylene and high-density polyethylene. *Colloid Polym. Sci.* **2013**, *291*, 773–787. [CrossRef] [PubMed]

33. Read, D.; Duckett, R.; Sweeney, J.; Mcleish, T. The chevron folding instability in thermoplastic elastomers and other layered material. *J. Phys. D Appl. Phys.* **1999**, *32*, 2087–2099. [CrossRef]

34. Krumova, M.; Henning, S.; Michler, G. Chevron morphology in deformed semicrystalline polymers. *Philos. Mag.* **2006**, *86*, 1689–1712. [CrossRef]

35. Nifant'ev, I.E.; Laishevtsev, I.P.; Ivchenko, P.V.; Kashulin, I.A.; Guidotti, S.; Piemontesi, F.; Camurati, I.; Resconi, L.; Klusener, P.A.A.; Rijsemus, J.J.H.; et al. C_1-symmetric heterocyclic zirconocenes as catalysts for propylene polymerization, 1: Ansa-zirconocenes with linked dithienocyclopentadienyl-substituted cyclopentadienyl ligands. *Macromol. Chem. Phys.* **2004**, *205*, 2275–2291. [CrossRef]

36. Mitchell, G.R. *XESA Reference Manual*; Institute Polytechnic Leira: Leiria, Portugal, 2016.

37. Nitta, K.-H.; Yamana, M. *Rheology*; De Vicente, J., Ed.; InTech: London, UK, 2012; ISBN 978-953-51-0187-1. Available online: http://www.intechopen.com/books/rheology/poisson-s-ratio-and-mechanical-nonlinearity-undertensile-deformation (accessed on 7 February 2018).

38. Geddes, A.J.; Parker, K.D.; Atkins, E.D.T.; Beighton, E. "Cross-β" conformation in proteins. *J. Mol. Biol.* **1968**, *32*, 343–358. [CrossRef]

39. Auriemma, F.; De Rosa, C. Stretching Isotactic Polypropylene: From "cross-β" to Crosshatches, from α Form to γ Form. *Macromolecules* **2006**, *39*, 7635–7647. [CrossRef]

40. Alexander, L.E. *X-ray Diffraction Methods in Polymer Science*; Wiley: New York, NY, USA, 1979.

41. Humbert, S.; Lame, O.; Chenal, J.; Rochas, C.; Vigier, G. Small strain behavior of polyethylene: In situ SAXS measurements. *J. Polym. Sci. B* **2010**, *48*, 1535–1542. [CrossRef]

42. Séguéla, R. Dislocation approach to the plastic deformation of semicrystalline polymers: Kinetic aspects for polyethylene and polypropylene. *J. Polym. Sci. B* **2002**, *40*, 593–601. [CrossRef]

43. Roe, R.-J. *Methods of X-ray and Neutron Scattering in Polymer Science*; Oxford University Press: New York, NY, USA, 2000; ISBN-10 0195113217.

44. Krogh-Moe, J. A Method for Converting Experimental X-ray Intensities to an Absolute Scale. *Acta Cryst.* **1956**, *9*, 951–953. [CrossRef]

45. Norman, N. The Fourier Transform Method for Normalizing Intensities. *Acta Cryst.* **1957**, *10*, 370–373. [CrossRef]

46. Brandrup, J.; Immergut, E.H. (Eds.) *Polymer Handbook*, 3rd ed.; John Wiley & Sons: New York, NY, USA, 1989.

47. Meille, S.V.; Brückner, S. Non-parallel chains in crystalline *g*-isotactic polypropylene. *Nature* **1989**, *340*, 455–457. [CrossRef]

48. De Rosa, C.; Auriemma, F. *Crystals and Crystallinity in Polymers: Diffraction Analysis of Ordered and Disordered Crystals*; John Wiley & Sons: New York, NY, USA, 2013.

polymers

MDPI

Article

An Investigation and Comparison of the Blending of LDPE and PP with Different Intrinsic Viscosities of PET

Shi-Chang Chen, Li-Hao Zhang, Guo Zhang, Guo-Cai Zhong, Jian Li, Xian-Ming Zhang * and Wen-Xing Chen

National Engineering Laboratory for Textile Fiber Materials and Processing Technology (Zhejiang), Zhejiang Sci-Tech University, Hangzhou 310018, China; scchenzstu@hotmail.com (S.-C.C.); zhanglihao0829@163.com (L.-H.Z.); 15858143135@163.com (G.Z.); zhongguocai1995@163.com (G.-C.Z.); 15857126685@163.com (J.L.); wxchen@zstu.edu.cn (W.-X.C.)
* Correspondence: joolizxm@hotmail.com; Tel.: +86-0571-8684-3623

Received: 5 December 2017; Accepted: 1 February 2018; Published: 5 February 2018

Abstract: The blending of aliphatic polyolefins and aromatic poly(ethylene terephthalate) (PET) based on different intrinsic viscosities (IV) was conducted in a torque rheometer. The comparison of blend components in terms of low density polythene (LDPE) and polypropylene (PP) in blending with PET was investigated, and the effects of the IV and proportion of PET on polymer blends are discussed in detail. Polymer blends with or without compatibilizer were examined by using a differential scanning calorimeter, thermogravimetric analyzer, rotary rheometer, field-emission scanning electron microscopy and a universal testing machine. It was found that the blending led to an increase in processability and a decrease in thermal stability for blends. The morphological analysis revealed that the incompatibility of blends was aggravated by a higher IV of PET, while this situation could be improved by the addition of compatibilizer. Results showed that there was an opposite effect for the tensile strength and the elongation at break of the polymer blend in the presence of a compatibilizer, wherein the influence of IV of PET was complicated.

Keywords: polymer blend; poly(ethylene terephthalate); intrinsic viscosity; polyolefin; compatibilizer

1. Introduction

Polythene (PE) and polypropylene (PP), as representatives of aliphatic polyolefins, and poly(ethylene terephthalate) (PET), as a representative of aromatic polymers, are the three most common commercialized synthetic polymers, owing to their excellent overall performance and affordable price. The yield of the three polymers in China reached more than 14, 20, 40 million tons in 2016, respectively. PET can be categorized into three types with different intrinsic viscosities (IV) or molecular weights according to the application field: fiber grade, bottle grade and industrial yarn, whose IV are about 0.65 dL/g, 0.85 dL/g and 1.0 dL/g and whose corresponding aims are to be applied in the field of textiles, containers bottles and belts, respectively [1]. Occasionally, these three polymers can be used simultaneously in one product, for example, in beverage bottles. It also can be seen that recycled PET from different sources is basically a mixture of PETs with different properties [2].

The blending of polyolefins and PET can be an alternative to improve cost-effectiveness for industrial applications [3]. However, the thermodynamically immiscible nature of blend roots with large differences in their polarities, which leads to insufficient interfacial adhesion between blend components, result in poor mechanical properties [4–6]. Therefore, the polyolefin/PET compatibilization has already attracted a great deal of attention in the past decades.

Generally, the compatibilization of two immiscible polymers can be achieved by implementing a suitable grafting modification of polyolefin chains with reactive groups. The modification is achieved by chemical reactions between polar functional groups and polymer chains in melting blending [7]. By enhancing the interactions between grafted functional groups of polyolefin and PET end groups, the phase dispersion and adhesion at the interface can be improved [8]. For example, the functional groups like maleic anhydride (MAH) or glycidyl methacrylate (GMA) can be easily grafted onto polyolefins, and advantageously react with the carboxyl and hydroxyl end groups of PET. This method presents an excellent compatibilization effect with respect to chemical interactions between polyolefin and PET [9,10]. However, the conversion of functional groups grafted onto polyolefins is usually insufficient even though some residual active ingredients get involved in crosslinking reactions [11]. More importantly, it is not difficult to imagine that this route goes against the large-scale application due to its high cost. From this perspective, the grafted functional groups of polyolefins can be used as compatibilizers, with a small amount added into the PET/polyolefin blend during extrusion processes, then the desirable compatibilization effect can be obtained. The compatibilizer is concentrated at the interface of two phases in the blending system, and the co-crystallization/molecular chains' entanglement between the polymer component of the compatibilizer and polymer matrix would be generated to facilitate the interfacial adhesion between two immiscible polymers. Consequently, two immiscible blending components are effectively compatible by forming a dispersed structure in one unity [12–14].

In the previous literature, GMA, MAH, acrylic acid (AA), ethylene-vinyl acetate copolymer (EVA), and maleimide (MI) are the common functional groups that can be either grafted onto polyolefins or copolymerized into compatibilizer. Recently, Jazani et al. [15] improved the mechanical properties of bottle grade PET by grafting GMA onto PP with the aid of styrene (St) comonomer, and found that the increase of impact strength was attributed to the enhanced interaction between epoxy groups of GMA and carboxyl end groups of PET, as well as the hydrogen bonds formed between PP-g-GMA and PET. In yet another approach, a compatibilizer composed of styrene–ethylene–butylene–styrene (SEBS) grafted with MAH was used as the compatibilizer in the blending of PP and PET spun fibers [16]. Results showed that the compatibilizer significantly affected the elongation at break of the blend, while no relevant effect could be observed in the anisotropically-oriented fibers. Using PP-g-AA as a compatibilizer, a good interaction between the PP and PET surfaces was achieved as PET islands was deformed in micro-fibers after hot stretching [17].

In most published works, PP is the dominant continuous phase in the blend system of polyolefins/PET. There is little research focusing on PE/PET blends, let alone on the compatibilizer, by grafting the functional groups onto PE. Lei et al. [18] studied the compatibility between PET and HDPE using PE-g-MAH, SEBS-g-MAH and E–GMA, respectively. It was found that the compatibilizing effectiveness of E–GMA was the best. Raffa et al. [19] indicated the chemical reactions among polymer and additives had a significant effect on the ultimate melt rheology and mechanical properties of recycled PET/polyolefin blends.

In most cases, a variety of nanomaterials as a third component are added into two immiscible polymers to improve the compatibilizing effect [20]. As a matter of fact, the functional groups grafted onto polyolefins have been commercialized and applied gradually, such as PP-g-MAH and PE-g-MAH, which offer the best convenience for designing polymer blend systems.

Based on the successful surface functionalization of polypropylene/polystyrene and their application in polymer blends, as described in our previous work [7,21–24], the motivation has turned to developing possible industrialization polymer blends by exploring the relationships between the structure and properties in the processable polymers. Consequently, this work concentrates on the comparative investigation of the blending of PP and LDPE with different IV of PET using commercialized polyolefin-g-MAH as a compatibilizer. A systematic performance investigation of the polymer blends with or without compatibilizer was performed to elucidate the processable correlation between several major polymer commodities.

Polymers **2018**, *10*, 147

2. Materials and Methods

2.1. Materials

Three poly(ethylene terephthalate) (PET) chips with different intrinsic viscosities (IV) of 0.7 dL/g, 0.85 dL/g and 1.0 dL/g, were purchased from Zhejiang Guxiandao Polyester Dope Dyed Yarn Co., Ltd. (Shaoxing, China). The moisture and carboxyl end group content of the PET chips are also listed as Table 1. LDPE (F401) and PP (DNDB7441) were purchased from Sinopec Yangzi Petrochemical Co., Ltd. (Nanjing, China), LDPE-*g*-MAH and PP-*g*-MAH were purchased from Fine-blend Compatibilizer Jiangsu Co., Ltd. (Nantong, China).

Table 1. The properties of the PET chips.

PET	[η] (dL/g)	[COOH] (mol/t)	H_2O (wt %)
PET1	0.70	22	0.35
PET2	0.85	18	0.26
PET3	1.00	15	0.10

In order to decrease the hydrolytic decomposition, the PET chips were dried at 130 °C for 9 h in a vacuum oven before being blended into either LDPE or PP. The polymer blends were processed in a RM-200C torque rheometer (Hapro, Haerbin, China). The processing temperatures were set to 260 °C for all blends in the rheometer, and the processing time was set to 5~10 min and the rotor speed was set to 50 r/min. The total weight of polymer blend was no more than 50 g. The weight ratio of the components in the blending are listed in Table 2. The compatibilizer usually accounts for five percent of the total weight of the two polymers (5 wt %). The blank control experiments were also carried out in the absence of compatibilizer.

Table 2. The weight ratio of the components in the polymer blends.

No.	PET1	PET2	PET3	PP	PP-*g*-MAH	LDPE	LDPE-*g*-MAH
1	20			80			
2		20		80			
3			20	80			
4	30			70			
5		30		70			
6			30	70			
7	20			80	5		
8		20		80	5		
9			20	80	5		
10	30			70	5		
11		30		70	5		
12			30	70	5		
13	20					80	
14		20				80	
15			20			80	
16	30					70	
17		30				70	
18			30			70	
19	20					80	5
20		20				80	5
21			20			80	5
22	30					70	5
23		30				70	5
24			30			70	5

2.2. Measurements

The thermal behavior of the blends was investigated using differential scanning calorimeter (DSC1, Mettler Toledo, Schwerzenbach, Switzerland) and thermogravimetric analyzer (TGA, Mettler Toledo, Schwerzenbach, Switzerland) under nitrogen atmosphere. The drying samples were initially heated from 25 to 300 °C at a rate of 50 °C/min. After that, the samples were kept at the final temperature for 3 min to eliminate the thermal history, and then cooled to 50 °C at a rate of 20 °C/min to determine the crystallization temperature (T_c) and before being kept for 3 min. Finally, the samples were heated to 300 °C at a rate of 10 °C/min to determine the melt point (T_m). About 5 mg of the samples was heated from 25 to 600 °C at a rate of 10 °C/min with a nitrogen flow rate of 45 mL/min to determine the decomposition temperature (thermal gravimetric rate 5%, $T_{5\%}$).

The rheological properties of the polymer blends as functions of angular frequency were investigated via plate-plate geometry (plate radius = 25 mm; gap = 1 mm) using a stress-controlled rheometer (MCR 301, Anton Paar, Graz, Austria) at 260 °C. The measurements of the granules and sheet samples were performed in air atmospheres. The loaded frequency sweep was undirectional from the angular frequency of ω = 500 1/s to 0.1 1/s.

The banded samples of polymers were prepared from a blending system so that the morphological characteristics of the polymer blends could be examined using an ULTRATM 55 field-emission scanning electron microscopy (SEM Ultra 55, Zeiss, Jena, Germany) at an acceleration voltage of 1 kV. The samples were cooled in liquid nitrogen for 15 min and then taken out quickly to be brittle fractured. The fracture surface of PET was etched by using the mixed solution of phenol and tetrachloroethane (1/1 (wt %)), then the fracture surface was gilded because of the poor electrical conductivity.

To test the mechanical property of the polymer blends, the round rod samples with diameter of 1.5 mm were stretched on a universal testing machine (Instron 3367, Instron, Canton, USA) at speed of 10 mm/min. Generally, the effective stretch length was about 2 mm, and then the tensile strength and elongation at break could be achieved.

3. Results and Discussion

3.1. Thermal Properties

Figure 1 shows differential scanning calorimeter (DSC) scans of polymers and polyolefin/PET blends. The crystallization could be observed in the cooling scan and the melting process was recorded in the subsequent heating scans. The evident cold crystallization peaks (T_{c1}) of the polymer blends, observed in Figure 1a, was between polyolefins and PET, while the second cold crystallization peaks (T_{c2}) was almost invisible in most cases. In contrast, there were two melting peaks in the heating DSC scans, which indicated the melting point of the polyolefin component (T_{m1}) and PET component (T_{m2}). Noticeably, the T_{c1}, T_{m1} and T_{m2} for all polyolefin/PET blends were listed in Table 3. It also could be seen that blending caused a gradual increase in the crystallization temperature (T_{c1}) from LDPE to LDPE/PET1, to LDPE/LDPET-g-MAH/PET, however, the T_{c1} of compatibilized PP/PET1 blends was lower than that of incompatible PP/PET1 blends. This might be ascribed to the plasticizing effect of compatibilizer on the PP/PET blend and the facilitation of crystallization [25]. On the other hand, it is well known that the regularity of the LDPE chain is better than PP, which increases the crystallization temperature of blending system.

Generally, the T_{c1} of incompatible LDPE/PET (Figure 1b) increased slightly, yet the T_{m1} and T_{m2} presented a gradual decline either with the increase of the content of PET or with the increase of the IV of the PET, but there was less significant difference among PP/PET blends. When the compatibilizer was added to two immiscible polymers (Figure 1c), the T_{c1} of LDPE/PET blends presented a minor depression, whereas the changing trends for the PP/PET blends is just opposite. It is worth stressing that the difference in the T_{c1}, T_{m1} and T_{m2} of the polymer blends was quite inconspicuous, and that the thermal behaviors need further verification. This is perhaps explained by the fact that both the content and property of PET have no significant influence on the cold crystallization performance of polyolefins/PET. These allow us to conduct a good deal of the blending of polyolefin with PET from different sources.

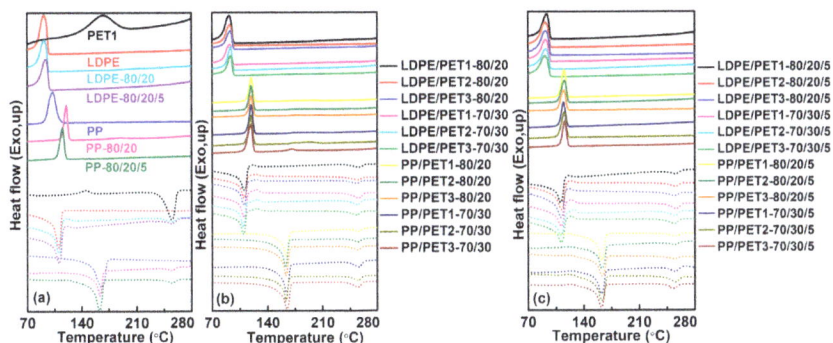

Figure 1. Cooling (solid line) and heating (short dot line) DSC scans of polymers and polymer blends, (**a**) polymers and blends; (**b**) incompatible polyolefin/PET blends; (**c**) compatibilized polyolefin/PET blends.

Table 3. The crystallization temperature and melting point of polymer blends.

Samples	T_{c1} (°C)	T_{m1} (°C)	T_{m2} (°C)
LDPE	91.09	110.12	—
PP	102.51	165.33	—
PET1	167.69	—	254.33
LDPE/PET1-80/20	91.85	110.68	255.67
LDPE/PET2-80/20	92.96	110.58	254.81
LDPE/PET3-80/20	93.43	110.23	253.32
LDPE/PET1-70/30	92.49	110.77	2557.5
LDPE/PET2-70/30	93.44	109.68	254.34
LDPE/PET3-70/30	94.44	109.86	252.88
PP/PET1-80/20	119.61	163.31	255.28
PP/PET2-80/20	119.14	162.96	255.25
PP/PET3-80/20	119.21	163.33	254.87
PP/PET1-70/30	119.13	163.70	255.75
PP/PET2-70/30	118.67	162.96	254.83
PP/PET3-70/30	118.65	164.42	254.81
LDPE/PET1-80/20/5	93.61	110.63	255.18
LDPE/PET2-80/20/5	91.39	111.66	255.96
LDPE/PET3-80/20/5	92.13	111.13	253.88
LDPE/PET1-70/30/5	91.78	111.40	256.22
LDPE/PET2-70/30/5	92.32	111.63	254.93
LDPE/PET3-70/30/5	91.23	111.66	254.66
PP/PET1-80/20/5	114.91	162.62	255.96
PP/PET2-80/20/5	115.61	162.86	254.93
PP/PET3-80/20/5	115.07	163.39	254.66
PP/PET1-70/30/5	114.36	162.09	255.44
PP/PET2-70/30/5	115.45	162.87	255.71
PP/PET3-70/30/5	116.00	162.09	253.88

The comparison of thermogravimetric (TG) and derivative thermogravimetric (DTG) scans of polymer with polymer blends is reported in Figure 2. PET presented decomposition at lower temperatures than LDPE and PP, as a result of the relatively poor stability of the ester group in PET chains. A small difference in the decomposition rate could be seen between PP and its blends based on the TG and DTG curves. Nevertheless, it was clear that the LDPE/PET blend showed worse thermal stability than that of LDPE and that the compatibilizers promote the thermal stability of two immiscible polymers. This was mainly attributed to the improvement of compatibility with the addition of grafting maleic anhydride; the interface adhesion was enhanced by the chemical reactions between the grafted functional groups of polyolefin and PET end groups [26].

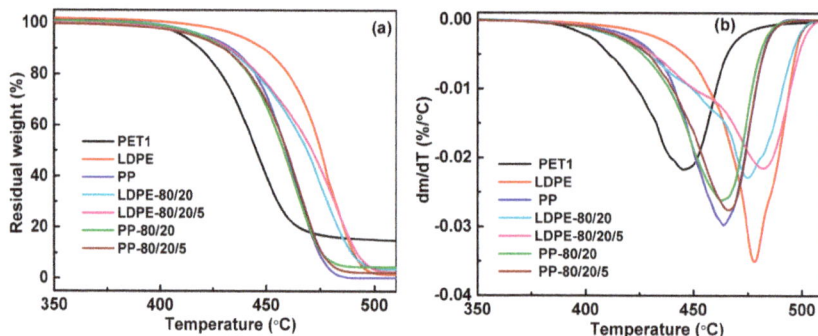

Figure 2. Comparison of TG (**a**) and DTG (**b**) curves of polymers and polymer blends.

Figures 3 and 4 show the effect of the content and IV of PET on the thermal stability of incompatible and compatibilized polyolefin/PET blends, respectively. The DTG curves of polymer blends presented asymmetric peaks, which indicated the whole decomposition of the different components. Accordingly, the decomposition temperature at thermal gravimetric rate of 5% ($T_{5\%}$) and the corresponding temperature of the maximum decomposition rate (T_{cmax}) of all polymer blends are also listed in Table 4. Regarding LDPE/PET blends without a compatibilizer (Figure 3), the decomposition temperature was slightly increased with the increase in the IV of PET, which was validated by the temperature of maximum decomposition rate as shown in the DTG graphs and Table 4, but the difference was unremarkable under the conditions of more PET content in the blends. On the contrary, it almost stayed at a constant thermal gravimetric rate for PP/PET blends in a ratio of 80 to 20 as the IV of PET increased, while it was a little better later with more PET content in the blends. When the compatibilizer was added to the blends (Figure 4), the thermal stability was slightly increased. As already mentioned, the interface could be enhanced. It was also noticed that both the maximum decomposition rate and the corresponding temperature of the compatibilized polyolefin/PET blends at a ratio of 70 to 30 were lower than that of 80/20 PP/PET blends in view of the poorer thermal stability of PET. Furthermore, the polymer blends with a higher IV of PET generally suggested better thermal stability and higher value of $T_{5\%}$ and T_{cmax}. These results are easy to understand if it is remembered that PET with a higher IV has a longer polymer chain.

Table 4. The decomposition temperature ($T_{5\%}$) and the corresponding temperature of the maximum decomposition rate (T_{cmax}) of polymer blends.

Blend	0/100		80/20		70/30		80/20/5		70/30/5		100/0	
	$T_{5\%}$ (°C)	T_{cmax} (°C)	$T_{5\%}$ (°C)	T_{cmax} (°C)	$T_{5\%}$ (°C)	T_{cmax} (°C)	$T_{5\%}$ (°C)	T_{cmax} (°C)	$T_{5\%}$ (°C)	T_{cmax} (°C)	$T_{5\%}$ (°C)	T_{cmax} (°C)
LDPE/PET1	410.8	444.3	419.7	473.4	420.5	478.7	419.6	482.8	419.4	481.7	437.5	477.7
LDPE/PET2	407.5	445.7	421.3	475.9	416.3	476.0	423.2	481.7	420.3	480.9	-	-
LDPE/PET3	407.2	447.7	422.5	480.3	416.4	475.8	423.6	480.8	419.5	479.2	-	-
PP/PET1	410.8	444.3	419.5	463.5	420.3	463.3	416.3	465.9	419.7	465.7	420.8	461.3
PP/PET2	407.5	445.7	417.1	463.4	414.3	461.4	421.1	463.3	411.5	465.0	-	-
PP/PET3	407.2	447.7	420.7	463.2	412.1	460.5	417.5	462.9	409.9	461.8	-	-

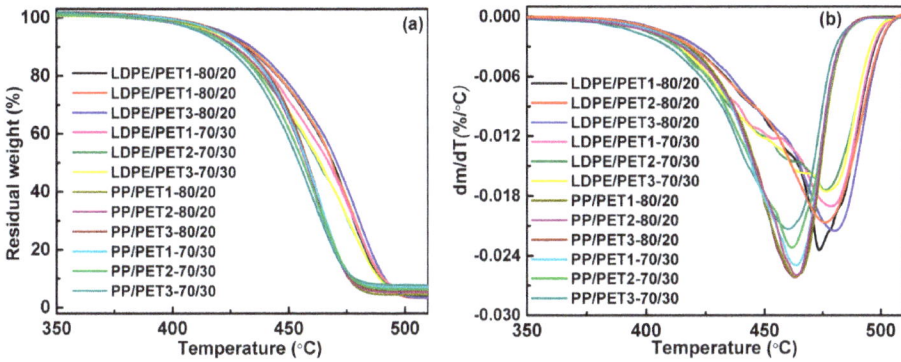

Figure 3. TG (**a**) and DTG (**b**) curves of incompatible polymer blends.

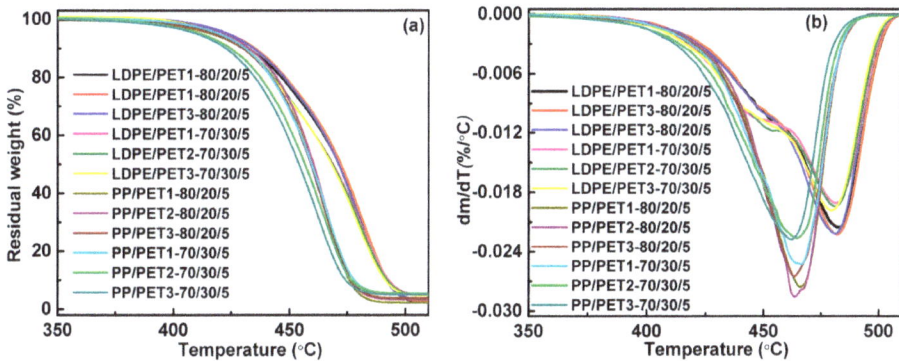

Figure 4. TG (**a**) and DTG (**b**) curves of compatibilized polymer blends.

3.2. Analysis of Rheological Behavior

Figure 5 shows the complex viscosity as a function of frequency at 260 °C for polymers and polymer blends. The rheological curves of polymer blends showed similar trends with the neat constituents of polymers. Considering that the T_m of PET was approximately equal to 260 °C, all the rheological measurements of polymers and polymer blends were conducted in air without inert gas protection, which was consistent with the blending process. Three kinds of PET displayed a pseudo-Newtonian behavior for the whole frequency range. The complex viscosity significantly increased with the IV of PET due to the longer polymer chain [1]. Both the polyolefins and polymer blends presented remarkable pseudoplastic behavior, and LDPE and its blends revealed a stronger shear thinning effect than PP and its blends because of the better chain mobility.

Compared with LDPE, the incompatible LDPE/PET blend showed a higher viscosity at low frequency and a lower viscosity at high frequency. However, the complex viscosity of compatibilized LDPE/PET blend was always lower than that of LDPE except at an extremely low frequency, and also lower than that of incompatible LDPE/PET blend within the whole frequency range. However, once the compatibilizer was formally incorporated into the system, the viscosity of the system decreased, which was due to the role of the compatibilizer in reducing the dispersity and inducing interaction at the interface between blend components, eventually increasing the mobility of the compatibilized system [27,28]. PP showed the lowest viscosity at low frequency, and the compatibilized PP/PET blend presented intermediate behavior between the incompatible blend and pure PP. As mentioned above, the carbonyl group in PET could interact with the pendant group in the PP chain so that the blending system gets less viscous, whereas the compatibilizer played a role of softening agent [29].

Figure 6 shows the complex viscosity of incompatible polymer blends. It can be seen that the viscosity of both the LDPE/PET and PP/PET blends present a slight increase as the IV of PET increases. This is probably due to the fact that the higher IV of PET has a longer polymer chain so that the PET itself possesses a higher viscosity. However, the blends with a higher content of PET did not exhibit a higher viscosity but displayed a slight decline. This suggested that the PET dispersed in polyolefins had some lubrication effect in a way due to the presence of flexible chain segments in PET.

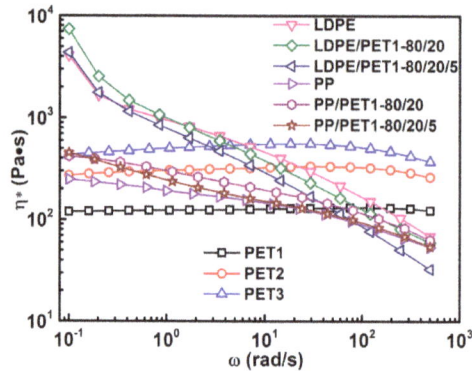

Figure 5. Complex viscosity for polymers and polymer blends.

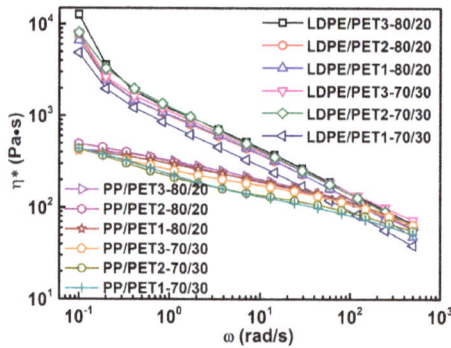

Figure 6. Complex viscosity for polymer blends with different IV of PET without compatibilizer.

Figure 7 shows the complex viscosity for polymer blends with different IV of PET in the presence of compatibilizer. It was in good agreement with the viscosity sequence of incompatible polymer blends with different IV of PET. Moreover, it was noticed that the viscosity of blends with a higher content of PET was significant increased. Several studies have shown evident results that polyolefin/PET blends are completely immiscible. By adding compatibilizer into the blending system, the interaction between the compatibilizer and blend components led to a reduction of interfacial tension and the suppression of coalescence [3,30–32]. Hence, it could be inferred that with the increase of PET content, the compatibilizer provided increasingly favorable support for an enhanced interaction between blend components.

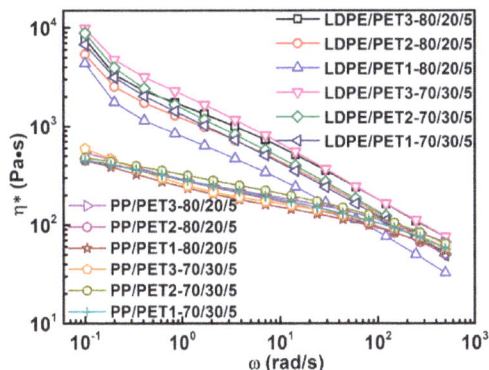

Figure 7. Complex viscosity for polymer blends with different IV of PET in the presence of compatibilizer.

3.3. Morphological Characterization of Polymer Blends

Figure 8 shows the morphology of incompatible and compatibilized polyolefin/PET1 polymer blends. It can be seen that the blending of two immiscible components results in an entirely clear phase interface. The cross profile of LDPE/PET was a brittle fracture and PET dispersed unevenly in LDPE phase in the form of particles. In yet another way, the large size of the circular hole indicated that the PET-dispersing phase appeared in the PP continuous phase, resulting in a typical sea-island structure. Figure 8 also shows the scanning electron microscopy (SEM) micrography of different amounts of PET with or without compatibilizer. With the increase of PET content, the holes of the LDPE/PET and PP/PET blends were increased in quantity, size and dispersion uniformity. The blends with a higher content of PET were more likely to be assembled in the blending process, resulting in the enlargement of the dispersed phase size. Compared to incompatible polymer blends, it was observed that compatibilized LDPE/PET blends had a better dispersity of holes appearing in the cross profile, the fracture surface was no longer clear but presented a significant layered hippocampal structure, and the circular hole in LDPE/PET blends with PP-*g*-MAH tended to be dispersed uniformly with diminishing size. All these confirmed that the compatibilizer had indeed improved the compatibility between polyolefins and PET. As mentioned above in the analysis of rheological behavior, the compatibilizer enhanced the interaction between the blend components which decreased the interfacial tension and suppressed the coalescence [3,30–32].

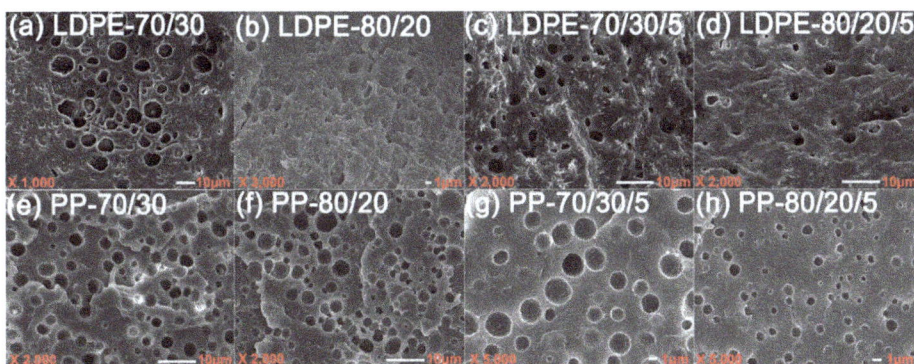

Figure 8. SEM micrographs of incompatible (**a,b,e,f**) and compatibilized (**c,d,g,h**) polymer blends.

The effect of the addition of different IV of PET on the blending system with or without compatibilizer was visualized by morphological characterization, as shown in Figures 9 and 10, respectively. In Figure 9, the dispersed phase in PP/PET blends was getting more and more clear with the increase in the IV of PET; by contrast, the phase separation of the two blend components was attenuated gradually, so that a rough interface and more crack propagation could be seen in PP/PET blends. This could be explained in that it was more difficult to diffuse into the polyolefin phase for a longer polymer chain of PET, as a consequence, the thinner interface layer suggested poor compatibility. With the addition of compatibilizer (Figure 10), the PET dispersed uniformly and the interface coalescence was strengthened. Not surprisingly, in this situation, the effect of the IV of PET on the dispersed phase size was weakened.

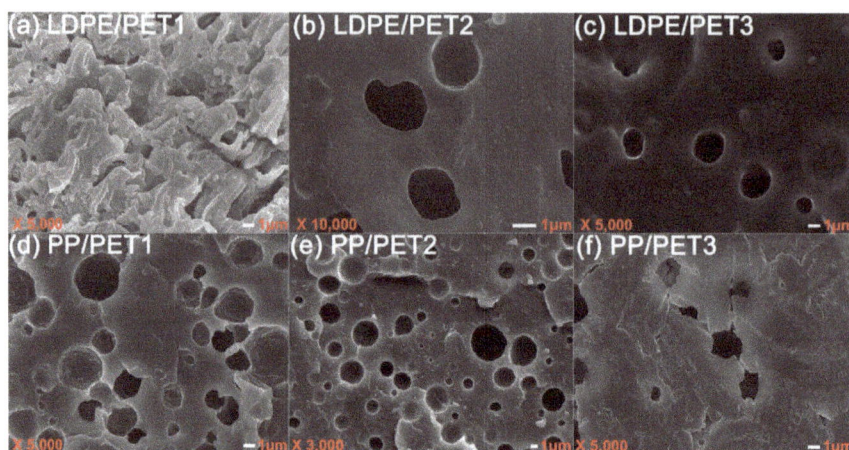

Figure 9. SEM micrographs of LDPE/PET (**a–c**) and PP/PET (**d–f**) blends with different IV of PET without compatibilizer (80/20).

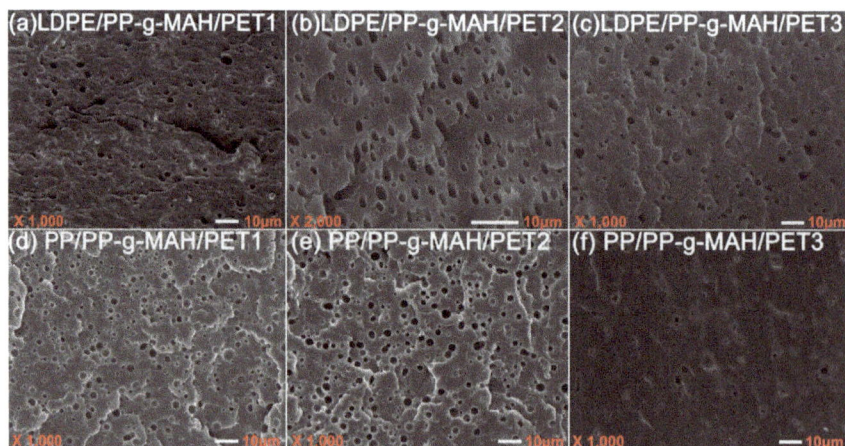

Figure 10. SEM micrographs of LDPE/PET (**a–c**) and PP/PET (**d–f**) blends with different IV of PET in the presence of compatibilizer (80/20/5).

3.4. Mechanical Properties of Polymers before and after Blending

Tables 5 and 6 show the elongation at break and the tensile strength of different polymer blends, respectively. The pure PET exhibited a decrease of elongation at break and an increase of tensile strength with the increase of IV. The polymer blends with a higher content of PET had a relatively lower elongation at break. As an immiscible polyolefin/PET blend, the interphase adhesiveness of two blending components was insufficient, which resulted in an evident interphase division, and therefore, worse interphase adhesiveness existed in polymer blends with higher content of PET.

Without compatibilizer, the elongation at break of both LDPE/PET blends and PP/PET blends declined with the increase of the IV of PET. Yet it increased remarkably after the compatibilizer was added to polyolefin/PET blends, especially for the highest IV of PET3 in blends. With longer polymer chains and higher steric effects, the PET with a higher enlarged interphase between the polyolefins and the PET, showed a decrease of elongation at break. The compatibilized polymer blends were expected to have a higher elongation at break because of the increased interphase adhesiveness and enhanced interaction. In this case, PET3 with better mechanical properties had a positive effect on the elongation at break of blends. This is different from the PE/PET blend with organoclay as a compatibilizer, as there may be a possible degradation of the clay surfactant during melt compounding [33]. It was noted that the elongation at break of the LDPE blending system was higher than that of the PP blending system, which was consistent with the observation of the rheological behavior which found that LDPE/PET blends had the benefit of higher viscosity. In addition, the elongation at break of the polyolefin/PET blend increased with the increase of PET dispersed phase content, which was in good agreement with the results of the research by Tavanaie et al. [34].

Table 5. The elongation at break of polymer blends.

Blend	0/100 (%)	80/20 (%)	70/30 (%)	80/20/5 (%)	70/30/5 (%)	100/0 (%)
LDPE/PET1	8.55	51.13	25.51	67.45	37.52	223.45
LDPE/PET2	8.45	44.67	23.50	78.25	41.47	-
LDPE/PET3	6.41	44.16	22.21	148.54	58.97	-
PP/PET1	8.55	15.1	11.7	38.5	30.5	428
PP/PET2	8.45	14.7	11.5	34.8	28.0	-
PP/PET3	6.41	13.7	10.2	31.3	24.3	-

Table 6. The tensile strength of polymer blends.

Blend	0/100 MPa	80/20 MPa	70/30 MPa	80/20/5 MPa	70/30/5 MPa	100/0 MPa
LDPE/PET1	26.0	16.71	23.52	14.32	18.25	12.31
LDPE/PET2	28.8	15.62	17.25	13.88	12.74	-
LDPE/PET3	37.5	14.45	17.01	12.90	12.24	-
PP/PET1	26.0	17.1	24.2	11.3	12.5	35.1
PP/PET2	28.8	20.6	25.0	11.9	12.41	-
PP/PET3	37.5	24.4	25.2	12.0	12.2	-

The variation of the tensile strength of polymer blends with PET content was largely different from the elongation at break, in that it was somewhat confused. In such a situation, the polymer blends were brittle fractured to subdue the tenacity of the materials. With the increase of the IV of PET, the tensile strength of the LDPE/PET blends was decreased, while it was gradually increased for PP/PET blends. According to the SEM micrographs, the interphase adhesiveness of incompatible LDPE/PET blends was relatively weaker than that of PP/PET blends. However, with the addition of compatibilizer, LDPE/PET blends showed better tensile properties since the shorter chain segment of LDPE-*g*-MAH was more likely to diffuse successfully into two blend components. It was also noticed that the tensile strength of all polymer blends generally declined, and the difference of tensile strength caused by either changing the IV of PET or changing the PET content also decreased. Therefore, it can

Polymers **2018**, *10*, 147

be stated that the compatibilizer played a role in increasing interphase adhesiveness to prevent brittle fracture, thus increasing the mechanical properties of polymer blends.

4. Conclusions

The blending of LDPE and PP with different intrinsic viscosities (IV) of PET was performed in the absence or presence of compatibilizer. The impact of the blending components on the thermal, rheological, morphological and mechanical properties was discussed in detail, and the following conclusions were drawn from the discussion:

(1) The blending leads to a higher crystallization temperature of the polyolefin component, yet the melting peak both of the polyolefin component and of the PET component in the blends are little different from their pure polymer. While the thermal stability of polymer blends is generally better than PET but worse than LDPE, and blending of PP with PET had no significant effect on the thermal stability of PP. The thermal performance of polyolefin/PET was slightly decreased by either the increase of the IV of PET or the decrease of PET content. It was found that the compatibilizer can improve the thermal stability of the blending system by enhancing the interface adhesion.

(2) The complex viscosity of incompatible polymer blends is slightly higher than that of compatibilized blends. The results show that it is increased with the increase of the IV of PET, yet decreased with the increase of PET content increasing in blends. The complex viscosity of the blend with compatibilizer can be similarly increased by either increasing the IV of PET or increasing the PET content in blend.

(3) PET disperses nonuniformly in LDPE in the form of granules; the cross-section of the incompatible blend shows brittle fracture. The number of holes in the blend increases with the increase of PET content. In contrast, the cross-section tends to be indistinct in the presence of LDPE-*g*-MAH as a compatibilizer. As the IV of the PET increases in the polymer blends, PET can be easily gathered to enlarge the scale of the dispersed phase, resulting in worse nonuniform dispersity, while it can be improved by adding compatibilizer.

(4) It was demonstrated that with the increase of the IV of PET, the elongation at break of the polymer blend decreases slightly and the tensile strength of the LDPE/PET blend also decreased accordingly, while the tensile strength of the PP/PET blend gradually increased. It is worth noting that the higher content of PET in the blend results in a smaller elongation at break and lower tensile strength. When the compatibilizer was added to the blends, an increase of elongation at break and a decrease of tensile strength could be seen simultaneously, and the difference of tensile strength caused by either changing the IV of PET or changing the PET content declined.

Acknowledgments: Gratitude is expressed to the National Key Research and Development Program of China (No. 2016YFB0303000) and Zhejiang Provincial Natural Science Foundation of China (No. LQ18E030011), Science Foundation of Zhejiang Sci-Tech University (ZSTU) (No. 17012072-Y, 2017YBZX01) for their financial support.

Author Contributions: Xian-Ming Zhang and Wen-Xing Chen conceived and designed the experiments; Guo-Cai Zhong and Jian Li performed the blending experiments and thermal analysis testing; Li-Hao Zhang conducted the SEM, rheology and mechanical properties measurement; Shi-Chang Chen and Guo Zhang contributed reagents/materials/analysis tools; Shi-Chang Chen analyzed the data and wrote the paper.

Conflicts of Interest: The authors declare no conflict of interest.

References

1. Chen, S.C.; Zhang, X.M.; Liu, M.; Ma, J.P.; Lu, W.Y.; Chen, W.X. Rheological characterization and thermal stability of different IV poly(ethylene terephthalate) in air and nitrogen. *Int. Polym. Proc.* **2016**, *31*, 292–300. [CrossRef]
2. Raffa, P.; Coltelli, M.B.; Savi, S.; Bianchi, S.; Castelvetro, V. Chain extension and branching of poly(ethylene terephthalate) (PET) with di- and multifunctional epoxy or isocyanate additives: An experimental and modelling study. *React. Funct. Polym.* **2012**, *72*, 50–60. [CrossRef]

3.   Pracella, M.; Chionna, D.; Pawlak, A.; Galeski, A. Reactive mixing of PET and PET/PP blends with glycidyl methacrylate-modified styrene-*b*-(ethylene-*co*-olefin) block copolymers. *J. Appl. Polym. Sci.* **2005**, *98*, 2201–2211. [CrossRef]

4.   Khonakdar, H.A.; Jafari, S.H.; Mirzadeh, S.; Kalaee, M.R.; Zare, D.; Saeb, M.R. Rheology-morphology correlation in PET/PP blends: Influence of type of compatibilizer. *J. Vinyl Addit. Technol.* **2013**, *19*, 25–30. [CrossRef]

5.   Bettini, S.H.P.; De Mello, L.C.; Munoz, P.A.R.; Ruvolo-Filho, A. Grafting of maleic anhydride onto polypropylene, in the presence and absence of styrene, for compatibilization of poly(ethylene terephthalate)/(ethylene–propylene) blends. *J. Appl. Polym. Sci.* **2013**, *127*, 1001–1009. [CrossRef]

6.   Zhang, X.; Li, B.; Wang, K.; Zhang, Q.; Fu, Q. The effect of interfacial adhesion on the impact strength of immiscible PP/PETG blends compatibilized with triblock copolymers. *Polymer* **2009**, *50*, 4737–4744. [CrossRef]

7.   Li, H.; Zhang, X.M.; Zhu, S.Y.; Chen, W.X.; Feng, L.F. Preparation of polypropylene and polystyrene with -NCO and -NH$_2$ functional groups and their applications in polypropylene/polystyrene blends. *Polym. Eng. Sci.* **2015**, *55*, 614–623. [CrossRef]

8.   Marcincin, A.; Kormendy, E.; Hricova, M.; Rusnak, A.; Aneja, A.P. Rheological behavior of polyester blend and mechanical properties of the polypropylene-polyester blend fibers. *J. Appl. Polym. Sci.* **2006**, *102*, 4222–4227. [CrossRef]

9.   Pracella, M.; Pazzagli, F.; Galeski, A. Reactive compatibilization and properties of recycled poly(ethylene terephthalate)/polyethylene blends. *Polym. Bull.* **2002**, *48*, 67–74. [CrossRef]

10. Wang, C.; Dai, W.; Zhang, Z.; Mai, K. Effect of blending way on b-crystallization tendency of compatibilized b-nucleated isotactic polypropylene/poly(ethylene terephthalate) blends. *J. Therm. Anal. Calorim.* **2013**, *111*, 1585–1593. [CrossRef]

11. Esmizadeh, E.; Moghri, M.; Saeb, M.R.; Mohsen, N.M.; Nobakht, N.; Bende, N.P. Application of Taguchi approach in describing the mechanical properties and thermal decomposition behavior of poly(vinyl chloride)/clay nanocomposites: Highlighting the role of organic modifier. *J. Vinyl Addit. Technol.* **2016**, *22*, 182–190. [CrossRef]

12. Li, S.C.; Lu, L.N.; Zeng, W. Thermostimulative shape-memory effect of reactive compatibilized high-density polyethylene/poly(ethylene terephthalate) blends by an ethylene-butyl acrylate-glycidyl methacrylate terpolymer. *J. Appl. Polym. Sci.* **2009**, *112*, 3341–3346. [CrossRef]

13. Nonato, R.C.; Bonse, B.C. A study of PP/PET composites: Factorial design, mechanical and thermal properties. *Polym. Test.* **2016**, *56*, 167–173. [CrossRef]

14. Li, H.; Xie, X.M. Morphology development and superior mechanical properties of PP/PA6/SEBS ternary blends compatibilized by using a highly efficient multi-phase compatibilizer. *Polymer* **2017**, *108*, 1–10. [CrossRef]

15. Jazani, O.M.; Rastin, H.; Formela, K.; Hejna, A.; Shahbazi, M.; Farkiani, B.; Saeb, M.R. An investigation on the role of GMA grafting degree on the efficiency of PET/PP-*g*-GMA reactive blending: Morphology and mechanical properties. *Polym. Bull.* **2017**, *74*, 4483–4497. [CrossRef]

16. Mantia, F.P.L.; Ceraulo, M.; Giacchi, G.; Mistretta, M.C.; Botta, L. Effect of a compatibilizer on the morphology and properties of polypropylene/poly(ethylene therephthalate) spun fibers. *Polymers* **2017**, *9*, 47. [CrossRef]

17. Si, X.; Guo, L.; Wang, Y.; Lau, K. Preparation and study of polypropylene/poly(ethylene terephthalate) composite fibers. *Compos. Sci. Technol.* **2008**, *68*, 2943–2947. [CrossRef]

18. Lei, Y.; Wu, Q.; Zhang, Q. Morphology and properties of microfibrillar composites based on recycled poly(ethylene terephthalate) and high density polyethylene. *Compos. Part A Appl. Sci. Manuf.* **2009**, *40*, 904–912. [CrossRef]

19. Raffa, P.; Coltelli, M.; Castelvetro, V. Expanding the application field of post-consumer poly(ethylene terephthalate) through structural modification by reactive blending. *J. Appl. Polym. Sci.* **2014**, *131*, 5829–5836. [CrossRef]

20. Yousfi, M.; Soulestin, J.; Vergnes, B.; Lacrampe, M.F.; Krawczak, P. Morphology and mechanical properties of PET/PE blends compatibilized by nanoclays: Effect of thermal stability of nanofiller organic modifier. *J. Appl. Polym. Sci.* **2013**, *128*, 2766–2778. [CrossRef]

21. Zhu, S.Y.; Li, H.; Zhang, X.M.; Chen, W.X.; Feng, L.F. Synthesis, copolymer composition, and rheological behavior of functionalized polystyrene with isocyanate and amine side groups. *Des. Monomers Polym.* **2015**, *132*, 232–241. [CrossRef]

22. Sun, M.H.; Zhang, X.M.; Chen, W.X.; Feng, L.F. Surface functionalization of polypropylene-bearing isocyanate groups in solid state and their cyclotrimerization with diisocyanates. *J. Appl. Polym. Sci.* **2015**, *132*, 42186. [CrossRef]

23. Sun, M.H.; Zhang, X.M.; Chen, W.X.; Feng, L.F. The effect of tunable modification parameters on the structure and properties of polypropylene with cyclotrimerization surface of isocyanates. *J. Appl. Polym. Sci.* **2016**, *133*, 43327. [CrossRef]

24. Zhang, X.M.; Sun, M.H.; Chen, W.X. Synergistic effects of silica nanoparticles and reactive compatibilizer on the compatibilization of polystyrene/polyamide 6 blends. *Polym. Eng. Sci.* **2017**, *57*, 24511. [CrossRef]

25. Chiu, F.C.; Yen, H.Z.; Lee, C.E. Characterization of PP/HDPE blend-based nanocomposites using different maleated polyolefins as compatibilizers. *Polym. Test.* **2010**, *29*, 397–406. [CrossRef]

26. Tao, Y.; Mai, K. Non-isothermal crystallization and melting behavior of compatibilized polypropylene/recycled poly(ethylene terephthalate) blends. *Eur. Polym. J.* **2007**, *43*, 3538–3549. [CrossRef]

27. Yee, M.; Souza, A.M.C.; Valera, T.S.; Demarquette, N.R. Stress relaxation behavior of PMMA/PS polymer blends. *Rheol. Acta* **2009**, *48*, 527–541. [CrossRef]

28. Souza, A.M.C.; Calvao, P.S.; Demarquette, N.R. Linear viscoelastic behavior of compatibilized PMMA/PP blends. *J. Appl. Polym. Sci.* **2013**, *129*, 1280–1289. [CrossRef]

29. De Souza, A.M.C.; Caldeira, C.B. An investigation on recycled PET/PP and recycled PET/PP-EP compatibilized blends: Rheological, morphological, and mechanical properties. *J. Appl. Polym. Sci.* **2015**, *132*, 41892. [CrossRef]

30. Hale, W.; Keskkula, H.; Paul, D.R. Compatibilization of PBT/ABS blends by methyl methacrylate-glycidyl methacrylate-ethyl acrylate terpolymers. *Polymer* **1999**, *40*, 365–377. [CrossRef]

31. Sanchez, S.A.; Calderas, F.; Manero, O. Influence of maleic anhydride grafting on the rheological properties of polyethylene terephthalate–styrene butadiene blends. *Polymer* **2001**, *42*, 7335–7342. [CrossRef]

32. Akbari, M.; Zadhoush, A.; Haghighat, M. PET/PP blending by using PP-*g*-MA synthesized by solid phase. *J. Appl. Polym. Sci.* **2007**, *104*, 3986–3993. [CrossRef]

33. Yousfi, M.; Lepretre, S.; Soulestin, J.; Vergnes, B.; Lacrampe, M.F.; Krawczak, P. Processing-induced degradation of nanoclay organic modifier in melt-mixed PET/PE blends during twin screw extrusion at industrial scale: Effect on morphology and mechanical behavior. *J. Appl. Polym. Sci.* **2013**, *131*, 1001–1007. [CrossRef]

34. Tavanaie, M.A.; Shoushtari, A.M.; Goharpey, F.; Mojtahedi, M.R. Matrix-fibril morphology development of polypropylene/poly(butylenesterephthalate) blend fibers at different zones of melt spinning process and its relation to mechanical properties. *Fiber Polym.* **2013**, *14*, 396–404. [CrossRef]

polymers

MDPI

Article

Effect of Curing Rate on the Microstructure and Macroscopic Properties of Epoxy Fiberglass Composites

Ammar Patel [1], Oleksandr Kravchenko [2] and Ica Manas-Zloczower [1,*]

[1]   Department of Macromolecular Science and Engineering, Case Western Reserve University, Cleveland, OH 44106, USA; aap89@case.edu

[2]   Department of Mechanical and Aerospace Engineering, Old Dominion University, Norfolk, VA 23529, USA; okravche@odu.edu

*   Correspondence: ixm@case.edu; Tel.: +1-216-368-3596

Received: 11 December 2017; Accepted: 24 January 2018; Published: 27 January 2018

Abstract: Curing rates of an epoxy amine system were varied via different curing cycles, and glass-fiber epoxy composites were prepared using the same protocol, with the aim of investigating the correlation between microstructure and composite properties. It was found that the fast curing cycle resulted in a non-homogenous network, with a larger percentage of a softer phase. Homogenized composite properties, namely storage modulus and quasi-static intra-laminar shear strength, remained unaffected by the change in resin microstructure. However, fatigue tests revealed a significant reduction in fatigue life for composites cured at fast curing rates, while composites with curing cycles that allowed a pre-cure until the critical gel point, were unaffected by the rate of reaction. This result was explained by the increased role of epoxy microstructure on damage initiation and propagation in the matrix during fatigue life. Therefore, local non-homogeneities in the epoxy matrix, corresponding to regions with variable crosslink density, can play a significant role in limiting the fatigue life of composites and must be considered in the manufacturing of large scale components, where temperature gradients and significant exotherms are expected.

Keywords: curing rate; epoxy microstructure; fatigue; composites; critical gel

1. Introduction

High performance thermoset composites are ubiquitous all around us. They are selected due to their light weight, high glass transition temperature, high strength and great chemical resistance [1–3]. The majority of the strength in a structural composite comes from the fiber being used for reinforcement, while the resin acts as a binder for the fiber mat holding the mat together [4,5]. Epoxies mixed with a curing agent or hardener represent a large percentage of thermoset resins, largely due to their excellent wettability and strong adhesion to the fiber. Epoxy composites are currently used in the manufacture of boat hulls, wind turbine blades, aircrafts and cars, to serve as structural materials. Despite the strength of the composite, stemming from the fiber reinforcement, the matrix has a large role to play in the failure of composites. The primary role of the matrix is to provide the stress transfer between the fibers. Therefore, failure of the composite usually occurs through matrix cracking, interface debonding, delamination or fiber breakage [6–8]. The microscopic damage in the matrix deteriorates the composite's mechanical properties and eventually results in coalescence of microscopic cracks and formation of transverse ply cracks and delaminations, which can lead to catastrophic failure of the composite. Thus, it is essential to study the microstructure of the matrix and understand how local non-homogeneities in the epoxy can affect the overall performance of the composite.

On a molecular scale, the epoxy amine reaction involves the opening up of the epoxide ring by reaction with an amine hydrogen [9–13]. The initial cure is largely dominated by the primary amine reaction taking place since the secondary amines are sterically hindered [14]. A post cure is thus required to ensure complete conversion of the secondary amines. On a macro scale, these reactions manifest themselves as the epoxy resin and curing agent mixture transforming from a fairly viscous liquid to a gel, through the formation of a three dimensional network, and finally, to a cross-linked intractable solid [15,16]. The onset of gelation was shown to be important for predicting residual stress development in composites. Previous studies have shown that residual stress build up in composites occurs past the gel point and is due to cure induced shrinkages in the epoxy matrix [17,18]. Cure rates past the gel point were shown to significantly affect residual deformation in composite laminates [19,20], while residual stresses alone can be significant enough to result in processing-induced damage in composites after cure. However, the method by which the cure conditions prior to gelation affect composite mechanical performance still remains not fully understood. Several studies have examined the microstructure of the epoxy network to get a better idea of the way gelation occurs [21–24]. In each of these studies, a two-phase structure was observed, consisting of a highly cross-linked, high density nodular phase, interspersed with a low density, softer phase. Small highly cross-linked microgel particles, in the order of ~10 nm, were the characteristic feature of the nodular phase, with larger nodules formed by the aggregation of these microgels. These investigative studies provided a model for gelation wherein:

1. Microgel particles, made up of uniformly cross-linked networks, nucleate and grow from the reacting polymeric network.
2. These microgels begin to coalesce, to form less coherent clusters, which continue to interact with other microgel clusters.
3. At gelation, the microgel clusters pack together to form a non-uniform, incoherent gel network with the partially reacted networks having a lower cross-link density, forming the dispersed phase.

Previous studies have manipulated the microstructure of the epoxy network by changing the stoichiometry of the epoxy amine reaction [22]. It was found that increasing the epoxy or amine content decreased the connectivity of the microgel clusters, resulting in a higher fraction of a dispersed phase. Studies on varying the stoichiometry of the epoxy amine reaction show either a decrease in mechanical properties or a decrease in the glass transition temperature (T_g), due to the presence of excess amine or epoxy [22,25,26]. Therefore, this does not present an effective way of tuning the microstructure of the epoxy, especially in the case of structural composites where high strength and T_g are required. In this work, we propose changing the microstructure by modifying the curing rate of the epoxy amine reaction. There are two major questions that need to be answered: What is the effect of increasing the reaction rate up to gelation? Can the reaction proceed so quickly that the microgels are unable to coalesce completely?

To the best of our knowledge, varying the curing temperatures, and thus, the curing rates, to study the microstructure of the resulting matrix, and observing the effect of this change on the composite properties, has not been studied. Increasing the curing rate, leading to quicker cycle times, also has the benefit of expediting composite manufacturing, which is of practical importance to industry. This work involves studying the effect of three different curing cycles on the evolution of the epoxy network and the implication it has on the final composite properties.

2. Experimental

2.1. Materials

Lab grade Diglycidyl ether of Bisphenol A (DGEBA) resin and the curing agent, isophorone diamine (IPD) are a product of Sigma–Aldrich® (St. Louis, MO, USA). Bi-directional glass fiber mats were procured from Fiberglast® (Brookville, OH, USA).

The epoxy resin and curing agent were mixed together using a glass rod for 5 min. The Epoxide Equivalent Weight (EEW) of DGEBA was 170 g/equivalent, obtained directly from the manufacturer mentioned above. The curing agent was added such that the ratio between the EEW and the Amine Hydrogen Equivalent Weight (AHEW) was 1:1. The Amine Hydrogen Equivalent Weight was calculated by the following equation.

$$\text{Amine hydrogen equivalent weight} = \frac{\text{Molecular weight of Amine}}{\text{number of active hydrogens}} = \frac{170.30}{4} = 42.575 \quad (1)$$

The amount of amine added in parts per hundred (phr) was thus found by

$$phr = \frac{AHEW}{EEW} \times 100 = \frac{42.575}{170.3} = 25 \text{ g per hundred grams of epoxy} \quad (2)$$

The composites were prepared by the hand layup method. Five layers of glass fiber mats were each impregnated with the resin curing agent mixture and stacked one on top of the other. Three different curing rates were used to cure the composite samples. After impregnation, one composite was allowed to cure at room temperature (RT) for 12 h, following which it was hot pressed, at a pressure of 2 metric tons, at 70 °C for 6 h, and then 140 °C for 6 h. This sample will be referred to as the "RT sample" in the rest of the manuscript. Another impregnated composite, termed the "70 sample", was hot pressed, with a pressure of 2 metric tons, at 70 °C for 6 h, and 140 °C for 6 h. A third composite, labeled the "140 sample", was directly hot pressed, at 140 °C, at a pressure of 2 metric tons, for 6 h following impregnation. The fiber mats used per composite were weighed prior to impregnation and the composite was weighed again post-curing, to obtain the fiber to resin weight ratio. For all the 3 curing cycles, the weight ratio of the composite was 65 parts fiber and 35 parts epoxy. In order to study the microstructure of the epoxy network, control samples without any glass fiber mats were also prepared. The epoxy resin and hardener were mixed together, and the mixture was poured into rectangular shaped molds and cured via the 3 curing cycles. Similar to the composite samples, these were labelled as "RT sample", "70 sample" and "140 samples", respectively.

2.2. Instrumental Methods

2.2.1. Soxhlet Extraction

Weighted samples obtained from the 3 different curing cycles were refluxed in Tetrahydrofuran (THF) for 72 h, following which the samples were dried to remove all solvent and weighed again to measure the difference in weight. The experiment was repeated from different samples at least twice for reproducibility. It was expected that all unreacted or partially reacted monomers of the epoxy amine reaction would be washed away during the THF reflux.

2.2.2. Differential Scanning Calorimetry (DSC)

DSC was performed on test samples using a TA Instrument DSC Q100 (New Castle, DE, USA) under nitrogen flow, using aluminum hermetic pans obtained from TA Instruments. Samples were subjected to a temperature ramp, from −60 to 200 °C, at a ramp rate of 3 °C/min. Samples were then equilibrated to −60 °C and ramped once more to 200 °C, at a rate of 3 °C/min. The glass transition temperature (T_g) of the samples was obtained as the midpoint of the drop in heat capacity during the heating cycle. Experiments were repeated 2–3 times to ensure reproducibility.

2.2.3. Atomic Force Microscopy (AFM)

Phase images to probe the microstructure of the epoxy samples were obtained using the Veeco Dimension 3100 Atomic Force Microscope (Bruker, Billerica, MA, USA) in Tapping mode with a Si-Carbide tip with a length of 125 mm, a spring constant of 40 Nm^{-1} and a resonant frequency of 372 KHz. Samples were microtomed, using a diamond knife, prior to imaging. Scan sizes of 250 nm and 1 μm were probed at a scan rate of 0.513 Hz. The contrast in phase images is affected by the viscoelastic behavior of the matrix, leading to brighter areas having a higher cross-link density.

The images obtained from the AFM were analyzed using ImageJ to find the percentage of bright and dark areas. During this analysis, care was taken to neglect those areas observed in the phase images that were clearly an artifact derived from the valleys and ridges obtained from the height imaging. The maximum surface roughness through the entire scan size did not exceed 35 nm. The thickness of these samples was ~1.5 mm.

2.2.4. Thermal Gravimetric Analysis (TGA)

TGA was performed on test samples using TA Instruments Q500 under nitrogen flow with a ramp rate of 10 °C/min until 600 °C. It is to be mentioned here that due to difficulties in cutting the glass fiber mat, the small amount of sample used primarily consists of the epoxy network and thus, does not reflect the fiber to resin ratio used in the composite.

2.2.5. Dynamic Mechanical Analysis (DMA)

TA instruments Q800 DMA, operating in dual cantilever mode, with an amplitude of 10 μm at a frequency of 1 Hz, was used to determine the Storage Modulus (E'). The glass transition temperature was obtained through the peak of the tan delta curve. Samples were run from 40 to 200 °C, at a ramp rate of 3 °C/min. Experiments were repeated 2–3 times for reproducibility.

2.2.6. Static and Dynamic Flexural Testing

Three-point bending tests were performed on the composite samples in static and fatigue modes. The static test was conducted on an Instron Tensile tester (Instron, Norwood, MA, USA) with a load cell of 5 kN at a speed of 2 mm/min and a span length of 50 mm. Dynamic three-point bending tests were performed on a servo-hydraulic test machine (MTS Model 20 kip, Eden Prairie, MN, USA) at the same span length, a frequency of 3 Hz of the loading sinusoidal wave and at a load that was 80% of the maximum load used to break the strongest samples determined from the static load test. The fatigue testing was conducted under load controlled mode with a load ratio, defined as the ratio between the minimum to maximum load, of $R = 0.2$. The number of cycles to failure was recorded and at least three specimens were tested for each sample, to ensure reproducibility.

3. Results and Discussion

3.1. Probing the Microstructure

The epoxy microstructure exposed to three different curing cycles was probed using the control samples without any glass fiber mats. Multiple pieces of the differently cured epoxies were subjected to Soxhlet extraction in THF. It was expected that a fully cured epoxy network would swell when exposed to THF but not dissolve. The unreacted parts of the network, i.e., the monomers or oligomers that had not cross-linked, would be washed away with the THF, observable by a weight change. A sample of the results for the Soxhlet extraction can be seen in Table 1. Since the initial sample weight changed when repeating the experiment, only the percentage change values reflect multiple tries.

Table 1. Soxhlet extraction of samples cured under the three curing cycles.

Curing Cycle	Weight before Drying (mg)	Weight after Drying (mg)	Percent Change in Weight (%)
Room temperature (RT) sample	258.6	258.4	0.077 ± 0.002
70 sample	176.7	176.2	0.275 ± 0.024
140 sample	294.8	286.1	2.886 ± 0.212

As can be seen in Table 1, the two samples (RT sample and 70 sample) that were allowed to gel completely before the post cure show almost no change in weight after extraction. Thus, the network was less homogenous in the 140 sample, since the microgels were not allowed to coalesce completely during the gelling stage.

AFM results show no significant differences between the samples on the 1 μm scale via the AFM, as can be seen in Supplementary Materials Figure S1. There were no nodular structures of 10–70 μm observed, as mentioned in other works [23,24]. At a scale of 250 nm, however, as seen in Figure 1, the inhomogeneous structure of the 140 sample is readily observable. The height images for all these samples can be seen in Supplementary Materials Figure S2. The fast cure of the 140 °C sample is expected to leave larger areas of partially reacted network that manifest themselves as a softer phase, with reduced mechanical properties, when compared with an otherwise hard epoxy network. These areas are regions of low cross-link density. The 140 sample is regularly interspersed with a softer phase material, indicating microgel clusters, meaning less connectivity between the highly cross-linked clusters. The RT sample, when compared to the 140 sample, shows a more homogenous network on either side of the height artifact. Significant dark spots in the 70 sample are mere artifacts, resulting from the differences in height. The 70 sample shows a similar structure to the RT in the bright spot dominated regions, with small amounts of a softer phase. It is interesting to note that the 140 sample has a similar structure to non-stoichiometric epoxy amine networks observed in other works [22], reinforcing the hypothesis that the fast cure in the 140 sample leaves portions of the system partially cured and thus, soft. Measuring of the relative areas of the bright and dark spots, i.e., The harder and softer phases of the epoxy network, respectively revealed that the 140 sample has a significantly higher percentage of softer phase regions when compared to the other samples (Supplementary Materials Table S1). The RT sample has ~15% of a softer phase region, whereas the 140 sample has ~31% of a softer phase region.

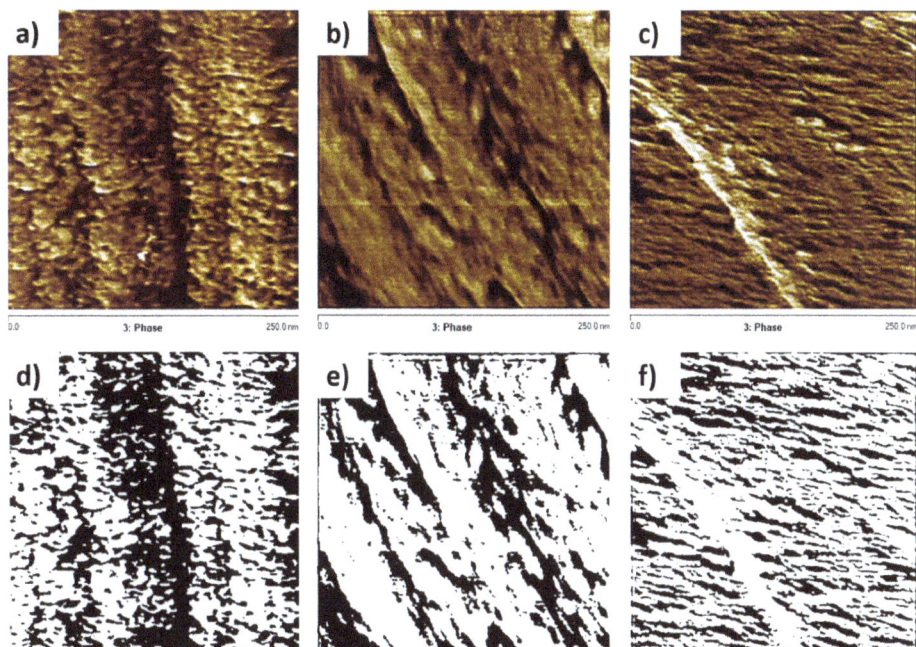

Figure 1. Phase images in tapping mode for (**a**) RT sample (**b**) 70 sample (**c**) 140 sample; figures (**d–f**) represent the threshold black and white images used to measure the percentage of bright and dark areas.

Differences in curing via the three cycles can also be observed by looking at the changes in glass transition via the DSC, as can be seen in Figure 2.

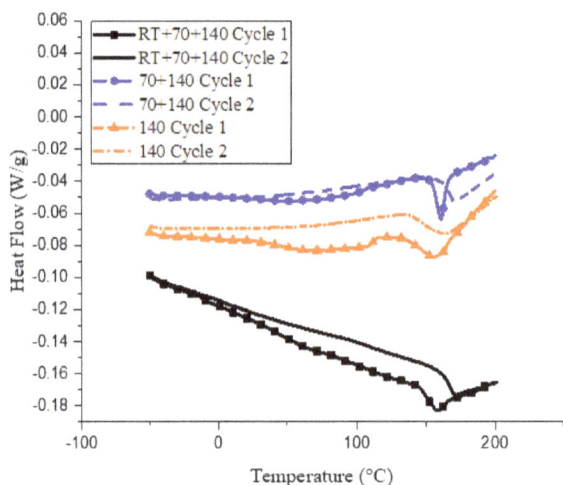

Figure 2. Heat capacity curves for samples cured under the three curing cycles.

All three samples, when subjected to a second curing cycle, show an increase in glass transition temperature. This is due to some amount of residual cure remaining during sample preparation, as is evident from the exothermic peaks observed immediately after T_g. For the 140 sample, a broad dip is observed in the 100–150 °C range, indicating a glassy to rubbery transition for the partially reacted oligomers in the network. This is apparent when comparing the T_g values of the three samples from the initial curing cycle, wherein the RT and 70 samples show similar T_g values of ~155 °C, whereas the 140 sample, due to the presence of these partially cured oligomers, has a lower T_g of 143 °C. However, the curing reaction for these oligomers is expected to persist, provided enough thermal activation is provided. This is exactly what is observed during the second curing cycle. During the second heating cycle, the bump observed in the 140 sample disappears, indicating that the oligomers have reacted. The T_g of the RT and 70 sample increase to similar temperatures of 165 °C during the second heating cycle, however, the T_g of the 140 cycle increases only up to 150 °C. Previous work [27] has noted that T_g is an exponential function of the degree of cure, indicating that close to full cure, a small increase in the degree of conversion causes a big increase in the T_g. This is similar to what is observed in the present DSC data. Within a particular microgel, the epoxy could be completely cured, but total coalescence of the individual microgels is essential for a complete degree of cure, which seems to be lacking in the 140 sample. The cross-linking reaction is still observed at the end of the second curing cycle for all samples because the reaction was not completed during the first cycle before the samples were cooled down, leading to vitrification.

Based on the microstructure, the 140 sample showed a distinct difference when compared to the RT and 70 samples. A closer look at the degree of conversions in the pre-curing stages of these RT and 70 samples provided an explanation for this behavior. As mentioned earlier, the T_g of the epoxy network increases as the curing reaction progresses [27] and if the T_g evolves to the temperature of the reaction, the epoxy transforms into a glass, leading to vitrification [28–30]. Also, gelation increases the viscosity by close to three orders of magnitude [31], after which the diffusion of the polymeric chains slows down, decreasing reaction kinetics [32]. Using an equation derived previously [27], the degree of conversion of the epoxy reaction can be obtained from the temperature at which the reaction proceeds (*T*), assuming enough time is given for the T_g of the network to reach the temperature of reaction.

$$\frac{T - Tg_0}{Tg_\infty - Tg_0} = \frac{\lambda \alpha}{1 - (1 - \lambda)\alpha} \tag{3}$$

Here, Tg_0 is the glass transition temperature of the epoxy monomer, which, in the case of DGEBA, is −25 °C; Tg_∞ is the glass transition temperature of the completely cured epoxy network, which was found to be 165 °C from DMA; α is the degree of conversion; and λ is a derived term that can be approximated to $\frac{Tg_0}{Tg_\infty}$ (both temperatures in Kelvin), which equates to 0.566. Both the RT and the 70 sample are given enough time to cure at 70 °C, which, when substituted into Equation (3), give a value of the degree of conversion of 0.63. According to the Flory Stockmayer theory, a bifunctional epoxy and a tetrafunctional amine, when mixed in a stoichiometric ratio, critically gel at

$$\alpha_{gel} = \left(\frac{1}{(f_a - 1)(f_b - 1)}\right)^{0.5} \rightarrow \alpha_{gel} = \left(\frac{1}{(2 - 1)(4 - 1)}\right)^{0.5} \rightarrow \alpha_{gel} = 0.577 \tag{4}$$

This value is very similar to that obtained for the RT and 70 samples. Thus, enough time is given for the microgels to diffuse and coalesce to reach their critical gel point and stabilize. The 140 sample, on the other hand, cures so quickly with the rise in reaction temperature, and thus, reaction rate, that even though microgel clusters form, the reaction medium has become viscous enough to prevent complete coalescing of these microgels. This leaves a large amount of partially unreacted chains in the network, exhibiting a pseudo-off stoichiometric network structure.

3.2. Composite Characterization

After the matrix was characterized, it was important to investigate how the observed changes in the microstructure correlate with macroscopic properties of the composite. The thermal stability of three types of sample was studied by TGA. Weight loss curves as a function of temperature show a one-step degradation process, as can be seen in Figure 3. The onset of thermal degradation at 5 weight percentage loss ($T_{d5\%}$) and the peak decomposition temperature (T_{peak}), shown in Supplementary Materials Table S2, were used to observe any differences in thermal decomposition. Values for $T_{d5\%}$ varied between 345 to 350 °C for all samples, while the difference between the T_{peak} values was minimal. Consequently, it is apparent that considered changes in microstructure, due to the differences in heating rates, did not translate into changes in the decomposition temperature of the composite.

A dynamic mechanical analysis, obtained from DMA and static flexural testing, revealed that macroscopic composite properties were not affected by post-curing conditions. As can be seen in Table 2, the macroscopic properties of storage modulus, above and below the glass transition temperature (Supplementary Materials Figure S3), flexural modulus (calculated from the linear region of the load-displacement curve), flexural strength (measured as the stress required for failure on the mid span of the outermost layer of the composite,) and strain at break, do not change significantly between samples (Supplementary Materials Figures S4–S6). Flexural stress was calculated using Equation (5)

$$\text{Flexural stress} = \frac{3FL}{2bd^2} \tag{5}$$

where *F* is the force applied, *L* is the span length (50 mm), *b* is the width of the sample and *d* is the thickness of the sample.

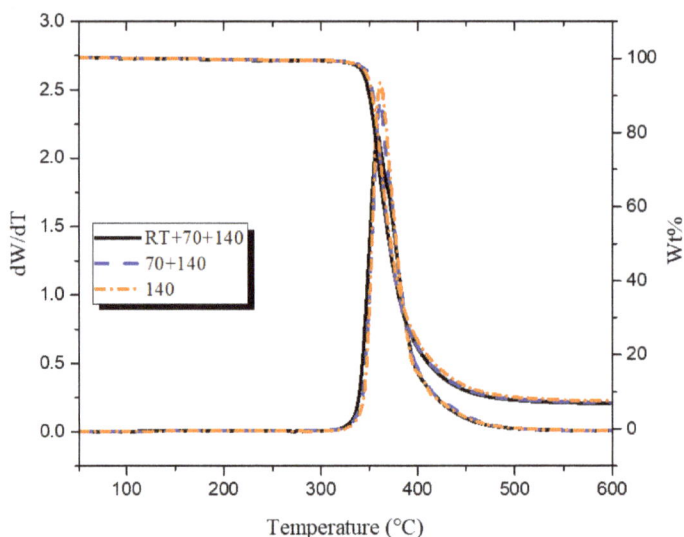

Figure 3. Thermal decomposition of epoxy glass-fiber composites.

Table 2. Storage modulus, glass transition temperature (T_g) and flexural properties of epoxy glass-fiber composites.

Sample	Storage Modulus at 40 °C (GPa)	Storage Modulus at 200 °C (GPa)	T_g (°C)	Flexural Modulus at 25 °C (GPa)	Flexural Strength at 25 °C (MPa)	Strain at Break (%)
RT + 70 + 140	13.8 ± 0.1	3.8 ± 0.3	165	20.6 ± 1.4	354 ± 17.7	1.9 ± 0.1
70 + 140	13.4 ± 0.1	3.7 ± 0.2	164	20.1 ± 0.6	350.4 ± 35.1	2 ± 0.2
140	12.5 ± 0.3	3.6 ± 0.4	165	20.2 ± 1.2	343.1 ± 41.0	2.1 ± 0.2

Similar values for fiberglass epoxy composites have been observed in previous works [33]. This behavior was expected, since the strength of the fibers will be dominant during quasi-static loading [34]. It is known that the presence of the hydroxyl groups on the glass fiber surface can accelerate the cure behavior of the epoxy amine reaction. It was hypothesized that this acceleration might be playing a role in changing the microstructure, leading to similar macroscopic properties. Thus, control samples used to probe the microstructure were also observed on the DMA under the same conditions. The results which can be seen in Supplementary Materials Table S3 show that the changes in curing cycle still do not affect the macroscopic properties. The values for storage modulus, above and below the glass transition, for the neat epoxy samples, are similar irrespective of the curing cycle.

The only difference that could be observed was in the tan delta peak height (Supplementary Materials Figure S7), which showed that the 140 sample had a lower peak height, when compared to the similar peak heights observed for the RT and 70 samples. Since tan delta is qualitatively known to reflect damping or relaxation of the matrix, fatigue testing was done, to determine if the microstructural changes developed during curing would affect the wear and relaxation of the composites. As mentioned previously, the common failure modes in composites are matrix cracking, interface debonding and delamination [6–8], which develop from coalescence of microscopic cracks throughout the fatigue life of a composite, thus, changes in the microstructure of the matrix were expected to have noticeable differences during cyclic testing under subcritical loads. The maximum load during the cyclic test was kept constant for the three types of composites and was selected as 80% of the maximum flexural force required to fail the RT sample (88.2 N). The number of cycles required

Polymers **2018**, *10*, 125

for the 140 sample to fail was significantly lower than those required for the RT and 70 samples, as shown in Figure 4.

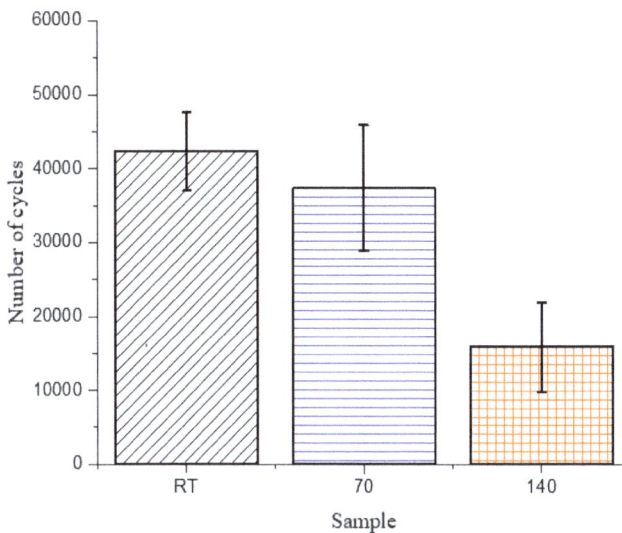

Figure 4. Number of cycles to failure in three-point bending fatigue tests for composite samples.

The change in the matrix microstructure, as depicted in the tan delta peak, is exactly reflective of the results obtained from the fatigue testing. The larger region of softer phase in the 140 sample would act as a stress concentration site and break prematurely, leading to early failure in those samples. The 70 and RT samples, having similar microstructures and tan delta peak heights, undergo a similar number of cycles to fail.

4. Conclusions

The epoxy samples were cured under different conditions, to obtain variations in their microstructures. It was found that those samples that were allowed to gel completely before post-curing showed similar microstructures and mechanical properties, whereas the sample that was cured significantly faster showed a larger region of a softer (lower crosslink density) phase, observable in the Soxhlet extraction, DSC and AFM. When using these curing cycles in epoxy glass fiber composites, it was found that static flexural tests, storage modulus and thermal decomposition remained unchanged, irrespective of microstructure. This result was explained by the fact that the corresponding properties are governed by the homogenized mechanical response of the composite, rather than the local non-homogeneous microstructure in epoxy matrix. The presence of microstructural variance in the matrix was found to play a significant role under fatigue loading, where the composite response is determined by the local "hot" spots in the microstructure, rather than the overall homogenized properties of the composite. The presence of soft sub-microscopic regions of epoxy with lower cross-link density played the role of weak links in the material, which was evident under fatigue loading conditions, where material degradation was gradual, but not in quasi-static loading. Consequently, composites manufactured with faster cure rates, and as a result containing greater volume content of the soft epoxy phase, showed lower fatigue life. However, if given enough time to critically gel, samples showed similar microstructures and thus, similar fatigue properties, irrespective of the rate of cure prior to gelation. Curing times can be expedited by pre-curing the composite at a temperature which vitrifies the network at the critical gel point degree of conversion, without having any consequences on the properties, as is apparent in the RT and 70 samples.

The present result has important practical significance for predicting the fatigue life of composite structures, which normally show a distribution of the cure rate profiles, as a result of the heat transfer and heat generated from the exothermic reaction during epoxy cure. Therefore, the varying cure rates would result in locally different microscopic compositions of the hard-soft phases in the epoxy matrix, and can be expected to affect the fatigue life of composite structure. Thus, the consideration of the cure state prior to gelation is an important aspect of predicting fatigue life of composites, as is the consideration of residual stresses, which are largely affected by the cure reaction past the gel point.

Supplementary Materials: The following are available online at www.mdpi.com/xxx/s1, Figure S1: Phase images of (a) RT sample (b) 70 sample (c) 140 sample at 1μm scale size, Figure S2: Height images of a) RT sample b) 70 sample c) 140 sample at 250nm scale size, Table S1: Percentage area of softer phase interspersed between the hard epoxy network, Table S2: Onset of thermal degradation at 5 weight % loss and peak degradation temperature for the samples cured at different rates, Figure S3: DMA curves for epoxy glass fiber composites , Figure S4: Flexural 3 point bending curves for RT sample, Figure S5: Flexural 3 point bending curves for 70 sample, Figure S6: Flexural 3 point bending curves for 140 sample, Table S3: Dynamic Mechanical Analysis of sample without glass fiber mats, Figure S7: Tan delta vs temperature for the 3 curing cycles without glass fiber mats.

Acknowledgments: The authors are grateful for funding received from the National Science Foundation Partnerships for International Research and Education (PIRE) Program (Award #1243313). The authors would also like to thank Richard Tomazin from the Swagelok Center for Surface Analysis of Materials (SCSAM) for assisting with AFM imaging.

Author Contributions: The manuscript was completed through contributions of all authors. Ammar Patel conceived, designed and performed the experiments; Ammar Patel and Oleksandr Kravchenko analyzed the data and contributed to discussions; Ammar Patel wrote the manuscript; Ica Manas-Zloczower provided recommendations for the investigative process and useful suggestions; Ica Manas-Zloczower contributed reagents and materials and is the principal investigator and corresponding author of this work. All authors revised and finally approved the submission.

Conflicts of Interest: The authors declare no conflict of interest.

References

1. Fan, Z.; Santare, M.H.; Advani, S.G. Interlaminar shear strength of glass fiber reinforced epoxy composites enhanced with multi-walled carbon nanotubes. *Compos. Part A Appl. Sci. Manuf.* **2008**, *39*, 540–554. [CrossRef]

2. Brøndsted, P.; Lilholt, H.; Lystrup, A. Composite Materials for Wind Power Turbine Blades. *Annu. Rev. Mater. Res.* **2005**, *35*, 505–538. [CrossRef]

3. Pilato, L.A.; Michno, M.J. Phenolic Resins: A century of progress. In *Advanced Composite Materials*; Springer: Berlin/Heidelberg, Germany, 1994.

4. Thwe, M.M.; Liao, K. Effects of environmental aging on the mechanical properties of bamboo-glass fiber reinforced polymer matrix hybrid composites. *Compos. Part A Appl. Sci. Manuf.* **2002**, *33*, 43–52. [CrossRef]

5. Latha, P.S.; Rao, M.V.; Kumar, V.K.; Raghavendra, G.; Ojha, S.; inala, R. Evaluation of mechanical and tribological properties of bamboo–glass hybrid fiber reinforced polymer composite. *J. Ind. Text.* **2016**, *46*, 3–18. [CrossRef]

6. Akinyede, O.; Mohan, R.; Kelkar, A.; Sankar, J. Static and Fatigue Behavior of Epoxy/Fiberglass Composites Hybridized with Alumina Nanoparticles. *J. Compos. Mater.* **2009**, *43*, 769–781. [CrossRef]

7. Birger, S.; Moshonov, A.; Kenig, S. Failure mechanisms of graphite-fabric epoxy composites subjected to flexural loading. *Composites* **1989**, *20*, 136–144. [CrossRef]

8. Lucas, J.P. Delamination fracture: Effect of fiber orientation on fracture of a continuous fiber composite laminate. *Eng. Fract. Mech.* **1992**, *42*, 543–561. [CrossRef]

9. Patel, A.; Maiorana, A.; Yue, L.; Gross, R.A.; Manas-Zloczower, I. Curing Kinetics of Biobased Epoxies for Tailored Applications. *Macromolecules* **2016**, *49*, 5315–5324. [CrossRef]

10. Mijovic, J.; Fishbain, A.; Wijaya, J. Mechanistic modeling of epoxy-amine kinetics. 2. Comparison of kinetics in thermal and microwave fields. *Macromolecules* **1992**, *25*, 986–989. [CrossRef]

11. Pascault, J.P.; Moschiar, S.M.; Williams, R.J.J. Buildup of Epoxycycloaliphatic Amine Networks. Kinetics, Vitrification, and Gelation. *Macromelecules* **1990**, *23*, 725–731. [CrossRef]

12. Mijovic, J.; Fishbain, A.; Wijaya, J. Mechanistic modeling of epoxy-amine kinetics. 1. Model compound study. *Macromolecules* **1992**, *25*, 979–985. [CrossRef]

13. Mijovic, J.; Wijayat, J. Reaction Kinetics of Epoxy/Amine Model Systems. The Effect of Electrophilicity of Amine Molecule. *Macromolecules* **1994**, *27*, 7589–7600. [CrossRef]

14. Pollard, M.; Kardos, J.L. Analysis of epoxy resin curing kinetics using the Avrami theory of phase change. *Polym. Eng. Sci.* **1987**, *27*, 829–836. [CrossRef]

15. Ivankovic, M.; Incarnato, L.; Kenny, J.M.; Nicolais, L. Curing kinetics and chemorheology of epoxy/anhydride system. *J. Appl. Polym. Sci.* **2003**, *90*, 3012–3019. [CrossRef]

16. Cheng, K.C.; Chiu, W.Y.; Hsieh, K.H.; Ma, C.C.M. Chemorheology of epoxy resin Part I epoxy resin cured with tertiary amine. *J. Mater. Sci.* **1994**, *29*, 721–727. [CrossRef]

17. Bogetti, T.A.; Gillespie, J.W., Jr. Process-Induced Stress and Deformation in Thick-Section Thermoset Composite Laminates. *J. Compos. Mater.* **1992**, *26*, 626–660. [CrossRef]

18. Kravchenko, O.G.; Kravchenko, S.G.; Pipes, R.B. Chemical and thermal shrinkage in thermosetting prepreg. *Compos. Part A Appl. Sci. Manuf.* **2016**, *80*, 72–81. [CrossRef]

19. Kravchenko, O.G.; Kravchenko, S.G.; Pipes, R.B. Cure history dependence of residual deformation in a thermosetting laminate. *Compos. Part A Appl. Sci. Manuf.* **2017**, *99*, 186–197. [CrossRef]

20. Madhukar, M.S.; Genidy, M.S.; Russell, J.D. A New Method to Reduce Cure-Induced Stresses in Thermoset Polymer Composites, Part I: Test Method. *J. Compos. Mater.* **2000**, *34*, 1882–1904. [CrossRef]

21. Morgan, R.J.; O'Neal, J.E. The microscopic failure processes and their relation to the structure of amine-cured bisphenol-A-diglycidyl ether epoxies. *J. Mater. Sci.* **1977**, *12*, 1966–1980. [CrossRef]

22. Vanlandingham, M.R.; Eduljee, R.F.; Gillespie, J.W., Jr. Relationships between Stoichiometry, Microstructure, and Properties for Amine-Cured Epoxies. *J. Appl. Polym. Sci.* **1999**, *71*, 699–712. [CrossRef]

23. Cuthrell, R.E. Epoxy polymers. II. Macrostructure. *J. Appl. Polym. Sci.* **1968**, *12*, 1263–1278. [CrossRef]

24. Vanlandingham, M.R.; Eduljee, R.F.; Gillespie, J.W., Jr. Moisture diffusion in epoxy systems. *J. Appl. Polym. Sci.* **1999**, *71*, 787–798. [CrossRef]

25. Palmese, G.R.; McCullough, R.L. Effect of epoxy-amine stoichiometry on cured resin material properties. *J. Appl. Polym. Sci.* **1992**, *46*, 1863–1873. [CrossRef]

26. Palmese, G.R.; McCullough, R.L. Kinetic and thermodynamic considerations regarding interphase formation in thermosetting composite systems. *J. Adhes.* **1994**, *44*, 29–49. [CrossRef]

27. Hale, A.; Macosko, C.W.; Bair, H.E. Glass transition temperature as a function of conversion in thermosetting polymers. *Macromolecules* **1991**, *24*, 2610–2621. [CrossRef]

28. Gillham, J.K. Formation and Properties of Thermosetting and High T_g Polymeric Materials. *Polym. Eng. Sci.* **1986**, *26*, 1429–1433. [CrossRef]

29. López, J.; Ramírez, C.; Torres, A.; Abad, M.J.; Barral, L.; Cano, J.; Díez, F.J. Isothermal curing by dynamic mechanical analysis of three epoxy resin systems: Gelation and vitrification. *J. Appl. Polym. Sci.* **2002**, *83*, 78–85. [CrossRef]

30. Lange, J.; Altmann, N.; Kelly, C.T.; Halley, P.J. Understanding vitrification during cure of epoxy resins using dynamic scanning calorimetry and rheological techniques. *Polymer* **2000**, *41*, 5949–5955. [CrossRef]

31. Chiou, P.-L.; Letton, A. Modelling the chemorheology of an epoxy resin system exhibiting complex curing behaviour. *Polym. Eng. Sci.* **1992**, *33*, 3925–3931. [CrossRef]

32. Vyazovkin, S.; Sbirrazzuoli, N. Mechanism and Kinetics of Epoxy-Amine Cure Studied by Differential Scanning Calorimetry. *Macromolecules* **1996**, *29*, 1867–1873. [CrossRef]

33. Yue, L.; Maiorana, A.; Patel, A.; Gross, R.; Manas-Zloczower, I. A sustainable alternative to current epoxy resin matrices for vacuum infusion molding. *Compos. Part A Appl. Sci. Manuf.* **2017**, *100*, 269–274. [CrossRef]

34. Nightingale, C.; Day, R.J. Flexural and interlaminar shear strength properties of carbon fibre/epoxy composites cured thermally and with microwave radiation. *Compos. Part A Appl. Sci. Manuf.* **2002**, *33*, 1021–1030. [CrossRef]

![polymers logo] *polymers*

MDPI

Article

Toughening of Poly(lactic acid) and Thermoplastic Cassava Starch Reactive Blends Using Graphene Nanoplatelets

Anibal Bher [1,2,3], Ilke Uysal Unalan [1,4], Rafael Auras [1,*], Maria Rubino [1] and Carlos E. Schvezov [3]

[1] School of Packaging, Michigan State University, East Lansing, MI 48824, USA; anibalbher@gmail.com (A.B.);
    iuysalunalan@gmail.com (I.U.U.); mariar@msu.edu (M.R.)
[2] Instituto Sabato, UNSAM-CNEA, San Martin, Buenos Aires 1650, Argentina
[3] Instituto de Materiales de Misiones (IMAM), CONICET-UNaM, Posadas, Misiones 3300, Argentina;
    schvezov@gmail.com
[4] Department of Food Engineering, Faculty of Engineering, İzmir University of Economics,
    İzmir 35330, Turkey
* Correspondence: aurasraf@msu.edu; Tel.: +1-517-432-3254

Received: 21 November 2017; Accepted: 15 January 2018; Published: 19 January 2018

Abstract: Poly(lactic acid) (PLA) was reactively blended with thermoplastic cassava starch (TPCS) and functionalized with commercial graphene (GRH) nanoplatelets in a twin-screw extruder, and films were produced by cast-film extrusion. Reactive compatibilization between PLA and TPCS phases was reached by introducing maleic anhydride and a peroxide radical during the reactive blending extrusion process. Films with improved elongation at break and toughness for neat PLA and PLA-*g*-TPCS reactive blends were obtained by an addition of GRH nanoplatelets. Toughness of the PLA-*g*-TPCS-GRH was improved by ~900% and ~500% when compared to neat PLA and PLA-*g*-TPCS, respectively. Crack bridging was established as the primary mechanism responsible for the improvement in the mechanical properties of PLA and PLA-*g*-TPCS in the presence of the nanofiller due to the high aspect ratio of GRH. Scanning electron microscopy images showed a non-uniform distribution of GRH nanoplatelets in the matrix. Transmittance of the reactive blend films decreased due to the TPCS phase. Values obtained for the reactive blends showed ~20% transmittance. PLA-GRH and PLA-*g*-TPCS-GRH showed a reduction of the oxygen permeability coefficient with respect to PLA of around 35% and 50%, respectively. Thermal properties, molecular structure, surface roughness, XRD pattern, electrical resistivity, and color of the films were also evaluated. Biobased and compostable reactive blend films of PLA-*g*-TPCS compounded with GRH nanoplatelets could be suitable for food packaging and agricultural applications.

Keywords: PLA; reactive blending; biobased films; graphene; nanoreinforcement

1. Introduction

Poly(lactic acid) (PLA), an aliphatic polyester derived from lactic acid, is a biobased polymer that can be obtained from renewable sources, and in recent years has been used increasingly in food packaging and medical devices and in the agriculture, textile, and automotive industries [1]. PLA is recyclable, biodegradable, and compostable under industrial composting conditions, and it is considered to be an alternative to replace traditional fossil non-biodegradable polymers [2]. However, some drawbacks regarding PLA's mechanical and barrier properties hinder its more widespread application in areas such as food packaging [1,3,4].

The blending of PLA with other polymers and the incorporation of nanofillers into the PLA matrix have been pursued to overcome some of the polymer's shortcomings [5–9]. PLA has been blended with polyethylene glycol, ethylene vinyl alcohol, poly(butylene adipate-*co*-terephthalate), and other polymeric systems to improve crystallinity, biodegradability, and thermal and mechanical properties [10–12]. One of the main biopolymers used for reactive blending functionalization of PLA is a starch-based thermoplastic, in order to maintain PLA's inherent compostability [13,14]. Starches obtained from crops such as corn, cassava, and sugar beets are cheap, biodegradable, and promote a lower environmental footprint [15]. Cassava starch is a good option for the production of thermoplastic cassava starch (TPCS) since cassava is inexpensive and is produced on a massive scale [16]. TPCS and modified TPCS offer acceptable properties as a PLA component for creating flexible packaging films [17,18].

Reactive blending is the best option to produce compatible PLA and TPCS blends [13,17,19,20]. The use of maleic anhydride in the presence of peroxide initiators, such as 2,5-bis(tert-butylperoxy)-2,5-dimethylhexane and dicumyl peroxide, has been widely reported as an efficient method to achieve an adequate compatibilization between the PLA and TPCS phases [13,17,20,21]. Films produced by reactive blending show a significant improvement in mechanical properties. However, the non-compatibilized or physical blends between these two biopolymers have poor mechanical properties and lower interfacial adhesion [17]. Although reactive blends of PLA-*g*-TPCS (i.e., PLA grafted to TPCS) present a good compatibilization of phases, the mechanical and barrier properties must be improved for the commercial production of films [21].

The use of clays (e.g., montmorillonite, halloysite), nanocellulose, metal nanoparticles (e.g., titanium dioxide, zinc oxide, silver), silica, and carbon nanotubes have been widely reported as reinforcement particles for some single or blended polymers [22–24].

Nanocomposites of PLA obtained using carbon-based nanofillers have been extensively reported with the aim of improving PLA's properties by the incorporation of carbon-based materials with remarkable values of tensile strength, conductivity, thermal stability, etc. [25]. One of the most exciting novel nanomaterials used in polymers to improve properties is graphene. Graphene is produced from natural graphite and is considered a multifunctional material that can be introduced into polymers and composites. Low concentrations of graphene have been shown to improve polymer stiffness, electrical and thermal conductivity, and barrier properties due to its high surface area, high modulus, low density, etc. [26–28]. Graphene (GRH) nanoplatelets typically consist of a few graphene layers, with thicknesses between 1 and 10 nm, average particle diameters of 3 to 50 μm, and a high aspect ratio [29]. In comparison with the single layer, GRH nanoplatelets are low cost and can be used as a reinforcement nanomaterial with the potential for large-scale production [30].

The incorporation of GRH oxide into the matrix of PLA or starch composites results in mechanical and barrier properties improvements [31,32]. The diffusional pathway is influenced by the morphology of the nanosheets or nanoplatelets (size, aspect ratio, etc.) as well as the degree of exfoliation and orientation of the individual sheets or flakes in the polymer matrix [33]. However, the inclusion of GRH nanoplatelets in the matrix of PLA-*g*-TPCS blends has not yet been explored, and it could provide a unique opportunity for the production of a novel blend with unique properties. This work aimed to develop a reactive blend film of PLA-*g*-TPCS reinforced with a low level of GRH nanoplatelets (0.1 wt %), to understand the effect of adding this amount of GRH on PLA-*g*-TPCS reactive blends, and then to fully characterize the new material by molecular weight, microscopy, and mechanical, thermal, optical, electrical, and barrier properties.

2. Materials and Methods

2.1. Materials

Ingeo™ biopolymer 2003D poly (96% L-lactic acid) (PLA) was acquired from NatureWorks LLC (Minnetonka, MN, USA), with a weight (M_w) and number (M_n) average molecular weight of $2.2 \pm 0.2 \times 10^5$ Da and $1.2 \pm 0.6 \times 10^5$ Da, respectively. Cassava starch with amylose content of $25\% \pm 6\%$ wt/wt and ~12% of moisture content was purchased from Aldema LLC

(Cooperativa Agricola e Industrial San Alberto Ltda., Puerto Rico, Mnes, Argentina). Glycerol (>99.5%), maleic anhydride (MA), and 2,5-bis(tert-butylperoxy)-2,5-dimethylhexane (L101) were purchased from Sigma Aldrich (Milwaukee, WI, USA). Graphene nanoplatelets xGnP® M-25, with an average particle diameter of 25 µm, a thickness of around 10 nm, and a surface area between 120 to 150 $m^2 \cdot g^{-1}$, were procured from XG Sciences Inc. (Lansing, MI, USA).

2.2. Production of Master Batch

PLA-*g*-MA was produced in a Century ZSK-30 twin-screw extruder (Century, Traverse City, MI, USA) with a composition of 2 wt % MA and 0.65 wt % L101, based on PLA weight. PLA, MA, and L101 were vigorously premixed in a bag before processing. The twin-screw extruder residence time for a rate of feed of 70 $g \cdot min^{-1}$ was ~3 min. The extruded mass was pelletized in a BT 25 pelletizer (Scheer Bay Co., Bay City, MI, USA), held in an oven at 50 °C for 3 h to remove residual water, and stored in a freezer at −15 °C until use. PLA-c (considered as control) was also produced, pelletized, and stored as previously described. TPCS was produced by mixing cassava starch and glycerol (in a resealable bag) in a proportion of 70/30 *wt/wt*, and holding for 12 h before use; TPCS was extruded and pelletized under similar conditions to those mentioned above. PLA-*g*-TPCS was produced by mixing PLA-*g*-MA with TPCS and using the same processing method as for PLA-*g*-MA. The proportion used of PLA-*g*-MA and TPCS was 70/30 *wt/wt*.

To produce the PLA-*g*-TPCS-GRH blend, pellets of PLA-*g*-TPCS were held in a vacuum oven overnight and were then premixed in a bag with 0.1 wt % of GRH nanoplatelets; the mixture was fed into the twin-screw extruder. For the PLA-GRH blend, neat PLA was premixed with 0.1 wt % of GRH and extruded using the twin-screw extruder, with an approximate residence time of 4 min. The extrusion was pelletized, held for 4 h at 50 °C in an oven, and stored in a freezer at −15 °C. Table 1 shows the processing conditions.

Table 1. Conditions of processing of master batches in the twin-screw extruder.

Films	Temperature Profile from the Feeder to the Die (°C)	Screw Speed (RPM)
PLA-c	140/150/160/160/160/170/170/170/170/160	120
TPCS	25/100/105/110/115/120/120/120/115/115	125
PLA-*g*-MA	140/150/160/160/160/170/170/170/170/160	120
PLA-*g*-TPCS	140/150/160/160/160/170/170/170/170/160	120
PLA-GRH	140/150/160/160/160/170/170/170/170/160	100
PLA-*g*-TPCS-GRH	140/150/160/160/160/170/170/170/170/160	100

RPM: revolutions per minute; PLA: Poly(lactic acid); TPCS: thermoplastic cassava starch; MA: maleic anhydride; GRH: graphene.

2.3. Production of Films

Master batches obtained from the twin-screw extruder were used to produce cast films in a single extruder (Randcastle Extrusion Systems, Inc., Cedar Grove, NJ, USA). Before the cast film process, all of the materials were conditioned overnight in an oven at 50 °C. The fabricated films were stored at −15 °C for further characterization. The processing conditions are presented in Table 2.

Table 2. Processing conditions of the cast films.

Films	Temperature Profile from the Feeder to the Die (°C)	Screw-Nip Roller-Winding Roller Speed (RPM)
PLA-c	140/150/160/160/160/170/160	30-50-12
PLA-*g*-TPCS	140/150/160/160/160/170/160	30-50-12
PLA-GRH	140/150/160/160/160/170/160	30-50-12
PLA-*g*-TPCS-GRH	140/150/160/160/160/170/140	30-50-12

2.4. Molecular Weight

Weight average molecular weight (M_w), number average molecular weight (M_n), and polydispersity index (*PI*) of PLA-c and the fraction of PLA present in PLA-*g*-TPCS, PLA-GRH, and PLA-*g*-TPCS-GRH were determined by the method described in previous work [17,21] using a Waters gel permeation chromatograph (Waters Corp., Milford, MA, USA) equipped with a 1515 isocratic HPLC pump, a 717plus autosampler, four Styragel® columns (HR1, HR2, HR3, and HR4), and a 2414 refractive index detector interface (Waters Corp.). Experiments were conducted in triplicate.

2.5. Scanning Electron Microscopy (SEM) and Atomic Force Microscopy (AFM)

SEM was used to investigate the morphology of the samples. Bar specimens of PLA-c, PLA-*g*-TPCS, PLA-GRH, and PLA-*g*-TPCS-GRH blends were produced by compression molding using pellets from the twin-screw extruder. Bars of PLA-*g*-TPCS and PLA-*g*-TPCS-GRH were immersed in liquid nitrogen for ~3 min, then fractured by hand, treated with hydrochloric acid (6 N) for 6 h to remove the TPCS phase, and air dried for 12 h in a fume hood [17]. Finally, the samples were mounted on aluminum stubs using carbon suspension cement (SPI Supplies, West Chester, PA, USA). Samples of films evaluated before and after tensile testing were also mounted on aluminum stubs with high vacuum carbon tabs (SPI Supplies) and coated with iridium at a thickness of ~2.7 nm. Samples were examined in a JEOL 6610LV (tungsten hairpin emitter) and a JEOL 7500F (field emission emitter) scanning electron microscope (JEOL Ltd., Tokyo, Japan) at various magnifications at 10 and 3.0 kV, respectively.

AFM was conducted using a Cypher™ atomic force microscope (Oxford Instruments Asylum Research, Inc., Santa Barbara, CA, USA) in the contact mode. Roughness parameters, calculated as the root mean square (Rq) and average roughness (Ra), were determined for each type of film and were calculated from the Htr mode image. Images were obtained in the Dfr mode. The film area for the determination of roughness was 900 μm^2.

2.6. Profilometer

A profilometer NanoMap 500LS (AEP Technology, Santa Clara, CA, USA) was also used for roughness determination. Films were attached to a microscope slide for measuring, and scans were conducted at 50 $\mu m \cdot s^{-1}$ with a sample frequency of 100 pts$\cdot s^{-1}$, data resolution of 5 μm, 6000 number of points per scan, and a contact force of 5.4 mg. The scan area was 9×10^6 μm^2. Values of Ra (nm) and peak-to-peak (nm) were determined from every measurement. A minimum of two samples was scanned for each film.

2.7. Tensile Properties

Tensile strength, elongation at break, and Young's modulus were evaluated using an Instron® Universal Machine 5565 (Instron, Norwood, MA, USA) according to ASTM D882-12 [34]. Film samples were cut into 2.54 cm × 20 cm strips and conditioned for 48 h in an environmental chamber (Environmental Growth Chambers, Chagrin Falls, OH, USA) at 23 °C and 50% relative humidity. PLA-c was tested with an initial grip and a rate grip separation of 125 mm and 12.5 mm·min^{-1}, respectively. All of the other samples were tested with an initial grip and a rate grip separation of 100 mm and 50 mm·min^{-1}, respectively. Five samples were evaluated for each formulation. Film thickness was determined by averaging five measurements for every specimen using a digital micrometer (Testing Machines Inc., Ronkonkoma, NY, USA).

2.8. Thermal Properties

Differential scanning calorimetry (DSC) of the films was carried out with a Q100 differential scanning calorimeter (TA Instruments, New Castle, DE, USA) equipped with a mechanical cooling system. Specimens between 5 and 10 mg were cut, weighed, and sealed in an aluminum pan.

Samples were first cooled from room temperature to $-50\,°C$, heated from $-50\,°C$ to $200\,°C$ (first heating cycle), underwent an isothermal for 1 min at $200\,°C$, again cooled until $-50\,°C$, and finally heated to $200\,°C$ (second heating cycle). The sample purge flow rate was 70 mL·min^{-1} of nitrogen. Glass transition temperature (T_g), cold crystallization temperature (T_{cc}), melting temperature (T_m), and degree of crystallinity (X_c) of PLA and blends were determined using the TA Universal Analysis 2000 software, version 4.5A (TA Instruments, New Castle, DE, USA). The X_c of PLA was calculated from an equation modified from Detyothin et al. [17]:

$$X_c = \frac{\Delta H_m - \Delta H_c}{\Delta H_f \times (1-\alpha)} \tag{1}$$

where ΔH_m and ΔH_c are the enthalpies of melting and crystallization, respectively; ΔH_f is the enthalpy of fusion of a pure crystalline PLA with a value of 93 J·g^{-1} [35]; and α is the sum of the weight fraction of TPCS, MA, and GRH in the final blends. The values of X_c are reported from the first and second heating run. Samples were run in triplicate.

Thermogravimetric analysis (TGA) of the films was conducted in a Q50 thermal gravimetric analyzer (TA Instruments) under a nitrogen atmosphere with a flow rate of 60 mL·min^{-1}. The method used was a ramp of $10\,°C\cdot min^{-1}$ from room temperature until reaching $600\,°C$ in an aluminum pan. Between 5 and 10 mg of sample was used for every run. Samples were run in triplicate.

Dynamic mechanical analysis (DMA) was conducted using an RSA-G2 analyzer (TA Instruments). Film samples (64 mm × 12 mm) were conditioned at 50% relative humidity and $23\,°C$ for 48 h in an environmental chamber. Five specimens were evaluated for each type of blend. The storage modulus (E'), loss modulus (E''), and tan delta were evaluated. A loading gap of 15 mm for a rectangular geometry was used. A tension axial force of 400 g with a sensitivity of 10 g was used, with an oscillation temperature ramp of $3\,°C\cdot min^{-1}$ from 25 to $100\,°C$ with a frequency of 1.0 Hz and a strain of 2%. Five specimens of each type of blend were assessed.

2.9. UV-Visible Spectroscopy

Transmittance of the films was evaluated with a Lambda 25 UV/Vis spectrophotometer (PerkinElmer, Waltham, MA, USA) in a wavelength range of 190–880 cm^{-1} using one cycle, with a scan speed of 480 nm·min^{-1}.

2.10. Barrier Properties to Oxygen and Water Vapor

Oxygen and water vapor barrier performance of film samples were measured with Ox-tran$^{®}$ model 2/21 and Permatran$^{®}$ model 3/33 testing instruments (Mocon, Minneapolis, MN, USA), respectively. Samples were mounted in masks of aluminum foil with an exposed area of 3.14 cm^2. Test conditions were 30% relative humidity (RH) and $23\,°C$ for oxygen and water vapor. The oxygen test was conducted with air.

2.11. Statistical Analysis

Statistical analysis was conducted with MINITAB™ software (State College Park, PA, USA). ANOVA and Tukey's test were used to evaluate the comparison of means at $p \leq 0.05$.

3. Results and Discussion

Reactive blends of PLA and TPCS compatibilized with MA using L101 as peroxide initiator were produced and characterized. GRH nanoplatelets were introduced with the objective of increasing the toughening of the reactive blends. Subsequently, cast films were produced and characterized. PLA-c and PLA-GRH were produced as reference films, and their properties are also reported.

3.1. Molecular Weight

The M_n, M_w, and *PI* values of the PLA films are reported in Table 3. The M_n and M_w for the fraction of PLA measured decreased significantly for the reactive blend of PLA-*g*-TPCS with respect to PLA-c. The overall reduction of the PLA-*g*-TPCS is related to several factors: (1) the use of MA and the peroxide initiator; (2) the presence of water and glycerol on TPCS; (3) the use of a two-step processing method (twin-screw extruder for the production of blends and single extruder for the production of films); and (4) the thermal conditions associated with the process.

Table 3. M_n, M_w, and *PI* of PLA-c and the PLA portion of the various films produced.

Films	M_n	M_w	PI
PLA-c	103.5 ± 0.1 [a]	201.9 ± 0.4 [a]	1.9 ± 0.1 [a]
PLA-*g*-TPCS	71.6 ± 10.5 [b]	147.6 ± 2.5 [b]	2.0 ± 0.3 [a]
PLA-GRH	98.8 ± 3.1 [a]	200.5 ± 0.7 [a]	2.0 ± 0.1 [a]
PLA-*g*-TPCS-GRH	71.0 ± 1.3 [b]	138.1 ± 1.3 [c]	1.9 ± 0.1 [a]

Note: Within columns, values followed by a different letter are significantly different at $p \leq 0.05$ (Tukey's test). *PI*: polydispersity index.

The degradation of PLA in the reactive blends can be associated with the dominant side reaction occurring in the maleated PLA obtained using a free radical grafting initiator, such as L101, and a compatibilizer, such as MA [19].

The presence of water due to the incorporation of TPCS for the production of the reactive blend is a reasonable factor affecting the reduction of the M_n and M_w of PLA [36]. Hydrolytic degradation is a well-known degradation mechanism of PLA due to the presence of moisture. The hydrolysis of PLA begins by the diffusion of water molecules into the amorphous portion of PLA, resulting in the cleavage of the ester bonds [37]. Thermal degradation of PLA during processing is also related to the hydrolysis of residual water, main-chain scission, and intra and intermolecular transesterification. A reduction of M_n for PLA by hydrolysis was reported in the presence of water, even at low temperatures [38]. The presence of low molecular weight additives, such as glycerol, has a negative impact on the molecular weight of PLA due to thermal degradation or the hydrolysis of polyester chains.

The production of films for characterization required a two-step processing method. In this case, the use of different temperatures and shear and mixing conditions allow for a reduction of molecular weight. Taubner and Shishoo reported a reduction of M_n for PLA during melt extrusion using a double-screw extruder influenced by the processing temperature and residence time [39].

In the case of PLA-GRH, the values of M_n and M_w were similar to those of PLA-c, showing that the addition of GRH nanoplatelets did not affect the PLA's structure. In the case of the final reactive blend, PLA-*g*-TPCS-GRH, this resulted in a ~30% decrease of M_n and M_w with respect to PLA due to the dual processing steps of the composite blends: first, to produce the reactive blend and then due to the addition of the GRH nanoplatelets and reprocessing. This reduction is associated with thermal conditions inherent with the processing in the twin-screw and single extruders. Nonetheless, the values of *PI* showed a narrow distribution of the molecular weight for all the samples: $PI \leq 2$.

3.2. Morphology of the Films

SEM images of the PLA blends were obtained to understand the grade of compatibilization between PLA and TPCS and the incorporation of GRH nanoplatelets. An optimal distribution of GRH nanoplatelets should play an important role in enhancing the barrier and mechanical properties of nanocomposite blends. The morphology of the fracture surface for the evaluated blends is shown in Figure 1. Figure 1a shows the surface of PLA-c. Figure 1b shows the reactive blend of PLA-*g*-TPCS after removing the TPCS phase; good compatibilization between PLA and TPCS was achieved in the reactive blend by the incorporation of MA and L101. Similar results for these reactive blends were reported by Detyothin et al. [17]. The proportion of cavities observed due to the removal of TPCS is in accordance with the proportion used for the production of the reactive blend. The compatibilization achieved allowed for good homogenization during the processing step using a twin-screw extruder. The incorporation of GRH nanoplatelets into neat PLA is presented in Figure 1c, where a small distribution of GRH nanoplatelets in the PLA matrix is in accordance with the low load of GRH used in the production (0.1 wt %). Figure 1d shows the presence and distribution of GRH nanoplatelets in the matrix of the compatibilized blend obtained with PLA and TPCS. By comparing Figure 1c,d, it appears that the GRH nanoplatelets are mostly located as flakes in the interface of PLA and TPCS. Figure S1 (supporting information available online) shows the graphene nanoplatelets powder used as a reinforcement and the formation of agglomerates. The TPCS domain was removed in Figure 1d, and it is apparent that the GRH nanoplatelets are embedded in the cavities surrounded by the PLA matrix. This observation may help explain the mechanical property results whereby the GRH nanoplatelets increased the tensile strength of PLA-GRH with respect to neat PLA but also increased the elongation at break of the reactive blend produced with TPCS.

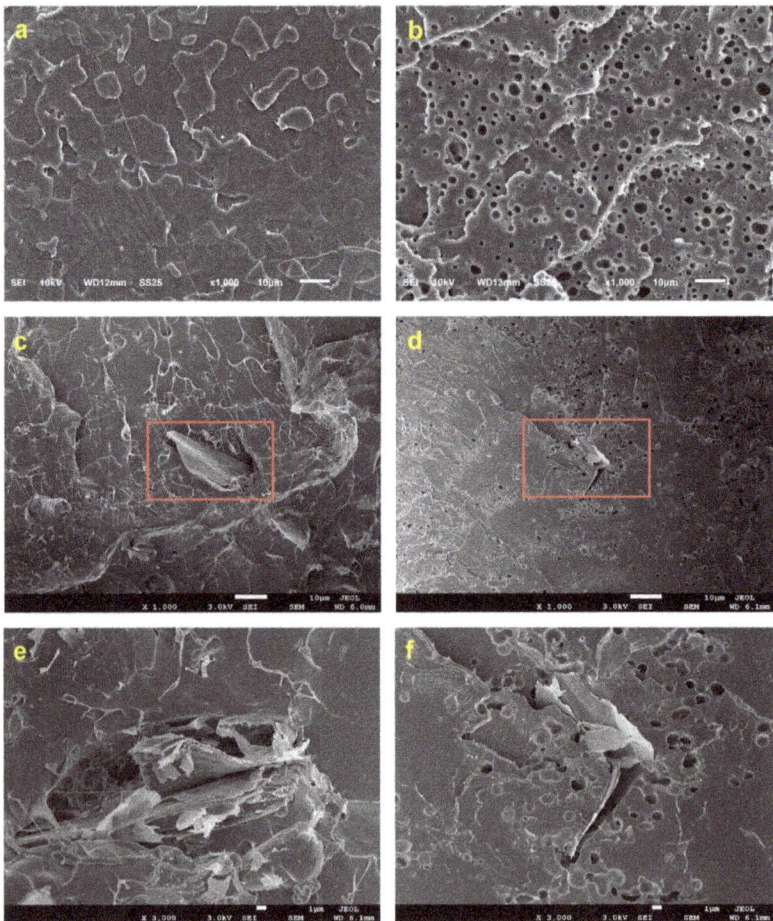

Figure 1. SEM images of film samples showing the polymer domains and distribution of GRH nanoplatelets: (**a**) PLA-c; (**b**) PLA-*g*-TPCS; (**c**) PLA-GRH (×1000); (**d**) PLA-*g*-TPCS-GRH (×1000); (**e**) PLA-GRH (×3000); (**f**) PLA-*g*-TPCS-GRH (×3000).

Figure 1e,f reveal further details of how the GRH nanoplatelets are inserted in both the PLA matrix and the PLA-*g*-TPCS-GRH. An uneven distribution of nanoplatelets (but good compatibilization and adhesion) in the matrix is evident. Figure 1e,f show the clustering or agglomeration of several GRH nanoplatelets, which could affect the barrier and mechanical properties. The presence of flakes of GRH in the matrix of PLA-c and PLA-*g*-TPCS exhibiting a surface structure with cracks was also reported by Gao et al. [40] working with nanocomposites of PLA and GRH nanoplatelets at concentrations between 5 and 15 wt %. Pinto et al. also reported deficient exfoliation of GRH nanoplatelets with aggregations of 5 to 10 sheets, as observed by SEM, and overlapping of nanoplatelets as observed by optical microscopy [26]. How the distribution of GRH nanoplatelets in the polymer matrix affected the mechanical properties of the films is described in the next section.

Table 4 and Figure 2 present the surface roughness of the PLA films as evaluated by AFM. The PLA-*g*-TPCS and PLA-*g*-TPCS-GRH films had higher values of surface roughness in comparison with the PLA-c and PLA-GRH films. This greater roughness could be attributed to the presence of the starch matrix in both reactive blends. Figure 2 shows a smooth surface for PLA-c and PLA-GRH. Roughness values obtained by profilometry are larger than those obtained by AFM due to the larger scan area; however, the values follow the same trend (i.e., roughness for PLA-c and PLA-GRH are lower than for PLA-*g*-TPCS and PLA-*g*-TPCS-GRH.).

Table 4. Roughness of PLA films as measured by AFM and profilometry.

Films	AFM		Profilometry	
	Rq (nm)	Ra (nm)	Ra (nm)	Peak-to-Peak (nm)
PLA-c	9.0 ± 2.1 [a]	6.3 ± 1.3 [a]	1046 ± 572 [a]	5140 ± 2023 [a]
PLA-*g*-TPCS	183.5 ± 0.3 [b]	143.6 ± 5.4 [b]	$35,167 \pm 12,176$ [b]	$126,667 \pm 23,283$ [b]
PLA-GRH	12.5 ± 2.3 [a]	9.8 ± 1.7 [a]	2811 ± 991 [a]	$13,986 \pm 4577$ [a]
PLA-*g*-TPCS-GRH	141.8 ± 33.5 [b]	104.4 ± 37.4 [b]	$17,100 \pm 7353$ [b]	$66,350 \pm 19,728$ [c]

Note: Within columns, values followed by a different letter are significantly different at $p \leq 0.05$ (Tukey's test).

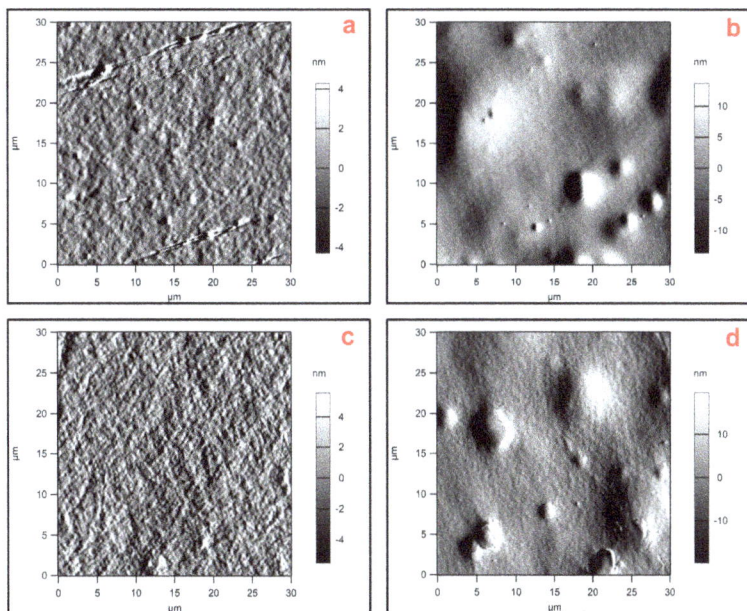

Figure 2. AFM images of films: (**a**) PLA-c; (**b**) PLA-*g*-TPCS; (**c**) PLA-GRH; (**d**) PLA-*g*-TPCS-GRH.

Figure 1a,b allow for a comparison of the difference observed in roughness by AFM due to the presence of the TPCS phase. An increase of roughness could be attributed to irregular shapes of starch granules as can be observed for the cavities where these granules were immersed. The creation of a topography with different values of roughness could be due to the presence of two different phases that were compatibilized.

The presence of GRH nanoplatelets could not be significantly affecting the roughness of the PLA films due to the low load used for film production, and its effect is masked in the reactive blends due to the effect of the TPCS phase. Pinto et al. evaluated the topography of PLA with GRH nanoplatelets as a nano-based material and reported an increase in roughness with a higher concentration of GRH nanoplatelets (0.4 wt %) [41].

3.3. Tensile Properties

An analysis of the tensile testing of the PLA films, evaluated in the machine direction, reveals a characteristic brittle behavior for PLA-c, with tensile strength values of ~25 MPa and elongation at break values around 9% (Table 5 and Figure 3). However, the introduction of GRH nanoplatelets to the PLA matrix resulted in improvements in both tensile strength and elongation at break of ~75% and 130%, respectively. Others have reported similar values of tensile strength when GRH nanoplatelets were used as a nanofiller for PLA [42]. Chieng et al. [43] reported tensile strength values of ~60 MPa for PLA with 0.3 wt % of GRH. Valapa et al. [27], using expandable graphite as reinforcement, found similar increments for elongation at break of PLA, but the increment for tensile strength was small with the same load (0.1 wt %).

Table 5. Tensile properties of films produced by cast-film extrusion.

Films	Thickness, µm	Tensile Strength, MPa	Modulus, GPa	Elongation at Break, %	Toughness, MJ·m^{-3}
PLA-c	24.2 ± 3.5 [a]	24.7 ± 0.7 [a]	1.2 ± 0.1 [a]	8.9 ± 5.1 [a]	1.3 ± 0.7 [a]
PLA-*g*-TPCS	46.0 ± 5.6 [b]	11.0 ± 1.5 [b]	0.7 ± 0.1 [b]	23.9 ± 2.1 [b]	2.3 ± 0.5 [a]
PLA-GRH	20.8 ± 2.9 [a]	43.2 ± 1.5 [c]	2.2 ± 0.3 [c]	21.2 ± 5.7 [b]	5.3 ± 1.1 [b]
PLA-*g*-TPCS-GRH	50.6 ± 8.4 [b]	13.7 ± 1.1 [b]	0.8 ± 0.1 [b]	103.4 ± 2.7 [c]	13.0 ± 1.4 [c]

Note: Within columns, values followed by a different letter are significantly different at $p \leq 0.05$ (Tukey's test).

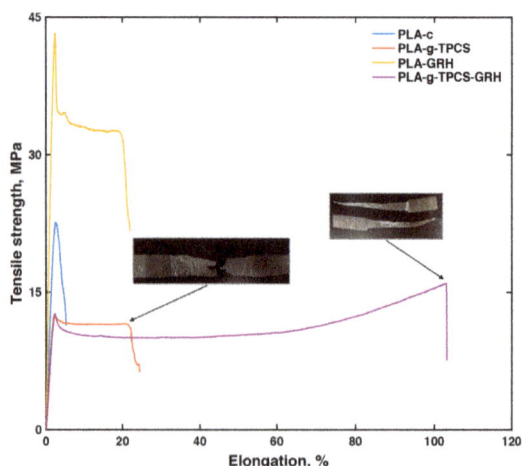

Figure 3. Tensile strength vs. elongation at break for films produced by twin-screw extrusion–cast-film extrusion.

The Young's modulus increased by ~100% with respect to neat PLA, indicating the stiffness behavior of the incorporated nanofiller. Reactive blending using MA and L101 to obtain better compatibilization between PLA and TPCS resulted in a blend with good elongation at break (~25%) but a ~50% reduction in tensile strength with respect to PLA-c. Similar values have been previously reported for these blends [17]. The addition of GRH nanoplatelets to the reactive blend of PLA-*g*-TPCS resulted in an important improvement in the elongation at break (values larger than 100%), showing an increment of ~300% with respect to PLA-*g*-TPCS. Values of toughness showed a significant improvement for PLA-*g*-TPCS-GRH, around ~900% with respect to PLA-c. Figure 3 insets are images of the films (PLA-*g*-TPCS and PLA-*g*-TPCS-GRH) after the tensile testing. The nanofiller incorporated did not improve the tensile strength of the reactive blend.

Figure 3 indicates that at least two mechanisms were acting when PLA-*g*-TPCS-GRH films were tested: a well-identified yield point and a strain hardening behavior. In Figure 1, the presence of gaps between the GRH nanoplatelets and the polymer matrix in the fractured surface of PLA-*g*-TPCS-GRH could be identified. Furthermore, Figure S2 shows the SEM images of the films after the tensile test. The cavities generated in the PLA-*g*-TPCS polymer matrix during the tensile test can be observed, which could be one of the main reasons for the improvement in the elongation at break (Figure S2b,e). Due to the incorporation of GRH nanoplatelets, when the material is under tension, the fractures created in the surface of the polymer matrix could find a free propagation path or a flake of GRH nanoplatelets. Since the GRH nanoplatelets are stiff materials, in the second case the fracture is forced to find an alternative path that continues with the propagation of the fracture and breaks the material during plastic deformation. Thus, an increase of the deformation energy and toughening is observed due to the addition of GRH, which is finally translated into high values of elongation at break [27,44].

The presence of GRH nanoplatelets, even at a low concentration, is enough to create a crack-bridging mechanism during tension [45]. This mechanism could increase the fracture toughness of a nanocomposite, and its efficiency is important when the nanofiller has a high value of aspect ratio [44]. One of the most significant properties of GRH nanoplatelets is their high aspect ratio. Since the adhesion of the GRH nanoplatelets to the PLA matrix is strong—as shown in the SEM images, Figure 1d and Figure S2f,l—and the nanoplatelets can act as a bridge between two fracture surfaces, they are avoiding or delaying the pullout and thus increasing the fracture energy, as was also demonstrated for other fillers [44–46].

On the other hand, it has been reported that when PLA was loaded with GRH nanoplatelets above 0.3 wt %, the elongation at break decreased and that was attributed to the large load of GRH nanoplatelets restricting the mobility of the polymer chains [42,47]. A high load of GRH nanoplatelets in a polymer matrix allows for the restack of nanosheets due to a Van der Waals force. As a consequence, under tension the primary effect will be the slippage of the graphene nanosheets with lower values of tensile strength [47].

Gao et al. [40] also concluded that the incorporation of graphene nanoplatelets (~15 nm) at a higher load (e.g., 5 to 10%) reduced the mobility of the PLA chains in the composite and increased its brittle behavior, which could be attributed to some aggregation of nanoplatelets due to the high concentration.

3.4. Thermal Properties

Figure 4 shows the TGA results for the evaluated samples. No significant difference in onset decomposition temperature was observed among neat PLA, PLA-*g*-TPCS, and samples with the GRH nanoplatelets incorporated into these matrixes; all samples had an onset temperature between 310–320 °C. An early change (before the onset temperature) was depicted, mainly due to the presence of the TPCS phase and the presence of glycerol. The incorporation of a nanofiller has been shown to enhance composite thermal stability; this was evident for PLA-GRH, but this was not evident for the addition of GRH nanoplatelets at low load (i.e., 0.1 wt %) in the PLA-*g*-TPCS blends. Chieng et al. [42], working with a plasticized PLA reinforced with GRH nanoplatelets, reported a similar value for the onset decomposition temperature. The addition of a load of GRH nanofiller above 0.5 wt % into the

PLA matrix was shown to significantly improve thermal stability [42]. The incorporation of GRH nanoplatelets at higher loads could work as a heat barrier, shifting the decomposition of the composite to higher temperatures, and also, due to its high aspect ratio, create a barrier for volatile degradation products present in the nanocomposite [48]. Similar values of onset decomposition temperature (T_{onset}) and maximum decomposition temperature (T_{dmax}) were observed for PLA and the reactive blends with and without the addition of GRH nanoplatelets (Table 6). Residual values observed are due to the decomposition of organic matter of TPCS and the formation of ash [49]. A lower residual was observed for PLA-GRH. This could be associated with the fact that GRH nanoplatelets are thermally conductive, and thus could improve the loss of residual mass of this blend at high temperature. Additionally, there is no presence of ash derived from the cassava starch.

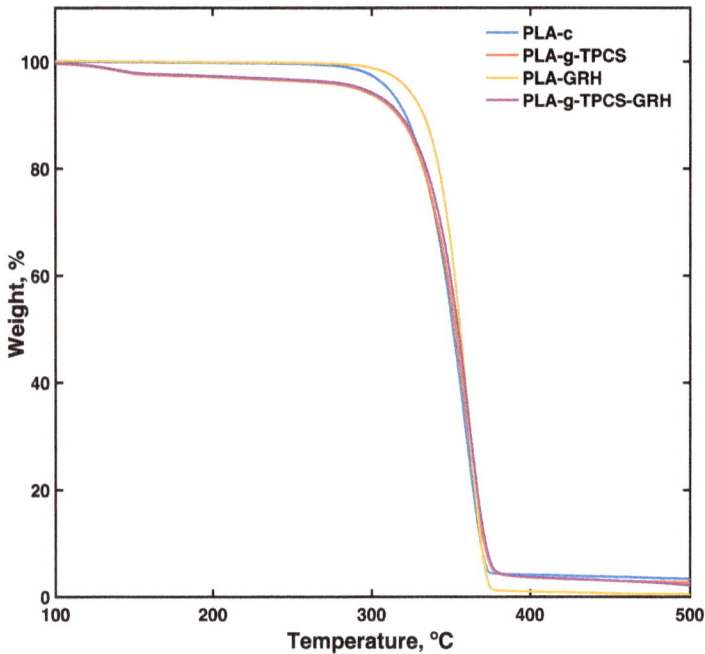

Figure 4. Thermogravimetric analysis (TGA) thermograms of the tested samples produced by twin-screw extrusion–cast-film extrusion.

Table 6. T_g, T_{cc}, T_m, and X_c from the second heating cycle of differential scanning calorimetry (DSC), and T_{onset} and T_{dmax} from TGA.

Films	T_g, °C	T_{cc}, °C	T_m, °C	X_c, %	T_{onset}, °C	T_{dmax}, °C
PLA	61.2 ± 0.7 [a]	93.2 ± 4.2 [a]	150.6 ± 0.2 [a]	3.1 ± 0.6 [a]	318.9 ± 8.4 [a]	360.8 ± 1.7 [a]
PLA-*g*-TPCS	52.3 ± 0.5 [b]	107.4 ± 0.5 [b]	143.4 ± 0.3 [b]	1.2 ± 0.6 [a]	310.4 ± 6.3 [a]	359.3 ± 0.4 [a]
PLA-GRH	61.0 ± 0.0 [a]	124.0 ± 0.8 [c]	149.5 ± 0.2 [c]	2.2 ± 0.6 [a]	320.1 ± 5.7 [a]	360.7 ± 0.8 [a]
PLA-*g*-TPCS-GRH	57.3 ± 0.1 [c]	118.8 ± 1.3 [c]	146.3 ± 0.2 [d]	2.0 ± 0.9 [a]	309.0 ± 2.7 [a]	359.9 ± 0.4 [a]

Note: Within columns, values followed by a different letter are significantly different at $p \le 0.05$ (Tukey's test).

The DSC results for the second heating cycle are presented in Table 6 and Figure 5. With respect to PLA-c, reductions in T_g of ~10 °C for the reactive blend PLA-*g*-TPCS and ~5 °C for the reactive blend with GRH nanoplatelets were observed. However, PLA-GRH had a similar T_g to PLA-c. Valapa et al. [27] also reported that a low content of expandable graphite in PLA composites did not affect the T_g, concluding that an incorporation of a low amount of reinforcement does not affect the mobility and reduction of PLA chains. On the other hand, lower T_g values have been reported for plasticized PLA nanoreinforced with GRH nanoplatelets [42], although the loads were higher (0.3 to 1.0 wt %). A similar reduction in T_g has been reported due to the presence of a TPCS phase and the use of MA and L101 [17]. The incorporation of GRH nanoplatelets in the PLA-*g*-TPCS increased its T_g compared with the unreinforced blends; this finding could be correlated with the tensile test results, whereby the GRH nanoplatelets acted as a reinforcement of the blends, reducing PLA chain mobility. The reactive blends had lower T_m values in comparison to PLA-c, and this decline could be associated with the reduction in M_n, which has a larger effect on T_m than on T_g. PLA-GRH exhibited a small decrease in T_m.

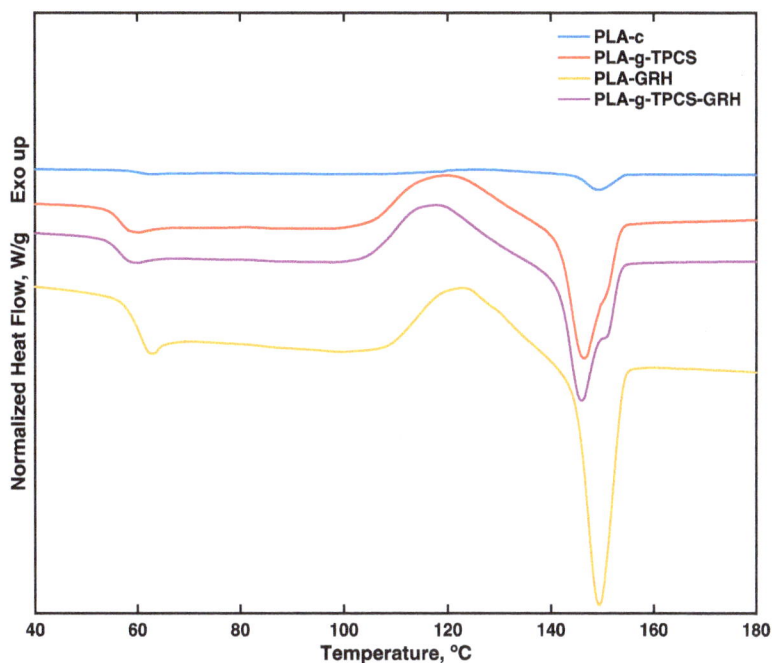

Figure 5. DSC thermograms of the second heating cycle of the tested films; samples were produced by twin-screw extrusion followed by cast-film extrusion.

The values of crystallization, as determined by the second heating cycle, remained stable for all of the blends, which are mostly amorphous. The presence of GRH nanoplatelets did not modify the crystallization behavior of PLA, as observed in Figure S3. A small crystallization peak is shown at 16.4° and assigned to the plane (200)/(110) of the α-crystal of PLA, confirming the presence of an ordered region due to the presence of the nanofiller [32]. However, the blending with TPCS disrupts the crystallization process, creating two types of crystal forms (α and α') as previously reported [50], which are difficult to crystallize. The XRD patterns (Figure S3) exhibited a fully amorphous behavior for all of the films produced, with a crystallization peak for PLA-GRH at 16.4° and a peak showing the presence of GRH nanoplatelets in PLA-GRH and PLA-*g*-TPCS-GRH at 26.5° (corresponding with the *d*002 spacing of graphite). The presence of this peak at 26.5° confirms that a high percentage of GRH nanoplatelets were intercalated in the polymer matrix. The thermal parameters obtained from the first heating cycle of the DSC (Table S1) showed that the crystallinity of all of the samples was mostly amorphous.

Figure 6a shows that the storage moduli, E', is reduced in the range of 35 to 60 °C due to the addition of TPCS phase concerning the PLA-c, indicating a reduction of the elastic region for PLA-*g*-TPCS. The reduction observed for E' upon the incorporation of GRH in PLA is due to the increase in toughening and the more plastic behavior of PLA-GRH. The E' represents the stiffness of the viscoelasticity of PLA and is proportional to the energy stored during the loading cycle. Since the addition of GRH improves the Young's Modulus of PLA-GRH samples and also the toughness of the samples through the crack-bridging mechanism as previously described, the increase of the Young's Modulus is due to the addition of the GRH nanoplatelets (the modulus is around 1 TPa for GRH), and the simultaneous reduction of E' is an indication of a less elastic and more plastic deformation with the temperature of the PLA-GRH sample. This incorporation of GRH into PLA-*g*-TPCS has less effect on the reduction of E' due to the presence of the plastic TPCS phase. With the increment of temperature after 60 °C, a slow decrease is observed for E' until it reaches the glass transition region where the drop of E' is important. Above the glass transition region, the values for E' are similar for all of the evaluated samples. An important reduction of E'' is observed for the reactive blends with respect to PLA. Thus, the reactive blends showed a tough behavior with less energy dissipated during deformation.

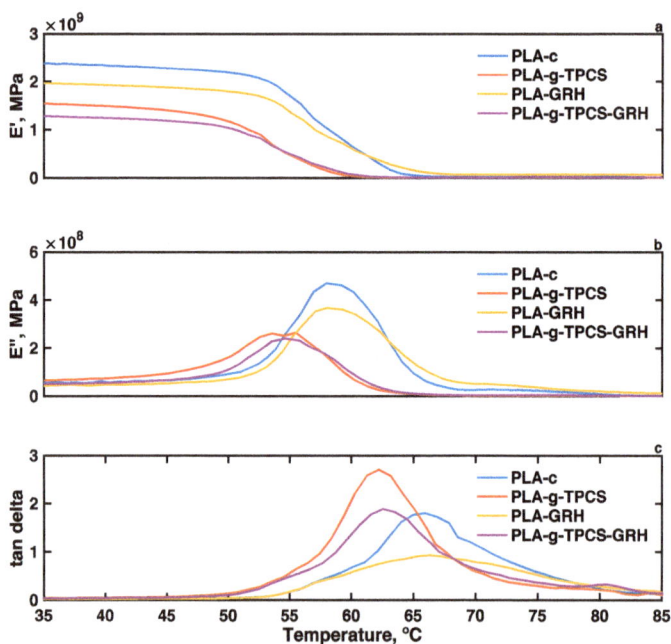

Figure 6. Dynamic mechanical analysis (DMA) thermograms of films produced by twin-screw extrusion–cast-film extrusion: (**a**) E', storage modulus; (**b**) E'', loss modulus; and (**c**) tan delta.

The T_g values have been reported to be higher when estimated by DMA than by DSC [51]. Nevertheless, the T_g trend observed in the DMA was the same as that observed in the DSC analysis, with similar values for PLA-c and PLA-GRH and a shift to lower temperatures for PLA-*g*-TPCS and PLA-*g*-TPCS-GRH. This shift is associated with the enhanced chain mobility of PLA due to the TPCS phase and the plasticizing effect achieved by reactive compatibilization. The plasticization effect could be described as the softening of the blend due to the presence of the glycerol. However, the decrease of the tan delta peak with the addition of GRH indicates that the polymer chains during the transition region are less restricted by the GRH nanoplatelets. Others have suggested that the reinforcement due to the addition of GRH nanoplatelets leads to a restriction of the chain mobility of the polymer [52]. A similar reduction of tan delta peaks was reported by Jonoobi et al. [53] when using cellulose nanofibers to reinforce PLA.

3.5. Optical Properties of Films

Transmittance of the characterized films is presented in Figure 7. The films can be divided into two groups. In the first group, comprising PLA-c and PLA-GRH, a ~10% reduction in transmittance between 250 and 880 nm was observed between the films due to the presence of GRH nanoplatelets. The films in the second group, PLA-*g*-TPCS and PLA-*g*-TPCS-GRH, had similar transmittance values, but these values were ~75% lower with respect to the first group. This phenomenon is mainly due to the presence of TPCS in the matrix of PLA-*g*-TPCS and PLA-*g*-TPCS-GRH, as can be observed in Figure S4, and which was discussed previously in terms of the increase of roughness in these films, as determined by AFM. A full characterization of the color and opacity of the film samples is provided in Table S2. The incorporation of 0.1% of nanofiller into PLA has shown similar characteristics as those reflected in this work [27].

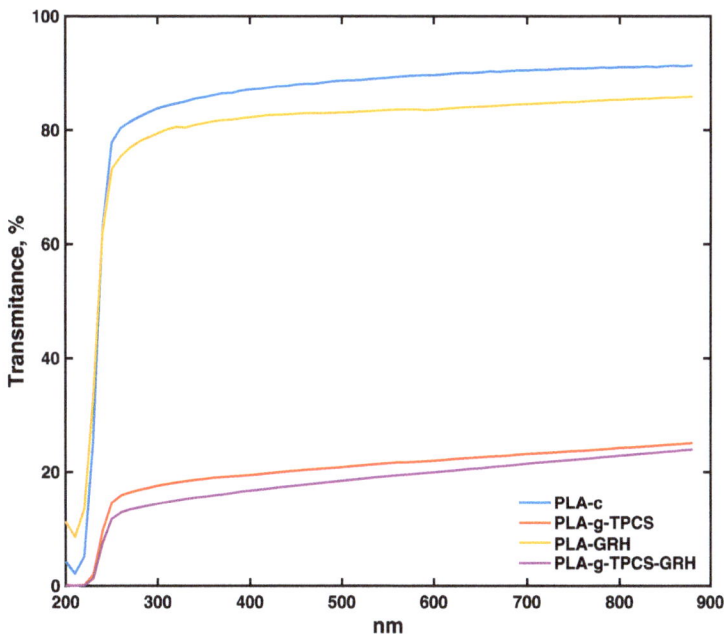

Figure 7. Transmittance (%) of film samples produced by twin-screw extrusion–cast-film extrusion.

It is also important to note that the presence of GRH in PLA-GRH and PLA-*g*-TPCS-GRH plays an important role in the electrical properties of these materials. The values of electrical resistivity for the films are provided in Table S3, and indicate that sample conductivity increases with the addition of very low amounts of GRH nanoplatelets. Higher values of conductivity were reported by Gao et al. [40] working with higher loads of GRH nanoplatelets (5–15%).

3.6. Barrier Properties to Oxygen and Water Vapor

Table 7 depicts the oxygen and water vapor permeability coefficients for the films tested under conditions of 30% RH and 23 °C. The oxygen permeability coefficients were lower in PLA films with GRH nanoplatelets, with a further improvement for the reactive blend with GRH nanoplatelets.

However, no reduction was observed due to the presence of TPCS. The water vapor permeability did not decline due to the addition of TPCS, perhaps due to the low RH testing condition. Furthermore, the water vapor permeability coefficient for the reactive blend did not decline due to the addition of the nanofiller. The presence of GRH nanoplatelets in the films tends to reduce the permeability to oxygen; nevertheless, due to the low load used, this improvement was not reflected in an important reduction in permeability coefficient value. The SEM images (Figure 1) showed the formation of flakes with several GRH nanoplatelets and poor dispersion. This could be one reason for the lack of oxygen barrier improvement. Research done in this field has demonstrated that the presence of graphene (nanosheets or nanoplatelets) improves the barrier properties of PLA and PLA blends. Huang et al. [54] presented three main factors affecting the "tortuosity" for the penetration and diffusion of molecules: (1) the volume fraction of nanoplatelets; (2) the morphology of the nanoplatelets (e.g., exfoliation, dispersion, and orientation perpendicular with respect to the direction of diffusion); and (3) the aspect ratio. Exfoliation of GRH nanosheets was reported to improve the barrier permeability to oxygen on PLA with a load of 0.1%, but had a negative effect with a load of 0.5%, which could be associated with the formation of aggregates at high loadings [27]. The presence of GRH nanosheets in PLA/starch blends was reported by Wu et al. [32], who demonstrated an important improvement in oxygen permeability for blends with ~30% starch and that were reinforced with 0.1% graphene nanosheets.

Table 7. Barrier properties of films to oxygen and water vapor at 30% relative humidity (RH) and 23 °C.

Films	O$_2$ Permeability Coefficient P $\times$ 10^{17} (kg·m·m^{-2}·s^{-1}·Pa^{-1})	Water Vapor Permeability Coefficient P $\times$ 10^{14} (kg·m·m^{-2}·s^{-1}·Pa^{-1})
PLA	2.2 $\pm$ 0.2 [a]	6.6 $\pm$ 0.8 [a]
PLA-*g*-TPCS	2.0 $\pm$ 0.1 [a]	6.6 $\pm$ 0.3 [a]
PLA-GRH	1.4 $\pm$ 0.1 [b]	6.1 $\pm$ 0.1 [a]
PLA-*g*-TPCS-GRH	0.9 $\pm$ 0.4 [c]	6.1 $\pm$ 0.7 [a]

Note: Within columns, values followed by a different letter are significantly different at $p \leq 0.05$ (Tukey's test).

4. Conclusions

Reactive functionalization of GRH nanoplatelets in the compatibilized matrix of PLA and thermoplastic cassava starch (TPCS) resulted in a significant improvement in elongation at break. The presence of GRH nanoplatelets in the PLA-*g*-TPCS reactive blend increased their toughness, with an improvement of ~900% and ~500% when compared to neat PLA and PLA-*g*-TPCS, respectively. GRH nanoplatelets in the reactive blend of PLA-*g*-TPCS were mainly located in the interface and were agglomerated in several flakes. The mechanism of crack bridging was identified as being responsible for the improvement in elongation at break for the reactive blends and the increase in the Young's modulus for PLA with the addition of GRH nanoplatelets. The use of two-step extrusion resulted in a large reduction, and hence a negative effect, on the molecular weight of PLA. Using a one-step processing condition could significantly improve the M_w for PLA-*g*-TPCS and PLA-*g*-TPCS-GRH. Production of the reactive blends by using a twin-screw extruder and a sheet film die adapted at the end of the extruder should be explored to scale this blend. Surface roughness was greater for the reactive blends with and without GRH. Transmittance decreased for the reactive blends with and without GRH in the UV-visible range, mainly due to the presence of the starch phase. Nevertheless, the reactive blends are still acceptable for see-through applications. Oxygen barrier properties improved with the addition of GRH nanoplatelets to PLA-c and PLA-*g*-TPCS. However, water barrier properties did not improve with the addition of GRH nanoplatelets; one of the main reasons could be the scarce exfoliation and orientation achieved. Moreover, the formation of flakes of GRH nanoplatelets in the polymer matrix could be diminishing the ability to create a tortuous path. Future work will focus on improving the barrier properties of these reactive blends with GRH nanoplatelets as a nanofiller reinforcement.

Polymers **2018**, *10*, 95

Supplementary Materials: The XRD characterization, color, opacity, and electrical resistivity of the films are available online at www.mdpi.com/2073-4360/10/1/95/s1, Table S1: T_g, T_{cc}, T_m, and X_c from the first heating cycle of DSC; Table S2: Values of optical properties of films produced by TSE-CF; Table S3: Resistivity of film samples; Figure S1: SEM images of graphene nanoplatelets powder used as reinforcement for production of the nanocomposites; Figure S2: SEM images of film samples before and after tensile test, evaluated in cross direction (CD) and machine direction (MD); Figure S3: XRD patterns obtained after tensile testing of film samples produced by twin-screw extrusion–cast-film extrusion; Figure S4: Picture of film samples produced by twin-screw extrusion followed by cast-film extrusion.

Acknowledgments: The authors thank the Center for Advanced Microscopy at Michigan State University (MSU) for assisting with the SEM images, the Department of Physics and Astronomy (at MSU) for the support with the AFM images, the Composites Materials and Structures Center (at MSU) for assisting with the resistivity tests, the Michigan Biotechnology Institute (at MSU) for providing access to the twin-screw extruder, the department of Electrical and Computer Engineering (at MSU) for assisting with the use of the profilometer, and the Center for Crystallographic Research (at MSU) for allowing the use of the XRD diffractometer. Anibal Bher thanks the National Scientific and Technical Research Council (CONICET) in Argentina for providing financial support through a Ph.D. fellowship, and the School of Packaging at MSU for partial financial support. Ilke Uysal Unalan thanks the Scientific and Technological Research Council of Turkey (TUBITAK) 2219-International Postdoctoral Research Fellowship Programme for providing financial support.

Author Contributions: Anibal Bher, Ilke Uysal Unalan, Rafael Auras, Maria Rubino, and Carlos E. Schvezov designed the study. Anibal Bher and Ilke Uysal Unalan conducted the experiments and processed the data. All the authors analyzed the experimental results, contributed to writing the manuscript, and approved the final version.

Conflicts of Interest: The authors declare no conflict of interest.

References

1. Farah, S.; Anderson, D.G.; Langer, R. Physical and mechanical properties of PLA, and their functions in widespread applications—A comprehensive review. *Adv. Drug Deliv. Rev.* **2016**, *107*, 367–392. [CrossRef] [PubMed]

2. Castro-Aguirre, E.; Iñiguez-Franco, F.; Samsudin, H.; Fang, X.; Auras, R. Poly(lactic acid)—Mass production, processing, industrial applications, and end of life. *Adv. Drug Deliv. Rev.* **2016**, *107*, 333–366. [CrossRef] [PubMed]

3. Jamshidian, M.; Tehrany, E.A.; Imran, M.; Jacquot, M.; Desobry, S. Poly-Lactic Acid: Production, Applications, Nanocomposites, and Release Studies. *Compr. Rev. Food Sci. Food Saf.* **2010**, *9*, 552–571. [CrossRef]

4. Kathuria, A.; Abiad, M.G.; Auras, R. Toughening of poly(L-lactic acid) with Cu3BTC2 metal organic framework crystals. *Polymer* **2013**, *54*, 6979–6986. [CrossRef]

5. Ahmed, J.; Arfat, Y.A.; Castro-Aguirre, E.; Auras, R. Mechanical, structural and thermal properties of Ag-Cu and ZnO reinforced polylactide nanocomposite films. *Int. J. Biol. Macromol.* **2016**, *86*, 885–892. [CrossRef] [PubMed]

6. Marra, A.; Silvestre, C.; Kujundziski, A.P.; Chamovska, D.; Duraccio, D. Preparation and characterization of nanocomposites based on PLA and TiO2 nanoparticles functionalized with fluorocarbons. *Polym. Bull.* **2017**, *74*, 3027–3041. [CrossRef]

7. Jalalvandi, E.; Majid, R.A.; Ghanbari, T.; Ilbeygi, H. Effects of montmorillonite (MMT) on morphological, tensile, physical barrier properties and biodegradability of polylactic acid/starch/MMT nanocomposites. *J. Thermoplast. Compos. Mater.* **2015**, *28*, 496–509. [CrossRef]

8. Bhardwaj, R.; Mohanty, A.K. Modification of brittle Polylactide by Novel Hyperbranched Polymer-Based Nanostructures. *Biomacromolecules* **2007**, *8*, 2476–2484. [CrossRef] [PubMed]

9. Murariu, M.; Dubois, P. PLA composites: From production to properties. *Adv. Drug Deliv. Rev.* **2016**, *107*, 17–46. [CrossRef] [PubMed]

10. Sheth, M.; Kumar, R.A.; Dave, V.; Gross, R.A.; McCarthy, S.P. Biodegradable Polymer Blends of Poly(lactic acid) and Poly(ethylene glycol). *J. Appl. Polym. Sci.* **1997**, *66*, 1495–1505. [CrossRef]

11. Lee, C.M.; Kim, E.S.; Yoon, J.S. Reactive Blending of Poly(L-lactic acid) with Poly(ethylene-co-vinyl alcohol). *J. Appl. Polym. Sci.* **2005**, *98*, 886–890. [CrossRef]

12. Ma, P.; Cai, X.; Zhang, Y.; Wang, S.; Dong, W.; Chen, M.; Lemstra, P.J. In-situ compatibilization of poly(lactic acid) and poly(butylene adipate-co-terephthalate) blends by using dicumyl peroxide as a free-radical initiator. *Polym. Degrad. Stab.* **2014**, *102*, 145–151. [CrossRef]

13. Huneault, M.A.; Li, H. Morphology and properties of compatibilized polylactide/thermoplastic starch blends. *Polymer* **2007**, *48*, 270–280. [CrossRef]

14. Ferri, J.M.; Garcia-Garcia, D.; Carbonell-Verdu, A.; Fenollar, O.; Balart, R. Poly(lactic acid) formulations with improved toughness by physical blending with thermoplastic starch. *J. Appl. Polym. Sci.* **2018**, *135*, 1–8. [CrossRef]

15. Jimenez, A.; Fabra, M.J.; Talens, P.; Chiralt, A. Edible and Biodegradable Starch Films: A Review. *Food Bioprocess Technol.* **2012**, *5*, 2058–2076. [CrossRef]

16. Zhu, F. Composition, structure, physicochemical properties, and modifications of cassava starch. *Carbohydr. Polym.* **2015**, *122*, 456–480. [CrossRef] [PubMed]

17. Detyothin, S.; Selke, S.E.M.; Narayan, R.; Rubino, M.; Auras, R.A. Effects of molecular weight and grafted maleic anhydride of functionalized polylactic acid used in reactive compatibilized binary and ternary blends of polylactic acid and thermoplastic cassava starch. *J. Appl. Polym. Sci.* **2015**, *132*, 1–15. [CrossRef]

18. Teixeira, E.D.M.; Curvelo, A.A.S.; Corrêa, A.C.; Marconcini, J.M.; Glenn, G.M.; Mattoso, L.H.C. Properties of thermoplastic starch from cassava bagasse and cassava starch and their blends with poly (lactic acid). *Ind. Crops Prod.* **2012**, *37*, 61–68. [CrossRef]

19. Detyothin, S.; Selke, S.E.M.; Narayan, R.; Rubino, M.; Auras, R. Reactive functionalization of poly(lactic acid), PLA: Effects of the reactive modifier, initiator and processing conditions on the final grafted maleic anhydride content and molecular weight of PLA. *Polym. Degrad. Stab.* **2013**, *98*, 2697–2708. [CrossRef]

20. Zhang, J.-F.; Sun, X. Mechanical Properties of Poly (lactic acid)/Starch Composites Compatibilized by Maleic Anhydride. *Biomacromolecules* **2004**, *5*, 1446–1451. [CrossRef] [PubMed]

21. Bher, A.; Auras, R.; Schvezov, C.E. Improving the toughening in poly(lactic acid)-thermoplastic cassava starch reactive blends. *J. Appl. Polym. Sci.* **2017**, 1–15. [CrossRef]

22. Raquez, J.M.; Habibi, Y.; Murariu, M.; Dubois, P. Polylactide (PLA)-based nanocomposites. *Prog. Polym. Sci.* **2013**, *38*, 1504–1542. [CrossRef]

23. Shayan, M.; Azizi, H.; Ghasemi, I.; Karrabi, M. Effect of modified starch and nanoclay particles on biodegradability and mechanical properties of cross-linked poly lactic acid. *Carbohydr. Polym.* **2015**, *124*, 237–244. [CrossRef] [PubMed]

24. Basu, A.; Nazarkovsky, M.; Ghadi, R.; Khan, W.; Domb, A.J. Poly(lactic acid)-based nanocomposites. *Polym. Adv. Technol.* **2017**, *28*, 919–930. [CrossRef]

25. Gonçalves, C.; Gonçalves, I.C.; Magalhães, F.D.; Pinto, A.M. Poly(lactic acid) composites containing carbon-based nanomaterials: A review. *Polymers* **2017**, *9*, 269. [CrossRef]

26. Pinto, A.M.; Cabral, J.; Tanaka, D.A.P.; Mendes, A.M.; Magalhaes, F.D. Effect of incorporation of graphene oxide and graphene nanoplatelets on mechanical and gas permeability properties of poly(lactic acid) films. *Polym. Int.* **2013**, *62*, 33–40. [CrossRef]

27. Valapa, R.B.; Pugazhenthi, G.; Katiyar, V. Effect of graphene content on the properties of poly(lactic acid) nanocomposites. *RSC Adv.* **2015**, *5*, 28410–28423. [CrossRef]

28. Kalaitzidou, K.; Fukushima, H.; Drzal, L.T. Mechanical properties and morphological characterization of exfoliated graphite-polypropylene nanocomposites. *Compos. Part A Appl. Sci. Manuf.* **2007**, *38*, 1675–1682. [CrossRef]

29. Drzal, L.; Fukushima, H. Expanded Graphite and Products Produced Therefrom. International Patent Application No. 7,550,529 B2, 23 June 2009.

30. Kalaitzidou, K.; Fukushima, H.; Drzal, L.T. Multifunctional polypropylene composites produced by incorporation of exfoliated graphite nanoplatelets. *Carbon N. Y.* **2007**, *45*, 1446–1452. [CrossRef]

31. Nuona, A.; Li, X.; Zhu, X.; Xiao, Y.; Che, J. Starch/polylactide sustainable composites: Interface tailoring with graphene oxide. *Compos. Part A Appl. Sci. Manuf.* **2015**, *69*, 247–254. [CrossRef]

32. Wu, D.; Xu, H.; Hakkarainen, M. From starch to polylactide and nano-graphene oxide: Fully starch derived high performance composites. *Rsc Adv.* **2016**, *6*, 54336–54345. [CrossRef]

33. Yoo, B.M.; Shin, H.J.; Yoon, H.W.; Park, H.B. Graphene and graphene oxide and their uses in barrier polymers. *J. Appl. Polym. Sci.* **2014**, *131*, 1–23. [CrossRef]

34. ASTM International. *ASTM D882: Standard Test Method for Tensile Properties of Thin Plastic Sheeting*; ASTM International: West Conshohocken, PA, USA, 2012.

35. Fischer, E.W.; Sterzel, H.J.; Wegner, G. Investigation of the structure of solution grown crystals of lactide copolymers by means of chemical reactions. *Polymere* **1973**, *251*, 980–990. [CrossRef]

36. Lim, L.-T.; Auras, R.; Rubino, M. Processing technologies for poly(lactic acid). *Prog. Polym. Sci.* **2008**, *33*, 820–852. [CrossRef]

37. Tsuji, H. Hydrolytic Degradation. In *Poly(Lactic Acid)*; Auras, R., Lim, L.-T., Selke, S.E.M., Tsuji, H., Eds.; John Wiley & Sons, Inc.: Hoboken, NJ, USA, 2010; pp. 343–381, ISBN 9780470293669.

38. Iñiguez-Franco, F.; Auras, R.; Burgess, G.; Holmes, D.; Fang, X.; Rubino, M.; Soto-Valdez, H. Concurrent solvent induced crystallization and hydrolytic degradation of PLA by water-ethanol solutions. *Polymer* **2016**, *99*, 315–323. [CrossRef]

39. Taubner, V.; Shishoo, R. Influence of processing parameters on the degradation of poly(L-lactide) during extrusion. *J. Appl. Polym. Sci.* **2001**, *79*, 2128–2135. [CrossRef]

40. Gao, Y.; Picot, O.T.; Bilotti, E.; Peijs, T. Influence of filler size on the properties of poly(lactic acid) (PLA)/graphene nanoplatelet (GNP) nanocomposites. *Eur. Polym. J.* **2017**, *86*, 117–131. [CrossRef]

41. Pinto, A.M.; Moreira, S.; Gonçalves, I.C.; Gama, F.M.; Mendes, A.M.; Magalhães, F.D. Biocompatibility of poly(lactic acid) with incorporated graphene-based materials. *Colloids Surf. B Biointerfaces* **2013**, *104*, 229–238. [CrossRef] [PubMed]

42. Chieng, B.W.; Ibrahim, N.A.; Wan Yunus, W.M.Z.; Hussein, M.Z.; Loo, Y.Y. Effect of graphene nanoplatelets as nanofiller in plasticized poly(lactic acid) nanocomposites: Thermal properties and mechanical properties. *J. Therm. Anal. Calorim.* **2014**, *118*, 1551–1559. [CrossRef]

43. Chieng, B.W.; Ibrahim, N.A.; Yunus, W.M.Z.W.; Hussein, M.Z.; Then, Y.Y.; Loo, Y.Y. Effects of graphene nanoplatelets and reduced graphene oxide on poly(lactic acid) and plasticized poly(lactic acid): A comparative study. *Polymers* **2014**, *6*, 2232–2246. [CrossRef]

44. Ahmadi-Moghadam, B.; Sharafimasooleh, M.; Shadlou, S.; Taheri, F. Effect of functionalization of graphene nanoplatelets on the mechanical response of graphene/epoxy composites. *Mater. Des.* **2015**, *66*, 142–149. [CrossRef]

45. Seshadri, M.; Saigal, S. Crack bridging in polymer nanocomposites. *J. Eng. Mech.* **2007**, *133*, 911–918. [CrossRef]

46. Shadlou, S.; Alishahi, E.; Ayatollahi, M.R. Fracture behavior of epoxy nanocomposites reinforced with different carbon nano-reinforcements. *Compos. Struct.* **2013**, *95*, 577–581. [CrossRef]

47. Zhao, X.; Zhang, Q.; Chen, D.; Lu, P. Enhanced mechanical properties of graphene-based polyvinyl alcohol composites. *Macromolecules* **2010**, *43*, 2357–2363. [CrossRef]

48. Chieng, B.W.; Ibrahim, N.A.; Yunus, W.M.Z.W.; Hussein, M.Z. Poly(lactic acid)/poly(ethylene glycol) polymer nanocomposites: Effects of graphene nanoplatelets. *Polymers* **2014**, *6*, 93–104. [CrossRef]

49. Beninca, C.; Colman, T.A.D.; Lacerda, L.G.; Carvalho Filho, M.A.S.; Bannach, G.; Schnitzler, G. The thermal, rheological and structural properties of cassava starch granules modified with hydrochloric acid at different temperatures. *Thermochim. Acta* **2013**, *552*, 65–69. [CrossRef]

50. Saeidlou, S.; Huneault, M.A.; Li, H.; Park, C.B. Poly(lactic acid) crystallization. *Prog. Polym. Sci.* **2012**, *37*, 1657–1677. [CrossRef]

51. Gracia-Fernández, C.A.; Gómez-Barreiro, S.; López-Beceiro, J.; Tarrío Saavedra, J.; Naya, S.; Artiaga, R. Comparative study of the dynamic glass transition temperature by DMA and TMDSC. *Polym. Test.* **2010**, *29*, 1002–1006. [CrossRef]

52. Petersson, L.; Oksman, K. Biopolymer based nanocomposites: Comparing layered silicates and microcrystalline cellulose as nanoreinforcement. *Compos. Sci. Technol.* **2006**, *66*, 2187–2196. [CrossRef]

53. Jonoobi, M.; Harun, J.; Mathew, A.P.; Oksman, K. Mechanical properties of cellulose nanofiber (CNF) reinforced polylactic acid (PLA) prepared by twin screw extrusion. *Compos. Sci. Technol.* **2010**, *70*, 1742–1747. [CrossRef]

54. Huang, H.D.; Ren, P.G.; Xu, J.Z.; Xu, L.; Zhong, G.J.; Hsiao, B.S.; Li, Z.M. Improved barrier properties of poly(lactic acid) with randomly dispersed graphene oxide nanosheets. *J. Membr. Sci.* **2014**, *464*, 110–118. [CrossRef]

polymers

MDPI

Article

Ionic Liquid as Surfactant Agent of Hydrotalcite: Influence on the Final Properties of Polycaprolactone Matrix

Luanda Chaves Lins [1,2], Valeria Bugatti [3], Sébastien Livi [1,2] and Giuliana Gorrasi [3,*]

[1]   University of Lyon, F-69003 Lyon, France; luandaqmc@gmail.com (L.C.L.); sebastien.livi@insa-lyon.fr (S.L.)
[2]   UMR 5223, Department of Engineering of Polymeric Materials, INSA Lyon, CNRS, F-69621 Villeurbanne, France
[3]   Department of Industrial Engineering, University of Salerno-via Giovanni Paolo II, 132,
      84084 Fisciano (Salerno), Italy; vbugatti@unisa.it
*   Correspondence: ggorrasi@unisa.it; Tel.: +39-089-964-146

Received: 30 October 2017; Accepted: 29 December 2017; Published: 5 January 2018

Abstract: This paper reports the surface treatment of layered double hydroxide (LDH) by using ionic liquid (IL) composed of phosphonium cation combined with 2-ethylhexanoate (EHT) counter anion as surfactant agent. Then, different amounts (1, 3, 5 and 7 wt %) of thermally stable organically modified LDH (up to 350 °C) denoted LDH-EHT were incorporated into polycaprolactone (PCL) matrix by mechanical milling. The influence of LDH-EHT loading has been investigated on the physical properties, such as the thermal and barrier properties, as well as the morphologies of the resulting nanocomposites. Thus, intercalated or microcomposite morphologies were obtained depending on the LDH-EHT loading, leading to significant reduction of the diffusion coefficient respect to water vapor. The modulation of barrier properties, using low functionalized filler amount, is a very important aspect for materials in packaging applications.

Keywords: polycaprolactone; layered double hydroxides; ionic liquids

1. Introduction

In the last decade, layered double hydroxides (LDHs) have received great attention from industry and academia. Their structure is similar to that of brucite where Mg atoms are octahedrally coordinated to six oxygen atoms belonging to six groups of –OH; each –OH group is, in turn, shared by three octahedral cations and points the hydrogen atom to the interlayer space. When cations of Mg(II) are isomorphously replaced by cations of Al(III), the substitution generates positive charges that are counterbalanced by the presence of counter ions, generally CO_3^{2-}, Cl^-, and NO_3^-, located in the interlamellar regions. The possibility to substitute these anions by ionic exchange procedures makes the hydrotalcites ideal solids to be used as host of potentially active molecules having a negative charge. LDHs have high level of purity and can be produced using simple procedures. They have been used as active filler for polymers and biopolymers for their capability of hosting active molecules [1–7], as catalysts for reactions [8,9], for the synthesis of polymers [10] and as surfactant adsorbents [11]. The exchange of the counter ions between the layers is not easy, therefore the modification of carbonated LDHs is generally carried out by calcination followed by immersion in a medium containing anions to be intercalated [12,13]. Very recently, ionic liquids (ILs) are attracting interest from several areas of basic and applied sciences, including chemistry, materials science, chemical engineering, and environmental science [14]. They are organic salts with low melting points (below 100 °C) based mainly on ammonium, imidazolium, pyridinium or phosphonium ILs combined with different counter anions such as halides, tetrafluoroborate (BF_4^-), hexafluorophosphate (PF_6^-), bistriflamide (TFSA), etc. ILs are considered as potentially environmentally friendly solvents having low volatility, low flammability, good chemical and thermal stabilities, and compatibility with organic and inorganic materials [15]. These characteristics make them a class of very versatile compounds. In fact, they have been used for organic synthesis [16], bio-processing operations, catalysis and gas separation [17], and inhibitors for the corrosion of magnesium alloys [18]. Recently, ILs have also been proposed as plasticizers, lubricants, structuring agents, or compatibilizing agents in several polymers [19–24]. They can also be used to overcome the low thermal stability of ammonium salts commonly used as organic modifiers of layered silicates, such as montmorillonites, that are used to enhance compatibility and dispersion of inorganic phases in organic media [25–28]. The hybrid organic-ionic nature of ILs and the ability to control their hydrophilic and lipophilic counterparts through proper selection of the constituent ions give rise to a complex set of phenomena, creating an area of study that is both fascinating and challenging. While attention has been devoted to ILs based on quaternary nitrogen, studies related to ILs based on quaternary phosphorous cations are still few. It has been demonstrated that some phosphonium based ILs have better properties than nitrogen-based ILs, [29,30]. In fact, research on nanocomposites containing clay minerals modified with phosphonium based ILs has recently increased [31–33]. In 2013, Bugatti et al. have developed a new method to deposit ionic liquid-modified LDHs on PLA films previously treated by plasma treatment leading to an increase of the water barrier properties of 30%. Recently, Kredatusova et al. have worked on the surface treatment of LDH by phosphonium ILs and they have used an environmentally friendly route based on microwave irradiation. Thus, they have demonstrated that the use of phosphonium ILs as surfactant agents of LDH induced an exfoliation morphology in polycaprolactone matrix. More recently, other authors have demonstrated that the incorporation of LDH-ILs into poly(butylene adipate-*co*-terephthalate) matrix led to a significant increase of its mechanical performances, especially of the stiffness without reducing the fracture behavior of the polymer matrix [34].

In a previous paper, we incorporated the LDH modified with trihexyl(tetradecyl)phosphonium 2-ethylhexanoate into a pectin matrix [35] for potential in biodegradable food packaging applications. Pectins are natural materials with no melting point, but high degradation temperature, and with high brittleness and poor elongation at break. These properties make them unable to be applied in flexible packaging. Poly(ε-caprolactone) (PCL) being a biodegradable polyester with low melting point and very high elongation at break but low elastic modulus appears one of the best candidate to be blended with pectins. The compatibilization between the two phases can be enhanced by the very wide chemistry of ionic liquids; in addition, also LDH can be considered an interesting filler hosting either ILs, either active molecules with peculiar functionalities. The aim of the present work is, then, to investigate the potential of layered double hydroxide modified with a phosphonium ionic liquid denoted trihexyltetradecylphosphonium 2-ethylhexanoate in PCL matrix for possible food packaging applications. In addition, the influence of the amount of modified LDHs will be evaluated on the morphologies but also in terms of thermal stability, surface analysis and barrier properties.

2. Experimental

2.1. Materials

Poly (ε-caprolactone) (PCL) Mn 80,000 was supplied from Sigma Aldrich (Milano, Italy). It was reduced in powder form in a mechanical mixer in presence of liquid nitrogen. A LDH (aluminum magnesium hydroxy carbonate), denoted PURAL® MG 63 HT, was chosen as pristine anionic clay and was provided by Sasol Performance Chemicals (Hamburg, Germany). The IL, coded as IL-EHT, based on trihexyltetradecylphosphonium cation associated with 2-ethylhexanoate counter anion was kindly provided by Cytec Industries Inc. (Thorold, ON, Canada).

2.2. LDH-Ionic Liquid Preparation

According to the literature, two steps are required: (i) calcination of the pristine LDH at 500 °C for 24 h; and (ii) LDH and 2 AEC (anion exchange capacity) of phosphonium ILs dispersion in 200 mL of deionized water/tetrahydrofuran (THF) mixture (300/100 mL) [33,36,37]. The suspensions were stirred and mixed at 60 °C for 24 h. The resulting precipitate was filtered and washed 5 times with THF. The residual solvent was removed by evaporation under vacuum and finally, the treated LDH was dried overnight at 80 °C. From the thermogravimetric analysis TGA (see inset of Figure 4) the amount of physisorbed and intercalated water (25–220 °C) is ≅15 wt %, the amount of carbonate anion and IL (250–500 °C) is ≅25 wt % and the amount of LDH (>500 °C) into the LDH-EHT is ≅60 wt %. The phosphonium ionic liquid used for the surface treatment and the following abbreviation used to designate the LDH-IL are summarized in Table 1.

Table 1. Designation of the phosphonium ionic liquid (IL) used for the modification of LDH.

Ionic Liquid	Chemical Structure	Code
Trihexyl(tetradecyl)phosphonium 2-ethylhexanoate		EHT

2.3. Composite Films Preparation

The incorporation of the filler into PCL was achieved by High Energy Ball Milling (HEBM) method. Powders composed of PCL and filler (vacuum dried for 24 h) were milled at different filler loading (i.e., 1, 3, 5, and 7 wt %) at room temperature in a Retsch (Haan, Germany) centrifugal ball mill (model S 100). The powders were milled in a cylindrical steel jar of 50 cm³ with 5 steel balls of 10 mm of diameter. The rotation speed used was 580 rpm. and the milling time was 60 min. The pure PCL, taken as reference, was milled in the same experimental conditions of the composites. The milled powders, were molded in a Carver laboratory press between two Teflon sheets, at 100 °C, followed by a quick cooling at ambient temperature. Films 150 μm thick were obtained and analyzed.

2.4. Methods of Investigation

Wide-angle X-ray diffraction spectra (WAXD) were collected on a Bruker D8 Advance X-ray diffractometer (Karlsruhe, Germany). A bent quartz monochromator was used to select the Cu Kα_1 radiation (k = 0.15406 nm) and run under operating conditions of 45 mA and 33 kV in Bragg-Brentano geometry.

Transmission electron microscopy (TEM) was carried out using a Philips CM 120 field emission scanning electron microscope (Philips, Eindhoven, The Netherlands) with an accelerating voltage of 80 kV. The samples were cut using an ultramicrotome equipped with a diamond knife, to obtain 60-nm-thick ultrathin sections. Then, the sections were set on copper grids.

Thermo Gravimetric Analysis (TGA) was carried out with a Mettler TC-10 thermobalance (Novate Milanese, Italy) from 25 °C to 600 °C at a heating rate of 10 °C/min under an air flow.

Differential scanning calorimetry (DSC) were carried out using by means of a DTA Mettler Toledo (DSC 30, GmbH, Greifensee, Switzerland) under nitrogen atmosphere. The films were heated from 25 °C to 100 °C at a heating rate of 10 °C/min.

Fourier transform infrared (FTIR) absorption spectra were recorded by a Perkin-Elmer spectrometer (Bruker Italia, Milano, Italy), model Vertex 70 (average of 32 scans, at a resolution of 4 cm^{-1}). Composite films, having the same thickness (≅150 μm) were analyzed. The LDH-P2 filler was analyzed in powder form after preparing a KBr based tablet (~1 mg of filler sample and ~100 mg of KBr).

Diffusion coefficients were evaluated, using the microgravimetric method, at different vapor activities (a = P/P_0), where P is the actual pressure to which the sample was exposed, and P_0 the saturation pressure at the test temperature. The penetrant was water vapor and the experiments were conducted at 30 °C. The transport properties were measured with water vapor in a range of activities

Polymers **2018**, *10*, 44

from 0.2 to 0.6. Measuring the increase of weight with time for the samples exposed to the vapor at a given partial pressure, P, both the equilibrium value of the sorbed vapor, C_{eq} (g/100 g), and the diffusion coefficient, D, were obtained.

Surface energies were determined with the sessile drop method using a Contact Angle System OCA, Dataphysics® (DataPhysics Instruments GmbH, Filderstadt, Germany). From contact angle measurements with water and diiodomethane as test liquids, the polar and dispersive components of surface energy were determined using the Owens-Wendt theory [38].

3. Results and Discussion

3.1. Morphologies of PCL/LDH Nanocomposites

Figure 1 shows the TEM micrographs evaluated on all composites. Two types of morphologies have been observed. For PCL containing 1 and 3 wt % of LDH-EHT, a homogenous dispersion of LDH has been obtained with the presence of small tactoïds. When enlargements at 500 and 200 nm have been performed, the majority of tactoïds have sizes less than 500 nm combined with the presence of few individual platelets. In the opposite, an increase of the amount of LDH-EHT (5 and 7 wt %) induced the formation of agglomerates but also a less homogeneous dispersion. In both cases, a microcomposite morphology is obtained for PCL containing 1, 3, 5 and 7 wt % of LDH-EHT.

Figure 2 reports the XRD spectra of PCL film and composites filled with LDH-EHT at different loading. The inset reports the spectrum of the LDH-EHT. In a previous work, we have observed by XRD that the surface treatment of LDH by phosphonium ILs highlighted only one diffraction peak at $2\theta = 11.6°$ corresponding to a basal spacing of 7.7 Å [33]. This result showed no influence of the intercalation of IL between clay layers and corresponded to the basal spacing of the pristine LDH. According to the literature, the absorption of the carbonate anions takes place during the surface treatment [33]. In addition, these results can be explained by the presence of short alkyl chains (< 12 carbons) on the anion leading to a planar configuration such as monolayer type. Indeed, Lagaly and Weiss have demonstrated on MMT that the surfactant chain length plays a key role on the different arrangements of organic salts between the clay layers [39,40]. In this work, the 2-ethylhexanoate counter anion is only composed of six carbons and one carbonyl bond (C=O). The carbonate anion also has only one carbonyl bond, and if IL-EHT assumes a planar configuration, no difference can be observed by XRD. The neat PCL matrix and the resulting nanocomposites containing different amounts of LDH-EHT i.e., 1, 3, 5 and 7 wt % exhibit different diffraction peaks. In fact, the diffraction peaks observed at $2\theta = 21.2°$, $21.8°$ and $23.5°$, corresponding to basal spacing of 4.2 Å, 4.1Å, and 3.8 Å, respectively, are attributed to the crystallinity of the polycaprolactone, whereas the diffraction peak observed at $2\theta = 11.6°$ is characteristic of the basal spacing of the pristine LDH, suggesting an aggregation of the LDH-EHT into the polymer matrix. These results clearly confirm the morphologies obtained by transmission electron microscopy (Figure 1).

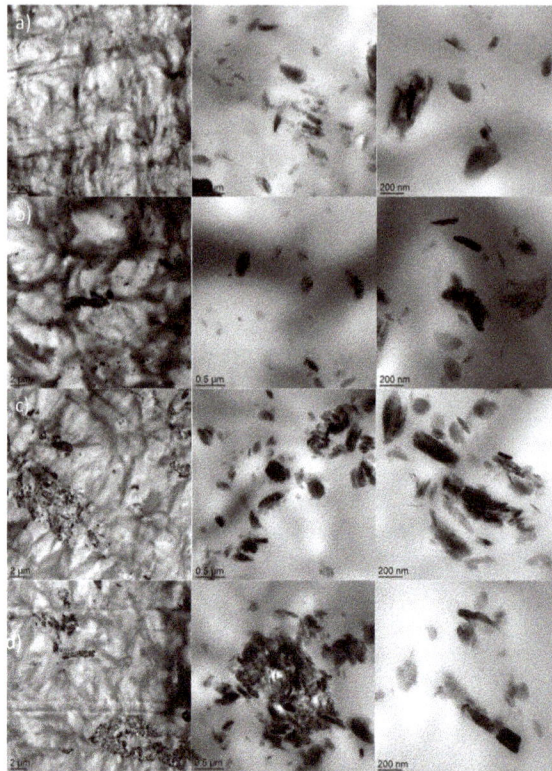

Figure 1. TEM micrographs of: (**a**) PCL + 1% LDH-EHT; (**b**) PCL + 3% LDH-EHT; (**c**) PCL + 5% LDH-EHT; and (**d**) PCL + 7% LDH-EHT.

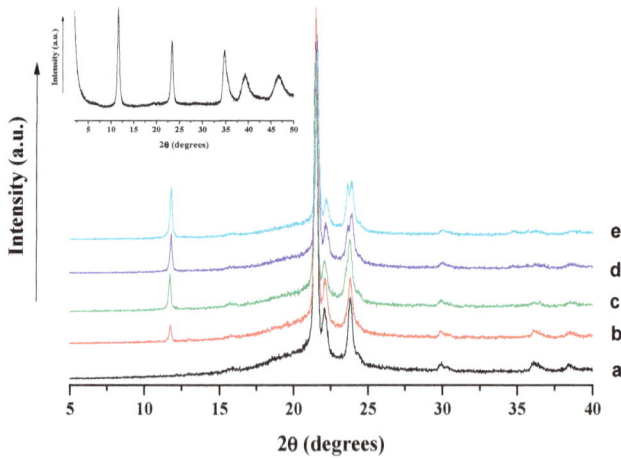

Figure 2. XRD spectra of: PCL (**a**); PCL/1% LDH-EHT (**b**); PCL/3% LDH-EHT (**c**); PCL/5% LDH-EHT (**d**); and PCL/7% LDH-EHT (**e**). Inset reports the XRD spectrum of LDH-EHT.

Figure 3A reports the FTIR spectra, in the range 530–580 cm^{-1}, of PCL and composites at different filler loading, evaluated on films having the same thickness (150 µm). The inset reports the spectrum of pure LDH-EHT. In this range, PCL does not show any peak, while, for LDH, the band at 553 cm^{-1} is assigned to translation modes of the hydroxyl groups mainly influenced by the trivalent aluminum [41]. Figure 3B reports the absorbance at 553 cm^{-1} as function of filler amount. It is evident that the absorbance at 553 cm^{-1} is linear with the filler loading, making this technique a useful tool for quantitative analysis.

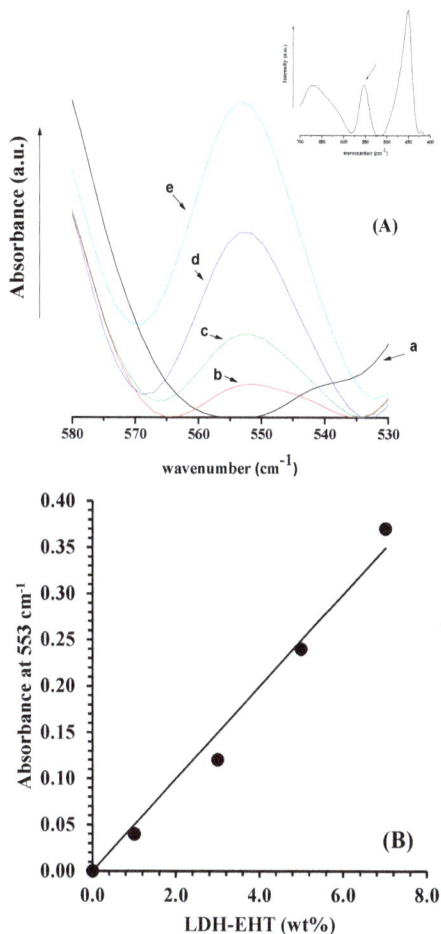

Figure 3. (**A**) FTIR spectra, in the range 530–580 cm^{-1}, of: PCL (a); PCL/1% LDH-EHT (b); PCL/3% LDH-EHT (c); PCL/5% LDH-EHT (d); and PCL/7% LDH-EHT (e). The inset reports the spectrum of pure LDH-EHT. (**B**) Absorbance at 553 cm^{-1} as a function of LDH-EHT (wt %).

3.2. Thermal Properties of PCL/LDH Nanocomposites

Figure 4 reports the TGA analysis evaluated on all the PCL composites. PCL is also reported for comparison. The inset of Figure 4 reports the TGA of the hybrid LDH-EHT.

The use of IL-EHT as surfactant agent of LDH generated different TGA profile of pristine LDH with the formation of two degradation peaks at 317 °C and 398 °C compared to 310 °C and 430 °C for LDH intercalated by carbonate counter anion. These results suggested the combined presence of carbonate and 2-ethylhexanoate counter anions which confirm XRD results. These results are consistent with the study where Kredatusova et al. showed by FTIR and XPS the presence of this IL onto the clay surface but also into the clay layers [33].

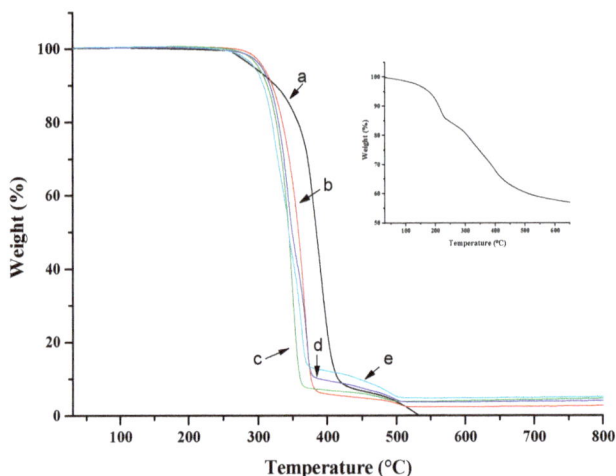

Figure 4. TGA analysis in air evaluated: PCL (**a**); PCL/1% LDH-EHT (**b**); PCL/3% LDH-EHT (**c**); PCL/5% LDH-EHT (**d**); and PCL/7% LDH-EHT (**e**). The inset reports the TGA on LDH-EHT.

The degradation of PCL occurs in two steps: The first one implies a statistical rupture of the polyester chains via ester pyrolysis reaction with production of 5-hexenoic acid, H_2O and CO_2. The second step leads to the formation of ε-caprolactone (cyclic monomer) as result of an unzipping depolymerization process [42]. The first degradation step is slightly anticipated in the composites, and the second step occurs around the temperature of degradation of unfilled PCL. This can be due either to the presence of LDH-EHT which starts to decompose at a lower temperature, or to the decrease of molecular weight of PCL for the milling treatment [43]. To better demonstrate this assumption, we submitted the PCL to the same milling processing of the composites and evaluated the Mw, by GPC analysis. We found a decreasing of Mw from 195 KDa of the unmilled PCL to 120 KDa of milled material. Table 2 reports the $T_{10\%}$ and $T_{50\%}$ of weight loss for all samples. In addition, different authors have demonstrated that the presence of water between LDH layers combined to the formation of metal oxides during the heating of hydrotalcite could accelerate the degradation of the polymer matrix [44,45]. In addition, the low thermal stability of the IL-EHT physisorbed on the LDH (around 340 °C) can also be explained an advanced degradation of the polymer matrix and highlighted by Xanthos et al. on a PLA matrix [46].

Table 2. Temperature at 10% and 50% of weight loss, extracted from TGA analysis (Figure 4).

Sample	$T_{10\%\text{weight loss}}$ (°C)	$T_{50\%\text{weight loss}}$ (°C)
PCL	327	385
PCL + 1% LDH-EHT	320	359
PCL + 3% LDH-EHT	315	343
PCL + 5% LDH-EHT	317	348
PCL + 7% LDH-EHT	307	344

3.3. Surface Analysis and Transport Properties of PCL/LDH Nanocomposites

According to the literature, it is well-known that the presence of ionic liquids into polymer matrix has a significant influence on the hydrophobic behavior of the polymer materials [21,24]. In a previous work, Livi et al. have demonstrated that the incorporation of only 2 wt % of ILs into PBAT matrix generated a significant increase of the water barrier properties due to the hydrophobic nature of phosphonium ionic liquids determined by Coutinho et al. [47,48]. In other works, the same authors have also demonstrated that the use of ILs as additives of epoxy networks induced a significant decrease of the polar component leading to more hydrophobic networks [49]. Thus, to determine the impact of the LDH-EHT on the surface energies of the polymer nanocomposites, the contact angles and surface energy on the neat polycaprolactone and the resulting nanocomposites containing 1, 3, 5 and 7 wt % of LDH-EHT determined by the sessile drop method are summarized in Table 3. Whatever the amount of LDH-EHT introduced into PCL matrix, similar values of surface energies were obtained between 30.8 and 33.5 mN/m. However, differences were observed concerning the values of the polar components. In fact, the incorporation of 1 and 3 wt % of organically modified LDH induced decreases in the polar component from 5 mN/m (neat PCL) to 0.4 mN/m (1 wt % LDH-EHT) and 0.8 mN/m (3 wt % LDH-EHT). These results confirm the presence of IL-EHT in the surface of LDHs and a homogeneous dispersion of the LDH-EHT, as seen on TEM micrographs (Figure 1). According to the literature, this decrease of the polar component is explained by the hydrophobic nature of IL-EHT functionalized by long alkyl chains [49,50]. Thus, different authors have demonstrated that the use of phosphonium IL as additives or surfactant agents of layered silicates such as montmorillonite (MMT) induced a significant reduction of the polar component [24]. Then, for PCL containing 5 and 7 wt % of the LDH-EHT, an increase in the polar component is observed (0.4 mN/m to 3.3 mN/m). These results can be explained by an increase of the amount of LDH-EHT. In fact, we have previously demonstrated by TGA that LDH-EHT contained 15 wt % of physisorbed and intercalated water. Consequently, an increase in the modified clays resulted in a slight decrease in the hydrophobicity of the PCL matrix through the polar component.

Table 3. Polar and dispersive components of the surface energy on the neat PCL and the resulting nanocomposites, from contact angles with water and diiodomethane at room temperature.

Sample	θ_{water} (°)	$\theta_{CH_2I_2}$ (°)	Surface Energy (mN/m)	Dispersive Component (mN/m)	Polar Component (mN/m)
PCL	77 ± 3	62 ± 0.4	30.8	26.1	4.7
PCL + 1% LDH-EHT	113 ± 11	62 ± 2.4	31.6	31.2	0.4
PCL + 3% LDH-EHT	106 ± 3	52 ± 1.8	33.1	32.3	0.8
PCL + 5% LDH-EHT	87 ± 4	49 ± 0.5	33.2	31.0	2.2
PCL + 7% LDH-EHT	85 ± 3	50 ± 3.6	33.5	30.2	3.3

The transport phenomena of small molecular weight molecules, such as water vapor, through polymeric multiphase systems are strictly influenced by the texture of the materials. In the case of composites, the dispersed phase is expected to play a significant role for the diffusion of the penetrating molecules [26]. Figure 5A reports the diffusion coefficient, D (cm²/s), as function of C_{eq} (g/100 g) of water vapour for PCL and the analyzed composites. A linear dependence of the diffusion with respect the equilibrium water sorption can be observed for all samples. It is, then, possible to extrapolate the thermodynamic diffusion coefficient, D_0, using the following empirical equation:

$$D = D_0 \exp\left(\gamma \, C_{eq}\right) \tag{1}$$

The thermodynamic diffusion coefficient is related to the fractional free volume and the tortuosity of the pathway. D_0 was extrapolated for all the samples according to Equation (1). The log D_0 versus filler loading is reported in Figure 5B. It is evident a significant decrease of the thermodynamic diffusion coefficient, even with 1 wt % of filler, a smooth decreasing going from 1 wt % to 3 wt % and a quite linear decreasing up to 7 wt % of filler loading. In XRD analysis, it was observed that the crystalline form of PCL is the same of the composites, and the degree of crystallinity, evaluated by DSC analysis, and here not reported is almost the same in unfilled sample and composites and ranges between 47% and 52%. The decreasing of the diffusion cannot be ascribed, then, to an increasing of crystallinity degree. The presence of the filler increases to a large extent the tortuosity of the pathway, leading to a decreasing in the diffusion. It is worth noticing that, already at 1 wt % of filler, the diffusion decreases of more than one order of magnitude. We have also attributed these results to the presence of IL into LDH. In fact, Coutinho et al. have investigated the effect of water on imidazolium and phosphonium ionic liquids combined with different counter anions [48,51]. For a small amount of water, the authors have highlighted a migration of the water to the near surface of IL inducing the formation of hydrogen bonding's with the counter anion. Oppositely, the use of extra water led to the formation of water clusters generating an increase of the surface tension of the neat ILs. By sessile drop method, we have observed an increase of the polar component from 3 wt % of LDH-EHT incorporated in the PCL matrix. Based on these different results and the obtained diffusion coefficients, we can suggest that the presence of only 1 wt % of IL modified-LDH in the polymer matrix can capture/slow water molecules through hydrogen-bonding interactions with the counter anion of IL-EHT. However, an increase in the amount of LDH-EHT containing up to 15 wt % of water generates a progressive formation of water clusters thus reducing the effect of the ionic liquid in favor of the dispersion of LDH, especially from 3 wt % to 5 wt % where the formation of well-dispersed aggregates into PCL matrix helps to increase the tortuosity of the pathway.

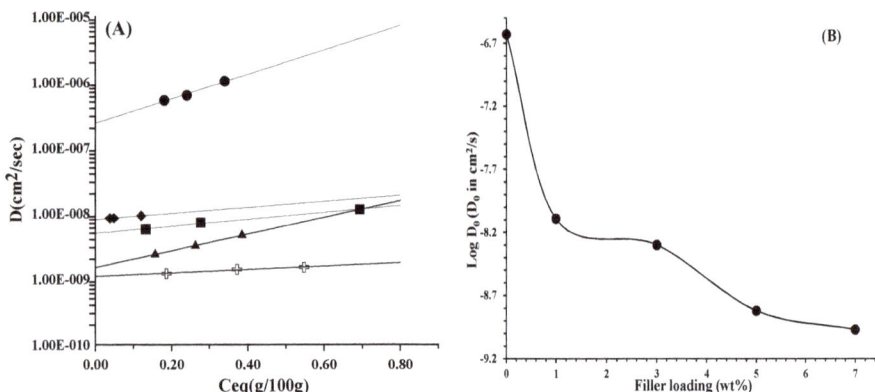

Figure 5. (**A**) The diffusion coefficient, D (cm^2/s), as function of C_{eq} (g/100 g) of water vapor for: PCL (●); PCL/1% LDH-EHT (◆); PCL/3% LDH-EHT (■); PCL/5% LDH-EHT (▲); and PCL/7% LDH-EHT (✛); and (**B**) the log D_0 as function of filler (LDH-EHT) loading (wt %).

4. Concluding Remarks

In this work, thermally stable organically modified hydrotalcite were prepared by using one phosphonium ionic liquid as modifier agent and were used as nanoparticles in PCL matrix. Then, different amounts of the LDH-EHT (1, 3, 5, 7 wt %) were introduced and their influence was investigated on the morphologies and the final properties of the polymer matrix. Two tendencies were observed in function of the quantity of LDH-EHT. For PCL containing 1 and 3 wt % of treated LDH, a homogeneous dispersion was obtained with the presence of small tactoïds leading to a significant decrease of the diffusion coefficient of water vapor, as well as a significant decrease of the polar component from 5 mN/m to 0.4 mN/m. Oppositely, an increase in the amount of LDH induced a microcomposite morphology characterized by the formation of large agglomerates. Moreover, whatever the amount of LDH-EHT incorporated into PCL matrix, a reduction of the thermal stability of the PCL nanocomposites was demonstrated. In summary, the use of small quantities of LDH has a significant impact on the water barrier properties of the polymer matrix opening up new potential prospects in food packaging.

Acknowledgments: This work was supported by the project "High Performing Advanced Material Platform for Active and Intelligent Food Packaging: Cronogard™" (H2020-SMEINST-2-2016-2017). Grant agreement No. 783696.

Author Contributions: Giuliana Gorrasi and Sebastien Livi conceived the paper and designed the experiments. Luanda Chaves Lins and Valeria Bugatti performed the experiments. All Authors analyzed the data. Giuliana Gorrasi and Sebastien Livi wrote the manuscript.

Conflicts of Interest: Authors declare no conflict of interest.

References

1. Costantino, U.; Ambrogi, V.; Perioli, L.; Nocchetti, M. Hydrotalcite-like compounds: Versatile layered hosts of molecular anions with biological activity. *Microporous Mesoporous Mater.* **2008**, *107*, 149–160. [CrossRef]
2. Costantino, U.; Bugatti, V.; Gorrasi, G.; Montanari, F.; Nocchetti, M.; Tammaro, L.; Vittoria, V. New Polymeric Composites Based on Poly(ε-caprolactone) and Layered Double Hydroxides Containing Antimicrobial Species. *ACS Appl. Mater. Interfaces* **2009**, *16*, 668–677. [CrossRef] [PubMed]
3. Bugatti, V.; Costantino, U.; Gorrasi, G.; Nocchetti, M.; Tammaro, L.; Vittoria, V. Nano-hybrid incorporation into poly(ε-caprolactone) for multifunctional applications: Mechanical and barrier Properties. *Eur. Polym. J.* **2010**, *46*, 418–427. [CrossRef]
4. Bugatti, V.; Gorrasi, G.; Montanari, F.; Nocchetti, M.; Tammaro, L.; Vittoria, V. Modified layered double hydroxides in polycaprolactone as a tunable delivery system: In vitro release of antimicrobial benzoate derivatives. *Appl. Clay Sci.* **2011**, *52*, 34–40. [CrossRef]
5. Gorrasi, G.; Bugatti, V.; Vittoria, V. Pectins filled with LDH-antimicrobial molecules: Preparation, characterization and physical properties. *Carbohydr. Polym.* **2012**, *89*, 132–137. [CrossRef] [PubMed]
6. Gorrasi, G.; Bugatti, V. Edible bio-nano-hybrid coatings for food protection based on pectins and LDH-salicylate: Preparation and analysis of physical properties. *LWT Food Sci. Technol.* **2016**, *68*, 139–145. [CrossRef]
7. Gorrasi, G.; Bugatti, V. Mechanical dispersion of layered double hydroxides hosting active molecules in Polyethylene: Analysis of structure and physical properties. *Appl. Clay Sci.* **2016**, *132*, 2–6. [CrossRef]
8. Choudary, B.M.; Lakshmi, M.K.; Venkat, C.R.R.; Koteswara, K.R.; Figueras, F. The first example of Michael addition catalyzed by modified Mg–Al hydrotalcite. *J. Mol. Catal. A Chem.* **1999**, *146*, 279–284. [CrossRef]
9. Guida, A.; Lhouty, M.H.; Tichit, D.; Figueras, F.; Geneste, P. Hydrotalcites as base catalysts. Kinetics of Claisen-Schmidt condensation, intramolecular condensation of acetonylacetone and synthesis of chalcone. *Appl. Catal. A Gen.* **1997**, *164*, 251–264. [CrossRef]
10. El-Toufaili, F.A.; Ahmadniana, F.; Dinse, A.; Feix, G.; Reichert, K.-H. Studies on hydrotalcite-catalyzed synthesis of poly(ethylene terephthalate). *Macromol. Mater. Eng.* **2006**, *291*, 1136–1143. [CrossRef]
11. Pavan, P.C.; Crepaldi, E.L.; Valim, J.B. Sorption of anionic surfactants on layered double hydroxides. *J. Colloid Interface Sci.* **2000**, *229*, 346–352. [CrossRef] [PubMed]
12. Crepaldi, E.L.; Tronto, J.; Cardoso, L.P.; Valim, J.B. Sorption of terephthalate anions by calcined and uncalcined hydrotalcite-like compounds. *Colloids Surf. A Physicochem. Eng. Asp.* **2002**, *211*, 103–114. [CrossRef]
13. Iyi, N.; Matsumoto, T.; Kaneko, Y.; Kitamura, K. Deintercalation of carbonate ions from a hydrotalcite like compound: Enhanced decarbonation using acid–salt mixed solution. *Chem. Mater.* **2004**, *16*, 2926–2932. [CrossRef]
14. Lei, Z.; Chen, B.; Koo, Y.; MacFarlane, D.R. Introduction: Ionic Liquids. *Chem. Rev.* **2017**, *117*, 6633–6635. [CrossRef] [PubMed]
15. Huddleston, J.G.; Visseer, A.E.; Reichert, M.W.; Willauer, H.D.; Broker, G.A.; Rogers, R.D. Characterization and comparison of hydrophilic and hydrophobic room temperature ionic liquids incorporating the imidazolium cation. *Green Chem.* **2001**, *3*, 156–164. [CrossRef]
16. Weyershausen, B.; Lehmann, K. Industrial application of ionic liquids as performance additives. *Green Chem.* **2005**, *7*, 15–19. [CrossRef]

17. Zhao, H.; Malhotra, S.V. Applications of ionic liquids in organic synthesis. *Aldrichim. Acta* **2002**, *35*, 75–83. [CrossRef]

18. Howlett, P.C.; Zhang, S.; Macfarlane, D.R.; Forsyth, M. An investigation of a phosphinate-based ionic liquid for corrosion protection of magnesium alloy AZ31. *Aust. J. Chem.* **2007**, *60*, 43–46. [CrossRef]

19. Rahman, M.; Brazel, C.S. Effectiveness of phosphonium, ammonium and imidazolium based ionic liquids as plasticizers for poly(vinyl chloride): Thermal and ultraviolet stability. In *Polymer Preprints (American Chemical Society; Division of Polymer Chemistry), Proceedings of the 227th ACS National Meeting, Anaheim, CA, USA, 28 March–1 April 2004*; American Chemical Society: Washington, DC, USA; Volume 45, pp. 301–302.

20. Park, K.; Xanthos, M. A study on the degradation of polylactic acid in the presence of phosphonium ionic liquids. *Polym. Degrad. Stab.* **2009**, *94*, 834–844. [CrossRef]

21. Livi, S.; Bugatti, V.; Soares, B.G.; Duchet-Rumeau, J. Structuration of ionic liquids in a poly(butylene-adipate-*co*-terephtalate) matrix: Influence on the water vapour permeability and mechanical properties. *Green Chem.* **2014**, *16*, 3758–3762. [CrossRef]

22. Livi, S.; Pham, T.N.; Gérard, J.-F.; Duchet-Rumeau, J. Polymer and Ionic Liquids: A successful wedding. *Macromol. Chem. Phys.* **2015**, *216*, 359–368. [CrossRef]

23. Livi, S.; Bugatti, V.; Marechal, M.; Soares, B.G.; Duchet-Rumeau, J.; Barra, G.M.O.; Gérard, J.-F. Ionic Liquids-Lignin combination: An innovative way to improve mechanical behaviour and water vapour permeability of eco-designed biodegradable polymer blends. *RSC Adv.* **2015**, *5*, 1989–1998. [CrossRef]

24. Lins, L.; Livi, S.; Duchet-Rumeau, J.; Gérard, J.-F. Phosphonium ionic liquids as new compatibilizing agents of biopolymer blends composed of poly(butylene-adipate-*co*-terephtalate)/poly(lactid acid) (PBAT/PLA). *RSC Adv.* **2015**, *5*, 59082–59092. [CrossRef]

25. Livi, S.; Duchet-Rumeau, J.; Pham, T.N.; Gérard, J.-F. A comparative study on different ionic liquids used as surfactants: Effect on thermal and mechanical properties of high density polyethylene nanocomposites. *J. Colloid Interface Sci.* **2010**, *349*, 424–433. [CrossRef] [PubMed]

26. Livi, S.; Duchet-Rumeau, J.; Gérard, J.-F. Supercritical CO$_2$—Ionic liquids mixtures for modification of organoclays. *J. Colloid Interface Sci.* **2011**, *353*, 225–230. [CrossRef] [PubMed]

27. Livi, S.; Duchet-Rumeau, J.; Pham, T.N.; Gérard, J.-F. Synthesis and physical properties of new surfactants based on ionic liquids: Improvement of thermal stability and mechanical behaviour of high density polyethylene nanocomposites. *J. Colloid Interface Sci.* **2011**, *354*, 555–562. [CrossRef] [PubMed]

28. Livi, S.; Dufour, C.; Gaumont, A.-C.; Levillain, J. Influence of the structure of the onium iodide salts on the properties of modified montmorillonite. *J. Appl. Polym. Sci.* **2013**, *127*, 4015–4026. [CrossRef]

29. Bonnet, L.G.; Kariuki, B.M. Ionic liquids: Synthesis and characterization of triphenylphosphonium tosylates. *Eur. J. Inorg. Chem.* **2006**, *2*, 437–446. [CrossRef]

30. Bradaric, C.J.; Downard, A.; Kennedy, C.; Robertson, A.J.; Zhou, Y. Industrial preparation of phosphonium ionic liquids. *Green Chem.* **2003**, *5*, 143–152. [CrossRef]

31. Ha, J.U.; Xanthos, M. Functionalization of nanoclays with ionic liquids for polypropylene composites. *Polym. Compos.* **2008**, *30*, 534–542. [CrossRef]

32. Calderon, J.U.; Lennox, B.; Kamal, M.R. Polystyrene/phosphonium organoclay nanocomposites by melt compounding. *Int. Polym. Process.* **2008**, *1*, 119–128. [CrossRef]

33. Kredatusova, J.; Benes, H.; Livi, S.; P-Georgievski, O.; Ecorchard, P.; Pavlova, E.; Bogdal, D. Influence of ionic liquid-modified LDH on microwave-assisted polymerization of ε-caprolactone. *Polymer* **2016**, *100*, 86–94. [CrossRef]

34. Livi, S.; Lins, L.; Peter, J.; Benes, H.; Kredatusova, J.; Donato, R.K.; Pruvost, S. Ionic Liquids as Surfactants for Layered Double Hydroxide Fillers: Effect on the Final Properties of Poly(butylene adipate-*co*-terephthalate). *Nanomaterials* **2017**, *7*. [CrossRef] [PubMed]

35. Lins, L.C.; Bugatti, V.; Livi, S.; Gorrasi, G. Phosphonium ionic liquid as interfacial agent of layered double hydroxide: Application to a pectin matrix. *Carbohydr. Polym.* **2018**, *182*, 142–148. [CrossRef] [PubMed]

36. Livi, S.; Bugatti, V.; Estevez, L.; Duchet-Rumeau, J.; Giannelis, E.P. Synthesis and physical properties of Layered Double Hydroxide based on Ionic Liquids: Application to a Polylactide matrix. *J. Colloid Interface Sci.* **2012**, *388*, 123–129. [CrossRef] [PubMed]

37. Livi, S.; Bugatti, V.; Hayrapeytan, S.; Wang, Y.; Estevez, L.; Vittoria, V.; Giannelis, E.P. Deposition of LDH on plasma treated polylactid acid to reduce water permeability. *J. Colloid Interface Sci.* **2013**, *396*, 47–52.

38. Owens, D.K.; Wendt, R.C. Estimation of the surface free energy of polymers. *J. Appl. Polym. Sci.* **1969**, *13*, 1741–1747. [CrossRef]

39. Lagaly, G.; Fernandez Gonzalez, M.; Weiss, A. Problems in Layer-Charge Determination of Montmorillonites. *Clay Miner.* **1976**, *11*, 173–187. [CrossRef]

40. Favre, H.; Lagaly, G. Organo-bentonites with quaternary alkylammonium ions. *Clay Miner.* **1991**, *26*, 19–32. [CrossRef]

41. Kloprogge, J.T.; Frost, R.L. Fourier Transform Infrared and Raman Spectroscopic Study of the Local Structure of Mg-, Ni-, and Co-Hydrotalcites. *J. Solid State Chem.* **1999**, *146*, 506–515. [CrossRef]

42. Persenaire, O.; Alexandre, M.; Degée, P.; Dubois, P. Mechanisms and Kinetics of Thermal Degradation of Poly(ε-caprolactone). *Biomacromolecules* **2001**, *2*, 288–294. [CrossRef] [PubMed]

43. Sorrentino, A.; Gorrasi, G.; Tortora, M.; Vittoria, V.; Costantino, U.; Marmottini, F.; Padella, F. Incorporation of Mg–Al hydrotalcite into a biodegradable Poly(ε-caprolactone) by high energy ball milling. *Polymer* **2005**, *46*, 1601–1608. [CrossRef]

44. Fan, Y.; Nishida, H.; Mori, T.; Shirai, Y.; Endo, T. Thermal degradation of poly(L-lactide): Effect of alkali earth metal oxides for selective L,L-lactide formation. *Polymer* **2004**, *45*, 1197–1205. [CrossRef]

45. Motoyama, T.; Tsukegi, T.; Shirai, Y.; Nishida, H.; Endo, T. Effects of MgO catalyst on depolymerization of poly-L-lactic acid to L,L-lactide. *Polym. Degrad. Stab.* **2007**, *92*, 1350–1358. [CrossRef]

46. Ha, J.U.; Xanthos, M. Novel modifiers for layered double hydroxides and their effects on the properties of polylactic acid composites. *Appl. Clay Sci.* **2010**, *47*, 303–310. [CrossRef]

47. Carvalho, P.J.; Neves, C.M.S.S.; Coutinho, J.A.P.J. Surface tensions of bis (trifluoromethylsulfonyl) imide anion-based ionic liquids. *J. Chem. Eng. Data* **2010**, *55*, 3807–3812. [CrossRef]

48. Almeida, H.F.D.; Lopes-da-Silva, J.A.; Freire, M.G.; Coutinho, J.A.P. Surface tension and refractive index of pure and water-saturated tetradecyltrihexylphosphonium-based ionic liquids. *J. Chem. Thermodyn.* **2013**, *57*, 372–379. [CrossRef]

49. Nguyen, T.K.L.; Livi, S.; Soares, B.G.; Benes, H.; Gérard, J.-F.; Duchet-Rumeau, J. Toughening of epoxy/ionic liquid networks with thermoplastics based on poly(2.6-dimethyl-1,4-phenylene ether) PPE. *ACS Sustain. Chem. Eng.* **2017**, *5*, 1153–1164. [CrossRef]

50. Koros, W.J.; Burgess, S.K.; Chen, Z. Polymer Transport Properties. *Encycl. Polym. Sci. Technol.* **2015**. [CrossRef]

51. Freire, M.G.; Carvalho, P.J.; Fernandes, A.M.; Marrucho, I.M.; Queimada, A.J.; Coutinho, J. Surface tensions of imidazolium based ionic liquids: Anion, cation, temperature and water effect. *J. Colloid Interface Sci.* **2007**, *314*, 621–630. [CrossRef] [PubMed]

MDPI

Article

Independent Evaluation of Medical-Grade Bioresorbable Filaments for Fused Deposition Modelling/Fused Filament Fabrication of Tissue Engineered Constructs

Mina Mohseni, Dietmar W. Hutmacher and Nathan J. Castro *

Institute of Health and Biomedical Innovation, Queensland University of Technology, Brisbane City 4059, QLD, Australia; mina.mohseni@hdr.qut.edu.au (M.M.); dietmar.hutmacher@qut.edu.au (D.W.H.)
* Correspondence: nathan.castro@qut.edu.au; Tel.: +61-7-3138-0055

Received: 9 December 2017; Accepted: 31 December 2017; Published: 2 January 2018

Abstract: Three-dimensional printing/additive manufacturing (3DP/AM) for tissue engineering and regenerative medicine (TE/RM) applications is a multifaceted research area encompassing biology, material science, engineering, and the clinical sciences. Although being quite mature as a research area, only a handful of clinical cases have been reported and even fewer commercial products have made it to the market. The regulatory pathway and costs associated with the introduction of bioresorbable materials for TE/RM have proven difficult to overcome, but greater access to 3DP/AM has spurred interest in the processing and availability of existing and new bioresorbable materials. For this purpose, herein, we introduce a series of medical-grade filaments for fused deposition modelling/fused filament fabrication (FDM/FFF) based on established and Federal Drug Administration (FDA)-approved polymers. Manufacturability, mechanical characterization, and accelerated degradation studies have been conducted to evaluate the suitability of each material for TE/RM applications. The comparative data serves to introduce these materials, as well as a benchmark to evaluate their potential in hard and soft tissue engineering from a physicochemical perspective.

Keywords: tissue engineering and regenerative medicine; bioresorbable polymers; 3D printing/ additive manufacturing; fused filament fabrication/fused deposition modelling; degradation; physicochemical characterization

---

1. Introduction

Three-dimensional printing (3DP) has established itself as a robust and effective technology for the manufacturing of highly controlled micro- and macro-scale structures that are suitable for use as tissue-engineered constructs (TECs). In combination with computer-aided drafting (CAD) and finite element modelling (FEM), custom TECs with tunable and pre-designed porous networks providing the necessary structural support that is required for defect repair and tissue regeneration can be realized. Rapid and precise fabrication of mechanically-analogous three-dimensional (3D) microenvironments make this technique most attractive. Not solely as a tool for controllable 3D architecture, but with the introduction and availability of high quality and traceable materials, spatiotemporal controlled delivery of biomolecular agents are feasible when compared to conventional manufacturing methods. Therefore, 3DP is a leading technology in both tissue engineering and drug delivery.

Fused-deposition modelling/fused filament fabrication (FDM/FFF) is one of the most common and accessible 3DP technologies. FDM/FFF is an additive manufacturing technology that is based on melt extrusion of thermoplastic polymers with ordered deposition in a layer-by-layer fashion.

With increasing interest in 3DP as a direct manufacturing technology, global demands for synthetic and natural materials with 3DP processability is a fast growing and untapped market. Although a plethora of single material and composite filaments are commercially available, the regulatory quality assurance/quality controls necessary for implantable 3DP medical devices from thermoplastic precursor filaments have limited translation to the clinic with few entities in pursuit of this burgeoning market. Therefore, the climate is ripe for the introduction of medical-grade 3DP thermoplastic filaments materials requiring thorough characterization of their potential in tissue engineering and regenerative medicine (TE&RM) applications. Thermoplastic polyesters and polyethers are good candidate materials for 3DP, owing to their high tunability with respect to melt properties, a unique property that is not typically seen in natural polymers where heating results in aloss of bioactivity and structural integrity. A uniform viscose flow and proper cooling rates is associated with the melting temperature, purity, molecular weight, and degree of crystallinity are important factors that are necessary for high resolution and reproducible printing of FDM/FFF 3D TECs.

TECs should exhibit physicochemical and mechanical properties suitable for initial tissue engraftment and maturation through the modulation of cell behavior, including mechanotransduction during TEC degradation. Cell-material interactions are complex where mechanical stimuli in the form of local deformation of binding receptors influence intracellular pathways leading to the promotion or inhibition of tissue growth [1,2]. Soft tissues, including adipose tissue, exhibit mechanoresponsiveness where adipogenesis is enhanced in the presence of static stretching, but impeded under static compression, as well as under dynamic loading [3]. In the case of bone TE&RM, bone deposition and resorption are mutually-exclusive events that are modulated by mechanical stimulation of the tissue [4,5]. According to the minimum effective strains (MES) hypothesis, applied strains greater than MES causes the adaptation of bone density and architectures, while strain below MES does not produce any change [6]. When considering the clinical application of 3D printed TECs, biomaterial selection and TEC architecture should be designed to withstand local deformation within the physiological range necessary for tissue deposition. The fundamental characteristics of the bulk material, the overall geometry of the TEC and the internal geometry synergistically regulate strain distribution on the TECs and, subsequently, the local deformation of microenvironments.

In addition to the structural contributions of 3D printed TECs, physicochemical changes of the structure, and, ultimately, the material during degradation, is another important aspect when designing TECs for TE&RM. Ideally, the rate of degradation and strength loss should be proportional to the rate of tissue growth [7]. As a TEC degrades, a loss in mechanical properties ensues and the rate of this action is a necessary consideration when designing TECs that inherently experience high stress. The rate of degradation and subsequent strength loss is an important design parameter for TE&RM research. Structural and mechanical integrity should remain during tissue remodelling, especially for large volume applications. As previously stated, the function of a scaffold is not only as a mechanical support, but also to serve as a substrate for tissue engraftment and remodelling.

Accelerated degradation of is a widely accepted method to study in vitro physiochemical changes of polymers [8]. The degradation rate of synthetic polyester and polyether polymers can vary from days-to-months, while accelerated degradation provides short-term degradation profiles, which makes it a time and cost effective method to evaluate and characterize materials in vitro [9–11]. In vivo degradation of large molecular chains predominantly begins and proceeds by the hydrolysis of amorphous regions and short chains in the polymer backbone, rather than through enzymatic degradation [12]. Utilizing alkaline medium, including sodium hydroxide (NaOH) solutions, which is rich in hydroxyl (–OH) groups, increases hydrolysis and serves as a comparable method for in vivo conditions [13].

In the current study, we have extensively characterized four new medical-grade filaments (Dioxaprene® 100 M, Max-Prene® 955, Lactoprene® 100 M, Caproprene™ 100 M kindly donated by Poly-Med, Inc., Anderson, SC, USA) for 3D printability. More importantly, the effects of degradation on physicochemical properties have been evaluated, with special emphasis on mechanical properties.

We have assessed these materials for printability via FDM/FFF by optimizing material-specific printing parameters, as shown in Table 1, as well as undertaken a comprehensive study characterizing physicochemical changes of 3D printed structures via accelerated degradation in alkaline conditions. Thermal analyses were also conducted to elucidate melt and flow properties for the optimization of printing temperature. In addition to melt behavior, thermal analysis allows for a better understanding of the morphology of macromolecular chains and re-crystallization of absorbable polymers. We conclude with a discussion wherein appropriate applications for each group of material is suggested based on degradation rate and strength loss.

Table 1. Optimized printing parameters.

Material	Temperature (°C)	Feed-Rate (mm/min)	Layer Height (mm)
DIO	120	1800	0.2
CAP	120	1800	0.2
LAC	192	1350	0.2
MAX	210	1800	0.16

2. Materials and Methods

2.1. Materials

Dioxaprene® 100 M (DIO), Caproprene™ 100 M (CAP), Lactoprene® 100 M (LAC), Max-Prene® 955 (MAX) were kindly donated by Poly-Med, Inc. (Poly-Med, Inc., Anderson, SC, USA). All of the materials were stored in a low humidity cabinet to minimize moisture absorption. Sodium hydroxide (NaOH, M_w = 40.00, Sigma-Aldrich, St. Louis, MO, USA) solutions of 1 M and 5 M were prepared with ultrapure water and used in all experiments.

2.2. Printing 3D Scaffolds

A series of 10 mm × 10 mm × 5 mm models were designed in SolidWorks (Dassault Systemes, Waltham, MA, USA) and the resultant computer-aided design (CAD) file was prepared for 3D printing by conversion to a computer numerical control file by Simplify3D® (Blue Ash, OH, USA). Next, the models were printed using a FlashForge Dreamer table-top FDM/FFF printer (FlashForge, Jinhua, China) at varying infill densities of: 10%, 20%, and 40%, respectively.

Infill density is a printing parameter that is related to the number of printed fibers comprising the internal architecture of the printed part. The porosity for each infill density was calculated as the ratio of actual volume to bulk volume using the following formula, where V_b is the bulk volume and V_t is the true volume.

$$\rho = \frac{V_b - V_t}{V_b} \times 100 \tag{1}$$

Actual volume was calculated as the mass ratio between the printed scaffold and bulk density of the material, as determined by the volume of the rectangular geometry (Table 2).

Table 2. The porosity and pore size for three infill densities. Data is represented as mean $\pm$ SD ($n = 5$).

Infill Density (%)	Porosity (%)	Pore Size (mm)
10	86.9 ± 2.5	2.19 ± 0.05
20	75 ± 4.5	1.01 ± 0.02
40	53.2 ± 3.1	0.37 ± 0.02

2.3. Degradation

Accelerated degradation was conducted for porous scaffolds using NaOH as the hydrolysis medium. A 1 M NaOH solution was used to study degradation for all four materials. Due to the slow degradation rate of PCL in 1 M NaOH, a higher concentration (5 M NaOH) was employed. The samples were immersed into NaOH and incubated at 37 °C, 5% CO_2. At each respective time point, degraded samples were rinsed with deionized water ($3\times$) and dried under vacuum at 40 °C for 8 h. Mass loss was calculated as the ratio of residual mass (W_r) and initial mass (W_0), according to the general formula.

$$W_L\% = \frac{W_0 - W_r}{W_0}\%$$

(2)

2.4. Scanning Electron Microscopy

The microstructure of pristine and degraded samples were analysed by scanning electron microscopy (SEM, JSM-7001F, JEOL Ltd., Tokyo, Japan). For cross-sectional views, samples were flash frozen in liquid nitrogen (5 min) and cut. All of the samples were gold sputter-coated using a JEOL fine sputter coater (JFC-1200, JEOL Ltd., Tokyo, Japan) for 75 s at 8 mA current and observed under vacuum at 2 KV accelerating voltage. Scaffold pore size was determined by measuring 10 fields of view using ImageJ (National Institutes of Health, Bethesda, MD, USA) [14].

2.5. Thermal Characterization

Melt and thermal properties of pristine and degraded materials were characterized via differential scanning calorimetry (Q100 DSC, TA Instruments, Newcastle, DE, USA) under non-isothermal conditions. Additionally, thermal transitions of non-printed monofilaments were determined to serve as a good estimation of the printing temperature. Samples weighing approximately 4 mg were sealed in an aluminium pan and exposed to non-isothermal heating at a ramp rate of 10 °C/min with kinetic analysis performed on the thermographs utilizing universal analysis 2000 software (TA Instruments, Newcastle, DE, USA).

2.6. Mechanical Testing

Unconfined, uniaxial compression was conducted on pristine and degraded samples using an Instron Micro Tester (5848, Instron, Melbourne, Australia). In an effort to characterize the effects of material composition and pore size on the mechanical properties, 3D printed TECs with three porosities were printed and evaluated. To study the effects of degradation on TEC stiffness, mechanical analysis was conducted for degraded samples. TECs of 20% infill density were chosen for strength loss characterization. Briefly, specimens were placed on a flat platen and were compressed at a rate of 0.6 mm/min up to 50% compression ($n = 5$) in deionized water at 37 °C. The elastic modulus was defined as the slope of the linear region (Range 4–10%), with the yield strength being defined as the peak stress of the linear region.

2.7. Contact Angle Analysis

Surface wettability of 3D printed TECs was evaluated by contact angle analysis. Thin films of the each polymer was prepared by heating to the respective melting point in an oven and cooled on a 20 mm circular glass cover slip at room temperature. A droplet with the volume of 2 μL was deposited on the films with images that were taken at the static condition using the FTA200 computer-controlled, video based instrument (First Ten Angstroms, Portsmouth, VA, USA). Five positions were randomly tested for each sample.

3. Results

3.1. Degradation

Thermal characterization of pristine, unprinted monofilaments by DSC (Figure 1) was used to determine the optimal temperature that was necessary for stable viscous flow and printability of the molten polymer. Table 1 shows optimized layer height, printing temperature, and material feed rate. Due to rapid cooling, interlayer fusion of MAX was inadequate at a layer height of 0.2 mm resulting in delamination. Therefore, a layer height of 0.16 mm was used. All of the materials were printed successfully, and SEM images were taken to quantify and validate the pore size of three different infill densities (Figure 2) with corresponding porosities, as determined using Equation (1).

Although DSC analysis is a good predictor of printing temperature, it is not reflective of the optimized printing temperature. For instance, PCL does not exhibit viscous flow up to 120 °C, while for all other materials, the printing temperature closely matches the melting point. This may be due to differences in the inherent viscosity and molecular weight of the material at the respective melt temperature.

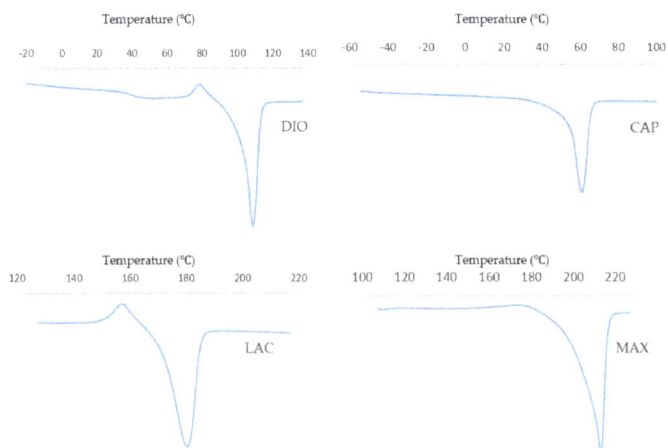

Figure 1. Non-isothermal DSC trace of non-printed monofilaments at a heating rate of 10 °C/min.

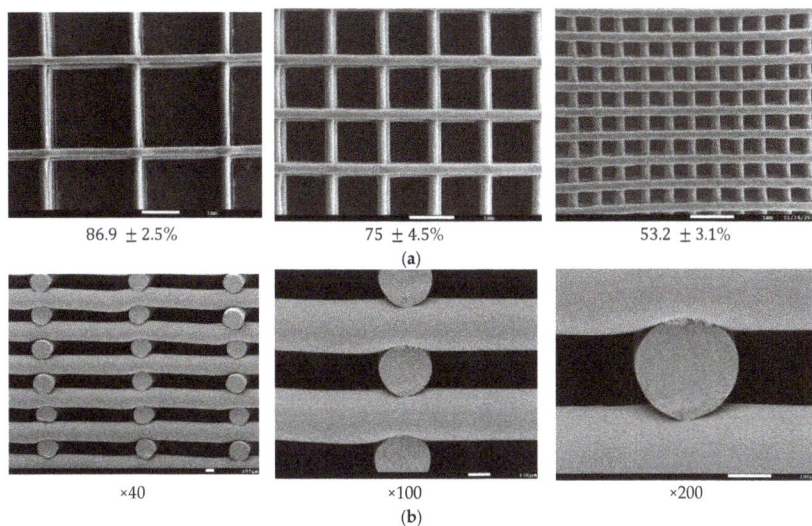

86.9 ± 2.5%   75 ± 4.5%   53.2 ± 3.1%

(a)

×40   ×100   ×200

(b)

Figure 2. Scanning electron microscopy (SEM) micrographs of three-dimensional (3D) printed scaffolds; (**a**) Representation of three porosities; (**b**) Representation of cross-sectional view at varying magnification.

3.2. Contact Angle Analysis

Figure 3 illustrates surface wettability, as determined by contact angle analysis of the four materials. Results show MAX as the most hydrophilic with a contact angle of $45.8° \pm 1.8°$, with increasing hydrophobicity of $58.4° \pm 0.42°$, $77.8° \pm 1°$, and $83.2° \pm 3.3°$ for DIO, LAC, and CAP, respectively.

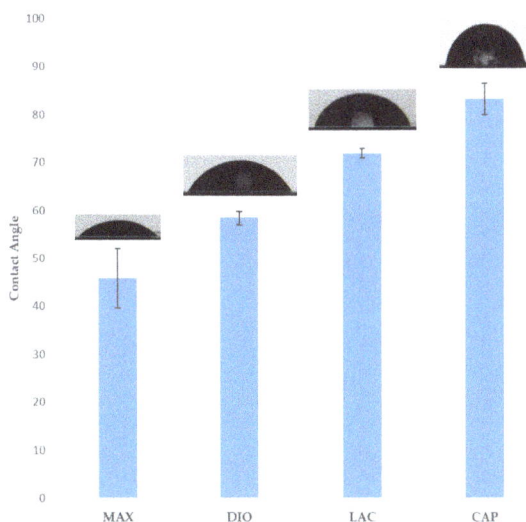

Figure 3. Water contact angle analysis of polymer films. Data is represented as mean $\pm$ SD ($n = 5$).

3.3. Mechanical Characterization

The mechanical behavior of 3D scaffolds under compression was characterized and is shown in Figure 4. All of the materials exhibit a toe region followed by linear and subsequent plastic deformation (Figure 4a). It is observed that mechanical strength of 3D printed structures is dependent on the inherent properties of the material, as well as pore size. When comparing the four materials, DIO with a pore size of 2.19 mm is the softest exhibiting a compressive modulus of 0.3 MPa which is 3.1, 25.8, and 32.6-fold less than CAP, LAC, and MAX, respectively. An increase in porosity leads to a significant increase in compressive modulus and yield strength, with DIO exhibiting the greatest change with respect to pore size as an increase in porosity from 53.2% to 86.9% produces a 53-fold increase in elastic modulus while similar porosities produce a 23, 8, and 9-fold increase for CAP, LAC, and MAX, respectively (Figure 4b). The inherent properties of the material and regulating pore size through precise 3D printing make it possible to achieve the desired mechanical properties.

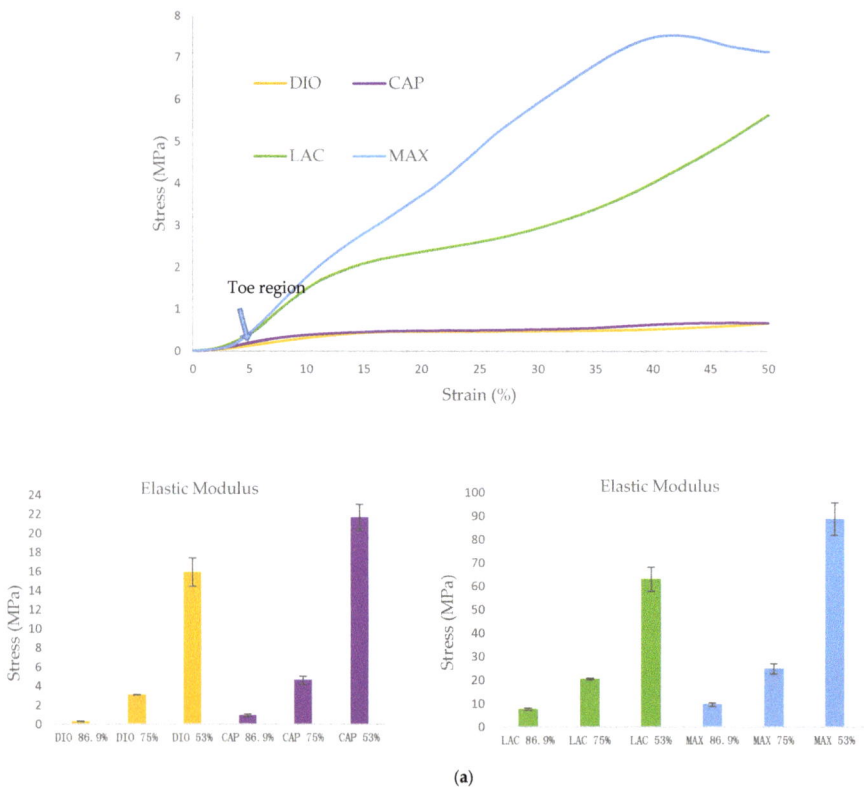

(a)

Figure 4. *Cont.*

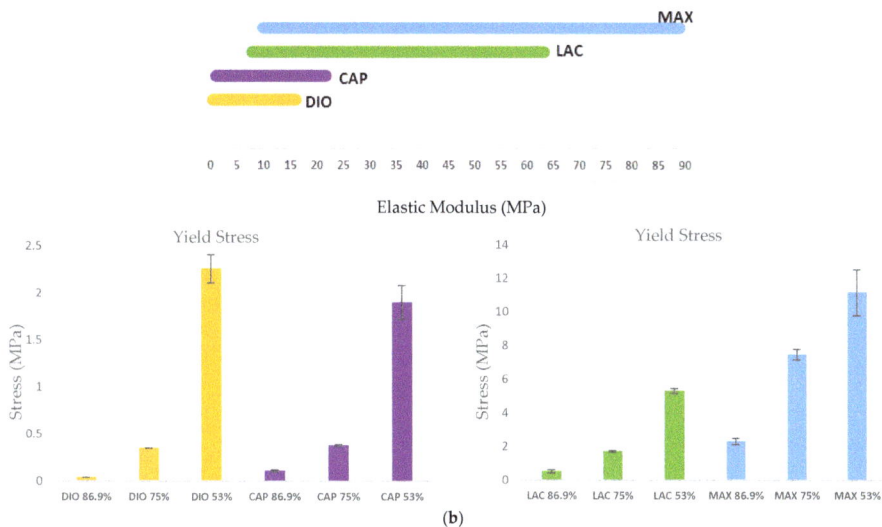

Figure 4. Unconfined, uniaxial compression. (**a**) Stress-strain curves for the scaffolds with different pore size; (**b**) Compressive modulus and yield strength for all materials with three porosities. Data is represented as mean ± SD ($n = 5$).

3.4. Degradation

Accelerated degradation was conducted under alkaline conditions to quantify the rate of mass loss, strength loss, and to evaluate morphological changes of printed fibers. 1 M NaOH was used to accelerate hydrolytic degradation of DIO, LAC, and MAX. Since the degradation rate of CAP in 1 M NaOH is considerably slower than all of the other materials (Figure 5a, less than 1% after 24 h), 5 M NaOH was used to accelerate CAP degradation within a comparable time frame. Hydrolytic degradation of polymers occurs by erosion of the polymer beginning with cleavage of hydrolytic bonds leading to the formation of water soluble components [15]. All of the materials show non-uniform mass loss under accelerated degradation at the early stages, which is associated with the cleavage of amorphous regions within the polymer. The degradation rate of amorphous regions is typically higher than crystalline regions as the accessibility of water molecules is less energetically costly. Single crystals in close proximity to the degraded amorphous region become unstable leading to cleavage and solubility. As a result, crystalline regions become more accessible and susceptible to hydrolytic cleavage and erosion.

MAX and DIO exhibit the fastest degradation rate, more than 75% mass loss after 2 h, while LAC shows less than 10% degradation (Figure 5b). When comparing LAC and CAP, after two days, LAC exhibited >90% mass loss, while CAP exhibited <1% degradation. In 5 M NaOH CAP shows accelerated degradation leading to ~95% mass loss after 19 days (Figure 5c).

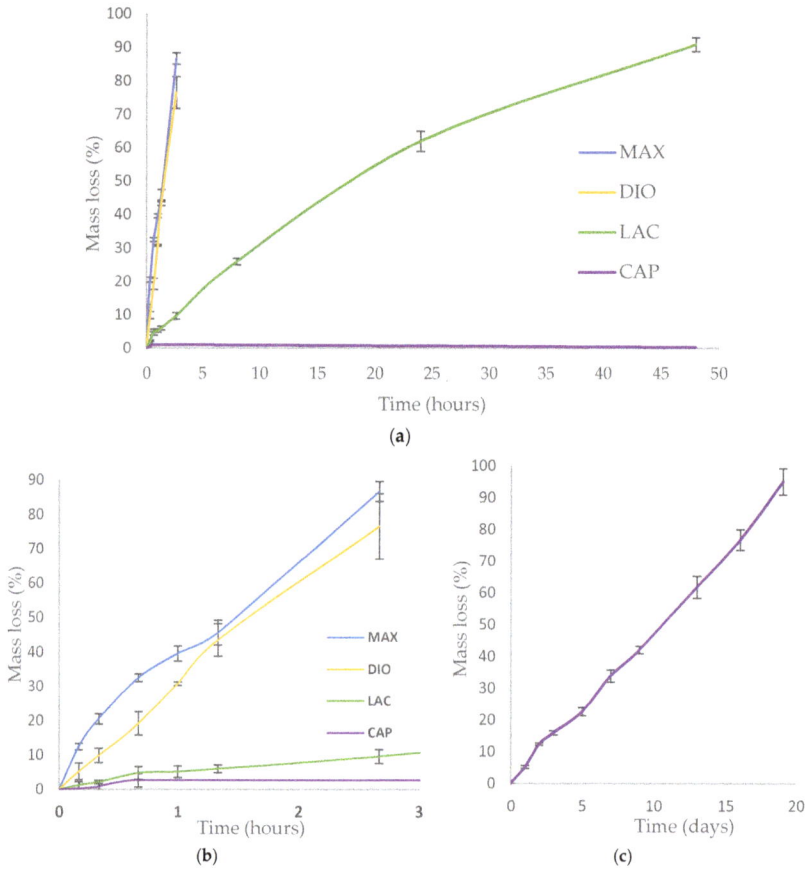

Figure 5. Accelerated degradation; (**a,b**) Degradation of MAX, DIO, LAC, and PCL in 1 M NaOH. (**c**) Degradation of CAP in 5 M NaOH. Data is represented as mean $\pm$ SD ($n = 5$).

Morphological studies were conducted via SEM of pristine and eroded samples. Erosion patterning and subsequent morphological changes of the printed structures are directly related to water accessibility and surface penetration. SEM images of MAX and LAC show a smooth surface at the top and cross section of the printed fibers, which indicates that the speed of water penetration into the core of the printed fiber is slower than the speed of hydrolysis of ester bonds. Subsequently, both amorphous and crystalline regions disintegrate prior to diffusion into the subsequent layers. A morphologically smooth surface and uniform decrease of fiber diameter and thickness leads to increased pore size (Figure 6 (DIO/CAP)). The results are indicative of a predominant surface erosion mechanism of degradation. For MAX, fusion points exhibit a similar erosion pattern with a regular smooth surface, while for LAC, fusion points have irregular surface morphology, producing a different morphological structure when compared to other regions of the 3D printed part. Therefore, the rate of hydrolysis at the fusion points of LAC is lower than water penetration. This observation indicates higher and larger crystalline structures that disintegrate in a slower rate when compared to amorphous sites leading voids and non-uniform surfaces. This can be related to the print speed and the overlaying of molten material wherein the material is cooled at a slower rate allowing for better recrystallization. MAX shows a uniform decrease of fusion points with circumferential cracks followed by longitudinal fractures (Figure 6 (LAC/MAX)). When compared to MAX, LAC exhibits a lower density of microfractures on the surface and fusion points, with only a few seen at the fusion points at 90% mass loss (Figure 6 (LAC)). CAP displays a roughened surface due to variations in hydrolysis speed of amorphous and crystalline regions. Crystallites exhibit greater resistance to hydrolysis and are morphologically noted as irregular grains on the surface. Initially, amorphous regions are cleaved, leading to higher crystallite density with more uniform degradation rate. Due to hydrophobic surfaces and high resistance of PCL toward water penetration, degradation proceeds slowly through cleavage of crystallites. The cross sectional view of printed fibers also displays a smooth surface absent of voids, which is indicative of a dominant surface erosion mechanism. Hydrolysis begins at the surface and proceeds very slowly due to high resistance of CAP to water penetration. The erosion pathway of the fusion points and surfaces are morphologically similar. DIO shows large pores on the surfaces and fusion points, which are attributed to fast hydrolysis and increased accessibility of water molecules into the interior of polymer. The cross section view shows that degradation begins from the edges causing non-uniform structures. Since no voids are seen in the core of the printed fiber, the degradation mechanism is predominately surface erosion with the fusion points exhibiting similar morphological changes when compared to other sites.

Strength loss of degraded scaffolds is presented as a factor of mass loss in an effort to evaluate the materials' potential for use in applications experiencing moderate to high levels of stress (Figure 7). For all of the samples, strength loss begins with a quick degradation rate followed by a rate decrease until complete disintegration. The rapid initial strength loss can be attributed to the initial attack and scission of amorphous regions by water molecules. This scission produces shortened polymer chains, which can be readily dissolved. At high mass loss, when the density of cracks is increased and the structure is susceptible to disintegration, the rate of strength loss increases again until complete fragmentation. When comparing the rates of mass loss to strength loss, the rate of strength loss exceeds that of mass loss due to the scission of amorphous regions and short chain polymers before the complete cleavage of the polymer chains ensues; mass loss appears as the chains are completely broken and dissolved in water. Before deviations in mass are detectable, strength loss is observed due to small scissions and cracks. MAX maintains structural and mechanical integrity up to 45% mass loss, while LAC preserves mechanical integrity up to 61% mass loss. Although these two polymers have a close mechanical behavior, the rate of strength loss of MAX is faster due to the higher density of cracks. Developed circumferential cracks at fusion points and longitudinal cracks on the surfaces increase stress concentration and the risk of failure. DIO shows an integrated structure of up to 30% mass loss, while PCL maintains its' integrity up to 42% mass loss. DIO is more fragile when compared to PCL and collapses sooner. As it is observed from SEM analysis, DIO shows large pores at fusion points due to faster hydrolysis speed and higher accessibility of water molecules to the interior of DIO. Higher stress concentration at irregular edges and smaller contact area at fusion points make DIO more fragile.

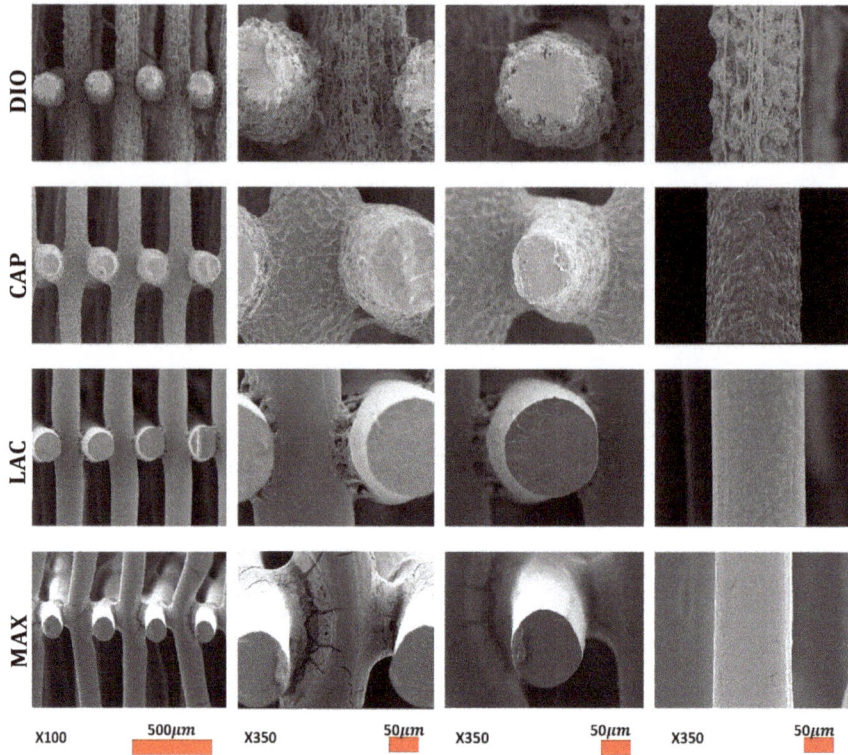

Figure 6. SEM micrographs of degraded samples after losing 30% mass loss at different magnifications.

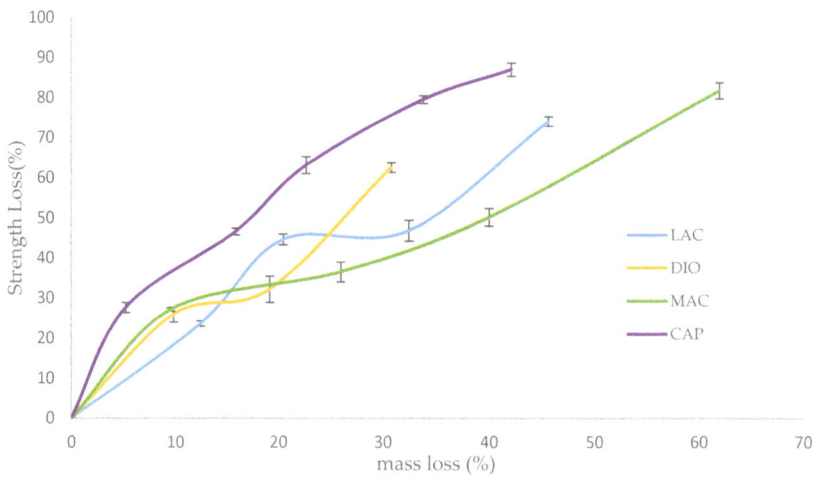

Figure 7. Strength loss versus mass loss for degraded scaffolds with porosity of 75%. Data is represented as mean $\pm$ SD ($n = 5$).

3.5. Thermal Analysis

Thermal transition of 3D printed non-degraded and degraded materials was monitored under non-isothermal analysis at a heating rate of 10 °C/min to evaluate the changes in the morphology and crystallinity of materials.

Thermal behavior of all materials is displayed in Figure 8. DIO shows an exothermic peak before melting which is related to the recrystallization of the polymer. For samples with higher degrees of degradation, the recrystallization temperature increases from 81.3 to 85.4 °C, which can be attributed to the larger lamellae requiring higher temperature and more energy for recrystallization.

Thermal transition of CAP shows an increase of enthalpy for degraded samples when compared to non-degraded samples. This observation illustrates a higher degree of crystallinity for degraded samples resulting in a slower degradation rate. CAP shows the highest melting peak and also the highest crystallinity (80.58 J/g) at 23% mass loss, which is the maximum resistance of crystalline structures toward hydrolysis and cleavage.

Thermal analysis of LAC shows a double melting behavior that can be explained by the melt-recrystallization model where the first endothermic peak is attributed to the melting of original lamellar structures, while the second endothermic peak is associated to the melting of newly formed crystallites during recrystallization. As the material melts during the first endothermic peak T_{m1}, larger lamellas begin to recrystallize leading to energy release. The newly formed crystalline structures melt at the higher temperature, T_{m2}, resulting in the second endothermic peak. As degradation proceeds, T_{m1} and T_{m2} decreases producing smaller original lamellas with lower molecule weights. The enthalpy decreases gradually from 54.91 to 54.10 J/g as materials lose their mass up to 61.94%. At 90.85% mass loss, enthalpy has reduced significantly to 50.30 J/g.

For MAX, during degradation the area of endothermic peaks increases from 59.59 to 95.98 J/g showing an increase in crystallinity. Additionally, the endothermic peak of the sample with the highest mass loss exhibits a narrowing which shows less crystallite size distribution. A small exothermic peak is observed before the melting peak that increases during degradation. This exothermic peak can be attributed to recrystallization before melting. Since higher levels of degradation produce a higher degree of ordered structures, there is greater probability of recrystallization as indicated by increasing exothermic peaks.

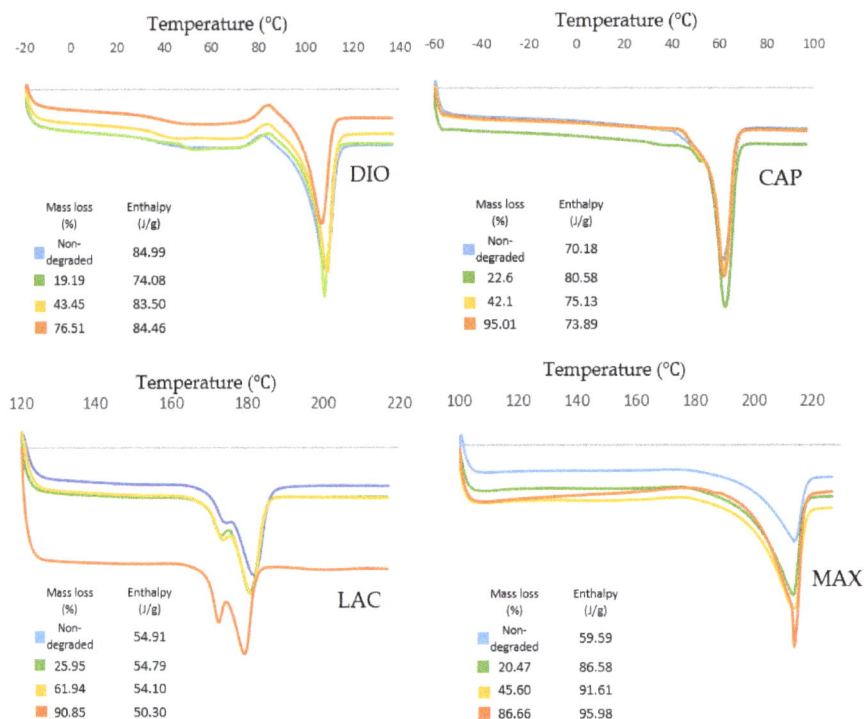

Figure 8. Thermal characteristics of non-degraded and degraded materials with different mass loss. Non-isothermal analysis was conducted at heating rate of 10 °C/min using DSC.

4. Discussion

In the current study, we have evaluated four new medical grade filaments for printability, physicochemical characteristics, and the potential for use in TE&RM applications. Comprehensive characterization was performed on non-degraded and degraded materials to assess their physical, mechanical, and morphological properties before and after degradation. These studies provide greater insight and empirical evidence to support proposed TE&RM applications.

Optimized printing conditions were determined after evaluating the heating effects on phase transition of monofilaments via DSC (Table 1, Figure 1). The materials were printed successfully with three different porosities and were assessed under unconfined, uniaxial compression to quantify the compressive modulus and yield strength. DIO with a pore size of 2.19 mm shows the softest behavior with an elastic modulus of 300 KPa (Figure 4b). DIO is composed of 100% Polydioxanone (PDO), which is a synthetic resorbable polymer with high flexibility supported by ether oxygen bonds and used extensively as an implantable suture. It exhibits excellent biocompatibility, degradability, and high flexibility, making it appropriate for a variety of biomedical applications. DIO is very sensitive to internal structures and provides a wide range of mechanical properties by tuning the pore size of the printed TEC, which makes it highly desirable for TE&RM applications (Figure 4b). This study is the first reported use of DIO in FFF/FDM 3D printing showing high proccessibility. This potential along with the inherent soft behavior makes it a good candidate for complex structures in soft tissue engineering. CAP, which is 100% caprolactone, is another appropriate candidate for soft tissue engineering. When compared to DIO, it shows a higher elastic modulus, however, as the pore size decreases, the difference is less dramatic. At a pore size of 2.19 mm, elastic modulus of PCL is approximately three times higher than DIO, while at a pore size of 0.37 mm, it is 1.3-fold higher. MAX, a copolymer of 95% glycolide and 5% lactide, shows the highest elastic modulus, at the pore size of 0.37 mm. Changing pore size from 0.37 to 2.19 mm gives a wide range of properties from 9.8 to 88.9 MPa. LAC is 100% Lactide and shows various elastic modulus from 7.79 to 63.17 MPa. By controlling the pore size of MAX and LAC, a wide range of materials properties can be achieved, which are suitable for hard tissue engineering including cartilage and bone tissue.

Accelerated degradation in alkaline medium was conducted to investigate hydrolytic degradation of all the monofilaments. Hydrolysis is a chemical reaction of water molecules with polymers that produces carboxylic acid, followed by cleavage of -OH bonds, resulting in the erosion of polymer matrix [16]. In this study, mass loss rate was evaluated in NaOH and depicted in Figure 5. By comparison of these results with contact angle analysis, it is realized that more hydrophilic surfaces shows higher degradation speed (Figure 3). SEM analysis shows surface erosion for four materials as the predominant degradation pathway. Figure 6 illustrates that degradation begins from the edges with no voids being seen radially towards the core. However, surface morphology of each polymer is different due to various rates of hydrolysis and water penetration into the polymer fibre with DIO exhibiting large voids on the surfaces. Hydrophilic surfaces for DIO, as well as fast hydrolysis, results in decreased stability and porosity. PCL exhibits a roughened surface due to the presence of highly water-resistant crystallites. MAX and LAC show smooth surfaces and uniform edges, with MAX exhibiting a high density of cracks at the fusion points and fibre surface leading to decreased mechanical stability and greater water accessibility.

In designing TECs, the bulk properties of materials as well as their degradation behavior and physicochemical changes should be considered. The inherent properties of materials and the internal structures of scaffolds, including pore size, are predominant factors in controlling the mechanical properties of designed TECs. Depending on tissue morphology and the cascade of biological phenomena during the regenerative process, an "optimized" pore size can vary. Feng. B showed that pore size <400 μm limits the growth and infiltration of blood vessels in a porous bioceramic scaffolds [17]. JP. Temple studied osteogenesis and angiogenesis of large craniomaxillofacial bone defects, with TEC exhibiting pore sizes of 0.1–2 mm [18]. The results show that uniform distribution of human adipose-derived stem cells with high density of vascular network was obtained at a pore size of ~800 μm. For larger pore sizes, cell aggregation ensues, leading to sedimentation and decreased activity, whereas for the smaller pore sizes, cell aggregation is minimized. In contrast to large bone defects, for barrier membrane in dental applications, smaller pores (<200 μm) is suggested to control the growth and infiltration of epithelial and gingival fibroblasts at early stages [19]. By decreasing the rate of cellular infiltration at early time points, bone tissue has more space to mature and remodel over time. Infection is one of the major concerns for applications involving barrier membranes. By employing TECs with small pore sizes, the risk of infection can be minimized [19]. By considering the limitation of pore size in a specific application, materials and internal structures should be chosen carefully to ensure suitable conditions for growth of tissue.

Based on the current study, MAX and LAC may be suitable candidates for musculoskeletal tissue engineering. These materials exhibit excellent printing fidelity allowing for the manufacture of complex geometries for large and small defects. By controlling pore size, a wide range of properties can be achieved. In addition, by printing composite structures of MAX and LAC, the unique properties of each material can be combined with predefined geometries leading to a higher degree of control on mechanical properties as well as degradation rate. Depending on the volume of defects and the speed of regeneration, the degradability of composite structures can be tuned. MAX degrades within 36 h in alkaline solution, while LAC exhibits 90% mass loss after 48 h. A composite structure of these two materials with different degradation rates is also feasible and merits further investigation in particular for drug delivery applications, where the release profile of drugs can be controlled by faster degrading material and the structure of scaffolds can be preserved by long-term degradable materials. As previously stated, infection is one of the major concerns with implantable TECs, which can be addressed by controlled release of antimicrobial drugs within the first few days of implantation when macrophages show the highest activity [20]. Utilizing dual extruding printers and controlling the number of layers and internal structures of each material shows great promise for controlled mechanical properties, degradation, and drug release.

With regards to soft tissue engineering, most available printing materials are natural hydrogels, which cannot be readily printed in complex geometries, especially for large volume tissue regeneration [21]. In this study, DIO has been introduced as a soft polymer with high flexibility along with high 3D printing fidelity for complex small and large structures. When compared to CAP, DIO exhibits a faster degradation rate, which makes it more appropriate for short-term regenerative applications. DIO may also be an appropriate polymer matrix for artificial skin patches owing to its mechanical and physicochemical properties. Additionally, skin does not require extended degradation for long-term regeneration and remodelling, which makes DIO a better candidate when compared to CAP or other materials. Degradation of PCL, which is the main component of CAP, is very slow (>2 years) [9,22], and a combination of DIO and CAP may provide the added capability of adjusting the degradation rate of TECs to be temporally compatible with tissue regeneration and remodelling. In addition to dual extrusion FDM/FFF printing, these polymers can be mixed during printing as they have similar printing temperature. The potential addition of DIO into the TEC allows for greater tunability with regards to degradation rate and mechanical properties. Recently, the concept of harnessing the body as a "bioreactor" has been investigated by our group and coupled with delayed fat injection [23] has shown promising results for large volumes. Delaying the administration of fat into the TEC leverages the initial cascade of biological events leading to fibrous tissue infiltration and neovascularization. By controlling the degradation rate of the internal TEC architecture, adequate space post-implantation can be modulated allowing for more uniform tissue ingrowth. One potential embodiment would utilize DIO as the internal structure, while providing the mechanical behavior that is necessary to ensure tissue viability by using PCL as a slow degradable polymer. Another important positive aspect of composite structures is in drug delivery where the degradation rate directly affects the release rate of incorporated drugs. The hydrophobic surface of CAP, as well as its slow degradation, limits sustained delivery. In most therapeutic/regenerative applications, the speed of tissue regeneration is slow, while drug delivery is required for a short period post-implantation. Therefore, for combinatorial TECs, including antibacterial embodiments, antimicrobial agents need to be released at the early stages of regeneration. Releasing antibiotics within the first weeks prior to complete formation of blood clot can decrease the risk of infection considerably. Therefore, employing DIO as a delivery system in soft tissue engineering is suggested.

5. Conclusions

The current study served to introduce a family of 3D printable bioresorbable medical-grade materials for use in TE/RM applications. With the combination of material traceability and accessible high-resolution 3D printing technologies, we may find ourselves at the forefront of commercially-viable 3DP/AM for TE/RM applications.

Acknowledgments: The authors would like to thank Poly-Med, Inc. for kindly supplying all materials used in the current study.

Author Contributions: Dietmar W. Hutmacher and Nathan J. Castro conceived and designed the experiments; Mina Mohseni performed the experiments; Mina Mohseni and Nathan J. Castro analyzed the data; Mina Mohseni wrote the paper.

Conflicts of Interest: The authors declare no conflict of interest.

Polymers **2018**, *10*, 40

References

1. Qin, Y.-X.; Lam, H. Intramedullary pressure and matrix strain induced by oscillatory skeletal muscle stimulation and its potential in adaptation. *J. Biomech.* **2009**, *42*, 140–145. [CrossRef] [PubMed]
2. Damaraju, S.; Matyas, J.R.; Rancourt, D.E.; Duncan, N.A. The effect of mechanical stimulation on mineralization in differentiating osteoblasts in collagen-I scaffolds. *Tissue Eng. Part A* **2014**, *20*, 3142–3153. [CrossRef] [PubMed]
3. Yuan, Y.; Gao, J.; Ogawa, R. Mechanobiology and Mechanotherapy of Adipose Tissue-Effect of Mechanical Force on Fat Tissue Engineering. *Plast. Reconstr. Surg. Glob. Open* **2015**, *3*, e578. [CrossRef] [PubMed]
4. Frost, H.M. Bone "mass" and the "mechanostat": A proposal. *Anat. Rec.* **1987**, *219*, 1–9. [CrossRef] [PubMed]
5. Frost, H.M. The mechanostat: A proposed pathogenic mechanism of osteoporoses and the bone mass effects of mechanical and nonmechanical agents. *Bone Miner.* **1987**, *2*, 73–85. [PubMed]
6. Frost, H. A determinant of bone architecture: The minimum effective strain. *Clin. Orthop. Relat. Res.* **1983**, *175*, 286–292. [CrossRef]
7. Ikada, Y. Challenges in tissue engineering. *J. R. Soc. Interface* **2006**, *3*, 589–601. [CrossRef] [PubMed]
8. Ginjupalli, K.; Shavi, G.V.; Averineni, R.K.; Bhat, M.; Udupa, N.; Upadhya, P.N. Poly(α-hydroxy acid) based polymers: A review on material and degradation aspects. *Polym. Degrad. Stab.* **2017**, *144*, 520–535. [CrossRef]
9. Sun, H.; Mei, L.; Song, C.; Cui, X.; Wang, P. The in vivo degradation, absorption and excretion of PCL-based implant. *Biomaterials* **2006**, *27*, 1735–1740. [CrossRef] [PubMed]
10. Felfel, R.; Poocza, L.; Gimeno-Fabra, M.; Milde, T.; Hildebrand, G.; Ahmed, I.; Scotchford, C.; Sottile, V.; Grant, D.M.; Liefeith, K. In vitro degradation and mechanical properties of PLA-PCL copolymer unit cell scaffolds generated by two-photon polymerization. *Biomed. Mater.* **2016**, *11*, 015011. [CrossRef] [PubMed]
11. Lam, C.X.; Savalani, M.M.; Teoh, S.H.; Hutmacher, D.W. Dynamics of in vitro polymer degradation of polycaprolactone-based scaffolds: Accelerated versus simulated physiological conditions. *Biomed. Mater.* **2008**, *3*, 034108. [CrossRef] [PubMed]
12. Therin, M.; Christel, P.; Li, S.; Garreau, H.; Vert, M. In vivo degradation of massive poly (α-hydroxy acids): Validation of in vitro findings. *Biomaterials* **1992**, *13*, 594–600. [CrossRef]
13. Htay, A.; Teoh, S.; Hutmacher, D. Development of perforated microthin poly (ε-caprolactone) films as matrices for membrane tissue engineering. *J. Biomater. Sci. Polym. Ed.* **2004**, *15*, 683–700. [CrossRef] [PubMed]
14. Schneider, C.A.; Rasband, W.S.; Eliceiri, K.W. NIH Image to ImageJ: 25 years of image analysis. *Nat. Meth.* **2012**, *9*, 671–675. [CrossRef]
15. Göpferich, A. Mechanisms of polymer degradation and erosion. *Biomaterials* **1996**, *17*, 103–114. [CrossRef]
16. Li, S. Hydrolytic degradation characteristics of aliphatic polyesters derived from lactic and glycolic acids. *J. Biomed. Mater. Res.* **1999**, *48*, 342–353. [CrossRef]
17. Feng, B.; Jinkang, Z.; Zhen, W.; Jianxi, L.; Jiang, C.; Jian, L.; Guolin, M.; Xin, D. The effect of pore size on tissue ingrowth and neovascularization in porous bioceramics of controlled architecture in vivo. *Biomed. Mater.* **2011**, *6*, 015007. [CrossRef] [PubMed]
18. Temple, J.P.; Hutton, D.L.; Hung, B.P.; Huri, P.Y.; Cook, C.A.; Kondragunta, R.; Jia, X.; Grayson, W.L. Engineering anatomically shaped vascularized bone grafts with hASCs and 3D-printed PCL scaffolds. *J. Biomed. Mater. Res. Part A* **2014**, *102*, 4317–4325. [CrossRef] [PubMed]
19. Rakhmatia, Y.D.; Ayukawa, Y.; Furuhashi, A.; Koyano, K. Current barrier membranes: Titanium mesh and other membranes for guided bone regeneration in dental applications. *J. Prosthodont. Res.* **2013**, *57*, 3–14. [CrossRef] [PubMed]
20. Johnson, C.T.; García, A.J. Scaffold-based anti-infection strategies in bone repair. *Ann. Biomed. Eng.* **2015**, *43*, 515–528. [CrossRef] [PubMed]

Polymers **2018**, *10*, 40

21. Hinton, T.J.; Jallerat, Q.; Palchesko, R.N.; Park, J.H.; Grodzicki, M.S.; Shue, H.-J.; Ramadan, M.H.; Hudson, A.R.; Feinberg, A.W. Three-dimensional printing of complex biological structures by freeform reversible embedding of suspended hydrogels. *Sci. Adv.* **2015**, *1*, e1500758. [CrossRef] [PubMed]
22. Lam, C.X.; Hutmacher, D.W.; Schantz, J.T.; Woodruff, M.A.; Teoh, S.H. Evaluation of polycaprolactone scaffold degradation for 6 months in vitro and in vivo. *J. Biomed. Mater. Res. Part A* **2009**, *90*, 906–919. [CrossRef] [PubMed]
23. Chhaya, M.P.; Balmayor, E.R.; Hutmacher, D.W.; Schantz, J.-T. Transformation of breast reconstruction via additive biomanufacturing. *Sci. Rep.* **2016**, *6*, 28030. [CrossRef] [PubMed]

polymers

MDPI

Article

Improved Processability and the Processing-Structure-Properties Relationship of Ultra-High Molecular Weight Polyethylene via Supercritical Nitrogen and Carbon Dioxide in Injection Molding

Galip Yilmaz [1,2], Thomas Ellingham [1,2] and Lih-Sheng Turng [1,2,*]

[1] Polymer Engineering Center, Department of Mechanical Engineering, University of Wisconsin–Madison, Madison, WI 53706, USA; gyilmaz@wisc.edu (G.Y.); ellingham@wisc.edu (T.E.)
[2] Wisconsin Institute for Discovery, University of Wisconsin-Madison, Madison, WI 53715, USA
* Correspondence: turng@engr.wisc.edu ; Tel.: +1-608-316-4310

Received: 8 December 2017; Accepted: 28 December 2017; Published: 30 December 2017

Abstract: The processability of injection molding ultra-high molecular weight polyethylene (UHMWPE) was improved by introducing supercritical nitrogen (scN_2) or supercritical carbon dioxide ($scCO_2$) into the polymer melt, which decreased its viscosity and injection pressure while reducing the risk of degradation. When using the special full-shot option of microcellular injection molding (MIM), it was found that the required injection pressure decreased by up to 30% and 35% when $scCO_2$ and scN_2 were used, respectively. The mechanical properties in terms of tensile strength, Young's modulus, and elongation-at-break of the supercritical fluid (SCF)-loaded samples were examined. The thermal and rheological properties of regular and SCF-loaded samples were analyzed using differential scanning calorimetry (DSC) and parallel-plate rheometry, respectively. The results showed that the temperature dependence of UHMWPE was very low, suggesting that increasing the processing temperature is not a viable method for reducing injection pressure or improving processability. Moreover, the use of scN_2 and $scCO_2$ with UHMWPE and MIM retained the high molecular weight, and thus the mechanical properties, of the polymer, while regular injection molding led to signs of degradation.

Keywords: ultra-high molecular weight polyethylene (UHMWPE); microcellular injection molding; supercritical fluid; supercritical N_2; supercritical CO_2

1. Introduction

Ultra-high molecular weight polyethylene (UHMWPE) is a linear homopolymer polyethylene that possesses many desirable solid-state characteristics [1]. When comparing UHMWPE to well-known polyethylene grades such as high-density polyethylene (HDPE), its distinct properties are directly associated with its ultra-high molecular weight, which ranges from 1 million g/mol to 6 million g/mol, compared to 0.05–0.25 million g/mol for HDPE [2–4]. Owing to its ultra-high molecular weight, UHMWPE has many high-performance properties, including high abrasion resistance, high toughness and fiber modulus, great chemical inertness, low friction coefficient, self-lubrication behavior, very low service temperature, noise dampening, and more.

As a result, UHMWPE has found successful applications in many industrial and medical sectors. For example, it has been employed in high-performance fiber products, artificial joint prosthesis, bearings, bumpers, and the sliding parts of production lines [4–6]. Its chemical inertness, low friction, and wear-resistant features make it a useful material for parts where lubricants could cause contamination or maintenance problems, such as in food and medical applications [1,7].

However, the ultra-high molecular weight that yields these desirable properties and applications also hinder the fabrication of UHMWPE by standard polymer processing methods, such as injection molding (IM), as it requires a very high processing pressure due to its high melt viscosity and lack of fluidity [1,4,8].

In the plastics processing industry, injection molding is one of the most important processes due to its wide range of processable materials, ready-to-use final products, capability to make complex parts, short cycle time, and high degree of automation [9]. It is estimated that injection molding is used to process approximately one-third of all thermoplastic materials, with the process encompassing around one-half of all plastic processing equipment [10].

At present, only a few UHMWPE grades in granular form are available for injection molding. Its original powder form is unsuitable for injection molding, even when using specially designed screws for powder resins. Due to its high viscosity, molding UHMWPE requires high injection rates and injection pressures (values such as 110 MPa compared to 55–70 MPa for most polymers) [3,11]. Although the special grades of UHMWPE allow for injection molding, a further understanding of their behavior is necessary to take advantage of all of the potential benefits and achieve optimal and reliable processing conditions. Moreover, if the processing of UHMWPE is not performed properly, the product quality can suffer due to thermal degradation.

Given its processing challenges, research literature on the injection molding of UHMWPE remains scarce. To the best of our knowledge, injection molding of UHMWPE is limited to Kuo's work [12–14]. In these studies, the injection molding of UHMWPE, process optimization for improved mechanical properties, and tribological characteristics were investigated.

There are two well-known approaches for successfully fabricating UHMWPE parts: (1) using high-pressure settings or (2) reducing the viscosity via solvents [1]. Solvents, including mineral oil, paraffin wax, decalin, and xylene, have been used successfully to process UHMWPE [6]. However, some tedious post-processing operations are necessary to clean parts and reclaim and recycle the solvents. Also, the high pressures required for processing increases tool wear and manufacturing costs.

An effective method for overcoming the high viscosity of UHMWPE is by using supercritical fluid (SCF) as a plasticizer. A compressed atmospheric gas, such as N_2 or CO_2 above its critical temperature and pressure values (cf. Table 1), is referred to as an SCF. An SCF exhibits the density of a liquid but has the high diffusivity and low viscosity of a gas. This specific phase allows not only accurate metering and pumping as a liquid, but also an increased diffusion rate into the polymer as a gas [15,16].

Once SCF is diffused into the polymer melt evenly, it can reduce the viscosity and the glass transition temperature of the polymer depending on the weight percentage of the SCF used [17,18]. The physical effect that SCF has on UHMWPE can be described by two mechanisms. The first and dominant mechanism is the increased free volume among entangled polymer chains [19,20]. The secondary mechanism is decreased chain entanglement due to the dilution of the polymer via the addition of the dissolved SCF [21]. The SCF also acts as a reversible plasticizer because it can leave the polymer matrix simply by diffusing out after processing is complete or when the thermodynamic instability is triggered due to changes in pressure or temperature. However, there are some drawbacks of SCFs, such as the need for high-pressure equipment. Nonetheless, SCF is still commonly used with favorable results because it can eliminate tedious solvent handling and all post-processing operations [22–24].

In this study, atmospheric gases, which are less expensive and unregulated—namely, N_2 and CO_2, were used in their supercritical state [15]. Several studies have been done on the plasticizing effects of supercritical CO_2 in various polymer processing methods [21,23–27]. Wilding et al. showed that the effect of supercritical CO_2 on viscosity intensifies as the molecular weight of the linear PE grade increases [28]. This suggests that the viscosity of UHMWPE can be reduced effectively via SCF. Moreover, supercritical N_2 has received some attention as an alternative to CO_2 for processing, especially for polyethylenes like low-density polyethylene (LDPE) [29]. In practice, injection molding using supercritical N_2 as a foaming agent is more common because it yields a finer and denser foamed microstructure [22,30]. The main advantage of N_2 over CO_2 is that the back pressure required to keep N_2 in its supercritical state is much lower (~54%) than the back pressure required for CO_2 [16]. A higher back pressure requires a longer dosage time, which is undesirable. On the other hand, the estimated maximum gas solubility of CO_2 is almost four times higher than N_2 in PE at 200 °C [15]. Therefore, CO_2 may have more potential to modify the melt properties when a higher amount of SCF is needed.

One of the commercially available technologies used to introduce SCF into the polymer is the microcellular injection molding (MIM) process. In this process, SCF is first dosed and injected into the injection molding barrel and mixed with the polymer melt during the dosage cycle. Thanks to the properties of SCFs, it diffuses to form small gas bubbles and eventually dissolves in the polymer matrix, thus creating a so-called "single-phase polymer/SCF solution" [15,16,29]. As the high pressure polymer/SCF solution enters the mold cavity, a sudden pressure drop causes the nucleation of bubbles, which expand to create a foamed product without the need for a packing cycle. This is because the expanding gas bubbles increase the volume of the foamed polymer, which compensates for material shrinkage as it cools, and thus fills the entire mold cavity [31–33]. The main usage of the MIM process is to create very fine and lightweight polymer foams with a typical average cell size of 3–100 µm [16,34] and a cell density of over 10^6 cells/cm^3.

Occasionally, solid products instead of foamed ones can also be produced while utilizing SCF in the MIM process. This process option is called high-pressure foam injection molding (FIM) [35]. In this case, the entire mold cavity is filled with a full shot volume's worth of material and then packing pressure is applied to suppress gas expansion and re-dissolve the gas into the polymer melt. In high-pressure FIM, the cells may or may not be eliminated completely depending on the processing parameters, such as the pressure in the cavity, SCF content, temperature, and polymer type [35,36].

In this study, the high-pressure foam injection molding option of the MIM process was used to produce solid parts rather than foamed ones in the UHMWPE injection molding experiments. Utilization of the plasticizing effect of supercritical N_2 and CO_2 was the aim. After the injection cycle, packing pressure was applied to block cell nucleation and re-solubilize the escaping gas. The injection molding of foamed UHMWPE parts using MIM was found to be a more involved project than their solid counterparts and remains an active on-going research subject. Herein, the processability in terms of injection pressure values and mold filling behavior of tensile bars, tensile properties, rheological properties, DSC data, and microstructure in forms of micro-computed tomography (µCT) images of the solid samples will be presented.

2. Materials and Methods

An injection molding grade of UHMWPE resin (GUR® 5129) was provided by Celanese Corporation (Irving, TX, USA). It had a viscosity average molecular weight of 4.7 million g/mol and a density of 0.93 g/cm^3. Industrial grade carbon dioxide (CO_2) and nitrogen (N_2) were purchased from Airgas (Greenville, SC, USA) and used as the SCF source for MIM.

2.1. Methods

A MIM-equipped injection molding machine, Arburg Allrounder 320S (Arburg, Lossburg, Germany) with a 25 mm diameter screw, was used to introduce scCO$_2$ and scN$_2$ into the melt. Here, the main purpose was to make a solid part rather than a foamed product by applying sufficient packing pressure to compress the full-size melt and suppress foaming. One critical requirement for the injection molding of UHMWPE is that the compression ratio of the screw must be less than 2.5:1, and preferably 2:1, due to the absence of melt flow based on the ISO 1133 standard. The compression ratio of the injection molding screw can be defined as the ratio of the feed section channel depth to the metering section channel depth [10]. The machine used in this study had a 2.5:1 compression ratio. Table 1 shows the injection-molded sample types and SCF loading percentages used in this study.

The percent of SCF loading used was determined experimentally to introduce a high and controllable amount of SCF.

Table 1. Sample name and SCF loading percent.

Sample Name	SCF	Critical Points [15]	Loading % by Weight
Regular	N/A	N/A	0
CO_2 sample	CO_2	31 °C, 7.37 MPa	1.5
N_2 sample	N_2	−147 °C, 3.40 MPa	1.5

The processing parameters used for injection molding are listed in Table 2 and were set according to the material's molding recommendations and the machine type used. The back pressure was adjusted for each gas to keep the SCF in its supercritical state. The cooling time was calculated as 25.2 s based on thermal diffusivity and target ejection temperature (85 °C) [37].

Table 2. Processing parameters used in injection molding.

Processing Parameters	Units	Value
Injection speed	cm^3/s	80
Injection vol.	cm^3	17.2
Cooling time	s	25
Back pressure	MPa	Regular: 0.5 N_2: 5 CO_2: 8
Packing pressure	MPa	110
Packing time	s	7
Nozzle temperature	°C	260
Mold temperature	°C	80

An aluminum mold designed and machined in-house was used to produce ASTM standard Type I tensile bars with a 2 mm thick fan gate. Figure 1 shows the mold cavity and the sprue location, which was located on top of the fan gate without a runner to minimize additional pressure requirements.

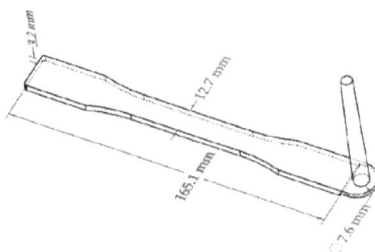

Figure 1. Schematic of the injection-molded specimen ASTM D638 Type I tensile bar mold cavity with its sprue.

Prior to any characterization experiments, the injection-molded UHMWPE samples were stored for at least one week to allow the dissolved gas to completely diffuse out.

2.2. Differential Scanning Calorimetry

Differential scanning calorimetry (DSC) was conducted according to the ASTM F2625-10 (2016) standard test method on a TA Instruments Q20 (TA Instruments, New Castle, DE, USA) to measure the enthalpy of fusion, percent crystallinity, and the melting point of the UHMWPE samples. The procedure was applied from ambient temperature to 200 °C, with a heat/cool/heat cycle at a 10 °C/min scanning rate. Four tests for each type of sample were performed with a weight range of 5 to 8 mg. Test samples were cut from the middle of the tensile bar. The enthalpy of melting per sample mass was normalized against the theoretical enthalpy of melting of 100% crystalline polyethylene to calculate the percentage of crystallinity in the test sample. The standard equation was used as follows:

$$\% \text{ Crystallinity} = 100 \times \frac{\Delta H_\text{s}}{\Delta H_\text{f}} \tag{1}$$

where ΔH_s was the sample's heat of fusion per unit mass during the freezing transition. The heat of fusion of 100% crystalline UHMWPE, ΔH_f = 289.3 J/g, was used in the calculation [38].

2.3. Rheology

A stress-controlled rheometer, TA Instruments AR 2000ex (TA Instruments, New Castle, DE, USA), was used to measure the complex viscosity of the samples. Tests were conducted with 25 mm diameter parallel plates with a 1 mm gap. Samples were cut from the tensile test bars, then compression molded at 200 °C for 10 s at 1 mm thickness. When room temperature was reached, the samples were punched into 25 mm diameter disks. First, a stress sweep test was run to determine the linear viscoelastic region and 0.5% strain was selected for the following tests. Frequency sweep tests were conducted from 0.01 Hz to 100 Hz at a temperature of 260 °C. For temperature dependency analyses, temperature ramp tests were run from 180 °C to 280 °C at two different frequencies, 0.1 and 100 Hz, respectively.

The temperature dependence of the materials was calculated using the following equation [39],

$$m(T) = m_\text{o} \cdot exp[-a(T - T_\text{o})] \tag{2}$$

where $m(T)$ is the consistency index, m_o is the consistency index at the reference temperature T_o, a is the temperature dependence or sensitivity coefficient, and T is the temperature at which the viscosity was measured.

2.4. Micro-Computed Tomography (µCT)

To examine if any bubbles formed in the samples, µCT scans were conducted with an industrial type Metrotom 800 µCT system from Carl Zeiss AG (Oberkochen, Germany). The scanning power was set at 32 kV and 120 µA for a 5 µm spot and voxel size. Image processing was performed using ImageJ software from the National Institutes of Health, Bethesda, MD, USA.

2.5. Tensile Tests

Tensile testing of the samples was performed on an Instron 5967 (Instron, Norwood, MA, USA) with a 30 kN load cell using ASTM D638-03 Type I tensile specimens. At least five samples of each type were tested with a crosshead speed of 50 mm/min as recommended in the standard. Before testing, all samples were examined with transmitted light to check for any remaining bubbles. Samples with defects were excluded. Moreover, the samples were weighed and compared to each other to eliminate any odd samples. A few samples were one standard deviation away from the mean and were also excluded.

2.6. Tensile Bar Images and Injection Pressure Measurements

To examine the mold-filing behavior of UHMWPE melt, the injection volume was reduced by 20% and the holding stage was canceled so that the injection pressure reduction and the effect of dissolved SCF on the shape of the short shots could be observed. Pressure data were collected at least five times for each injection speed setting from 10 cm^3/s to the maximum machine injection speed of 80 cm^3/s in 10 cm^3/s increments. All samples were weighed each time to ensure that the same amount of material was injected within 0.1 g.

3. Results and Discussion

3.1. Differential Scanning Calorimetry

The first and second heating DSC graphs of the neat and injection-molded (regular, CO_2, and N_2) samples can be seen in Figure 2, and the data derived from the DSC test is reported in Tables 3 and 4 for the first and second heating tests, respectively.

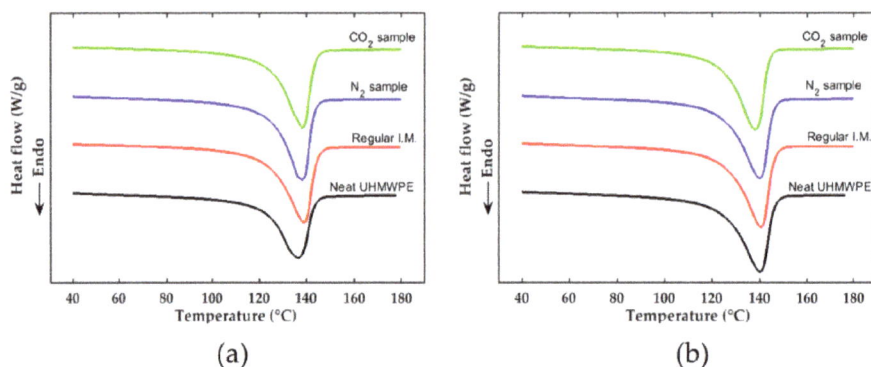

Figure 2. (a) The first heating and (b) the second heating of the differential scanning calorimetry (DSC) graphs of the injection-molded samples and neat ultra-high molecular weight polyethylene (UHMWPE).

Table 3. The average thermal behavior of injection-molded samples obtained from the first heating thermograms.

Sample	T_m (°C)	ΔH_s	% Crystallinity
Neat	136.9 ± 0.1	142.4 ± 1.9	49.2 ± 0.7
Regular	139.5 ± 0.4	158.2 ± 3.2	54.7 ± 1.1
N_2 sample	138.7 ± 1.0	161.1 ± 2.7	55.7 ± 0.8
CO_2 sample	138.6 ± 0.7	159.6 ± 2.2	55.1 ± 0.9

Table 4. The average thermal behavior of injection-molded samples obtained from the second heating thermograms.

Sample	T_m (°C)	ΔH_s	% Crystallinity
Neat	139.8 ± 0.6	166.7 ± 0.7	57.6 ± 0.2
Regular	140.3 ± 0.9	167.7 ± 2.7	58.0 ± 0.8
N_2 sample	140.2 ± 0.8	169.8 ± 1.5	58.7 ± 0.5
CO_2 sample	139.7 ± 1.2	167.6 ± 1.9	57.9 ± 0.6

In Figure 2a, the first heating curves for all injection-molded samples seemed identical and showed a higher melting point than the neat sample. It is known that different thermomechanical (temperature–pressure) histories, as well as material degradation, can alter the first heating thermograms. The distinct factor in the thermal history of the injection-molded samples is expected to be the presence of scN_2 and $scCO_2$ during processing. However, the degree of crystallinity (% crystallinity) of the samples tabulated in Table 3 was similar within the standard deviation. The degree of crystallinity of the neat sample was much less than the injection-molded samples, likely due to a different thermomechanical history from its production process.

The difference between the regular and SCF-loaded injection-molded samples was smaller than the standard error for each of them, thus indicating that SCF did not have a clear effect on the thermal properties. Even so, the N_2 and CO_2 samples had slightly higher-than-average degrees of crystallinity as compared to the regular sample—1.0% more for the N_2 samples and 0.4% more for the CO_2 samples. This behavior might have been due to the slightly increased molecular mobility and higher packing pressure on the N_2 and CO_2 samples. As mentioned earlier, scN_2 and $scCO_2$ can reduce the viscosity of the polymer by increasing the free volume between polymer chains. Therefore, it is expected to enhance the crystallization kinetics and degree of crystallinity. In addition, lower flow resistance on the polymer should transfer more pressure during the packing cycle. On the other hand, the foaming and melt fracture mechanisms during the filling stage caused some gas to escape and vent out just before the packing stage. This lowered the amount of SCF in the melt during the cooling phase and the packing cycle, which decreased crystallization in these phases.

Since the crystal structure is very rigid compared to the amorphous structure, a higher degree of crystallinity usually increases the Young's modulus of the final product [40]. As will be seen later, the Young's modulus in the tensile tests in this study followed the results of the DSC tests. In the absence of SCF, the degree of crystallinity for the regular sample was slightly lower than for the SCF samples. Meanwhile, the average melting points were similar among the samples.

Because the second heating had the same thermal history, the second heating thermograms should reflect the thermal degradation of the material rather than the processing history. It is known that a lower molecular weight PE, namely HDPE, tends to have a higher degree of crystallinity [41]. The data in Table 4 show that all injection-molded samples had slightly higher degrees of crystallinity than neat UHMWPE, although there were no significant differences in degrees of crystallinity or melting points among the injection-molded samples.

Yasuniwa et al. reported that, although the melting point of the PE grades increased linearly with the molecular weight, above 10^6 g/mol, the change was insignificant [42]. The results in our study agree with this finding. While the melting points of all injection-molded samples were similar, the rheology and mechanical test results to be discussed below showed that the regular sample had a lower viscosity and tensile properties than the SCF samples.

One possible source of error in the DSC tests could stem from a non-uniform distribution of material with different molecular weights in the final product. Within the injection-molding barrel, thermal degradation mainly occurs on the surface of the barrel rather than on the screw root. Once the material enters the cavity, material degradation takes place where the temperature (due to viscous heating) or stress (due to shear and elongational flows) is highest. Although four tests for each type of sample were performed, the degraded material may not have been distributed and dispersed evenly within the main matrix.

3.2. Mechanical Properties

The bar graphs in Figure 3a–c show the mechanical properties of the injection-molded samples. The same data are also tabulated in Table 5. The Young's moduli of the N_2 and CO_2 samples increased slightly from 8% to 9% compared to the regular sample. A similar slight increase was also observed for tensile strength, with about 1 MPa of difference.

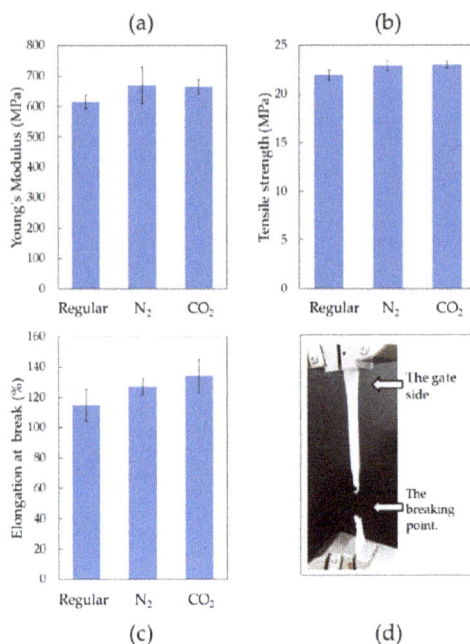

Figure 3. (a–c) Mechanical properties of UHMWPE samples and (d) an image of the breaking point.

Table 5. Tensile properties of injection-molded samples.

Samples	Young's Modulus (MPa)	Elongation at Break (%)	Tensile Strength (MPa)	Toughness (J/m^3)
Regular	615.2 ± 21.4	115 ± 10	21.9 ± 0.5	64.4 ± 7.1
N$_2$ samples	669.6 ± 59.9	127 ± 5	22.9 ± 0.5	73.2 ± 4.0
CO$_2$ samples	664.0 ± 22.7	134 ± 11	23.0 ± 0.3	77.8 ± 8.4

The largest improvement was observed on the elongation-at-break. The N$_2$ sample had a 12% elongation-at-break increase and the CO$_2$ sample had a 19% elongation-at-break increase, indicating that the samples with super-critical gas deformed more before fracturing than the regular sample.

During the mechanical tests, all parts ruptured in the same thin section on the side farthest from the gate (cf. Figure 3d). Elongation-at-break was mainly determined by the weakness of this point. This weakness was attributed to relatively low pressure at that point along the part during the packing cycle, as the pressure during the packing phase decreased from the gate to the end of the part. The dominant role of the melt pressure on the mechanical properties of injection-molded parts is a well-known phenomenon [43,44]. However, melt pressure values across the same part do not vary much for low viscosity materials. Here, the reason might be that the high viscosity of the UHMWPE caused a relatively high pressure variance, which resulted in weakness at the indicated far-end location on the tensile bar. While the N$_2$ samples exhibited a slightly higher Young's modulus, the CO$_2$ samples had higher values in terms of elongation-at-break and toughness. However, the differences between the N$_2$ and CO$_2$ samples were within the standard variation. The reason for this might have been that both gases had similar physical effects on the flow behavior.

3.3. Rheology

UHMWPE is normally selected as a material for its excellent impact strength and wear resistance based on its ultra-high molecular weight. This high molecular weight must be preserved to an acceptable level after thermal processing via a process like injection molding. The molecular weight of the final product should be checked to determine whether the process was successful or not. Due to the high molecular weight of UHMWPE and the difficulty of accessing a suitable solvent for the gel permeation chromatography (GPC) device, the molecular weight of the neat and injection-molded samples were analyzed in an indirect way by examining the change in melt viscosity. Moreover, a better understanding of the rheological characteristics of UHMWPE may help design processing parameters. Figure 4a displays the complex viscosity of neat and injection-molded UHMWPE samples.

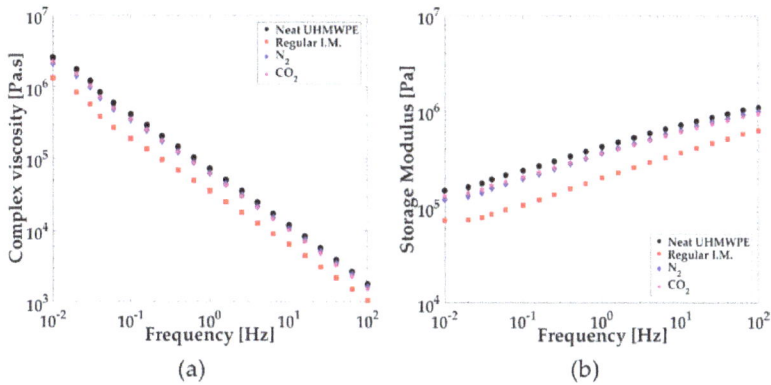

Figure 4. (**a**) Complex viscosity versus frequency; (**b**) Storage modulus versus frequency.

In the tested frequency range in Figure 4a, all UHMWPE samples fit the Power-Law model with a strong shear-thinning behavior and without a zero-shear rate Newtonian viscosity plateau. At the lowest frequency tested, 0.01 Hz, the complex viscosity of neat UHMWPE corresponded to 2.553×10^6 Pa·s. Regular injection-molded UHMWPE samples processed without any SCF had a complex viscosity of 1.324×10^6 Pa·s, which was about 48% lower. On the other hand, the viscosities of the N_2 and CO_2 samples were closer to the neat sample, with an 18% reduction for N_2 samples and a 12% reduction for CO_2 samples. This viscosity reduction was likely due to material degradation as a result of viscous heating, especially during the recovery or dosage cycles and the injection-molding process.

Thermal degradation lowers the molecular weight and broadens its distribution by breaking the polymer chains. Although increasing the amount of low-molecular-weight chains in the polymer typically improves flowability, it degrades the impact strength and wear resistance of the final products [39,45]. Based on our experimental observations, the thermal degradation of UHMWPE was minimized with proper machine and mold selection, such as using a smaller compression-ratio screw, small volume part, and a larger gate. Independent of this, the rheological results suggested that SCF helped minimize the thermal degradation of UHMWPE.

Figure 4b shows the storage moduli of neat and injection-molded UHMWPE samples. It can be seen that the storage moduli of the N_2 and CO_2 samples behaved similarly to the viscosity results, showing only a slight decrease compared to the neat sample, whereas the regular injection-molded sample displayed a comparatively large decrease.

Equation (3) shows the expression of viscous heating upon shear deformation:

$$\dot{Q}_{VH} = \eta \dot{\gamma}^2. \tag{3}$$

In the above equation, $\dot{Q}_{VH}$ is the rate of viscous heating, η represents the shear viscosity, and $\dot{\gamma}$ is the shear deformation rate [39,45]. Thermal degradation of a material that occurs during processing is largely the result of the viscous heating. Although all samples except the neat sample were produced using the same processing parameters, the viscous heating might have been lower for the N_2 and CO_2 samples due to the plasticizing effect of the SCF, which lowers viscosity [17,46].

The temperature dependence of the UHMWPE viscosity should be understood for successful manufacturing. For this purpose, a temperature sweep test was performed and reported in Figure 5 at two frequencies (0.1 and 100 Hz). Only a small temperature dependence was observed for the UHMWPE viscosity, as shown by the slope of the data points in Figure 5 for both frequencies tested. The temperature sweep tests suggested that increasing the temperature of the UHMWPE melt for processing ease was not an effective way of decreasing the viscosity of the polymer, despite its common practice.

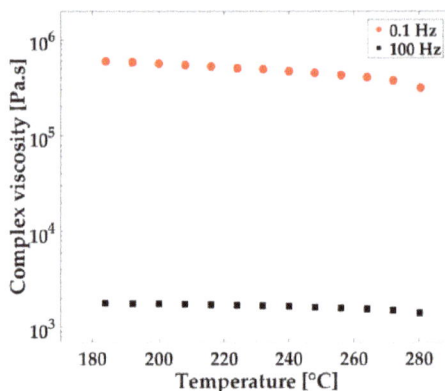

Figure 5. Complex viscosity versus temperature.

When temperature dependence, based on the temperature dependence relationship presented in Equation (2), was compared among common thermoplastics (HDPE, LDPE, Polystyrene (PS), Polypropylene (PP), Polyamide (PA66), Polycarbonate (PC), and Polyvinyl chloride (PVC)), from a general standpoint, the smallest temperature dependence coefficient, a, was reported as 0.002 °C^{-1} for HDPE [39]. According to the same model, the temperature dependence coefficient of UHMWPE was calculated as 0.001 °C^{-1}, about half that of HDPE's dependence. Low temperature dependence for a polymer can be a desirable property for a reliable injection-molding process. Therein, the polymer can maintain its viscosity within stable limits, despite factors like viscous heating or varying zone temperatures along the barrel. However, having a lower temperature dependence shifts the focus to the resistance to flow which, regardless of the melt temperature, becomes the major impediment in processing UHMWPE.

3.4. Tensile Bar Images and Injection Pressure Measurements

Images of complete and short-shot injection-molded UHMWPE parts are shown in Figure 6a,b. The gate location was on the right side of the parts. Although the appearance of the complete parts in Figure 6a was identical for all sample types, the short-shot images in Figure 6b showed a unique difference in flow behavior between the regular injection-molded samples and the SCF samples. It is well-known that many engineering plastics exhibit so-called "fountain flow" behavior [47]. However, in Figure 6b, it can be seen that the UHMWPE melt failed to demonstrate fountain flow behavior. This was because the material slipped at the mold surface as a result of its high viscosity and very low friction properties as well as the severe melt fracture that occurred as the melt entered the cavity under high stress. As a result, the flow front of the regular short-shot sample formed an irregular porous structure that became less porous near the gate. This might have been due to the huge pressure near the gate region where the part transitioned from a porous-like structure to a continuous-like structure. That is, the high-pressure development in the gate area caused the polymer to form a solid structure near the gate of the regular injection-molded part. Interestingly, it was observed that the N_2 and CO_2 samples in Figure 6b filled the entire cavity with a completely porous-like, foam-like structure, even near the gate side, as a result of gas expansion and venting. Such a porous structure was later compacted into solid-like parts (cf. Figure 6a) through packing pressure during the packing stage.

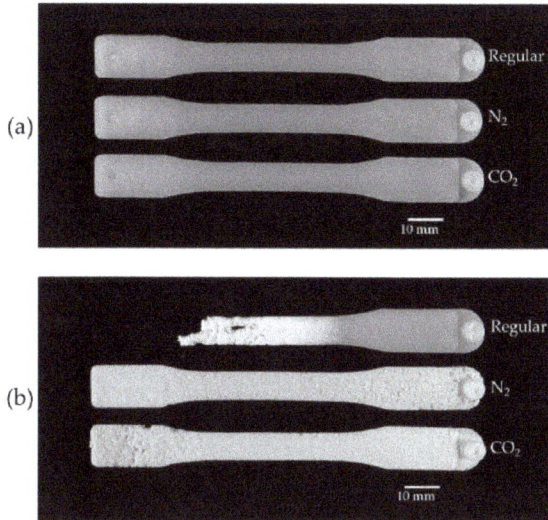

Figure 6. Images of injection-molded UHMWPE samples: (**a**) complete parts and (**b**) short-shot samples.

Figure 7 shows the injection pressure values versus flow rate on a linear scale axis. There was a noticeable decrease between the regular injection-molded and SCF-loaded samples. The N_2 samples yielded slightly lower pressure values compared to the CO_2 samples.

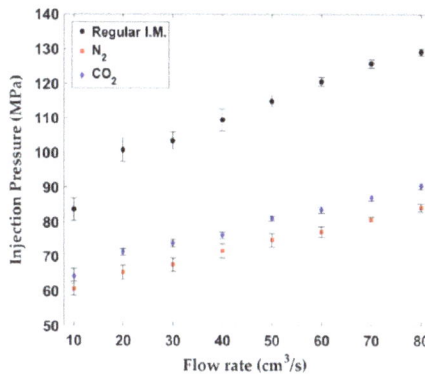

Figure 7. Injection pressure versus flow rate.

The shapes of the curves, especially from moderate to high flow rates, seemed linear. For the 80 cm^3/s flow rate used to create the injection-molded samples, the average pressure reduction was about 30% for CO$_2$ samples and 35% for the N$_2$ samples.

This pressure reduction might have been due to the viscosity reduction effects of the dissolved SCF in the polymer melt. Previous studies have reported this effect for scN$_2$ and scCO$_2$ on various PE grades [28,29,46].

Other contributions to the pressure drop may have been that the material changed its flow behavior as demonstrated in Figure 6b. It seems that SCF decreased the material's tendency to agglomerate and stick to the mold surface, thus resulting in less pressure development. Furthermore, the escaping high-pressure gas during the filling stage may have caused a lubrication effect and slipping between the UHMWPE part and the mold surface.

Another reason for the pressure reduction might have been undissolved gas bubbles in the screw chamber or emerging gas bubbles in the mold cavity, which may also explain why scN$_2$ reduced the pressure slightly more than scCO$_2$. In microcellular injection molding (MIM), until the injection cycle begins, it is assumed that finely distributed micro-gas bubbles completely dissolve in the polymer melt and create a single-phase solution. However, in practice, the single-phase polymer/gas solution may not form completely at certain molding settings over a typical recovery time [16]. As long as the back pressure is kept above the super critical pressure, the residual micro-bubbles will not affect the MIM process [16].

The gas solubility in polymers is a strong function of pressure and temperature. Based on our experimental observations, it was estimated that the residual microbubbles when scN$_2$ was used were more numerous than with the scCO$_2$ case because of the lower solubility of scN$_2$ in the UHMWPE melt. To the best of our knowledge, gas solubility and diffusion data are not available for UHMWPE. For one grade of HDPE, the gas solubility was reported as 1.2 wt % for scN$_2$ and 4.6 wt % for scCO$_2$ at 270 °C [16]. Similarly, as gas emerges from the polymer/gas solution during injection, numerous bubbles form due to thermodynamic instability. Since scN$_2$ has a lower solubility in UHMWPE, it will have a higher degree of supersaturation for scN$_2$ given the same gas loading at 1.5 wt % (cf. Table 1). Thus, more gas bubbles are expected to form for the N$_2$ samples. Either way, tiny bubbles with relatively negligible viscosity may reduce the average flow resistivity of the melt-gas mixture.

3.5. Micro-Computed Tomography (µCT)

Although µCT resolution (5 µm) is limited when compared to electron microscopy methods, as a nondestructive method, it enables us to scan large volumes conveniently instead of only a limited area of the fracture surface. It also allows one to effectively analyze and select the most representative cross-section images. µCT images of injection-molded samples are shown in Figure 8.

Remaining bubbles in injection-molded parts can cause some cosmetic and mechanical defects. When SCF is used, remaining bubbles can be a challenge to eliminate. During cooling, the shrinkage of the polymer can create some random bubbles in that part. However, the injection molding images in Figure 6b show that UHMWPE underwent foaming and allowed gas to emerge and escape. Even if there is a relatively small amount of gas remaining, it is easy to pack the part and eliminate the bubbles for both N$_2$ and CO$_2$ samples, resulting in solid injection-molded parts free of any residual bubbles as shown in Figure 8. In µCT images of all injection-molded samples, the corners of the cross sections images demonstrated unexpectedly sharp and light colored edges. This might have been due to the exponential edge-gradient effect observed as a defect of the µCT mathematical reconstruction algorithm whenever long sharp edges of high contrast are encountered [48].

Polymers **2018**, *10*, 36

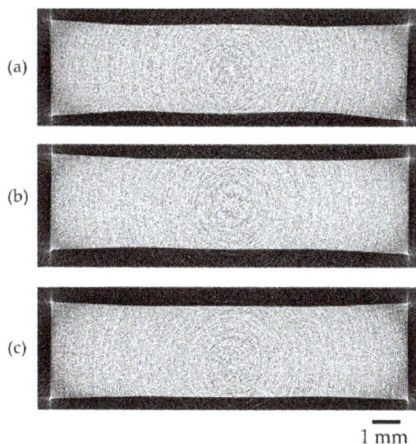

Figure 8. Micro-Computed Tomography (μCT) images of injection-molded samples: (**a**) regular injection-molded sample; (**b**) N_2 sample; and (**c**) CO_2 sample.

4. Conclusions

In summary, UHMWPE was processed via microcellular injection molding with supercritical N_2 or CO_2 as a reversible plasticizer and compared with regular injection-molded samples. It was found that both scN_2 and $scCO_2$ reduced the required injection pressure for the chosen material flowrate during injection. The flow behavior was also found to be different when scN_2 and $scCO_2$ were used. UHMWPE failed to demonstrate typical fountain flow behavior; instead, it slipped on the mold surface. While DSC results showed a small difference in terms of the degree of crystallization and melting temperature among all samples, the rheological tests showed that the complex viscosity and storage modulus of the N_2 and CO_2 samples were closer to that of the neat sample and higher than the regular injection-molded sample. This suggests that some thermal degradation may have occurred during processing, but scN_2 and $scCO_2$ helped reduce the overall degradation due to the plasticizing effect in the barrel and injection molding cycle. In addition, the rheological tests showed that increasing the melt temperature would not reduce the viscosity. From the mechanical tests, it was found that CO_2 samples had the most favorable elongation-at-break results while N_2 samples had the highest Young's modulus. Finally, μCT images showed that N_2 and CO_2 samples were free of any residual bubbles.

Acknowledgments: The authors gratefully acknowledge the support from the Wisconsin Institute for Discovery (WID), the Kuo K. and Cindy F. Wang Professorship, the College of Engineering, and Office of Vice Chancellor for Research and Graduate Education at the University of Wisconsin-Madison. The UHMWPE was generously donated by the Celanese Corporation (Irving, TX, USA).

Author Contributions: Lih-Sheng Turng directed the research and conceived and designed the experiments. Galip Yilmaz carried out all experimental work, analyzed the data, and wrote the manuscript. Thomas Ellingham participated in the interpretation of the research, data analysis, and manuscript writing.

Conflicts of Interest: The authors declare no conflict of interest.

References

1. Peacock, A.J. *Handbook of Polyethylene: Structures, Properties and Applications*; Marcel Dekker: New York, NY, USA, 2000; ISBN 0471166286.
2. Edidin, A.A.; Kurtz, S.M. Influence of mechanical behavior on the wear of 4 clinically relevant polymeric biomaterials in a hip simulator. *J. Arthroplast.* **2000**, *15*, 321–331. [CrossRef]
3. Osswald, T.A.; Baur, E.; Brinkmann, S.; Oberbach, K.; Schmachtenberg, E. *International Plastics Handbook*; Hanser: München, Germany, 2006; ISBN 9781569903995.

4. Kurtz, S.M. *The UHMWPE Biomaterials Handbook: Ultra-High Molecular Weight Polyethylene in Total Joint Replacement and Medical Devices*, 2nd ed.; Academic Press: Burlington, MA, USA, 2009; Volume 2, ISBN 978-0-12-374721-1.

5. McKellop, H.; Shen, F.; Lu, B.; Campbell, P.; Salovey, R. Development of an extremely wear-resistant ultra high molecular weight polythylene for total hip replacements. *J. Orthop. Res.* **1999**, *17*, 157–167. [CrossRef] [PubMed]

6. Schaller, R.; Feldman, K.; Smith, P.; Tervoort, T.A. High-Performance Polyethylene Fibers "Al Dente": Improved Gel-Spinning of Ultrahigh Molecular Weight Polyethylene Using Vegetable Oils. *Macromolecules* **2015**, *48*, 8877–8884. [CrossRef]

7. Mourad, A.-H.I.; Fouad, H.; Elleithy, R. Impact of some environmental conditions on the tensile, creep-recovery, relaxation, melting and crystallinity behaviour of UHMWPE-GUR 410-medical grade. *Mater. Des.* **2009**, *30*, 4112–4119. [CrossRef]

8. Xu, M.M.; Huang, G.Y.; Feng, S.S.; McShane, G.J.; Stronge, W.J. Static and dynamic properties of semi-crystalline polyethylene. *Polymers* **2016**, *8*, 77. [CrossRef]

9. Heim, H.-P. *Specialized Injection Molding Techniques*, 1st ed.; William Andrew: Oxford, UK, 2015; ISBN 9780323371216.

10. Osswald, T.A.; Turng, L.-S.; Gramann, P.J. *Injection Molding Handbook*, 2nd ed.; Hanser: München, Germany, 2008; ISBN 9781569904206.

11. Bryce, B.D.M. *Plastic Injection Molding Manufacturing Process Fundamentals*; Society of Manufacturing Engineers: Dearborn, MI, USA, 1996; Volume I, ISBN 0872634728.

12. Kuo, H.C.; Jeng, M.C. The influence of injection molding and injection compression molding on ultra-high molecular weight polyethylene polymer microfabrication. *Int. Polym. Process.* **2011**, *26*, 508–516. [CrossRef]

13. Kuo, H.-C.; Jeng, M.-C. The influence of injection molding on tribological characteristics of ultra-high molecular weight polyethylene under dry sliding. *Wear* **2010**, *268*, 803–810. [CrossRef]

14. Kuo, H.-C.; Jeng, M.-C. Effects of part geometry and injection molding conditions on the tensile properties of ultra-high molecular weight polyethylene polymer. *Mater. Des.* **2010**, *31*, 884–893. [CrossRef]

15. Okamoto, K.T. *Microcellular Processing*; Hanser: München, Germany, 2003; ISBN 1569903441.

16. Xu, J. *Microcellular Injection Molding*; Wiley: Hoboken, NJ, USA, 2010; ISBN 9781118057872.

17. Kikic, I. Polymer-supercritical fluid interactions. *J. Supercrit. Fluids* **2009**, *47*, 458–465. [CrossRef]

18. Hernández-Ortiz, J.C.; Van Steenberge, P.H.M.; Reyniers, M.F.; Marin, G.B.; D'hooge, D.R.; Duchateau, J.N.E.; Remerie, K.; Toloza, C.; Vaz, A.L.; Schreurs, F. Modeling the reaction event history and microstructure of individual macrospecies in postpolymerization modification. *AIChE J.* **2017**, *63*, 4944–4961. [CrossRef]

19. Vrentas, J.S.; Duda, J.L. Diffusion in Polymer-Solvent Systems. I. Reexamination of the Free-Volume Theory. *J. Polym. Sci. Polym. Phys. Ed.* **1977**, *15*, 403–416. [CrossRef]

20. D'Hooge, D.R.; Van Steenberge, P.H.M.; Reyniers, M.F.; Marin, G.B. The strength of multi-scale modeling to unveil the complexity of radical polymerization. *Prog. Polym. Sci.* **2016**, *58*, 59–89. [CrossRef]

21. Nalawade, S.P.; Picchioni, F.; Janssen, L.P.B.M. Supercritical carbon dioxide as a green solvent for processing polymer melts: Processing aspects and applications. *Prog. Polym. Sci.* **2006**, *31*, 19–43. [CrossRef]

22. Sun, X.; Turng, L.S. Novel injection molding foaming approaches using gas-laden pellets with N_2, CO_2, and $N_2 + CO_2$ as the blowing agents. *Polym. Eng. Sci.* **2014**, *54*, 899–913. [CrossRef]

23. Tomasko, D.L.; Burley, A.; Feng, L.; Yeh, S.-K.; Miyazono, K.; Nirmal-Kumar, S.; Kusaka, I.; Koelling, K. Development of CO_2 for polymer foam applications. *J. Supercrit. Fluids* **2009**, *47*, 493–499. [CrossRef]

24. Wingert, M.J.; Shen, J.; Davis, P.M.; Lee, L.J.; Tomasko, D.L.; Koelling, K.W. Rheological Studies of Polymers under High Pressure Carbon Dioxide. In Proceedings of the Society of Plastics Engineers Annual Technical Conference (ANTEC), Boston, MA, USA, 1–5 May 2005; pp. 1143–1147.

25. Garcia-Leiner, M.; Song, J.; Lesser, A.J. Drawing of ultrahigh molecular weight polyethylene fibers in the presence of supercritical carbon dioxide. *J. Polym. Sci. Part B Polym. Phys.* **2003**, *41*, 1375–1383. [CrossRef]

26. Kiran, E. Supercritical fluids and polymers—The year in review—2014. *J. Supercrit. Fluids* **2016**, *110*, 126–153. [CrossRef]

27. Ellingham, T.; Duddleston, L.; Turng, L.-S. Sub-critical gas-assisted processing using CO_2 foaming to enhance the exfoliation of graphene in polypropylene + graphene nanocomposites. *Polymer* **2017**, *117*, 132–139. [CrossRef]

28. Wilding, M.D.; Baird, D.G.; Eberle, A.P.R. Melt Processability and Foam Suppression of High Molecular Weight Polyethylenes Plasticized with Supercritical Carbon Dioxide. *Int. Polym. Process.* **2008**, *23*, 228–237. [CrossRef]
29. Hsu, C.-L.; Turng, L.-S.; Osswald, T.A.; Rudolph, N.; Dougherty, E.; Gorton, P. Effects of Pressure and Supercritical Fluid on Melt Viscosity of LDPE in Conventional and Microcellular Injection Molding. *Int. Polym. Process.* **2012**, *27*, 18–24. [CrossRef]
30. Sun, X.; Kharbas, H.; Peng, J.; Turng, L.S. A novel method of producing lightweight microcellular injection molded parts with improved ductility and toughness. *Polymer* **2015**, *56*, 102–110. [CrossRef]
31. Kharbas, H.A. *Developments in Microcellular Injection Molding Technology*; University of Wisconsin—Madison: Madison, WI, USA, 2003.
32. Turng, L.-S.; Kharbas, H. Development of a Hybrid Solid-Microcellular Co-injection Molding Process. *Int. Polym. Process.* **2004**, *19*, 77–86. [CrossRef]
33. Kharbas, H.; Ellingham, T.; Turng, L.-S. Use of core retraction to achieve low density foams in microcellular injection molded polypropylene parts. In Proceedings of the Technical Conference & Exhibition, Indianapolis, IN, USA, 23–25 May 2016; pp. 1285–1290.
34. Gong, S.; Yuan, M.; Chandra, A.; Kharbas, H.; Osorio, A.; Turng, L.S. Microcellular Injection Molding. *Int. Polym. Process.* **2005**, *20*, 202–214. [CrossRef]
35. Shaayegan, V.; Wang, G.; Park, C.B. Effect of foam processing parameters on bubble nucleation and growth dynamics in high-pressure foam injection molding. *Chem. Eng. Sci.* **2016**, *155*, 27–37. [CrossRef]
36. Shaayegan, V.; Wang, C.; Costa, F.; Han, S.; Park, C.B. Effect of the melt compressibility and the pressure drop rate on the cell-nucleation behavior in foam injection molding with mold opening. *Eur. Polym. J.* **2017**, *92*, 314–325. [CrossRef]
37. Liang, J.Z.; Ness, J.N. The calculation of cooling time in injection moulding. *J. Mater. Process. Technol.* **1996**, *57*, 62–64. [CrossRef]
38. Wunderlich, B.; Cormier, C.M. Heat of fusion of polyethylene. *J. Polym. Sci. Part A Polym. Phys.* **1967**, *5*, 987–988. [CrossRef]
39. Osswald, T.A.; Rudolph, N. *Polymer Rheology: Fundamentals and Applications*; Hanser: Munich, Germany, 2015; ISBN 978-1-56990-517-3.
40. Pang, W.; Ni, Z.; Chen, G.; Huang, G.; Huang, H.; Zhao, Y. Mechanical and thermal properties of graphene oxide/ultrahigh molecular weight polyethylene nanocomposites. *RSC Adv.* **2015**, *5*, 63063–63072. [CrossRef]
41. Ergoz, E.; Fatou, J.G.; Mandelkern, L. Molecular Weight Dependence of the Crystallization Kinetics of Linear Polyethylene. I. Experimental Results. *Macromolecules* **1972**, *5*, 147–157. [CrossRef]
42. Yasuniwa, M.; Tsubakihara, S.; Nakafuku, C. Molecular Weight Effect on the High Pressure Crystallization of Polyethylene. *Polym. J.* **1988**, *20*, 1075–1082. [CrossRef]
43. Kubáut, J.; Månson, J.-A.; Rigdahl, M. Influence of mold design on the mechanical properties of high-pressure injection—molded polyethylene. *Polym. Eng. Sci.* **1983**, *23*, 877–882. [CrossRef]
44. Kalay, G.; Sousa, R.A.; Reis, R.L.; Cunha, A.M.; Bevis, M.J. The enhancement of the mechanical properties of a high-density polyethylene. *J. Appl. Polym. Sci.* **1999**, *73*, 2473–2483. [CrossRef]
45. Osswald, T.A.; Menges, G. *Material Science of Polymers for Engineers*; Hanser: München, Germany, 2012; ISBN 9781569905142.
46. Sauceau, M.; Fages, J.; Common, A.; Nikitine, C.; Rodier, E. New challenges in polymer foaming: A review of extrusion processes assisted by supercritical carbon dioxide. *Prog. Polym. Sci.* **2011**, *36*, 749–766. [CrossRef]
47. Beaumont, J.P. *Runner and Gating Design Handbook: Tools for Successful Injection Molding*; Hanser: München, Germany, 2004; ISBN 9783446226722.
48. Joseph, P.M.; Spital, R.D. The exponential edge-gradient effect in X-ray computed tomography. *Phys. Med. Biol.* **1981**, *26*, 473–487. [CrossRef] [PubMed]

![polymers logo] *polymers*

MDPI

Article

Mechanical Properties Distribution within Polypropylene Injection Molded Samples: Effect of Mold Temperature under Uneven Thermal Conditions

Sara Liparoti [1], Vito Speranza [1,*], Andrea Sorrentino [2] and Giuseppe Titomanlio [1]

[1] Department of Industrial Engineering, University of Salerno, via Giovanni Paolo II, 132, 84084 Fisciano, Italy; sliparoti@unisa.it (S.L.); gtitomanlio@unisa.it (G.T.)

[2] Institute for Polymers, Composites and Biomaterials (IPCB-CNR), Via Previati, 1/C, 23900 Lecco, Italy; andrea.sorrentino@cnr.it

* Correspondence: vsperanza@unisa.it; Tel.: +39-089-964145

Received: 13 October 2017; Accepted: 3 November 2017; Published: 7 November 2017

Abstract: The quality of the polymer parts produced by injection molding is strongly affected by the processing conditions. Uncontrolled deviations from the proper process parameters could significantly affect both internal structure and final material properties. In this work, to mimic an uneven temperature field, a strong asymmetric heating is applied during the production of injection-molded polypropylene samples. The morphology of the samples is characterized by optical and atomic force microscopy (AFM), whereas the distribution of mechanical modulus at different scales is obtained by Indentation and HarmoniX AFM tests. Results clearly show that the temperature differences between the two mold surfaces significantly affect the morphology distributions of the molded parts. This is due to both the uneven temperature field evolutions and to the asymmetric flow field. The final mechanical property distributions are determined by competition between the local molecular stretch and the local structuring achieved during solidification. The cooling rate changes affect internal structures in terms of relaxation/reorganization levels and give rise to an asymmetric distribution of mechanical properties.

Keywords: indentation; Harmonix AFM; polymer morphology; mechanical properties

1. Introduction

Understanding the effect of the process on the internal morphology of polymeric artifacts and its relationship with the final mechanical properties is the key-point for a correct prediction of reliability, performance and durability of a product. The complexity of the internal features, developed in the micro and sub-micro scales during common processing of thermoplastic polymers, requires additional studies on the relationships between morphology and properties. Consistently, characterization tests on micro and sub-micro scales, such as micro and nano indentation, are emerging as new approaches to provide quantitative information about the mechanical properties and morphology of polymeric materials [1,2]. The ability to measure properties on the nanometric length-scale is particularly important for objects such as molded parts in which the localized internal structures can have a significant impact on the bulk properties [3–8].

Recently, a multiscale mechanical characterization of injection-molded samples was performed by combining dynamic mechanical analysis, micro Indentation, and HarmoniX Atomic Force Microscopy tests [9,10]. Those results show that the molded samples present a complex multilayer morphology in which, starting from the sample surface, less organized structures (globules), characterized by mechanical properties similar to the quiescent amorphous phase, coexist with well-organized and oriented structures (fibrils), characterized by a high mechanical modulus. Intermediate behavior was found in the central part of the sample, where well developed spherulitic structures are characterized by a lower level of orientation [11,12].

The morphology developed in the molded parts can be highly dependent on imperfect or wrong processing conditions. They can be due to problems related to a combination of poor material characterization, imperfect tooling design, and inadequate control of the processing variables [13–16]. Knowing how intentionally created molding defects impact on the final product quality can help to recognize and avoid them in real conditions. Probably, among other parameters, the mold temperature is the most critical parameter to be monitored and controlled [17,18]. Despite the extensive studies, a clear correlation between the product properties and the mold temperature is not yet completely established. The situation becomes still more complex when a complete and accurate temperature control is difficult or even impossible. As a matter, non-controlled asymmetrical temperature fields are quite common when geometrical constraints, non-balance (or failure) of the cooling systems, and local overheating are present. In some special cases, high-temperature gradients are imposed by specific production requests such as micro-injection, in-mold labeling and over-molding [19,20]. Regardless of the specific reason for the non-uniform temperature, the results may give rise to severe problems such as inhomogeneous filling, low cycle repeatability, warpage, a lack of part performances and poor surface finishing [21,22]. The uneven temperature field may produce a sample with strong mechanical unbalancing with the possibility of warpage and the formation of cracks.

Liparoti et al. [12,23] reported the effect of induced asymmetrical fast surface temperature evolution on the morphology of injection molded isotactic polypropylene (iPP) samples. They showed a strong effect of two factors on the morphology of the final part: increasing the cavity surface temperature and varying the temperature pulse duration [24]. Despite the continuous technological interest in this subject, only a few papers have tried to correlate the complex internal morphology produced in these conditions with the local mechanical properties within the molded parts.

The objective of this work is to provide an experimental morphological and mechanical characterization of molded samples obtained in highly asymmetrical mold temperature conditions. The mechanical properties along the sample thickness have been correlated to the mold temperature condition adopted. The correlation among processing conditions, mechanical properties and internal morphology of the moldings are also analyzed and discussed.

2. Materials and Methods

Thin bars of isotactic polypropylene (iPP) with a rectangular section were obtained by injection molding, using a mold cavity that can impose high temperature gradients in a very short time. The iPP adopted for the experiments was supplied by Montell (Ferrara, Italy), now Lyondell Basell Industries. It is a commercial grade (tradename Moplen T30G). Moplen T30G is a general purpose homopolymer for extrusion/molding applications, with a melt flow index of 3.6 (ASTM D1238/L). The molecular weight distribution was determined by a size exclusion chromatography, with weight-average molar mass M_w of 376,000 g·mol^{-1}, polydispersity index M_w/M_n of 6.7, and a meso pentads content 87.6%.

Figure 1 shows the geometry of the samples and the sketch of the heating device. In particular, the mold has a gate 1 mm thickness and a rectangular cavity of length $L = 70$ mm, width $W = 20$ mm, and thickness $S = 1$ mm.

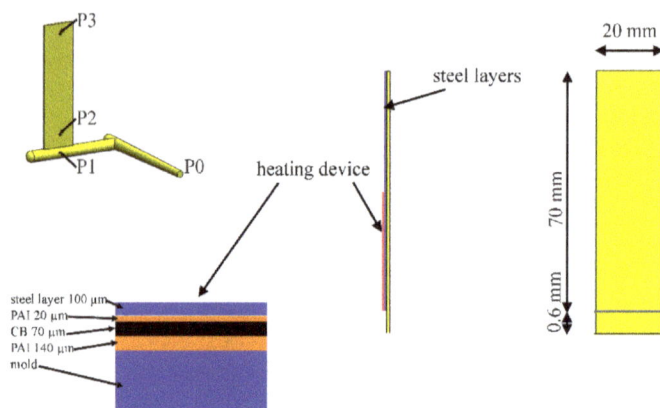

Figure 1. Sketch of the cavity geometry and of the heating device adopted. Positions P0 ÷ P3 where the pressure evolutions were measured during the injection molding tests are also indicated. (PAI = poly(amide-imide) insulating layer; CB = electrical conductive layer, made of carbon black loaded PAI).

The injection molding tests were carried out with a 70-ton Canbio reciprocating screw, injection-molding machine (Negri-Bossi, Cologno Monzese, Milan, Italy). The experiments were performed with an average volumetric flow rate of 3 cm$^3 \cdot$s^{-1} (the cavity filling time was about 0.5 s), a melt injection temperature of 220 °C and a mold temperature of 20 °C. A holding pressure of 300 bar was adopted with a holding time of 2 s. The pressure was measured in four different positions along the flow path by means of piezoelectric pressure transducers. In particular, one transducer is located in the injection chamber, P0, one just before the gate, P1, and two in the cavity, P2 and P3 (20 and 50 mm downstream from the gate). Their locations are indicated in Figure 1. As also shown in Figure 1, a thin heater is located below the cavity surface and it is protected by a thin steel layer (100 μm). The thin heater was adopted for quickly and accurately controlling the temperature on one side of the cavity (on the moving plate of the mold). A detailed description of the heating device is reported elsewhere [25]. The melt temperature in position P2 was measured with a temperature probe located on the protective steel layer.

The molding conditions adopted in the tests considered in this work are summarized in Table 1. During the injection molding tests, the heating device was supplied by a constant electrical power, P_e. The heating device was activated for two seconds, t_a, before the local contact with the polymer in position P2, where pressure and temperature were measured. After that, the electrical power was supplied for additional heating times, t_h, 0.5 s, 8 s and 18 s. The asymptotic temperature, T_{level}, reached on the cavity surface when the heater is supplied with 7 W/cm^2, for longer heating time, is 120 °C. For comparison, a non-conditioned experiment, T_off, was performed without activating the heater (in this case T_{level} is the ejection temperature, 28 °C).

Table 1. Experimental molding conditions.

Test Run	P_e (W/cm^2)	T_{level} (°C)	t_h (s)	t_a (s)	P_{hold} (bar)
T_off	0	28	0	0	300
120-05	7	120	0.5	2	300
120-8	7	120	8	2	300
120-18	7	120	18	2	300

2.1. Sample Preparation

The specimens were cut in position P2 from the cross section of the sample and carefully polished with a predefined procedure [9]. For the Atomic Force Microscopy (AFM) characterizations, an additional chemical etching procedure was carried out on the specimens in order to eliminate local micro deformation induced by the cutting procedures. In particular, the Basset's etched procedure [12,23,26] was adopted to prepare selected samples to the AFM analysis. A solution of potassium permanganate, in a mixture of 10:4:1 volumes of concentrated sulphuric acid, orthophosphoric acid and distilled water (1 g of potassium permanganate in 100 mL of mixture), was adopted as etchant. A 2 h period of etching at room temperature was generally sufficient to reveal the surface topography.

2.2. Optical Microscopy Tests

In order to gather information on sample morphology, cross section slices with 0.1 mm thickness were cut from injected samples by means of a Leica slit microtome in position P2. These samples were observed in Polarized Optical Microscopy (POM) with an optical microscope (model Olympus BX41) equipped with a digital camera by using a magnification level of 20X. Micrographs of the slices were taken so that flow direction was rotated of 45° with respect the analyzer direction.

2.3. Indentation Tests

Indentation tests were performed with a Nano Test™ Platform (Micro Materials Ltd., Wrexham, UK). Tests were carried out with an initial load of 0.02 mN, a load rate of 1 mN/s and a maximum load of 120 mN held for 60 s. A three-sided Berkovich pyramidal diamond tip with a radius of about 100 µm was adopted to indent the surface of the sample. Elastic modulus were evaluated from each imprint, directly by Nano Test™ Platform software 4, according to the Oliver and Pharr method [27].

2.4. AFM Analysis

AFM analyses were performed by a NanoScope MultiMode V scanning probe microscope (Veeco, Santa Barbara, CA, USA) equipped with HarmoniX tool. Tests were performed with HMX probe silicon cantilevers (Bruker, Billerica, MA, USA) with nominal radii of c.a. 10 nm. The cantilever oscillation is composed by two different movements, torsional and vertical. These movements have different frequencies, in particular the amplitude frequency of torsional movement is higher than the tapping frequency [28,29]. The reconstruction of sample morphology is due to the vertical movements in standard tapping mode, whereas, the reconstruction of elastic modulus maps is due to the tip sample force interactions during the torsional movement [28]. Lastly, the AFM elastic modulus distribution is obtained averaging the elastic moduli on windows of 20 × 20 µm by the Nanoscope software version 7.30.

HarmoniX measurements were done in air. Cantilevers were calibrated using a standard polystyrene/low density polyethylene (PS/LDPE) sample. The adopted vertical frequency was 44 kHz and the torsional frequency was 989 kHz. The cantilever vibration free amplitude was of 750 mV in air. The force level was modulated adopting the amplitude set point. The amplitude set point was used for feedback control, as a reference value; in particular ca. 60% of the free amplitude was selected. Imaging was performed with 0.5 Hz scan rates, considering 20 harmonics.

The calibration method reported by Gojzewski et al. [4] allows evaluating the deflection sensitivity and cantilever spring constants that were 360.6 nm/V and 1.58 N/m respectively.

Image processing and data analysis were performed with the NanoScope software version 7.30 and NanoScope Analysis version 1.20. The NanoScope software gives elastic modulus maps by elaborating HarmoniX AFM imaging through Derjaguin-Muller-Toporov (DMT) model [30].

3. Results

3.1. Molding Experiments

An example of the temperature evolutions recorded during the molding experiments is reported in Figure 2. In the same figure, the temperature evolution recorded for the experiment T_off is also reported. The activation of the electric heater starts two seconds before the melt reaches position P2 ($t < 3.7$ s), during such a time the temperature rises up to about 120 °C. At this point, the temperature increase up to about 150 °C, is due to the contact of the hot melt with the cavity surface. Obviously, the temperature recorded for the experiment T_off only shows the temperature increase due to the hot melt contact. In all cases, after a peak the temperature starts to decrease down to the selected T_{level}, i.e. 120 °C for tests performed with the heater activated and 28 °C for the experiment T_off. For the tests performed with the heater activated, soon after the heater deactivation, the temperature decreases quickly to 28 °C. The cooling phase duration depends only slightly on the duration of the heating stage.

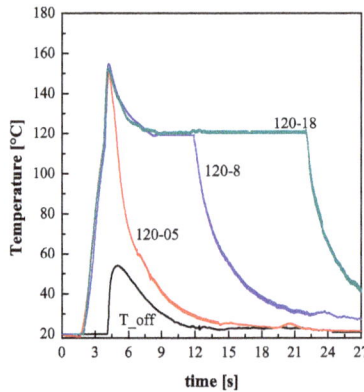

Figure 2. Temperature evolutions measured during tests reported in Table 1.

Figure 3 shows the pressure evolutions recorded in different positions along the flow path of two selected experiments, T_off and 120-18. After the filling stage (that ends after 4.5 s), the pressure undergoes a sudden increase upwards, reaching a sharp peak followed by a quick decrease toward the holding value. The peak in the pressure evolutions is higher in the T_off experiment because the pressure required to fill the cavity is higher at ambient temperature. As soon as the holding pressure in P0 is released ($t = 6.5$ s), the pressure in P1 quickly reduces. Half a second after the holding pressure release, the screw rotates again and moves back to its starting position, applying a back-pressure for about 0.5 s (batching step). As the melt viscosity in the sprue and runners is low enough, the pressure evolution in P1 follows the pressure in P0 showing a similar back-pressure spike. On the opposite side, the viscosity in the cavity is so high to prevent any back-pressure effect on the pressure evolutions in position P2 and P3.

Apart from some minor differences, the high temperature reached on one side of the sample surface seems to have only marginal effect on the pressure evolution recorded on the unheated side. A similar finding was described in previous works [12]. The pressure evolution in a certain position depends on the temperature distribution along the flow path. Comparing the pressure evolution of the T_off test with the 120-18 test it is evident that in the T_off case the pressure drop between position P2 and P3 is significantly higher. This is due to the fact that the melt cools down faster since it is in contact with a colder surface.

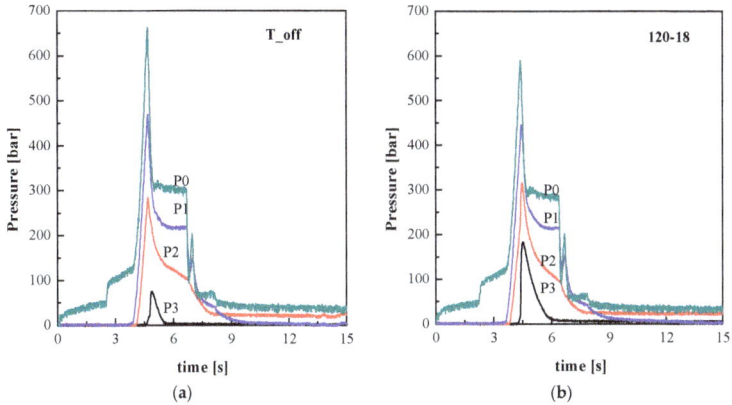

Figure 3. Pressure evolutions measured in different positions along the flow path for the tests (**a**) T_off and (**b**) 120-18.

3.2. Optical Morphology

A preliminary analysis of morphology developed along the sample thickness allows observing some of the effects of an asymmetric temperature field. An example of cross-section morphology obtained by the optical micrograph of the sample T_off is shown at the top of Figure 4. The optical image in polarized light turns out to be symmetric for that sample and it is characterized by alternating colored bands. Each band is characteristic of a polymer layer with different levels of crystallinity and orientation. These layers were formed at different times of the process and thus experienced different thermomechanical histories before solidification. The competition between solidification and the relaxation process produces a characteristic distribution of orientation through the thickness of the sample. In the literature, the presence of three distinct morphologies is reported: a thin, oriented "skin layer" at the sample surface, formed by elongated structures at the sample wall and by less oriented "globular" elements soon after [12], an highly oriented fibrillar "shear layer", just adjacent to the skin, and a less oriented layer, the "spherulitic" one, in the sample core [12,23,31,32].

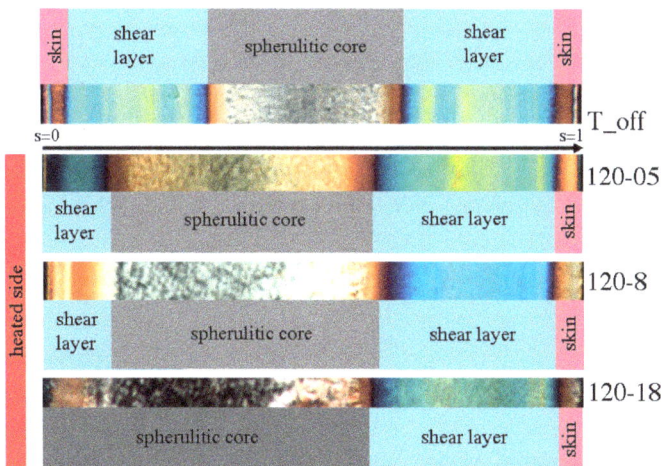

Figure 4. Optical micrographs of the samples from test runs T_off, 120-05, 120-8 and 120-18. The heated side is reported on the left most of the figure.

The morphologies of the samples obtained keeping the cavity surface temperature at 120 °C for 0.5, 8 and 18 s are compared in Figure 4. The different layers developed along the sample thickness are strongly influenced by the temperature of the cavity surface. In particular, the shear layer moves toward the sample surface and its thickness decreases by increasing the heating time. The skin layer disappears for longer heating times, while with a heating time of 0.5 s the presence of the skin layer can not be clearly evidenced from the POM micrograph. Figure 4 shows that a 0.5 s heating time is sufficient to lose the symmetric morphology of the sample.

3.3. AFM Morphology

A deeper investigation of the local morphology has been carried out by means of atomic force microscopy. Since HarmoniX AFM allows the elastic modulus on the sample surface to be mapped, it is possible to attempt a correlation between the morphology distribution and the mechanical properties along the sample thickness. The elastic modulus has been obtained from force-indentation curves applying the Derjaguis-Muller-Toporov (DMT) model [31]. Examples of the sample characterization are reported in Figures 5 and 6. These figures show both the POM image of the whole thickness of the sample and some of the height and the elastic modulus maps acquired by HarmoniX AFM for the tests 120-8 and 120-18, respectively. The dimension of each AFM map has been selected to take into account the characteristic dimension of the morphological structures that are observed in the sample.

The micro structures developed in injection-molded parts are generally anisotropic and largely dependent on stretch distribution, molecular orientation, and cooling rate induced by the molding cycle. Comparing Figures 5 and 6 one can observe that the morphology developed on the unheated side, where the temperature was at 28 °C, is essentially independent of the heating time. In particular, the AFM maps show that in the shear layer the fibrils are well packed with a mean thickness of about 250 ± 30 nm. On the heated side, the skin is not detectable for both selected heating times; the shear layer, detectable for the sample 120-8 of Figure 5, is made up of fibrils that have a mean thickness (500 ± 50 nm) that is bigger than the thickness of the fibrils on the unheated side. Figure 6, which refers to a longer heating time, shows that, on the heated side, the fibrils of the shear layer completely disappear and the spherulitic layer impinges on the sample surface. Evidently, in that case, the stretch due to the flow has sufficient time to relax the molecular orientation. Thus, the molecules crystalize into a spherulitic morphology during cooling to ambient temperature. Additionally, Figure 6 shows that there is a significant difference of the spherulite dimensions within the spherulitic layer. The spherulites that are formed close to the heated side appear smaller (15 μm diameter near the heated wall and 25 μm in the inner part of the layer). The reason for this difference can be due to the different cooling rates experienced by the melt when the heater is deactivated: close to the sample wall the cooling rate is higher; this favors higher nucleation density that means smaller spherulites. On the other border of the spherulitic layer, for both heating times, the spherulites appear elongated, consistent with the fact that they are in the transition zone. Finally, the morphology of the skin layer is shown by the AFM height map at the bottom of Figure 5. The fibrils of the shear layer tend to disappear and are replaced by un-oriented globular structures as shown by the magnification of the area enclosed in the green square reported in the AFM map.

Figures 5 and 6 also show the elastic modulus maps acquired in the same positions where morphology was detected. The comparison among the elastic modulus maps shows that the shear layer on the unheated side is characterized by highest values of elastic modulus whereas the spherulitic layer shows the smaller values. The skin layer is characterized by the smallest modulus values (see Figure 5). This is consistent with the fact that in this layer globular elements have been found [12]. Moreover, the shear layer on the heated side is characterized by smaller modulus values with respect to the shear layer on the unheated side. This suggests that the fibrils found in these zones have a different mechanical behavior. In sample 120-18 (Figure 6) the spherulitic core impinges the sample wall on the heated side. The spherulites close to the heated side show modulus comparable with the elastic modulus of the spherulites in the inner part of the sample.

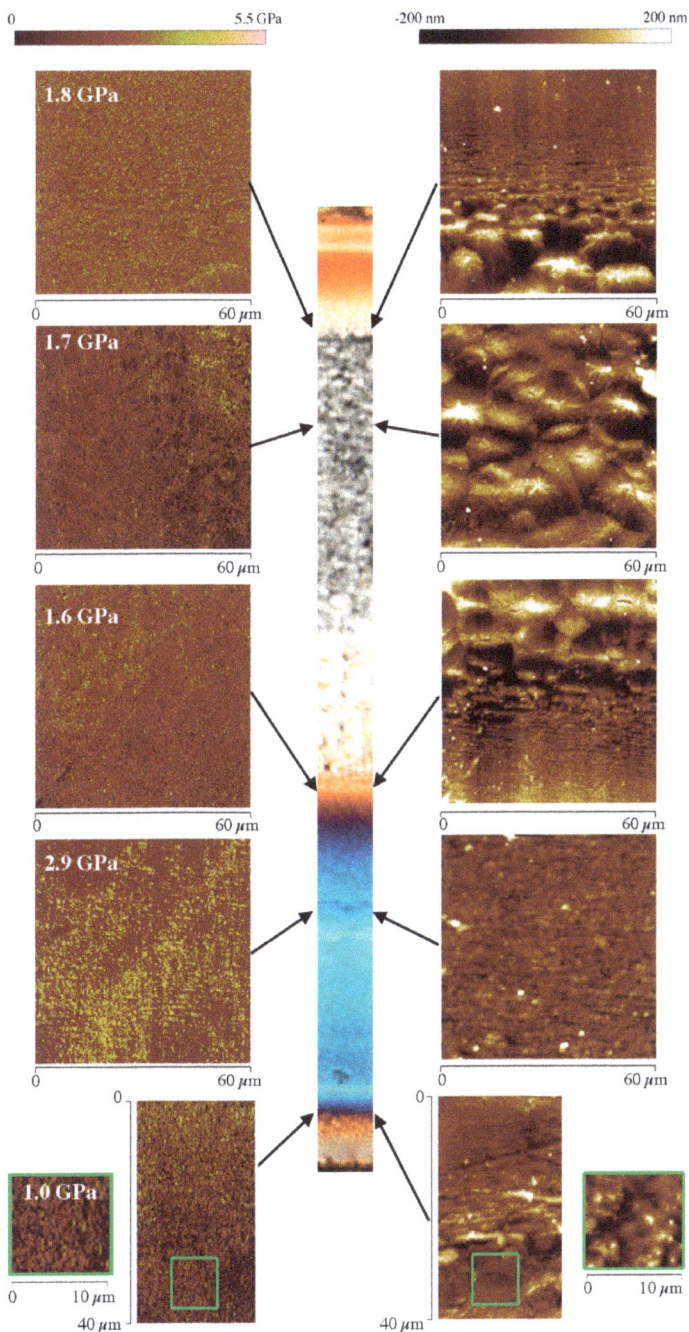

Figure 5. Atomic force microscopy (AFM) maps, height and modulus, obtained in different zones along the sample thickness, for the sample 120-8. The average moduli are reported in the map of each zone. Polarized Optical Microscopy (POM) micrographs and AFM maps have all the same alignment with flow direction in the thickness plane.

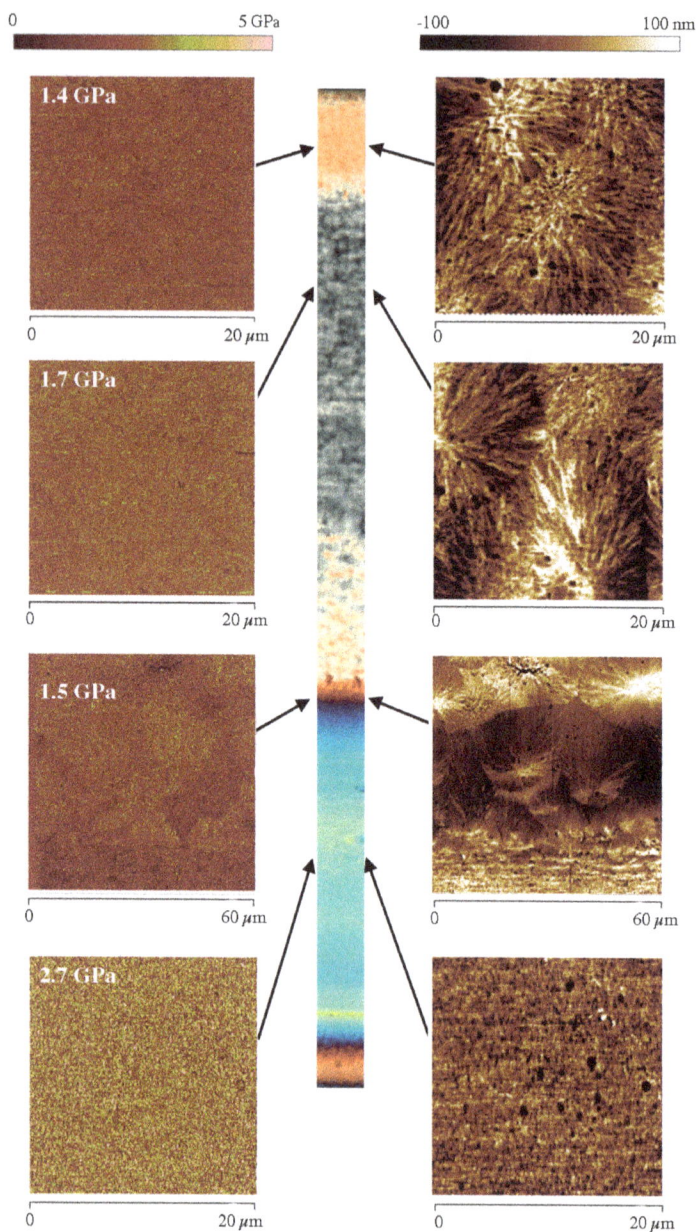

Figure 6. AFM maps, height and modulus, obtained in different zones along the sample thickness, for the sample 120-18. The average moduli are reported in the map of each zone. POM micrographs and AFM maps have all the same alignment with flow direction in the thickness plane.

For further investigation of mechanical modulus distributions, indentation tests have also been performed in different positions along the sample's thickness. Figure 7 shows the load displacement curves obtained in different positions along the thickness of the sample 120-18. In particular, three different values of the dimensionless distance from the heated side, d, where considered: d = 0.1, d = 0.4 and d = 0.75. Following the Oliver–Pharr model [27], the elastic modulus mainly depends on the initial unloading slope in the load displacement plot. In particular, the elastic modulus increases with the unloading slope.

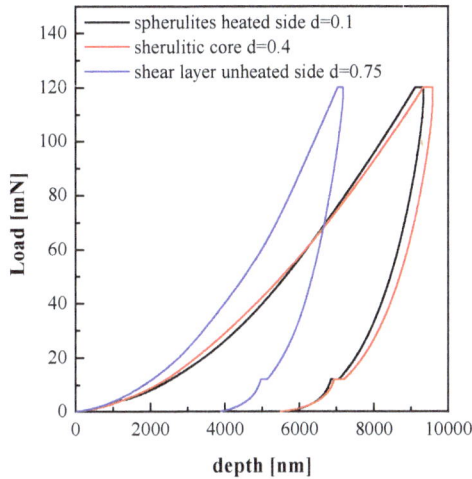

Figure 7. Load-displacement curves of three different positions along the thickness of sample 120-18.

The indentation analyses confirm that the highest moduli were found in the shear layer on the unheated side, the spherulites show lower moduli.

Figure 8 shows the distributions of elastic moduli evaluated by both HarmoniX AFM and indentation analyses in several positions along the thickness of the samples 120-05, 120-8 and 120-18. In order to further study the effect of the cavity surface temperature on the mechanical properties of the samples, the experimental results obtained by the indentation analysis of the T_off sample are also reported in each plot.

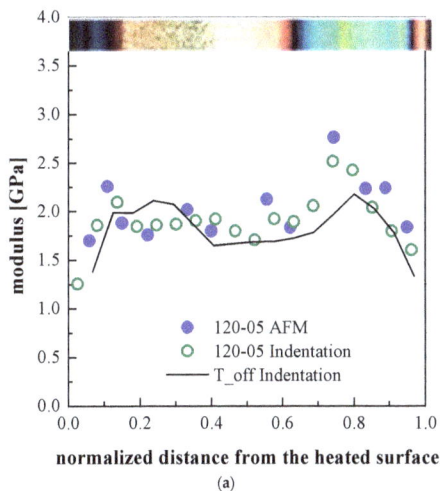

(a)

Figure 8. *Cont.*

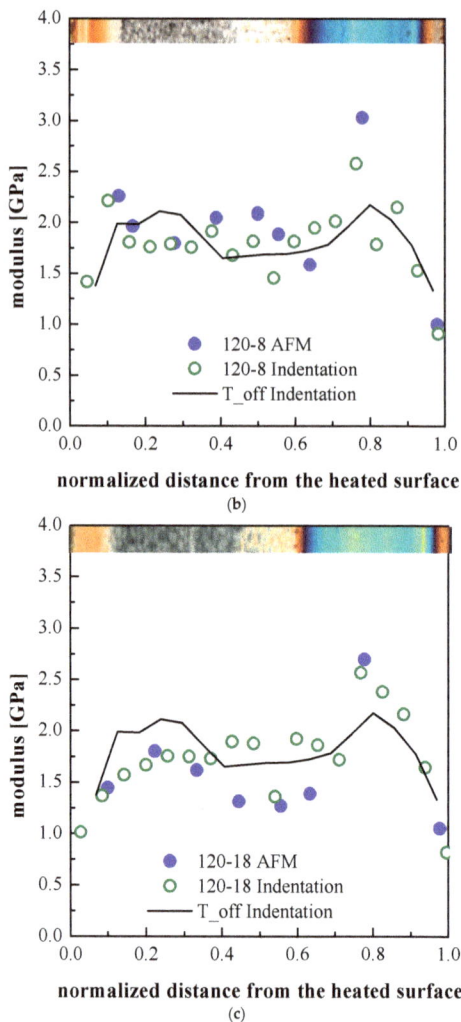

Figure 8. Distributions of elastic modulus along the sample thickness obtained by indentation and HarmoniX AFM tests for the samples (**a**) 120-05; (**b**) 120-8 and (**c**) 120-18.

The experimental results for the values of the modulus of the T_off sample reported in Figure 8 show a characteristic symmetric distribution with two maxima located in the shear layers, as shown in Figure 4. The skin layer shows the lowest moduli and the spherulitic core has intermediate modulus values. By increasing the mold temperature, the modulus distribution becomes asymmetric. This is due to both the uneven temperature field evolutions and to the asymmetrical flow developed during the filling stage. Because of the reduction of the shear layer thickness, on the heated side, the maximum value of the modulus approaches the sample surface. Moreover, the value of the maximum is lower with respect to the value of the maximum found on the unheated side. With 18 s heating time, the shear layer disappears on the heated side and the elastic modulus shows the characteristic values of the spherulitic layer.

Figure 9 shows a comparison among the elastic modulus distribution obtained from indentation analyses of the samples T_off, 120-05, 120-8 and 120-18. On the unheated side the modulus appears significantly affected by the heating only close to the internal border of the shear layer; however, the modulus appears to be almost unaffected by the heating duration.

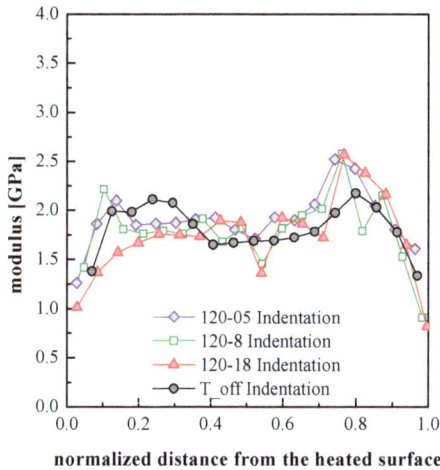

Figure 9. Elastic modulus distributions (obtained from indentation analyses) along the sample thickness for the samples T_off, 120-05, 120-8 and 120-18.

4. Discussion

The mechanisms that govern the temperature evolution inside the polymer part during the molding process are: fountain flow, heat convection, heat diffusion, viscous and latent heat generation [32]. Fountain flow occurs when the melt is forced, with elongational flow, outwards from the center of the flow front toward the mold wall. The fountain flow is recognized to be the main reason for the high level of molecular orientation found at the skin layer [33–35]. It is also responsible for a temperature profile that is quite uniform along the thickness of the flow front. This situation changes quickly soon after the first contact of the polymer with the surface by the effect of the heat transfer with the wall.

The evolution of temperature distribution inside an asymmetrically heated sample is given by Equation (1) [36], where convective contributions of the melt flow are not accounted for, or it could be neglected during the post-filling stages of the injection molding process. It considers a slab of thickness s starting from uniform temperature T_i (220 °C in the considered cases), with one surface held at constant temperature T_1 and the other surface at T_2 (120 °C and 28 °C, respectively).

$$\frac{T(y,t) - T_1}{T_2 - T_1} = \frac{y}{S} + \sum_{n=1}^{\infty} \frac{Z_n}{T_2 - T_1} \left[exp\left(\frac{-\alpha n^2 \pi^2 t}{S^2} \right) * sin\left(\frac{n\pi y}{S} \right) \right] \tag{1}$$

where, α is the thermal diffusivity, t is the time and

$$Z_n = 2\frac{1 - (-1)^n}{n\pi}(T_i - T_1) + \frac{2(-1)^n}{n\pi}(T_2 - T_1) \tag{2}$$

In order to take into account the heat of crystallization, the thermal diffusivity α is corrected with the Stefan number, *Ste*, according to Equation (3).

$$\alpha = \frac{\alpha_1}{1 + Ste} \tag{3}$$

with:

$$\alpha_1 = \frac{k}{\rho\,C_p}\,Ste = \frac{X_c \cdot \Delta h_\infty}{C_p \cdot \left(T_i - T_f\right)} \tag{4}$$

where Δh_∞ is latent heat of crystallization, X_c is the crystallinity and T_f is the final temperature of the slab. For asymmetrical boundary conditions, T_f is not constant along the slab thickness and depends on s. The values of the parameters adopted for the calculation are reported in Table 2.

Table 2. Parameters adopted to calculate the temperature distribution along the sample thickness.

Parameter	Value
α_1	$8.25 \times 10^{-8}\ \mathrm{m^2 \cdot s^{-1}}$
C_p	$2.62\ \mathrm{J \cdot g^{-1} \cdot K^{-1}}$
Δh_∞	$114\ \mathrm{J \cdot g^{-1}}$
X_c	0.5
T_f at midplane	$70\ ^\circ\mathrm{C}$

The evolutions of the temperature distribution along the sample thickness, calculated by the Equations (1)–(4) at different times, are reported in Figure 10. The temperature distributions asymptotically approach the equilibrium distribution T_{eq} (the dotted line in Figure 10).

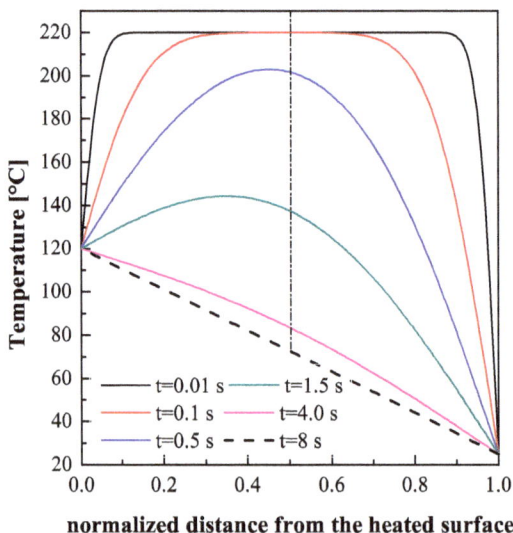

Figure 10. Predictions of temperature distribution along the cavity thickness at different times in an asymmetrically heated sample calculated by Equations (1)–(4).

Figure 11 shows the time to cool down each layer in the temperature range 70–100 °C, that corresponds to the temperature range within which the iPP crystallization takes place. The time distributions (down to 70 °C and to 100 °C) are reported for both symmetric thermal conditions and asymmetric thermal conditions.

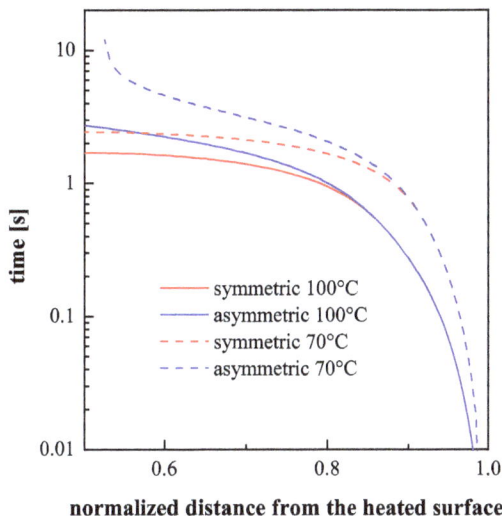

Figure 11. Distributions of the times taken to cool down to 100 °C and to 70 °C on the unheated side of the sample, calculated both for the T_off sample and for the asymmetrically heated sample.

The morphology distributions depend on the molecular stretch, which is determined, essentially, by the filling flow and on the subsequent relaxation time that, in turn, depends on the temperature evolution. As reported above, the surface heater was held active for three different times (namely, 0.5 s, 8 s and 18 s) after the contact of the melt with the cavity in position P2. The time 0.5 s corresponds to the cavity filling time, 18 s corresponds to the dominant relaxation time at 150 °C evaluated under processing conditions [37,38] and 8 s corresponds to about the time needed to reduce the distance from the equilibrium temperature distribution by about 95 percent.

The thicknesses of the shear layers on the side held at 28 °C show a small dependence upon the heating time (Figure 4). However, the asymmetric heating affects the evolution of temperature on the whole thickness. Figure 10 shows that there is a significant thickness (of about 0.15 mm) on the unheated mold surface where the cooling time results essentially unaffected by the mold heating on the other side (Figure 11). As expected, on this layer the morphologies of all samples correspond to that of the sample T_off, which experienced symmetrical cooling. Vice versa, the skin layer is not observed on the heated side where cooling rate, although high, does not achieve the values of the other side.

Deeper inside the unheated side, as shown in Figure 11, the cooling time during asymmetric heating experiments becomes larger with respect to the symmetrical cooling. The packing flow in the sample core, because of the higher temperature, remains active for a time longer than in the symmetrical case. Consequently, the molecular stretch reaches values sufficient to achieve shear layer thicknesses slightly wider than the thickness of the shear layer of the T_off case.

The micrographs reported in Figure 8 show that the morphology at the heated side changes with the heating time. In particular, with 0.5 s heating time a small shear layer was detected, which becomes thinner with 8 s and disappears with 18 s heating time. A parallel growth of the spherulitic layer toward the surface takes place due to surface heating. These effects must be determined by the gradual relaxation of the molecular stretch inside the polymer, while it is still in the molten state. As a consequence, it can be concluded that at the heated surface the polymer crystallizes during the fast cooling stages which follow the shutdown of the heating device.

The final mechanical properties are determined by the local structure, which obviously depends upon the local molecular stretch and by the local structuring achieved during solidification.

The structuring from a melt (where the molecules are randomly distributed and deeply entangled) to a crystalline structured solid is a kinetic process and the structuring level enhances if more time is available. The time available to the polymer for crystalline structuring increases as the cooling rate (through the crystallization and related phenomena) decreases, thus for a given molecular stretch, structuring improves as the distance from the cold surface increases. On the other hand, the intensity of the molecular stretch (mainly generated by the shear intensity during the cavity filling) increases from the sample center toward the sample surface. Applying such a reason to the whole cross section, the moduli are expected to increase with the distance from the surface, due to the cooling rate decrease. At the same time, they are expected to decrease with the distance from the surface, due to the stretch decrease. The combination of these two factors is consistent with a maximum in between the symmetry plane and the sample surface, as it can actually be observed in the Figures 8 and 9.

In the T_off conditions, the maximum within the shear layer is located closer to the spherulitic layer rather than to the external border of the shear layer. The structuring is expected to be very poor at both surfaces of the sample because of the extremely high cooling rate. On approaching the surfaces, the modulus decreases especially where skin layers are observed. When the heating stage is active the position of the maximum of the elastic modulus, on the unheated side, ends up being essentially unaffected by the heating stage.

What has still to be discussed is the effect of the heating time on the values of the maximum of the modulus within the shear layers. It is not a surprise that the moduli within the two shear layers on the heated side (of the samples 120-05 and 120-8) follow similar paths (with clear small maxima) since both of those layers crystallize during the fast cooling which follows the shutdown of the heating device. The sample 120-18 does not show any maximum on the heated side since in this case the polymer chains have enough time to relax and the shear layer is not observed.

On the unheated side, it has to be considered that the three samples (120-05, 120-8, 120-18) experienced cooling histories that are very different one from the other. Furthermore, within each test, the cooling histories through the crystallization are a function of the distance from the surface, also within the shear layer. The temperature evolutions within the sample shown in Figure 10 were calculated under the condition that the temperature on the heated side is held at 120 °C. However, for the sample 120-05 the heating was interrupted soon after the end of the filling, i.e., 0.5 s. At that time, the temperature distribution on the sample cross-section was already significantly different with respect to the symmetrical cooling case (see the temperature distribution at $t = 0.5$ s, in Figure 10). Furthermore, the sample surface will take a small but significant time to decrease to 28 °C, as shown in Figure 2. As a consequence, the subsequent cooling through the crystallization on the unheated side will be slowed down. This determines a larger time for structuring and thus it justifies the increase (with respect to the symmetric case) of the maximum of the modulus inside the shear layer.

The other two samples, 120-8 and 120-18, undergo a first cooling toward the temperature of the dotted line in Figure 10, and afterward their temperature remains essentially constant until the heating shutdown and then they undergo a second cooling to the mold temperature after the heating shutdown. At the end of the first cooling step the temperature becomes sufficiently small (45–60 °C at 0.6–0.8 mm distance from the heated wall) to assure that, in both conditions, the polymer crystallizes in the time available.

It is well known that crystallization temperature during cooling decreases with the increase of the cooling rate and increases by the effect of the flow, both effects increase toward the sample surface and partially compensate each other. The crystallization of the α-phase of polypropylene has a maximum rate at about 70 °C with a half crystallization time of about 2 s [39,40]. That time decreases of 1–2 orders of magnitude by the effect of the flow, as a consequence the α crystallization temperature in a shear layer has to take place in the interval 70–100 °C. The permanence time within that temperature interval determines the time available, during cooling, for structuring during solidification, which in its turn (for a given stretch) determines the material properties (the modulus in our case). The permanence times within that temperature interval can be evaluated from Figure 11 also for the sample T_off.

In particular, Figure 11 refers only to the unheated side. The permanence times are only slightly dependent on the distance from the sample surface in the shear layer. However, in the internal zone of the shear layer, the permanence times in the crystallization temperature range for the samples 120-8 and 120-18 show values clearly larger than those calculated for the sample T_off. This is consistent with the findings related to the mechanical properties: the sample obtained with 120 °C mold surface temperature shows higher maximum values of elastic modulus in the shear layer of the unheated side with respect the shear layer of the T_off sample.

5. Conclusions

In this work, iPP samples were obtained by injection molding adopting a cavity with a surface kept at a high temperature for different time periods (0.5 s, 8 s and 18 s) during the process. Resulting molded samples have been mechanically characterized at different length-scales by indentation and AFM HarmoniX.

The asymmetry of the thermal conditions on the mold surface completely changes the classical skin-core structure shown by the samples. The three characteristic morphological layers, namely skin, shear and core layer are affected in a different way by external heating. The core layer, generally characterized by a spherulitic morphology with intermediate modulus, increases in thickness and moves up to the heated sample side. The shear layers show different behaviors with respect to the half-thickness position. The one closer to the heated side moves in the surface direction and gradually decreases its thickness up to completely disappear for 18 s heating time. On the opposite, the layer closer to the unheated surface slightly increases in dimension with respect to those formed in a symmetrically cooled sample. The skin layer, characterized by loosely structured crystalline elements, is practically unaffected on the unheated side, yet it disappears even for a very short heating time (0.5 s) on the heated side.

Higher values of the modulus were found in the shear layers (especially those close to the unheated side). Minimum and intermediate values were found in the skin and in the spherulitic layers, respectively.

The variations of the modulus were interpreted on the basis of variations of both the molecular stretch and the time available for molecular structuring: the cooling rate through the crystallization determines the time available and thus affects the level of structuring achieved inside the solid. The modulus increases with the increase of local structuring and molecular stretch.

Acknowledgments: Andrea Sorrentino acknowledges financial support from the "VINMAC" project, ID 139455, Decreto Regionale Lombardia n. 9559 del 02/08/2017, CUP: E67H16000980009.

Author Contributions: Giuseppe Titomanlio and Sara Liparoti conceived and designed the experiments; Sara Liparoti performed the molding and POM tests; Sara Liparoti and Giuseppe Titomanlio analyzed the acquired data; Vito Speranza and Sara Liparoti performed and analyzed the AFM tests; Andrea Sorrentino performed and analyzed the Indentation tests; Vito Speranza wrote the paper.

Conflicts of Interest: The authors declare no conflict of interest.

References

1. Giró-Paloma, J.; Roa, J.J.; Díez-Pascual, A.M.; Rayón, E.; Flores, A.; Martínez, M.; Chimenos, J.M.; Fernández, A.I. Depth-sensing indentation applied to polymers: A comparison between standard methods of analysis in relation to the nature of the materials. *Eur. Polym. J.* **2013**, *49*, 4047–4053. [CrossRef]
2. Flores, A.; Naffakh, M.; Díez-Pascual, A.M.; Ania, F.; Gómez-Fatou, M.A. Evaluating the reinforcement of inorganic fullerene-like nanoparticles in thermoplastic matrices by depth-sensing indentation. *J. Phys. Chem. C* **2013**, *117*, 20936–20943. [CrossRef]
3. Lu, Z.; Zhang, K.F. Morphology and mechanical properties of polypropylene micro-arrays by micro-injection molding. *Int. J. Adv. Manuf. Technol.* **2009**, *40*, 490–496. [CrossRef]

4. Gojzewski, H.; Imre, B.; Check, C.; Chartoff, R.; Vancso, J. Mechanical mapping and morphology across the length scales unveil structure-property relationships in polycaprolactone based polyurethanes. *J. Polym. Sci. Part B* **2016**, *54*, 2298–2310. [CrossRef]
5. Kalay, G.; Bevis, M.J. Processing and physical property relationships in injection-molded isotactic polypropylene. 2. Morphology and crystallinity. *J. Polym. Sci. Part B* **1997**, *35*, 265–291. [CrossRef]
6. Guadagno, L.; Longo, P.; Raimondo, M.; Naddeo, C.; Mariconda, A.; Sorrentino, A.; Vittoria, V.; Iannuzzo, G.; Russo, S. Cure behavior and mechanical properties of structural self-healing epoxy resins. *J. Polym. Sci. Part B* **2010**, *48*, 2413–2423. [CrossRef]
7. Sorrentino, A.; Pantani, R.; Brucato, V. Injection molding of syndiotactic polystyrene/clay nanocomposites. *Polym. Eng. Sci.* **2006**, *46*, 1768–1777. [CrossRef]
8. Wieme, T.; Tang, D.; Delva, L.; D'hooge, D.R.; Cardon, L. The relevance of material and processing parameters on the thermal conductivity of thermoplastic composites. *Polym. Eng. Sci.* **2017**. [CrossRef]
9. Liparoti, S.; Sorrentino, A.; Speranza, V.; Titomanlio, G. Multiscale mechanical characterization of iPP injection molded samples. *Eur. Polym. J.* **2017**, *90*, 79–91. [CrossRef]
10. Liparoti, S.; Sorrentino, A.; Speranza, V. Micromechanical characterization of complex polypropylene morphologies by HarmoniX AFM. *Int. J. Polym. Sci.* **2017**, *2017*, 9037127. [CrossRef]
11. Liparoti, S.; Sorrentino, A.; Titomanlio, G. Fast cavity surface temperature evolution in injection molding: Control of cooling stage and final morphology analysis. *RSC Adv.* **2016**, *6*, 99274–99281. [CrossRef]
12. Liparoti, S.; Sorrentino, A.; Guzman, G.; Cakmak, M.; Titomanlio, G. Fast mold surface temperature evolution: Relevance of asymmetric surface heating for morphology of iPP molded samples. *RSC Adv.* **2015**, *5*, 36434–36448. [CrossRef]
13. Huang, M.-C.; Tai, C.-C. The effective factors in the warpage problem of an injection-molded part with a thin shell feature. *J. Mater. Process. Technol.* **2001**, *110*, 1–9. [CrossRef]
14. Ozcelik, B.; Sonat, I. Warpage and structural analysis of thin shell plastic in the plastic injection molding. *Mater. Des.* **2009**, *30*, 367–375. [CrossRef]
15. Wang, C.; Huang, M.; Shen, C.; Zhao, Z. Warpage prediction of the injection-molded strip-like plastic parts. *Chin. J. Chem. Eng.* **2016**, *24*, 665–670. [CrossRef]
16. Vietri, U.; Sorrentino, A.; Speranza, V.; Pantani, R. Improving the predictions of injection molding simulation software. *Polym. Eng. Sci.* **2011**, *51*, 2542–2551. [CrossRef]
17. Kurt, M.; Saban Kamber, O.; Kaynak, Y.; Atakok, G.; Girit, O. Experimental investigation of plastic injection molding: Assessment of the effects of cavity pressure and mold temperature on the quality of the final products. *Mater. Des.* **2009**, *30*, 3217–3224. [CrossRef]
18. De Santis, F.; Pantani, R. Development of a rapid surface temperature variation system and application to micro-injection molding. *J. Mater. Process. Technol.* **2016**, *237*, 1–11. [CrossRef]
19. Jiang, J.; Wang, S.; Sun, B.; Ma, S.; Zhang, J.; Li, Q.; Hu, G.H. Effect of mold temperature on the structures and mechanical properties of micro-injection molded polypropylene. *Mater. Des.* **2015**, *88*, 245–251. [CrossRef]
20. Larpsuriyakul, P.; Fritz, H.G. Warpage and countermeasure for injection-molded in-mold labeling parts. *Polym. Eng. Sci.* **2011**, *51*, 411–418. [CrossRef]
21. Nian, S.-C.; Wu, C.-Y.; Huang, M.-S. Warpage control of thin-walled injection molding using local mold temperatures. *Int. Commun. Heat Mass Transf.* **2015**, *61*, 102–110. [CrossRef]
22. Wang, X.; Zhao, G.; Wang, G. Research on the reduction of sink mark and warpage of the molded part in rapid heat cycle molding process. *Mater. Des.* **2013**, *47*, 779–792. [CrossRef]
23. Liparoti, S.; Titomanlio, G.; Sorrentino, A. Analysis of asymmetric morphology evolutions in iPP molded samples induced by uneven temperature field. *AIChE J.* **2016**, *62*, 2699–2712. [CrossRef]
24. Liparoti, S.; Sorrentino, A.; Titomanlio, G. Rapid control of mold temperature during injection molding process: Effect of packing pressure. *AIP Conf. Proc.* **2015**, *1695*, 020052–020057. [CrossRef]
25. Liparoti, S.; Landi, G.; Sorrentino, A.; Speranza, V.; Cakmak, M.; Neitzert, H.C. Flexible poly(amide-imide)-carbon black based microheater with high-temperature capability and an extremely low temperature coefficient. *Adv. Electron. Mater.* **2016**, *2*, 1600126. [CrossRef]
26. White, H.M.; Bassett, D.C. On row structures, secondary nucleation and continuity in alpha-polypropylene. *Polymer* **1998**, *39*, 3211–3219. [CrossRef]
27. Pharr, G.M.; Oliver, W.C.; Brotzen, F.R. On the generality of the relationship among contact stiffness, contact area, and elastic modulus during indentation. *J. Mater. Res.* **1992**, *7*, 613–617. [CrossRef]

28. Gòmez, C.J.; Garcia, R. Determination and simulation of nanoscale energy dissipation processes in amplitude modulation AFM. *Ultramicroscopy* **2010**, *110*, 626–633. [CrossRef] [PubMed]

29. Sahin, O.; Magonov, S.; Su, C.; Quate, C.F.; Solgaard, O. An atomic force microscope tip designed to measure time-varying nanomechanical forces. *Nat. Nanotechnol.* **2007**, *2*, 507–514. [CrossRef] [PubMed]

30. Derjaguin, B.; Muller, V.; Toporov, Y. Effect of contact deformations on the adhesion of particles. *J. Colloid Interface Sci.* **1975**, *53*, 314–326. [CrossRef]

31. Pantani, R.; Coccorullo, I.; Speranza, V.; Titomanlio, G. Modeling of morphology evolution in the injection molding process of thermoplastic polymers. *Prog. Polym. Sci.* **2005**, *30*, 1185–1222. [CrossRef]

32. Kamal, M.R.; Isayev, A.I.; Liu, S.-J. Injection molding: Technology and fundamentals. In *Progress in Polymer Processing*; Carl Hanser Verlag GmbH & Co.: Munich, Germany, 2009; p. 737.

33. Coyle, D.J.; Blake, J.W.; Macosko, C.W. The kinematics of fountain flow in mold-filling. *AIChE J.* **1987**, *33*, 1168–1177. [CrossRef]

34. Lafleur, P.G.; Kamal, M.R. A structure-oriented computer simulation of the injection molding of viscoelastic crystallinepolymers part I: Model with fountain flow, packing, solidification. *Polym. Eng. Sci.* **1986**, *26*, 92–102. [CrossRef]

35. Mavridis, H.; Hrymak, A.N.; Vlachopoulos, J. Finite element simulation of fountain flow in injection molding. *Polym. Eng. Sci.* **1986**, *26*, 449–454. [CrossRef]

36. Latif, M.J. Transient conduction. In *Heat Conduction*; Springer: Berlin, Germany, 2009; p. 128.

37. Pantani, R.; Speranza, V.; Titomanlio, G. Evolution of iPP relaxation spectrum during crystallization. *Macromol. Theory Simul.* **2014**, *23*, 300–306. [CrossRef]

38. Pantani, R.; Speranza, V.; Titomanlio, G. Simultaneous morphological and rheological measurements on polypropylene: Effect of crystallinity on viscoelastic parameters. *J. Rheol.* **2015**, *59*, 377–390. [CrossRef]

39. De Santis, F.; Adamovsky, S.; Titomanlio, G.; Schick, C. Scanning nanocalorimetry at high cooling rate of isotactic polypropylene. *Macromolecules* **2006**, *39*, 2562–2567. [CrossRef]

40. Pantani, R.; Speranza, V.; Titomanlio, G. Effect of Flow-Induced Crystallization on the distribution of spherulite dimensions along cross section of injection molded parts. *Eur. Polym. J.* **2017**, *97*, 220–229. [CrossRef]

![polymers logo] *polymers*

MDPI

Article

Deformation-Induced Phase Transitions in iPP Polymorphs

Harm J. M. Caelers [1,†], Enrico M. Troisi [1,†], Leon E. Govaert [1,2] and Gerrit W. M. Peters [1,*]

1   Department of Mechanical Engineering, Materials Technology Institute, Eindhoven University of
    Technology, P.O. Box 513, 5600 MB Eindhoven, The Netherlands; Harm.Caelers@SABIC.com (H.J.M.C.);
    Enrico.Troisi@SABIC.com (E.M.T.); l.e.govaert@tue.nl (L.E.G.)
2   Faculty of Engineering Technology, University of Twente, P.O. Box 217, 7500 AE Enschede, The Netherlands
*   Correspondence: g.w.m.peters@tue.nl; Tel.: +31-040-247-4840
†   Current address: Sabic T&I, STC Geleen, P.O. Box 319, 6160 AH Geleen, The Netherlands.

Received: 21 August 2017; Accepted: 18 October 2017; Published: 24 October 2017

Abstract: This detailed study reveals the relation between structural evolution and the mechanical response of α-, β- and γ-iPP. Uni-axial compression experiments, combined with in situ WAXD measurements, allowed for the identification of the evolution phenomena in terms of phase composition. Tensile experiments in combination with SAXS revealed orientation and voiding phenomena, as well as structural evolution in the thickness of the lamellae and amorphous layers. On the level of the crystallographic unit cell, the WAXD experiments provided insight into the early stages of deformation. Moreover, transitions in the crystal phases taking place in the larger deformation range and the orientation of crystal planes were monitored. At all stretching temperatures, the crystallinity decreases upon deformation, and depending on the temperature, different new structures are formed. Stretching at low temperatures leads to crystal destruction and the formation of the oriented mesophase, independent of the initial polymorph. At high temperatures, above $T\alpha_c$, all polymorphs transform into oriented α-iPP. Small quantities of the initial structures remain present in the material. The compression experiments, where localization phenomena are excluded, show that these transformations take place at similar strains for all polymorphs. For the post yield response, the strain hardening modulus is decisive for the mechanical behavior, as well as for the orientation of lamellae and the evolution of void fraction and dimensions. β-iPP shows by far the most intense voiding in the entire experimental temperature range. The macroscopic localization behavior and strain at which the transition from disk-like void shapes, oriented with the normal in tensile direction, into fibrillar structures takes place is directly correlated with the strain hardening modulus.

Keywords: polymorphism; isotactic polypropylene; deformation; phase transitions; uniaxial compression; uniaxial tensile deformation; temperature; in situ X-ray; cavitation

1. Introduction

Isotactic polypropylene (iPP) shows several crystal modifications (polymorphs) [1,2]. Depending on chain architecture [3], additives [4] or the conditions experienced during processing [5] the formation of either one of them can be enhanced. The structures most often found are the monoclinic α-, the pseudo-hexagonal β- [6–8], the orthorhombic γ- [2,3,9,10] and the metastable meso-phase [11]. These crystals form a morphology in which the crystalline layers (lamellae) are alternated with amorphous layers. In a recent study, the authors presented remarkable similarities in the yield and failure kinetics of the α-, β- and γ-iPP, obtained from uni-axial compression experiments, even though the crystal structures are very different [12]. For the intrinsic material response (true stress versus true strain), however, some interesting differences were found, mainly in the post-yield behavior. It is well known that the softening and hardening of iPP, displayed in the region after yielding, involves a number of structural modifications among which are crystal destruction/melting, orientation of the surviving crystallites and the amorphous network and recrystallization (possibly in other crystal-phases [13–15]). Furthermore, the typical alternating heterogeneous structure of a relatively soft amorphous domain and a stiff crystal part introduces stress and strain concentrations on a local scale [16]. In the case of uni-axial extension, this gives rise to an additional phenomenon. Due to the negative hydrostatic pressure that develops while stretching, cavitation or voiding is initiated when the cavitation strength of the amorphous phase, which is lower than that of the crystals, is exceeded [17].

From an engineering point of view, it is of great importance to understand the relation between the various (micro- and meso-scale) structural features of the material and the behavior observed on a macroscopic scale, and therefore, this has been the subject of several studies. The structural evolution of α- [13,15,18–20], β- [21–24] and γ-iPP [14,21,25] was studied for several loading geometries. In general, cavitation is observed in the case of uni-axial tensile deformation, and it was found to affect yield stress and the properties at large strain tremendously.

Because of this strong coupling, Humbert et.al. investigated the initiation of voiding extensively [26] and concluded that, depending on molecular topology and the micro-structural parameters, cavitation can either proceed or follow the onset of plasticity. They related the type of cavitation, i.e., homogeneous or heterogeneous, to a critical lamellar thickness l_{cc}. Below this thickness, the critical shear stress for plastic crystal deformation is exceeded before cavitation takes place, while at higher thickness, the critical stress for cavitation is exceeded before the crystals deform plastically. The effect of lamellar thickness on cavitation found by Wang et al. [27] can be explained with this framework. The influence of temperature was further investigated by Xiong et al. [28], who also modeled the cavitation/shear competition successfully [29].

In all these studies, the amorphous material plays a crucial role, for example because it determines the critical cavitation stress. Starting from that perspective, some studies were devoted to develop techniques to quantify the Young's modulus of the amorphous phase [30] and the strain hardening modulus from tensile experiments [31]. The Young's modulus of the amorphous phase in the initial stages of deformation can be determined as shown by Xiong et al. [30], who used a procedure based on in situ SAXS and WAXD tensile experiments. They found a decreasing Young's modulus of the amorphous network with increasing crystallinity and temperature. With respect to the strain hardening modulus it is concluded that at large strains the role of the network is dominant over the residual crystalline structures, particularly in the case of tensile deformation [32,33]. In this perspective disentanglement induced cavitation at large strains, and void propagation via network relaxation were investigated, as well [19,20], and found to be able to reduce the strain hardening modulus to a large extent. Even in strongly oriented material, whitening as a result of cavitation can still occur. Subsequent to voiding and the structure transformation into a fibrillated system, the amount of tie molecules is decisive for the final strength before fracture/failure takes place, as is shown by Ishikawa et al. [34].

Besides the structural features (nanometer scale) and the loading conditions discussed before, the relation between the crystal unit cell structure itself (Angstrom scale) and the cavitation process is investigated for α- and β-iPP [35]. Using SEM techniques, they showed that α-spherulites show micro-cracking in tensile loading. Cavitation takes place at the early stages of deformation and starts at the spherulite boundaries or in the equatorial region [35]. Contrarily, the β-phase deforms plastically up to high deformations. Cavitation starts in the center of the spherulite and subsequently propagates to the equatorial regions [35]. Basic mechanisms like lamellar separation, plastic slip and crazing seem to be involved. Due to the cross-hatched structures of α-iPP lamellae, an increased stiffness is reported since crystalline domains are considered to form a connected physical network [35,36]. The resistance against plastic flow is significantly less for β-iPP when compared to α-iPP, even at deformation where the junctions of the cross hatched structures are destroyed. Therefore, intrinsic mechanical properties of crystals are more likely to be involved in plastic behavior than their micro-structural arrangement [13,36,37]. Due to reduced chain interactions and lower packing density in β-crystals, an enhanced chain mobility is obtained [36,37]. An important limitation of the SEM observation [35] is that the assessed length scale is relatively large, especially compared to the data generated with X-ray techniques.

Despite the large number of studies in this field, a detailed comparison between α-, β- and γ-iPP, tested in the same conditions, is still missing. The deformation and resulting phase transitions, onset of voiding, lamellar growth or destruction and the orientation are revealed in this work. Moreover, the relation of events taking place at the nano- and micro-scale level with the macroscopically intrinsic mechanical response is still unclear. Therefore, in this study in situ WAXD and SAXS experiments are performed in combination with compression and tensile experiments. The compression experiments allow one to investigate the phase transitions as a function of true stress and strain. Cavitation does not occur in compression tests due to a positive hydrostatic pressure. Due to the experimental restrictions, related to the compression setup blocking the diffracted X-ray, the maximal achievable compressive strain is restricted to a true strain of about 80%. The tensile experiments allow one to deform the material up to large strains and investigate voiding. The disadvantage is that it is difficult to relate the phase transitions to true strain. The combination of the two mechanical tests, together with a comparison of the different polymorphs deformed at the exact same conditions, offers a complete overall picture.

This work provides insight on the sequence in which all these events, at the level of crystal structure (Å-scale), lamellae (nm-scale) and voids (μm-scale), take place as a function of deformation and temperature. It deals with the structural changes occurring in uni-axial compression and extension of α-, β- and γ-iPP. This detailed picture of what happens upon deformation allows us to explain the material behavior on a macroscopic scale in terms of micro-scale events. Since the amount of generated data is rather large, a small overview of how the results are presented is given first:

- First, the mechanical response obtained from the tensile experiments is shown, together with the calculated true stress as a function of the true strain. The behavior is discussed and compared with the intrinsic material response measured in uni-axial compression.
- The structural evolution in terms of phase transitions, crystal plane orientation and slip is revealed by WAXD experiments combined with uni-axial tensile tests. The crystal phase transitions are measured in compression experiments, as well, allowing one to compare these results with the ones obtained from tensile experiments (phase transitions) and couple them to the true strain.
- The combination of SAXS experiments with uni-axial tensile tests provided insights on the onset of voiding, lamellar orientation and evolution of amorphous and lamellar thickness. Furthermore, a transition of initially disk-like average void structure to a fibrillar structure was found.

2. Materials and Methods

2.1. Material and Sample Preparation

In this work, an isotactic polypropylene homo polymer was used (Sabic) with a weight averaged molar mass Mw of 320 kg/mol and a polydispersity index Mw/Mn of 5.4. Sheets of iPP material with a thickness of approximately 1 mm, containing monoclinic α-crystals, were prepared by compression molding. The granular material was placed in a mold and heated to 230 °C, after which a pressure of about 5 MPa was applied stepwise. The material was kept under these conditions for 10 min to erase the thermo-mechanical history. Then, the solidification was induced by putting the mold in a cold press (25 °C) for three minutes. The preparation of sheet material containing pseudo-hexagonal β-crystals was done following the same procedure but with a material containing 0.1 wt % β-specific nucleating agent (NJSTAR NU100, New Japan Chemical Group, Osaka, Japan). To obtain the highest possible β-phase content, solidification was induced by cooling for 3 min in a press set at a temperature of 90 °C, after which the sample was cooled further in ambient conditions to room temperature. For the preparation of samples containing γ-crystals with an orthorhombic unit cell structure, a custom build hydrostatic compression tool is used [12]. The temperature and pressure history, applied during processing, is given in Figure 1. The preparation procedure is based on the work of Mezghani et al. [10]. First the temperature is increased to melt the iPP granulate. After erasing the thermo-mechanical history and the remaining air, a pressure of 180 MPa is applied. Subsequently the sample is slowly cooled to 150 °C, which is sufficiently low to crystallize (The melting temperature increases approximately 30 °C for a pressure increase of 100 MPa [38] and thus the undercooling increases with 30 °C). Finally, when the crystallization process is completed, the sample is cooled further by blowing compressed air through the cooling channels. After reaching room temperature cylindrical sheets of approximately 1 mm in thickness were taken out of the mold. All the iPP-sheets were stored at room temperature for two months before tensile experiments were performed. Thick plates of α-, β- and γ-iPP were prepared following the same sample preparation procedure as described in [12].

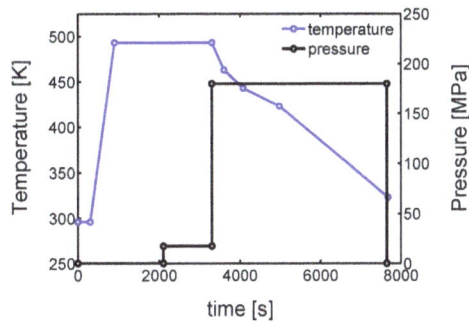

Figure 1. Schematic of the pressure-temperature protocol, used to prepare γ-iPP samples.

2.2. Mechanical Testing

Two loading geometries were used for mechanical testing procedures, i.e., uni-axial compression and uni-axial tensile. Compression experiments were performed with a well defined constant true strain rate. In this way true stress as a function of the true strain can be related to structural changes. A special tool was developed to prevent the compression setup from blocking the diffracted intensity during in situ X-ray measurements to the highest possible strain; see Figure 2.

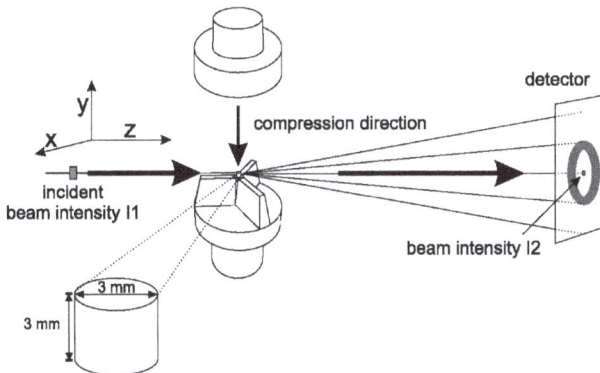

Figure 2. Schematic representation of the compression setup, combined with in situ X-ray experiments.

The in situ mechanical experiments were performed with a Zwick Z5.0 (Ulm, Germany) and the cylindrically-shaped samples, with dimensions of $\varnothing\, 3 \times 3\ \mathrm{mm}^2$, were machined from the thick plates. With a custom built oven, temperatures of 110 °C were applied. In the ex situ compression experiments, performed on β- and γ-iPP, samples of $\varnothing\, 6 \times 6\ \mathrm{mm}^2$ were used. Since in uni-axial compression a true strain can be applied at a constant rate, and the true stress is measured, no geometrical affects are expected. After the application of a true strain of about 80%, the experimental setup starts to block the diffracted or scattered intensity. In order to study structural evolutions at higher strain, tensile experiments were performed. In this way, structural changes, even the ones taking place far beyond yielding, could be monitored. Furthermore, the onset and the evolution of the shape and dimensions of voids can be measured during tensile deformation. A disadvantage is that it is not possible to control the true strain rate. In tensile deformation the material deforms inhomogeneously after reaching the yield point. To force the tensile sample to neck in front of the beam, dumbbell shaped tensile bars were cut from the iPP-sheets, using a punch according to Figure 3. Another important drawback is that the local strain of the material in the beam area, related to the observed patterns, cannot be determined directly. However, an alternative approach will be explained in Section 2.3.4. A Linkam TST 350 tensile stage (Tadworth, UK) is used to stretch the samples. Temperatures were chosen at 25, 50, 80 and 110 °C respectively, while a stretching speed of 12 μm/s was applied, being equivalent to an engineering strain rate of 8×10^{-4} s^{-1}. The samples were stretched in horizontal direction, schematically illustrated in Figure 3. In this schematic, the incident beam intensity I_1 and the intensity measured with the photo diode I_2 are also indicated.

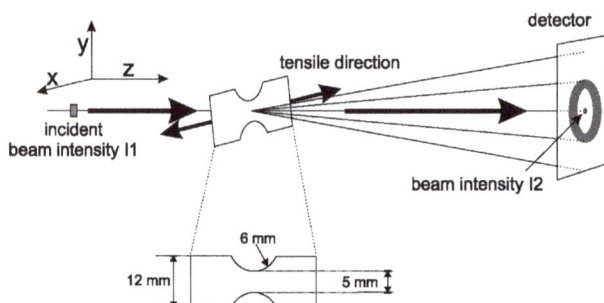

Figure 3. Schematic representation of the tensile setup, combined with in situ X-ray experiments.

Stretching the γ-iPP samples as prepared, resulted in brittle failure at temperatures below the α-transition. Obviously, this is not desired from an experimental point of view. Therefore, the ductility of the γ-iPP samples is improved with a thermal rejuvenation treatment just before the tensile experiment. This turned out to be a successful method to obtain the formation and growth of a stable neck. By keeping the sample for 300 s at a temperature of 85 °C, the constrained amorphous domains in the vicinity of the crystals were rejuvenated, after which the samples were quenched to room temperature. The drop in the yield stress resulting from this treatment is sufficient to obtain ductile behavior [12,39,40].

2.3. X-ray Techniques and Analysis

X-ray experiments were carried out at the Dutch-Belgian beamline BM26 (DUBBLE) in the European Synchrotron and Radiation Facility in Grenoble, France [41]. To monitor the structural evolution upon stretching at the level of the unit cell, lamellae and voids, the same set of experiments was repeated three times:

- The 2D WAXD patterns where recorded with a Frelon detector. This detector has a pixel size of 97.65×97.65 μm^2 and was placed at approximately 140 mm from the sample. The acquisition time was 2.5 s.
- The 2D SAXS patterns where recorded with a Pilatus 1 M detector. Simultaneously with these patterns, a Pilatus 3 K was used to record WAXD patterns. From now on, we will refer to these patterns as 1D WAXD-patterns. The pixel size of the Pilatus 1 M and 1 K was 172×172 μm^2, with a sample to detector distance of approximately 2707 mm (for SAXS) and 283 mm (for

WAXD). At this distance the Porod region is included in the SAXS data. Again the acquisition time was 2.5 s.

- Another set of 2D SAXS experiments were performed with a sample to detector distance of approximately 7258 mm, allowing one to obtain scattering data at low q-values. Again these frames where acquired simultaneously with 1D WAXD patterns, recorded with the Pilatus 3 K placed at a distance of approximately 183 mm, and an acquisition time of 2.5 s.
- During the compression experiments, in situ WAXD patterns where recorded using a Frelon detector. This detector is the same as the one used for the 2D WAXD experiments in tensile deformation, but was now placed at approximately 200 mm from the sample. In this case the acquisition time was 1 s.

The wavelength in all the experiments was $\lambda = 1.033\text{Å}$. The size of the beamspot was approximately 0.3×1 mm^2. The incident beam intensity I_1 is measured using an ionization chamber and the transmitted beam intensity I_2 was measured with a photo diode

2.3.1. Data Reduction

To extract results from the X-ray experiments, the data have to be normalized and corrected. In the case of the WAXD experiments a Frelon detector was used and as a result dark current I_{dc}, i.e., the signal recorded with a closed shutter, has to be subtracted. This is done on both the patterns recorded with (I_m) and without the sample (I_{bkg}). This correction is not needed for the patterns taken with a Pilatus detector because of the absence of readout noise and dark current. To correct for the air scattering the background I_{bkg} is subtracted. For proper subtraction I_{bkg} was first corrected for the ratio $I_1/I_{1,bkg}$ between the incident beam intensity and the background. Next, a correction factor $C = I_{1,bkg}/I_{2,bkg}$ is introduced to compensate for the fact that different devices were used to detect the incident beam intensity I_1 and the intensity recorded by the photo diode in the beam stop I_2. The correction for the attenuation due to the presence of a sample is then $C \cdot I_2/I_1$. In the case of tensile experiments the sample thickness d_t decreases during the experiment and the transmission T is not constant. To correct for these contributions, Equations (1) and (2) have to be used,

$$d_t = \mu \cdot ln(I_{1,t}/(C \cdot I_{2,t})) \tag{1}$$

$$T = \frac{C \cdot I_2}{I_1} \tag{2}$$

where μ is the absorption coefficient of the sample that can be determined from the intensity measured for the initial sample thickness d_0. Combining all the corrections results in [42]:

$$I_{cor} = \frac{\frac{(I_m - I_{dc})}{I_1} - C \cdot \frac{I_2}{I_1}\frac{I_1}{I_{1,bkg}}\frac{(I_{bkg} - I_{dc})}{I_{1,bkg}}}{T \cdot d_t} \tag{3}$$

2.3.2. Phase Composition

The crystallographic structures of iPP display a unique diffraction pattern, with intense scattering at angles that are related to specific crystal plane d-spacings by Bragg's law. In Figure 4 typical 2D diffraction patterns of isotropic α-, β- and γ-iPP are shown.

Polymers **2017**, *9*, 547

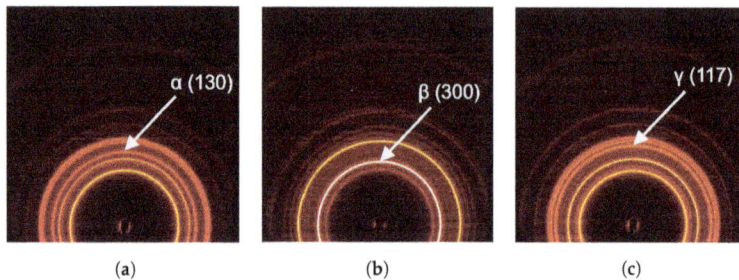

Figure 4. 2D WAXD patterns of iPP, characteristic for (a) α-iPP; (b) β-iPP and (c) γ-iPP, as prepared for this study . The diffraction peak unique for α-iPP is the third clear one going from the center towards outside of the pattern. The same holds for γ-iPP, whereas the diffraction ringspecific for β-iPP is the most clear one in the β pattern. All characteristic peaks are indicated in the figure.

The most clear diffractions of α-iPP come from the crystal planes corresponding to d-spacings of 6.26Å (110), 5.24Å (040) 4.78Å (130) and 4.17Å/4.05Å (111)/(041). For the γ-iPP the 2D-pattern looks similar and the biggest difference can be observed from the third diffraction ring. The crystal planes and corresponding d-spacings are 6.39Å (111), 5.20Å (008) 4.38Å (117) and 4.17Å/4.05Å (202)/(026). For β-iPP only two important diffraction peaks can be observed at d-spacings of 5.50Å (300), 4.19Å (301) respectively. The arrows in Figure 4 indicate characteristic reflections of α-iPP (Figure 4a), β-iPP (Figure 4b) and γ-iPP (Figure 4c). When these structures orient, the intensity migrates azimuthally to specific angles where the diffracted intensity concentrates.

To obtain quantitative information about the phase composition upon stretching, the patterns are radially integrated over an azimuthal angle of 180°. Voigt functions were fitted to the integrated 1D patterns to quantify crystal fractions as a function of strain and temperature. As a first step of the automated peak fitting procedure, the amorphous halo was fitted. After subtraction of the halo, Voigt functions were fitted on the first frame of the series. Here, the sample consisted of almost solely α-, β- or γ-iPP. Upon deformation the diffraction peaks can move to slightly higher or lower q-values. Therefore, the initial peak positions were not fully constrained, and allowed to move over a range of ± 0.25 nm^{-1}. Peaks that are present at the initial peak positions (± 0.25 nm^{-1}), were then subtracted from the intensity pattern. After sufficient deformation, the residual signal can contain additional peaks, resulting from the phase transitions. The newly formed peaks, that are subsequently fitted, represent the characteristic diffraction of other crystal phases. Note that these are the only reflections that can be clearly distinguished. The mesophase is defined as everything that is left after the subtraction of the amorphous phase and all the crystal diffractions, and was also fitted with Voigt functions. Finally, after fitting all the peaks of an integrated WAXD pattern, an additional optimization routine was used to fit the exact height and position of the crystal diffraction peaks. In this optimization run, the shape of the peaks was constrained to the values determined in the first peak fitting routine. The expected error that is made for this kind of fitting procedures can in general easily be in the order of 5% to 10%. Within this study, the data are treated the same way every time, and the fitting is done according to a consistent procedure. When comparing between the different results, this consistency is believed to be the most important feature for reducing the relative errors, made in the determination of the phase composition.

In Figure 5a,b two typical examples of a deconvoluted signal of γ-iPP are shown. The first one, Figure 5a, is the sample at room temperature prior to deformation and the second one, Figure 5b, is the sample after deformation (last image in top row of Figure 15).

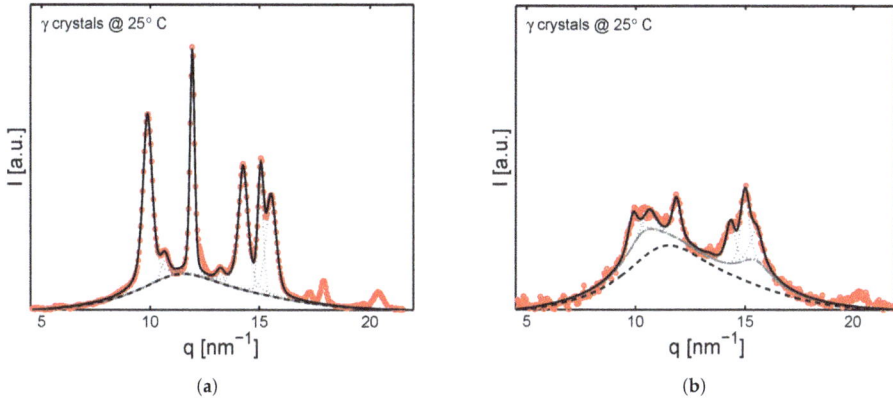

Figure 5. (**a**) An example of a deconvoluted undeformed γ-iPP sample and (**b**) a deformed γ-iPP sample, stretched at 25 °C. The dashed black line is the amorphous halo, the dotted gray lines are the peak fittings, the solid gray line represents the mesophase and the solid black line is the sum of the fitted peaks. The red markers represent the radially integrated pattern obtained from the experiments.

The dashed black line in this figure is the amorphous halo that was fitted to the pattern to determine the weight fraction of the crystallinity according to:

$$\chi_w = \frac{A_{tot} - A_a}{A_{tot}} \qquad (4)$$

where A_{tot} is the total diffracted intensity (integrated area) and A_a is the integrated area of the scaled amorphous halo (determined on quenched low tacticity iPP with negligible crystallinity). The dotted gray lines represent Voigt functions that were fitted on the diffraction peaks and used to determine the weight fraction of α-, β-, γ- or meso-iPP according to:

$$\chi_i = \chi_w \cdot \left(\frac{A_i}{A_\alpha + A_\beta + A_\gamma + A_{meso}} \right) \qquad (5)$$

where A_α, A_β and A_γ are the surfaces corresponding to the characteristic α-,β- and γ-reflections of the (110), (300) and (111) planes respectively. A_{meso} refers to the area of the two mesophase peaks. In Figure 5, the gray lines correspond to this latter phase and, together with the amorphous part (black dashed lines) and the crystal reflections (dotted gray lines), the total (black line) fits the measured patterns very well. The weight fraction of α-, β- and γ-iPP, as well as the crystallinity χ_w, is a mass percentage because the diffraction is proportional to the number of diffracting centers, i.e., the number of crystals, and hence the mass.

2.3.3. Determination of lp, lc and la

Two different methods are used to obtain the long period, the lamellar thickness and the thickness of the amorphous layer. The first one is based on Bragg's law and follows from the integrated Lorentz corrected scattering (SAXS) intensity, given by:

$$I_1(q) = I(q)q^2 \qquad (6)$$

where I is the intensity and q is the scattering vector. The long period then follows via:

$$l_p = \frac{2\pi}{q_{I_1,max}} \qquad (7)$$

with $q_{l_1,max}$ the maximum of the Lorentz corrected intensity. Since the long period is constructed of an amorphous volume and a crystalline volume , it is straightforward that the lamellar thickness l_c follows from:

$$l_c = \frac{2\pi}{q_{l_1,max}} \chi \tag{8}$$

where χ is the volume percentage of the crystallinity. This percentage is obtained from WAXD experiments via [32] :

$$\chi = \frac{\frac{\chi_w}{\rho_c}}{\frac{\chi_w}{\rho_c} + \frac{1-\chi_w}{\rho_a}} \tag{9}$$

In this equation the mass fraction of crystals, χ_w, follows from Equation (4). The amorphous layer thickness can then easily be obtained by subtraction of the lamellar thickness from the long period:

$$l_a = l_p - l_c \tag{10}$$

An alternative way to determine these quantities is via the 1D auto-correlation function $\gamma_1(r)$ [43,44]. For spherical symmetry this can be calculated with:

$$\gamma_1(r) = \frac{1}{Q} \int_{q_0}^{q_\infty} I_1(q) \cos(qr) \, dq, \tag{11}$$

where r is the real space and Q is the scattering invariant defined by:

$$Q = \int_{q_0}^{q_\infty} I_1(q) \, dq, \tag{12}$$

To extrapolate the experimentally assessed range of the scattering vector, Debye-Bueche [45] and the Porod law [46] are used, respectively. The long period, amorphous layer thickness and lamellar thickness were then determined as described by Stein et al. [47]. The most important difference between the determination of these quantities via Bragg's law and the auto-correlation function is that the latter one gives higher values for the lamellar thickness, and thus lower values for the amorphous domain thickness. This originates from the interface between the two phases which is more dense than the bulk amorphous parts and therefore considered to be part of the lamellae in the approach using the autocorrelation function.

Lamellar quantities like the long period l_p can be determined from the scattering in either the meridional region, or the equatorial region, parallel and perpendicular to the tensile direction respectively (see Figure 6).

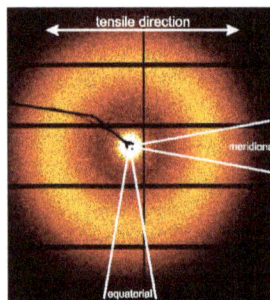

Figure 6. Regions used for radial integration of the SAXS patterns.

This allows one to distinct between these features in either the tensile direction (integration of the meridional region) or the transverse direction (integration of the equatorial region). Obviously, this is only of interest at relatively low strains before lamellae start to break, voiding takes place or, depending on the temperature, the material recrystallizes.

2.3.4. Determination of the Strain

Due to the shape of the tensile samples and the inhomogeneous deformation after yielding, the true strain of the sample volume in front of the X-ray beam cannot be calculated directly from the applied engineering strain. However, the X-ray data can also be used to determine the draw ratio of the sample volume in front of the beam [48]. This allows us to obtain all structural parameters as a function of actual deformation. It should be noticed that, although we intended to create an uni-axial deformation, this was not guaranteed due to the initial sample shape, which was forced by the experimental conditions (i.e., the start of a neck in a known position where the beam is positioned), and the complex plastic deformation mechanisms accompanied with localization phenomena and crystal phase transitions. Moreover, the deformation mode (anything between uni-axial and planar) can change during the deformation path and these changes can be temperature, deformation rate, void formation and phase dependent. To deal with this complex situation we will analyze our experimental data by considering two limiting cases: pure uni-axial and pure planar deformation.

To determine the true strain, the draw ratio λ_l in the tensile direction, defined as:

$$\lambda_l = \frac{l_t}{l_0} \tag{13}$$

should be calculated first. Here, l_0 is the initial length of an arbitrary volume in front of the beam at $t = 0$, and l_t is the length at time t. If we now assume incompressibility, we get:

$$l_0 w_0 d_0 = l_t w_t d_t \tag{14}$$

Combining this with Equation (13), and with the assumption of uni-axial deformation, i.e., the contraction in the tensile-, width- and thickness direction is equal, the width at $t = 0$ (w_0) and at time t (w_t) can be substituted by the thickness d_0 and d_t respectively, and Equation (15) can then be applied to obtain the actual draw ratio. This calculated ratio is resulting from the decrease of polymeric material in the beam area. Throughout the entire stretching process and the associated deformation of the test specimen, the sample remains larger than the X-ray beam. The latter one has dimensions of 0.3×1 mm^2, which is clearly smaller than the final sample dimensions.

$$\lambda_{l,uni} = \frac{1}{\lambda_d^2} = \left(\frac{d_0}{d_t}\right)^2 \tag{15}$$

The sample thickness at time t, defined as d_t, can be straightforwardly calculated using Equation (1), and at time $t = 0$ the initial thickness d_0 follows from [48]:

$$d_0 = \mu ln(I_{1,t=0}/(C \cdot I_{2,t=0})) \tag{16}$$

Substitution of Equations (1) and (16) into Equation (15) allows us to determine the draw ratio on the local level of the beam spot, without determination of μ. This ratio is determined on a homogenized volume since the beam dimensions are much larger than the spherulites and the lamellae. To calculate the draw ratio for plane strain conditions, Equation (17) is used.

$$\lambda_{l,pl} = \frac{1}{\lambda_d} = \left(\frac{d_0}{d_t}\right) \tag{17}$$

In this case, the draw ratio in tensile direction equals the draw ratio in thickness direction, but the sample width remains constant upon stretching. The true strain and true stress then follow via:

$$\epsilon_{true,n} = \ln(\lambda_{l,n}) \tag{18}$$

and:

$$\sigma_{true,n} = \sigma_e \cdot \lambda_{l,n} \tag{19}$$

for both the assumptions of uni-axial deformation and plane strain conditions. In the latter equation, σ_e is the engineering stress. The subscript n in this equation denotes either uni-axial deformation or plane strain conditions.

The approach presented here, which is based on the intensity drop as a result of the sample presence, can be used if the following assumption holds:

$$I_1 = I_{trm} + I_{rfl} + I_{abs} + I_{sct} \approx I_{trm} + I_{abs} \tag{20}$$

where I_1 is the incident beam intensity, I_{trm} is the transmitted intensity which is measured with the photo-diode, I_{rfl} is the reflected intensity, I_{abs} is the absorbed intensity and I_{sct} is the scattered intensity. However, if voids appear, large density differences are introduced and as a result the scattered part of the intensity is no longer negligible with respect to the transmitted part. Since the scattering from the voids takes place under a very small angle (due to the relatively large length scales involved), the intensity measured in the photo diode during the WAXD experiments is the sum of the scattering due to the voids, and the transmission. This means that for the WAXD experiments the transmitted intensity can be determined even if voids appear, and Equations (1) and (16) can be applied. To obtain the strain for the SAXS experiments, where this is not the case, simultaneously recorded WAXD patterns (Pilatus 3 K) were linked to the 2D-WAXD patterns via superposition, indicated with the rectangular section in Figure 7a. The white lines indicate the slice-shaped area used for radial integration. In Figure 7b, an example of the relation of the SAXS frames as a function of the WAXD frames is shown. The color represents the relative difference between the scaled integrated intensity as a function of the scattering vector q of the 2D- and the 1D WAXD patterns. The red markers highlight the minimum. With this coupling, determined for all sets of SAXS-, and their corresponding WAXD experiments, the evolution in sample thickness can be obtained.

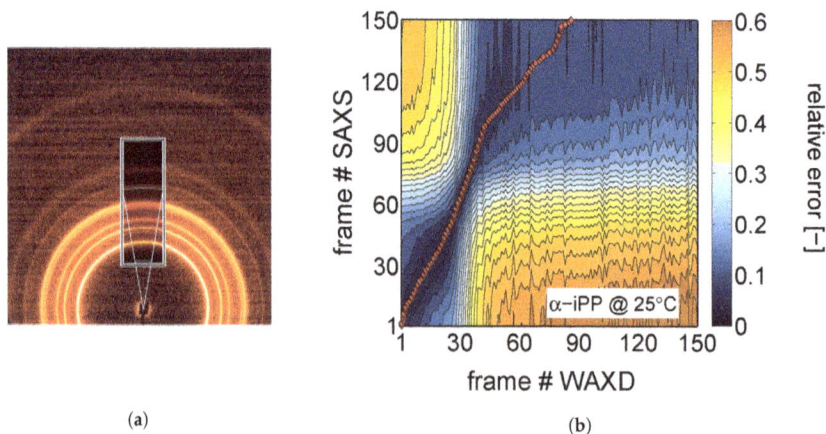

Figure 7. (a) Example of a WAXD pattern measured with the Frelon detector. In the inset the WAXD pattern measured simultaneously with SAXS is put into the 2D WAXD pattern; (b) An example of a figure used to find the minimal difference between the two WAXD patterns. Numbers on the axes indicate the frame number.

On a very local level, an estimate for the strain can be deduced from the changing distance between the crystal planes. Upon stretching, the evolving d-spacing can be used, similar to Xiong et al. [30] according to:

$$\epsilon_{d-space} = \frac{d_{hkl} - d_{hkl,0}}{d_{hkl,0}} \tag{21}$$

to express the evolution in terms of a strain. In this equation d_{hkl} is the d-spacing of a specific plane hkl in time, and $d_{hkl,0}$ is the spacing at time $t = 0$.

2.3.5. Void Fraction

Initially the scattering is caused by the crystals and amorphous phase, but upon deformation, voids appear, and the scattering resulting from these voids becomes dominant. This is accompanied by a strong increase in the scattering invariant due to the large difference in the density of the material and a void. For the calculation of the scattering invariant, cylindrical symmetry is assumed in the loading direction. The normalized scattering invariant is then given by [42]:

$$\frac{Q}{Q_0} = \frac{\int\limits_{-\infty}^{\infty} \int\limits_{0}^{\infty} I_{cor}(q_x, q_y) q_y dq_x dq_y}{\int\limits_{-\infty}^{\infty} \int\limits_{0}^{\infty} I_{cor,t=0}(q_x, q_y) q_y dq_x dq_y} \tag{22}$$

where q_y and q_x are the scattering vector in vertical and horizontal direction respectively. Q_0 is the invariant at time $t = 0$, prior to deformation.

Under the assumption that the we start from a situation without cavities, the void fraction can be calculated directly form the invariant according to [42]:

$$\phi_v = \left[\frac{Q}{Q_0} - 1 \right] \cdot \left[\frac{\chi \rho_c^2 + (1-\chi)\rho_a^2}{\chi(1-\chi)(\rho_c - \rho_a)^2} - 1 \right]^{-1} \tag{23}$$

with ρ_c the crystal density, given by:

$$\rho_c = \frac{\chi_\alpha}{\chi} \cdot \rho_\alpha + \frac{\chi_\beta}{\chi} \cdot \rho_\beta + \frac{\chi_\gamma}{\chi} \cdot \rho_\gamma + \frac{\chi_{meso}}{\chi} \cdot \rho_{meso} \tag{24}$$

where χ_i is the mass fraction of crystal phase i, with density ρ_i ($i \sim \alpha, \beta, \gamma, meso$). The values used for the density of the specific crystal phases are given in Table 1, as well as the density of the amorphous phase. The fractions can be obtained from Equation (5).

Table 1. List of densities used in Equation (24).

	α	β	γ	*meso*	*amorphous*
ρ (kg/m^3)	946	921	938	880	850

3. Results and Discussion

Several strain and stress definitions are used in this study depending on the type of results presented. The engineering stress σ_e and strain ϵ_{app} (also referred to as apparent macroscopic strain) are calculated with the initial gauge length and the initial cross sectional area. These are used when we intend to link structural transitions to the macroscopically observed tensile behavior. When we want to clarify transitions, the true stress σ_{true} and strain ϵ_{true} are sometimes used. To obtain the strain hardening modulus we use $\lambda^2 - 1/\lambda$ as a strain measure [39] and finally, for the strain determined at the local scale of the lamellae, ϵ_l is used.

3.1. The Mechanical Response

3.1.1. Tensile Tests (Mechanical)

In Figure 8a,c,e, the engineering stress as a function of macroscopic apparent strain is shown for tensile tests, performed on different iPP-polymorphs and at multiple temperatures. Initially, linear elastic behavior can be observed, but with increasing strain and stress the deformation becomes plastic, ultimately leading to yield. Temperature facilitates the mobility and, therefore, has a reducing effect on the resistance against yielding. This can also be seen in Table 2, where an overview of the obtained yield stresses at different temperatures is given for α-, β- and γ-iPP.

Table 2. Yield stresses obtained in the tensile experiments.

	25 °C	50 °C	80 °C	110 °C
σ_y of α-iPP (MPa)	32.0	22.3	12.6	8.4
σ_y of β-iPP (MPa)	28.5	20.4	13.2	7.9
σ_y of γ-iPP (MPa)	32.3	22.8	13.6	7.9

The two limiting cases of uni-axial extension and plane strain of α-, β- and γ-iPP are shown in Figure 8b,d,f. Up to yielding, where the deformation is close to homogeneous, the discrepancy between uniaxial and planar deformation is mainly observed in the initial elastic stiffness. After yielding a cross-over takes place and, depending on the amount of softening seen in the engineering stress, the uni-axial response increasingly deviates from the plane strain response. For β-iPP, with almost no softening, the difference is rather small.

Although in agreement with the findings presented by G'Sell et al. [49], the true stress-true strain results obtained in the tensile experiments are different from the ones from uni-axial compression, where softening is observed and the yield stress is much higher (see Figures 9a and 10a). First of all, this discrepancy can result from the formation of cavities, that can develop due to negative hydrostatic stresses in the case of the tensile experiments. The cavities appear on a local level and cause macroscopic softening to (partially) disappear [50], even though locally the softening is maintained. This phenomenon is called sequential yielding and can start prior to macroscopic yield, as will be shown in the following sections. Second, the stress and strain fields in a tensile experiment can become highly inhomogeneous after yield, and cause the sample to deform with a variable strain rate. As soon as the sample softens, locally the strain rate strongly increases. This has a reducing effect on the amount of softening observed in the macroscopic mechanical response. Finally, the sample preparation procedure is slightly different for compression and tensile experiments, also having a minor effect. In order to prepare compression samples, slightly thicker plates were compression molded. Because of that, the average cooling rates upon quenching in a cold press are slightly lower. This could lead to small differences in crystallization temperature and lamellar thickness, which manifests itself in the mechanical response in terms of the observed yield stress [51]. Furthermore, extreme differences in cooling rate and crystallization temperature can also have an affect on the strain hardening modulus [32]. In this case, however, the differences in processing are so minor, that it is assumed that it can be neglected.

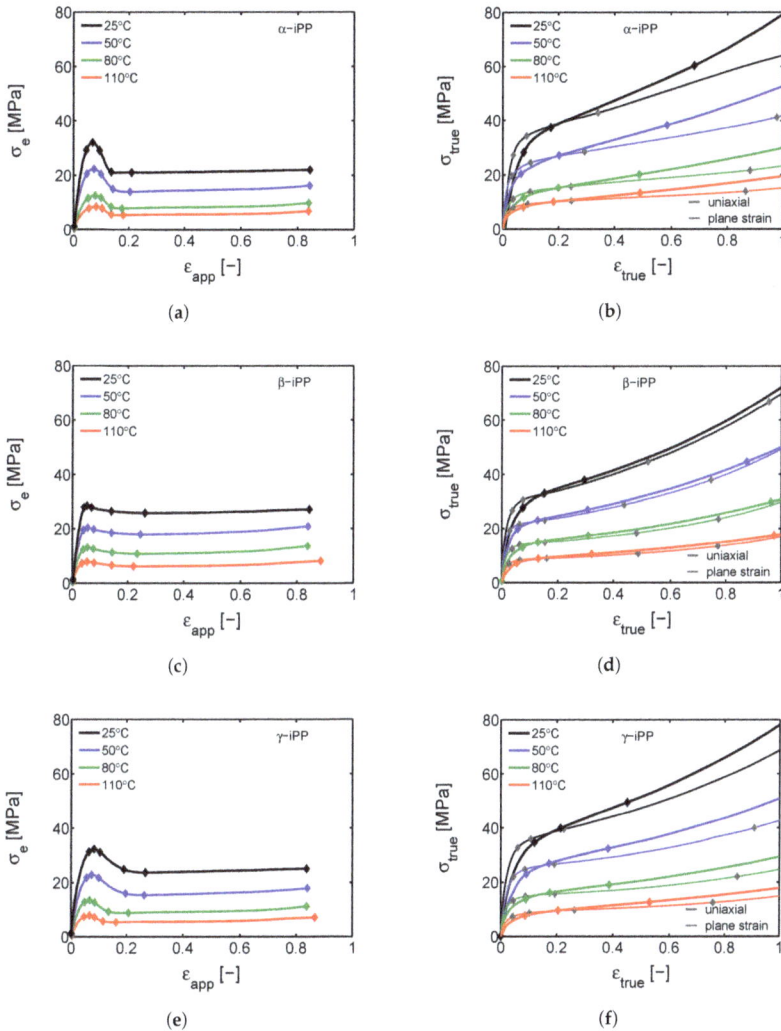

Figure 8. (**a**,**c**,**e**) The engineering stress as a function of the apparent strain for α-, β- and γ-iPP tensile experiments respectively, stretched at different temperatures and a rate of 12 μm/s; (**b**,**d**,**f**) The true stress as a function of true strain, corresponding to figure (**a**,**c**,**e**). The thick solid lines are calculated for the assumption of uni-axial deformation and the thin lines for plane strain. The markers in the figures correspond to the 2D WAXD and SAXS images, shown in the following sections.

3.1.2. Compression Tests (Mechanical)

Since the amorphous network is crucial for the mechanical response at high strains [33], the strain hardening modulus G_r is determined from compression experiments. From tensile experiments it is not possible to find a qualitative value for the hardening modulus due to the difficulties discussed before. The compression experiments, on the other hand, are perfectly suitable for this goal since localization phenomena are excluded and a well defined constant true strain rate can be applied. Consequently, the intrinsic true stress and true strain can be obtained and the strain hardening modulus can straightforwardly be determined according to Haward [39]. In Figure 9b a Gaussian plot, i.e., the true stress as a function of $\lambda^2 - 1/\lambda$, is shown, for α-, β- and γ-iPP compressed at room temperature with a true strain rate of 10^{-3} s^{-1}. The dashed lines are compression experiments performed under the same loading conditions, but on thermally rejuvenated samples as explained in our previous work [12]. Linear fits, determined in the hardening regime, are represented by the dotted lines. These slopes give an estimation for the strain hardening modulus G_r. It is clear that β-iPP has the highest strain hardening modulus (4.2 MPa) and α-iPP the lowest (2.2 MPa). The strain hardening modulus of the γ-iPP is in between (3.5 MPa). Moreover, the thermal rejuvenation treatment does not seem to affect the hardening (in this strain regime), but reduces the softening significantly. Good agreement is found when comparing the hardening modulus of α-iPP with the results reported by Schrauwen et al. [32] and Haward [39]. To the knowledge of the authors no research has been devoted to the determination of hardening moduli for β- and γ-iPP.

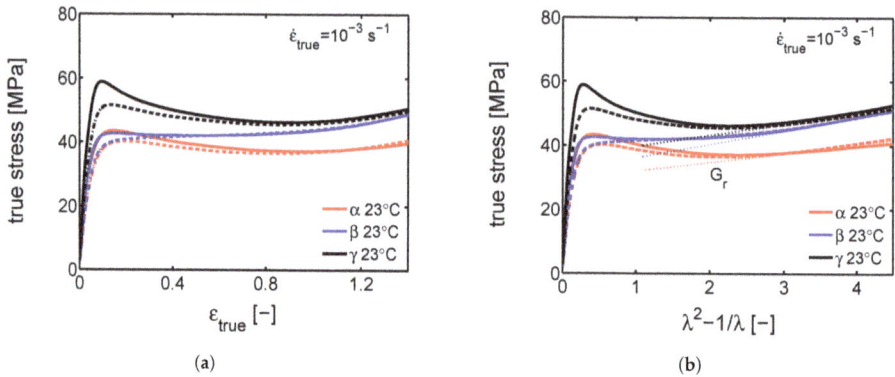

Figure 9. (**a**) True stress-true strain curves for α-, β- and γ-iPP. The results are obtained from uni-axial compression experiments at a strain rate of 10^{-3} s^{-1} and a temperature of 23 °C on samples with a dimension of Ø6 × 6 mm^2 (β- and γ-iPP) or Ø3 × 3 mm^2 (α-iPP). The dashed lines are the true stress-true strain response obtained on thermally rejuvenated samples; (**b**) The corresponding Gaussian plots for α-, β- and γ-iPP. The dotted lines represent the strain hardening moduli G_r.

In our previous work [12] we showed that the contribution of the constrained amorphous phase to the effect of thermal rejuvenation on the yield stress, is almost similar for α- and β-iPP, independent of the strain rate. This thermal rejuvenation treatment has no effect on the hardening modulus, see Figure 9b, but mainly leads to a reduction in the true stress-true strain response at strains around yielding. After yielding the thermally rejuvenated samples coincide again with the aged samples. This observation is in good agreement with the behavior typically observed in aged and, either thermally or mechanically, rejuvenated amorphous polymers [52].

Softening after yielding is a property of an amorphous glassy material. The degree of softening is a direct consequence of the amount of aging a sample has experienced, prior to the mechanical testing. In the case of iPP, at a temperature above the glass transition of the bulk material, and below the melting temperature of the crystals, softening is related to the constrained amorphous phase (which is considered to be in a glassy state). In the vicinity of the relatively immobile crystalline domains, the mobility of the amorphous material is strongly reduced. The extent of the constraints can depend on a number of morphological and crystallographic features. The strength of the secondary interactions in the crystal is related to the density, which is the lowest in case of β-iPP (see Table 1). The lamellar thickness and the density of crystal defects also affect the strength of the constraints. Furthermore, the crossed stacking of chains within the lamellae of γ-iPP is also positively contributing to the constraints. All these constraints are directly coupled to the properties of the crystal. With thermal rejuvenation softening partially disappears. Particularly in γ-iPP a large reduction in yield stress is found, but also in α- and β-iPP a decrease is observed, indicating that the iPP contains constrained amorphous material in all three crystal phases. At 110 °C, where the constrained amorphous material is mobile (above the α_c-relaxation temperature), both α- and γ-iPP still show some remaining softening, see Figure 10. This means that as a result of deformation, the structural integrity further deteriorates after yielding. As can be observed from Figure 20 in Section 3.2.2, this is not governed by a large reduction in crystallinity. Moreover, this reduction is very similar for the β-iPP. Based on these observations it could be hypothesized that the break-down of the cross hatched crystal network is responsible for the softening in α- and γ-iPP, without destroying the crystals themselves, since these structures are typically present in these polymorphic forms [2,53,54]. In this perspective, the absence of these structures in the case of β-iPP could explain why the mechanical response displays no softening. The trends in the strain hardening moduli at 110 °C are similar to that at 25 °C.

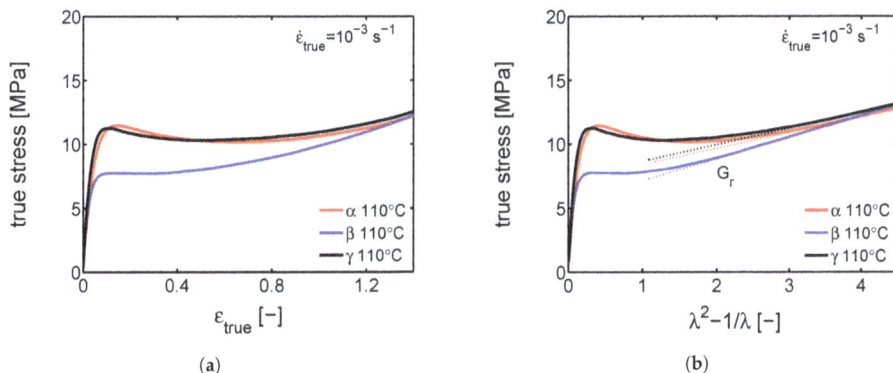

Figure 10. (a) The solid lines are the true stress as a function of the true strain obtained from uniaxial compression experiments on α-, β- and γ-iPP, measured at a strain rate of 10^{-3} s^{-1} and a temperature of 110 °C on samples with a dimension of Ø6 × 6 mm^2 (β- and γ-iPP) or Ø3 × 3 mm^2 (α-iPP); (b) The corresponding Gaussian plots for α-, β- and γ-iPP. The dotted lines represent the strain hardening moduli G_r.

To summarize, from the mechanical behavior in uni-axial compression it is hypothesized that the intrinsic softening observed in polypropylene is an effect, mainly resulting from the constrained amorphous domains in the vicinity of the crystals. However, as follows from the compression experiments performed at 110 °C, combined with the results of the WAXD experiments presented in Section 3.2.2, suggests that the constraints implied by the cross-hatched structures, present in α- and γ-iPP, also contribute to the softening.

3.2. WAXD Analysis

3.2.1. Tensile Tests (WAXD)

Figure 11 shows the evolution of the α-iPP crystal structure during tensile testing in terms of the normalized 2D WAXD patterns, as a function of strain for 4 different temperatures. Since most of the structural changes take place in the macroscopically observed softening regime, i.e., in a small apparent strain range, it is chosen to depict the changes as a function of uniaxial true strain for clarity reasons. The patterns depicted here correspond to the markers in Figure 8. Initially, before straining the sample, an isotropic diffraction pattern is observed which clearly shows the monoclinic α-phase reflections. Upon straining the sample, the pattern becomes slightly elliptical, indicating that the d-spacings of the crystals in the polar regions are extended whereas the distances in the equatorial regions reduce. As a result of further straining, the reflections become less pronounced and the scattering intensity migrates to certain angles, indicating (strong) orientation. At high elongation at room temperature a transition from α-iPP to oriented mesophase can be observed [55], whereas at high temperatures the isotropic α-iPP transforms to a strongly oriented α configuration.

This azimuthal orientation is investigated in more detail by integrating the (110) diffraction peak over an angle ranging from 0 to 180°. An angle of 90° represents the equator, while 0 and 180° are the polar regions. This integration was carried out for α-iPP stretched at the lowest and the highest temperature, i.e., corresponding to the top and bottom row in Figure 11, and the results are given in Figure 12. Due to the strong orientation and the high intensity at an azimuth of 90° the minor effects at lower strains are unclear. In the inset an enlarged plot of this region is shown. The intensity transforms from nearly isotropic in the initial stage of deformation to a slightly oriented state prior to yielding.

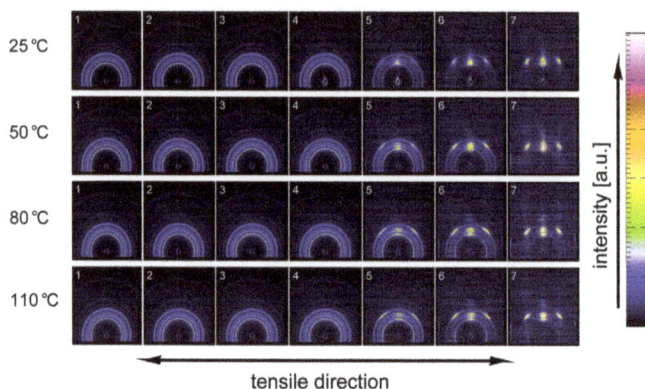

Figure 11. Normalized 2D WAXD patterns of α-iPP stretched at temperatures of 25, 50, 80 and 110 °C from top to bottom. The true strains, determined with the assumption of fully uni-axial deformation, are given as well. The macroscopic strains at which the patterns were taken are indicated by the markers in Figure 8. The stretching direction is horizontal.

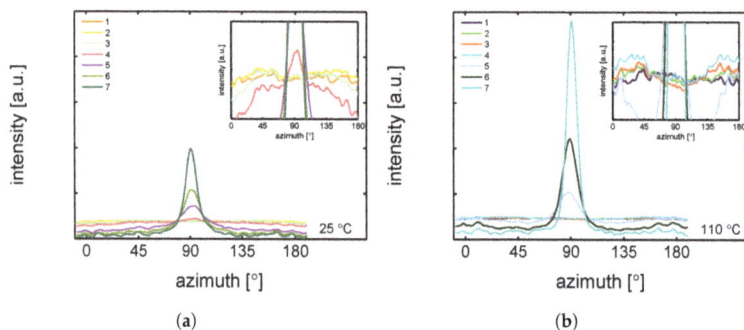

Figure 12. Azimuthal spread of the (110) diffraction of α-iPP at various strains; (**a**) uni-axial stretching at 25 °C and (**b**) 110 °C. The numbers in the legend correspond to the 2D patterns in Figure 11.

Straining the sample further leads to selective melting and recrystallization, and orientation of "old crystals" into a strongly oriented mesophase at 25 °C, and α-phase at 110 °C. The peak in the azimuthal integration, representative for these oriented structures shows up first in the yield point and evolves in the softening and hardening regions thereafter. Small intensity maxima at 10° and 170° degrees, observed in the material stretched at 110 °C, indicate the survival or appearance of cross-hatched structures in the stretched sample [56].

A similar figure can be made for β-iPP, see Figure 13. In the initial stages of deformation the isotropic diffraction rings resulting from the pseudo-hexagonal β-phase structure become elliptical due to the (elastic) deformation. Contrary to α-iPP at room temperature, the diffraction of the β-iPP crystals present in the initial isotropic crystal structure seems to be partially maintained during stretching. The diffraction of the β-iPP is more intense than the reflections of the α- and γ-iPP. The scaling of all 2D-WAXD images in this work is the same, and as a result, a larger part of the crystals has to be destroyed to get rid of the β-diffraction peaks. Moreover, due to different localization behavior, the uni-axial true strain belonging to β-iPP in front of the beam is lower than for α- and γ-iPP, even if the same amount of macroscopic strain is applied. Consequently, after finishing an experiment that is stopped at the same macroscopic strain for all iPP samples, the actual true strain reached by β-iPP in front of the beam is clearly lower. This makes a direct comparison complicated. The other part of the initial structure that is destroyed forms new oriented structures. At room temperature the diffracted intensity appears at positions typical for oriented mesophase. At elevated temperature and upon deformation, a transition from β-crystals to oriented and more stable α-crystals takes place.

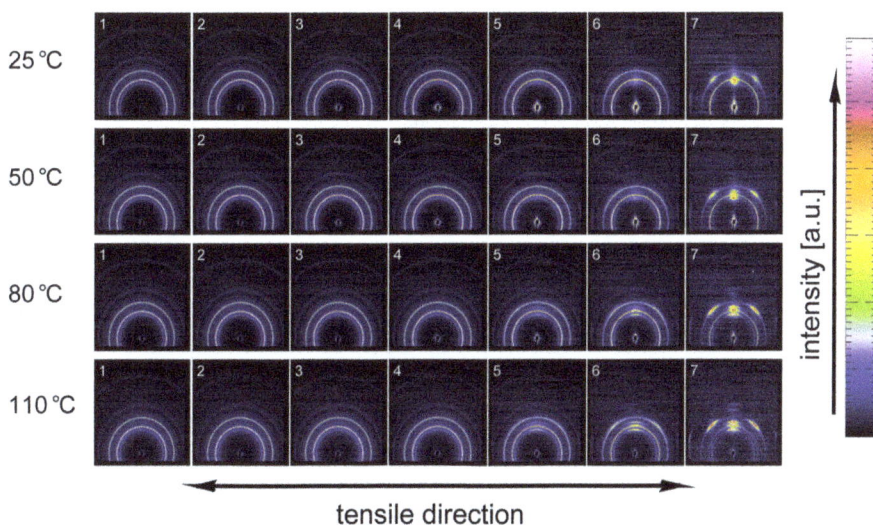

Figure 13. Normalized 2D-patterns of β-iPP stretched at temperatures of 25, 50, 80 and 110 °C from top to bottom. The true strains, determined with the assumption of fully uni-axial deformation, are given as well. The macroscopic strains at which the patterns were taken are indicated by the markers in Figure 8. The stretching direction is horizontal.

From an azimuthal integration over the (300) diffraction peak of β-iPP, deformed at 25 and 110 °C (Figure 14), it becomes clear that the amount of material involved in partial melting and/or orientation is much less (as a function of apparent macroscopic strain) as compared to α-iPP. Only in the last few frames a clear decrease in the intensity at an azimuth of 0 and 180° is observed. The deformation of the initial structure to preferred orientations on the other hand, continues also after the yield point (frame 3). At room temperature, as well as at 110 °C, the first evidence of recrystallization in a new oriented phase is found in frame 5, corresponding to deformation well beyond the macroscopic yield point.

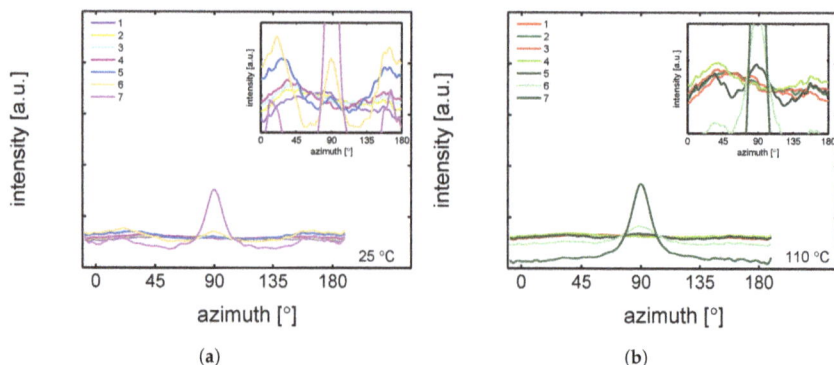

Figure 14. Azimuthal spread of the (300) diffraction of β-iPP at various strains. (a) uni-axial stretching at 25 °C and (b) 110 °C. The numbers in the legend correspond to the 2D patterns in Figure 13.

The reflections of the orthorhombic γ-phase become very vague upon deformation at 25 °C, and the scattering transforms into a pattern indicative for oriented mesophase, see Figure 15. Stretching at 50 °C result is a similar response, however, besides mesophase some small amounts of oriented α-iPP seems to be formed. Elongation at 80 and 100 °C results in a transition of isotropic γ-iPP into strongly oriented α-iPP. All these transitions take place after yielding. These 2D-patterns provide qualitative information about the orientation resulting from the stretching.

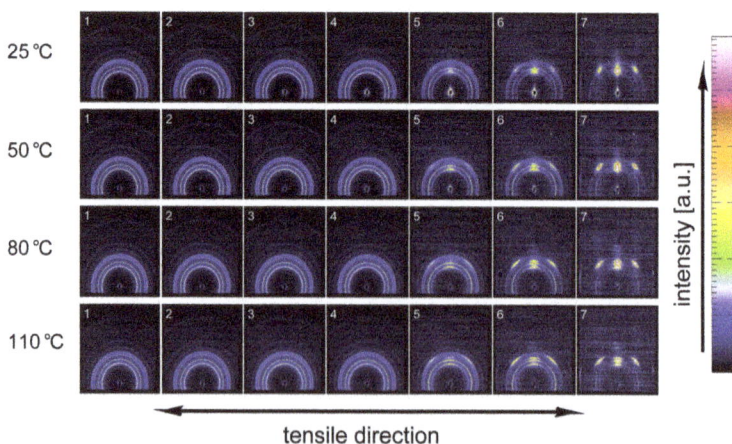

Figure 15. Normalized 2D-patterns of γ-iPP stretched at temperatures of 25, 50, 80 and 110 °C from top to bottom. The true strains, determined with the assumption of fully uni-axial deformation, are given as well. The macroscopic strains at which the patterns were taken are indicated by the markers in Figure 8. The stretching direction is horizontal.

The azimuthal scan of the (111) peak and the evolution as a result of the deformation is given in Figure 16, for the γ-samples deformed at 25 and 110 °C. Interestingly, the intensity remains rather isotropic until the onset of the softening after yielding (frame 5–7). The absence of the off-axis orientation means that the γ-crystals do not allow for partial orientation of the initial crystallites. After yielding, when selective melting and recrystallization takes place, the intensity of the (111) peak in the polar regions starts to decrease and the newly formed structures appear at an azimuth of 90°.

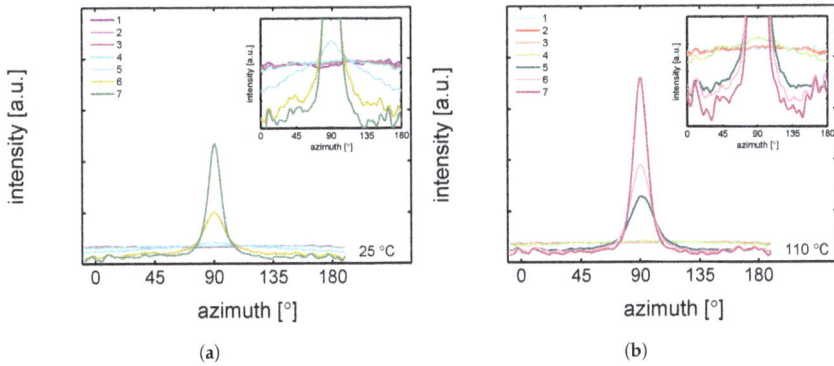

Figure 16. Azimuthal spread of the (111) diffraction of γ-iPP at various strains. (**a**) uni-axial stretching at 25 °C and (**b**) 110 °C. The numbers in the legend correspond to the 2D patterns in Figure 15.

In Figure 17 the results of radial integration of the patterns measured on all crystal phases, stretched at testing temperatures of 25 and 110 °C are plotted as a function of the scattering vector q and the true strain ϵ_{true}. At 25 °C the reflections of α- and γ-iPP become less intense and broaden, particularly in the macroscopic softening region, corresponding to true strains of about 0.2 to 2 [-]. The β-crystals, on the other hand, seem to maintain their crystallographic structure, and the biggest change observed is broadening of the peaks. At 25 °C and high strain, α- and γ-iPP eventually transform into a 1D intensity typical for the mesophase. In the figures corresponding to elongation at 110 °C it is evident that the α-reflection seems to be maintained rather well, whereas for β and γ-iPP the characteristic reflections disappear, simultaneously with the appearance of the α-reflection. The reflections remain sharp with well defined peaks up to high true strains.

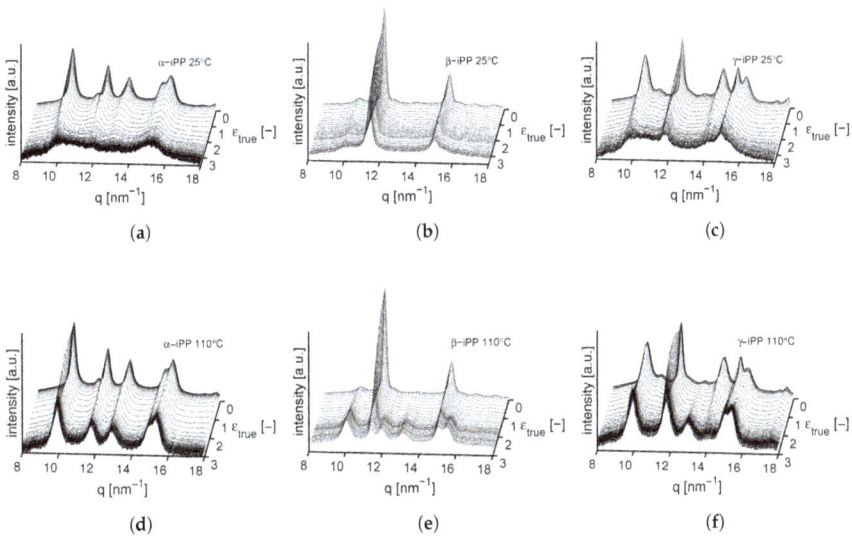

Figure 17. The integrated intensity as a function of the scattering vector q. From (**a–c**) we see the evolution of the crystallinity upon stretching at 25 °C of the α-, β- and γ-iPP respectively. Similar results, obtained from stretching experiments performed at 110 °C, are shown in (**d–f**). The structural evolution is given as a function of the true strain.

The results of the approach, discussed in Section 2.3.2, to quantify the crystal phase composition, are given in Figure 18. Here, together with the apparent macroscopic stress-strain response, the phase content of the samples deformed at 25 and 110 °C, is shown as function of the macroscopic strain. Results of tensile experiments performed at 50 and 80 °C can be found in the Appendix Figure A1.

The difference in the structural evolution at room temperature of the α-, β- and γ-samples that immediately draws the attention is the strong and drastic crystal destruction during the macroscopic softening for α- and γ-iPP, which is contrasting to the slow and gradual changes in β-iPP. In α-iPP, almost all the crystals that are initially present in the sample are being destroyed at low temperatures (25 °C), and either transform to oriented mesophase or amorphous phase. At higher temperatures, the amount of newly formed mesophase is lower and ultimately, at a drawing temperature of 110 °C reduces to zero. At the highest temperatures the α-iPP fraction slightly changes, however, based on the results presented in Figure 12 it is known that the crystals partially melt and orient, and that the sample subsequently recrystallizes. The gradual transitions observed upon deformation of β-iPP are also observed at elevated temperatures. Where the transition at low temperatures is mainly from β to amorphous or mesophase, the recrystallization at elevated temperature gives α-iPP. The overall crystallinity slightly decreases at all drawing temperatures. The γ-iPP transforms partially to α-iPP already at a temperature of 50 °C (see the Supporting Information). With increasing temperature the fraction of γ- transforming to α-phase increases, until a temperature of 110 °C where this transition takes place exclusively. To summarize, at low temperatures all polymorphs are being partially destroyed and form amorphous material or oriented mesophase, while at high temperatures all polymorphs transform to oriented α-iPP. In any case the crystallinity slightly decreases. The structural changes observed from X-ray and the coupling to the apparent macroscopic response are affected by geometric effects, rather than pure deformation, due to localization effects. Therefore, in Section 3.2.2, these transitions are investigated by means of uni-axial compression.

For crystal phase transitions to take place, the chains within the crystal structures need to have a certain amount of mobility, induced by either the applied temperature or stress. From the results presented in Figure 18 it is clear that this mobility is mainly achieved after the yield point. From that point on, the material is able to partially melt and form new oriented structures. The deformation at this stage is plastic. In order to determine the onset of plastic deformation in the crystal, the evolution of the d-spacing is used. The increasing distance between crystal planes in the polar regions of the diffraction pattern is reflected in the d-spacing and thus, the peak positions observed in the WAXD frames. In Figure 19, this evolution in d-spacing is depicted in terms of a strain obtained from Equation (21), and presented as a function of the macroscopically applied strain ϵ_{app}.

As expected for purely elastic behavior, a linear increase in the $\epsilon_{d-space}$ is obtained first. Subsequently this increase levels off, indicating that the further increasing stress, transmitted on the crystals, no longer results in the same increase in d-spacing. This leveling off is clearly observed prior to yielding and can be interpreted as the onset of collective stress-induced α-relaxation. Therefore, the onset of this deviation marks the beginning of crystal plasticity. In the case of β-iPP this transition takes place at low macroscopic strains compared to α- and γ-iPP. The transition correlates to a combination of morphological features like density of the crystal on one hand, and lamellar thickness on the other hand. In α- and γ-iPP, the crystals have a similar thickness (see Section 3.3) and density (Table 1). The crossed stacking of chains in the γ-iPP lamellae seems to slightly postpone plastic deformation as well. Note that prior to yielding this comparison can be made since the deformation is (close to) homogeneous.

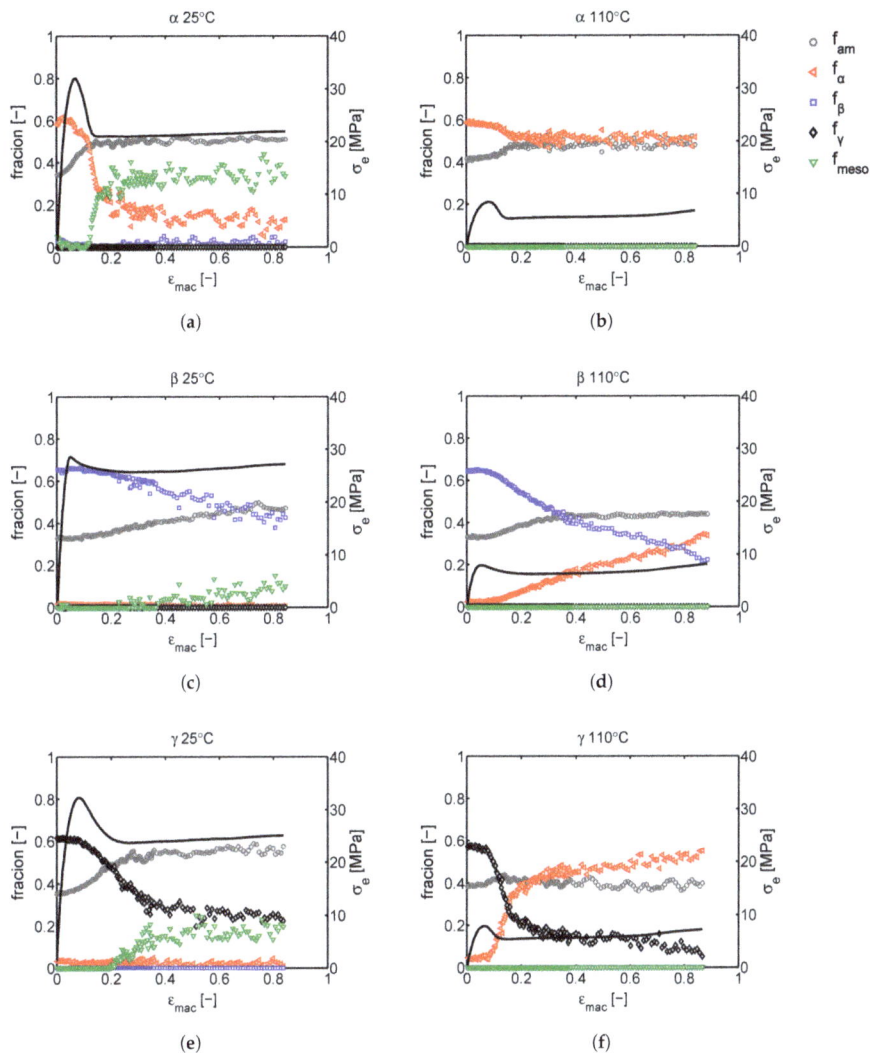

Figure 18. The crystal phase fractions obtained from the WAXD experiments. In (**a,c,e**) we see the evolution of the crystallinity and phase fractions upon stretching at 25 °C. Figure (**b,d,f**) show the result of the tensile experiments performed at 110 °C. From top to bottom we see α-, β- and γ-iPP. The solid lines are the macroscopic engineering stress as a function of the apparent macroscopic strain.

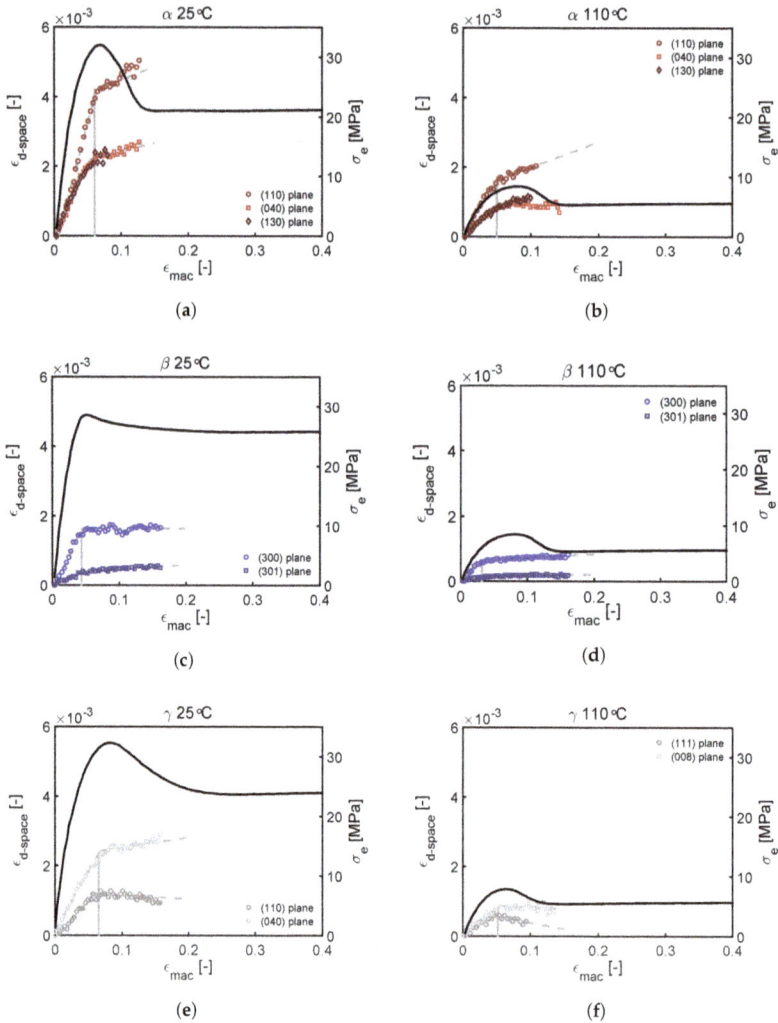

Figure 19. The increase of the distance between crystal planes of (**a,b**) α-iPP; (**c,d**) β-iPP and (**e,f**) γ-iPP. The strain at which the distance between the crystal planes no longer increases following the initial slope is associated with the onset of the plastic deformation. Figure (**a,c,e**) are obtained from tensile tests performed at 25 °C, while (**b,d,f**) are taken at 110 °C.

3.2.2. Compression Tests (WAXD)

Since all the phase transitions are depicted as a function of the macroscopic tensile strain, the question rises how big the effect of localization phenomena are. To answer this question, similar analysis are done on α-, β- and γ-iPP deformed in uni-axial compression experiments at temperatures of 25 and 110 °C, where the applied true strain rate is constant and thus the true stress as a function of the true strain can be obtained, see Figure 20.

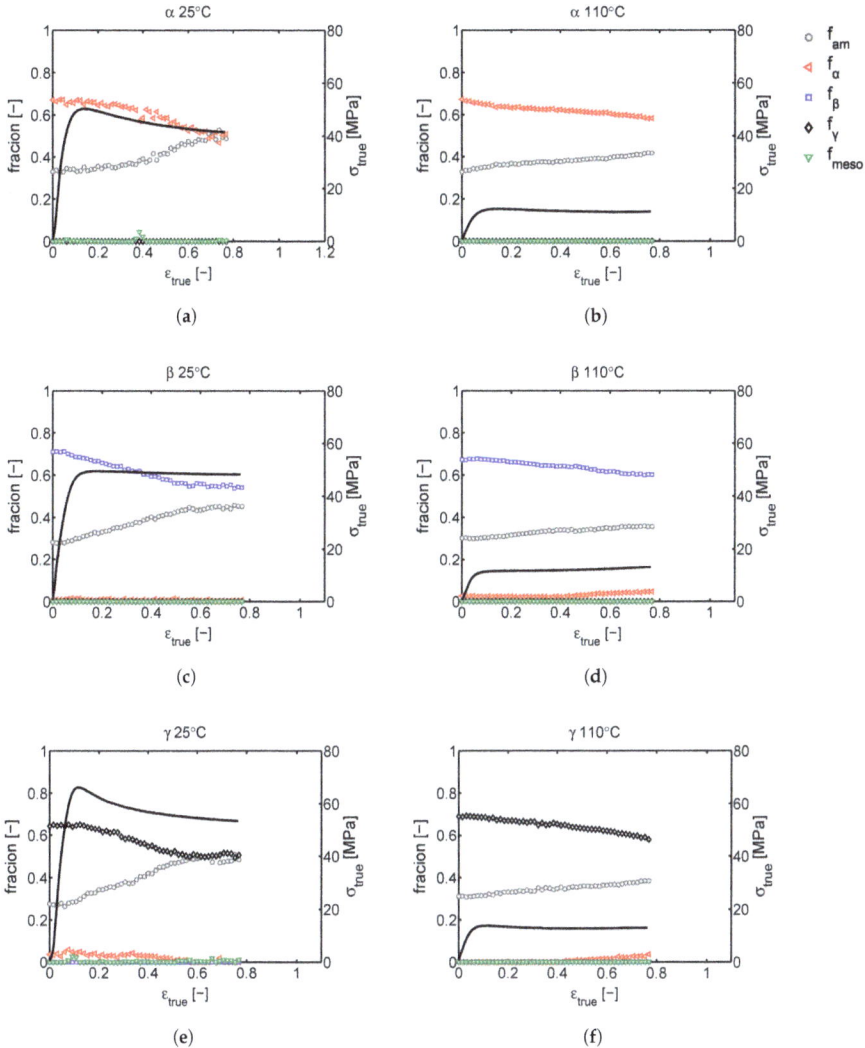

Figure 20. The crystal phase fractions obtained from the WAXD experiments. From top to bottom we see α-, β- and γ-iPP. From left to right we see the evolution of the crystallinity and phase fractions upon compressing at 25 and 110 °C. The true stress as a function of the true strain obtained at a true strain rate of 10^{-2} s^{-1} is shown with the solid black lines.

The apparent conservation of the structural integrity of β-iPP upon deformation as observed from the tensile experiments (Figure 18) is purely a result of reduced localization compared to α- and γ-iPP. In fact, during compression and at low temperatures it even seems as if the destruction of crystals already starts at relatively low strains, while for α- and γ-iPP the crystallinity starts to decrease mainly after yielding. Furthermore, the compression experiments show no softening in β-iPP, independent of the temperature, whereas in the tensile experiments the β-iPP samples do show softening as a result of geometrical effects. These findings compare well with the observations reported by Xu et al. [57]. Although this work only provides a comparison between α- and β-iPP, the observed phenomena are similar to what is found in this study. α-iPP shows more pronounced strain softening and smaller strain hardening compared to β-iPP. To the knowledge of the authors no publications exist that comment on the strain hardening and softening of γ-iPP in uni-axial compression experiments. The early onset of crystal destruction in β-iPP compared to α-iPP, and the slightly later onset in γ-iPP perfectly matches the findings of Lezak et al., who studied the deformation of α-, β- and γ-iPP in plane strain compression at room temperature [21,25] and at elevated temperature [22,58]. From their extended research they concluded that the initiation of plastic deformation is relatively easy in β-iPP as compared to α-iPP since both crystal slip and shear are easier. For γ-iPP on the other hand, the opposite conclusion was drawn, and plastic deformation was found to be more difficult than in case of α-iPP. This matches with the observations presented on crystal deformation in this work, even though in the tensile experiments the initiation and growth of voids can have a large effect on the local stress states present in the material.

3.3. SAXS Analysis

3.3.1. Tensile Tests (SAXS)

To investigate the structural evolution at the nanometer length scale, SAXS data was taken for the same set of experimental conditions as for the WAXD experiments. In Figure 21, the results of the experiments conducted on α-iPP are shown.

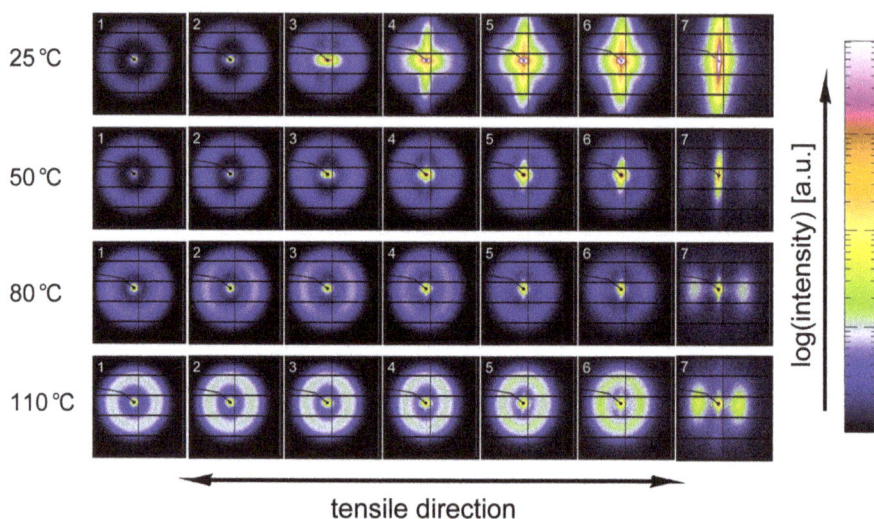

Figure 21. Normalized 2D SAXS patterns of α-iPP stretched at temperatures of 25, 50, 80 and 110 °C from top to bottom. The true strains, determined with the assumption of fully uni-axial deformation, are given as well. The macroscopic strains at which the patterns were taken are indicated by the markers in Figure 8. The stretching direction is horizontal.

The intensity scattered from the lamellae increases with temperature, which can be explained by the difference in thermal expansion between the crystals and the amorphous domains that increases the density contrast. The undeformed samples, with randomly oriented lamellar stacks, give rise to a scattering circle with a homogeneous distribution. Upon deformation, the sample transforms to a state in which the scattering of the lamellae concentrates in the polar regions. At 25 °C voids are already present at the yield point (frame 3), evidenced by the clear lobes near the beam center. From the scattering pattern it becomes clear that the dimensions of the voids initially have the largest dimension in the direction perpendicular to the tensile direction. During strain softening and hardening this gradually evolves into the opposite; the largest dimensions of the voids are now parallel to the stretching direction. This observation is accompanied by a transition from high aspect ratio craze-like features to a highly voided state with shear deformation zones. This transition to micro-necking is related to the entanglement network as was shown by for example Kausch et al. [59]

and Ishikawa et al. [34]. At high temperature no clear scattering as a result of voiding is observed at yield, however, close to the center some kind of non-cylindrical pattern is observed. This is an indication for the formation of voids, but with dimensions larger than the detectable length scales. Finally, it is evident that lamellae of the newly formed crystals are strongly oriented with the so-called lamellar normal of the stacks parallel to the tensile direction, meaning that the iPP chains orient along the tensile direction.

To quantify the lamellar stack orientation upon stretching, azimuthal integrations are performed in a similar way as for the WAXD patterns. In Figure 22 the intensity along the azimuth is shown for the α-sample stretched at room temperature. After yield, in frames 4 to 7, a strong intensity increase in the equatorial regions develops as a result of the transformation of the shape and the growth of fibrillar voids elongated in tensile direction. For this reason, only the first 3 or 4 frames are shown in the inset, i.e., depending on where voiding starts to affect the result. Although the last frame depicted in the insets already shows some features of voiding (development of a peak at 90°), the main intensity distribution during the early stages of deformation is a direct result of the orientation of the lamellar stacks.

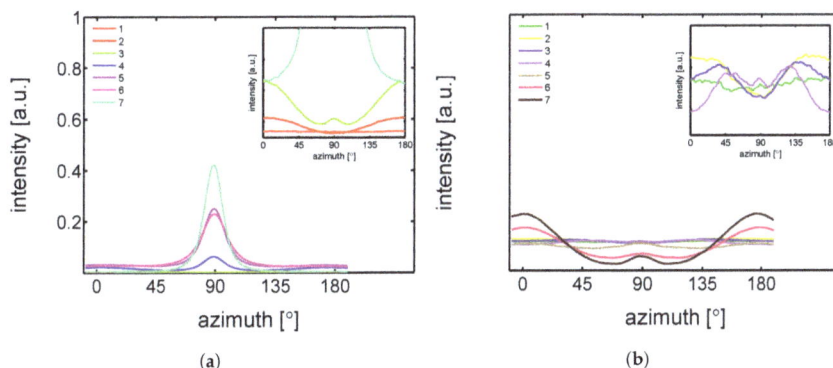

Figure 22. Azimuthal intensity of the lamellar scattering at various strains of α-iPP. (**a**) uni-axial stretching at 25 °C and (**b**) 110 °C. The numbers in the legend correspond to the 2D patterns in Figure 21.

An important observation done at 110 °C is that in the initial stages of deformation the intensity of the lamellae concentrates at angles of 45° and 135°. This phenomenon is often attributed to break-up in between lamellae [26,60–62], following after cavitation. Here, the scattering patterns show no evidence of voids with a length scale similar to that of the lamellae. Combined with the absence of this preferential orientation in tests performed at room temperature, where there is voiding prior to yielding, it is therefore more likely that the material adopts this orientation because it is preferential for plastic deformation. At room temperature, where we are below T_{α_c}, a different mechanism takes place which could indicate that the constrained amorphous network is strong enough with respect to the crystallites, to prevent the material to transform to this orientation.

The 2D SAXS patterns measured on β-iPP are shown in Figure 23. The scattering intensity close to the center is much stronger than that observed in the α-samples, meaning that the samples exhibit much more voids in the domain of detectable void sizes. This holds for all temperatures. The process of voiding starts already in the early stages of deformation.

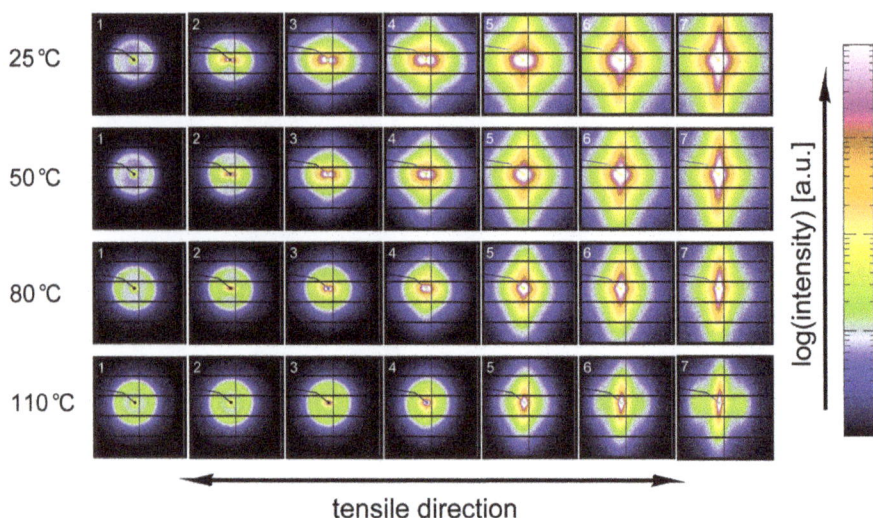

Figure 23. Normalized 2D SAXS patterns of β-iPP stretched at temperatures of 25, 50, 80 and 110 °C from top to bottom. The true strains, determined with the assumption of fully uni-axial deformation, are given as well. The macroscopic strains at which the patterns were taken are indicated by the markers in Figure 8. The stretching direction is horizontal.

The transition from cavities with the largest dimension perpendicular to the tensile direction to cavities with the largest dimension parallel to the tensile direction takes, compared to the α-phase, place at higher strains and seems less clear/unfinished, even in the final stages of the stretching experiment. The scattering from the newly formed crystals appears in the polar regions, indicating that the lamellar normals are oriented parallel to the tensile direction.

The intensities along the azimuth is shown in Figure 24 upon stretching the sample. As can be observed, the tensile experiment was started from a state which is not completely isotropic. Apparently the compression molding process led to minor orientation in the sample. This only becomes clear after zooming in extensively on the result of an azimuthal integration. Since the anisotropy level is so low, it is assumed that this artifact is not of any influence on the further results. At 25 °C the scattering of the voids affects the integrated intensity almost immediately. At 110 °C cavitation is observed first in the third frame, corresponding to the yield point. Although a small anisotropy seems to be present from the beginning of the experiment there is no evidence for a tendency of the formation of a pattern with a orientation at $+45°$ and $-45°$, which is different from α-iPP. These observations are similar to the results found by Men et al., who worked on Poly(1-butene) [63], and claimed cavitation through lamellar breakup.

Finally, Figure 25 contains the SAXS patterns taken from tensile experiments performed on γ-iPP. Based on the scattering intensity it is expected that the void fraction (within the detectable range) is in between that of α- and β-iPP. The onset of voiding is clearly before yielding at low temperatures, and also at high temperatures there is some evidence since the scattering close to the center deviates from circular. At 110 °C the scattering of the original lamellae becomes strongly elliptical, before the selective melting and recrystallization into oriented crystallites with the lamellar normal in tensile direction takes place. Similar to the α- and β-iPP the dimensions of the voids are initially larger perpendicular to the tensile direction and transform to shapes that have the largest dimensions parallel to the drawing direction. This happens beyond yielding and takes place more gradual than in case of α-iPP.

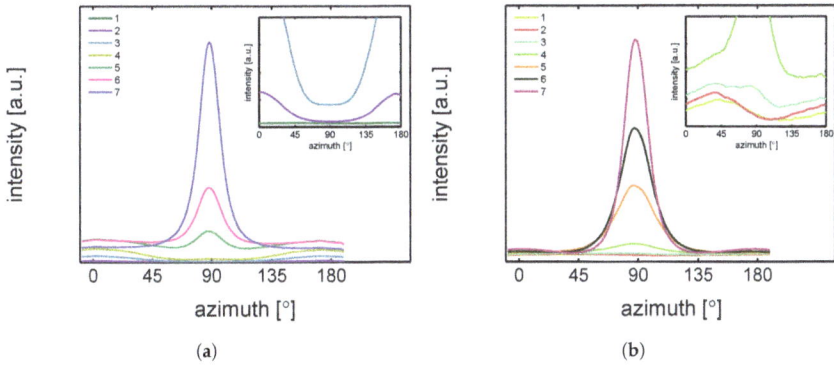

Figure 24. Azimuthal intensity of the lamellar scattering at various strains of β-iPP. (**a**) uni-axial stretching at 25 °C and (**b**) 110 °C. The numbers in the legend correspond to the 2D patterns in Figure 23.

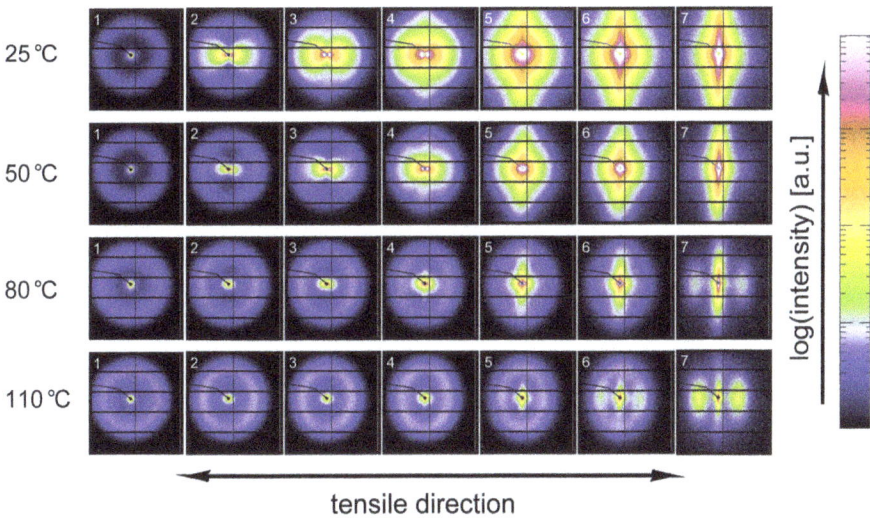

Figure 25. Normalized 2D SAXS patterns of γ-iPP stretched at temperatures of 25, 50, 80 and 110 °C from top to bottom. The true strains, determined with the assumption of fully uni-axial deformation, are given as well. The macroscopic strains at which the patterns were taken are indicated by the markers in Figure 8. The stretching direction is horizontal.

The intensity as a function of the azimuth is depicted in Figure 26 for the tests performed at 25 and 110 °C. The latter has a preferred orientation at angles of 45° and 135°, which indicates that also in γ-iPP the lamellar break-up or orientation takes place prior to yielding, similar to α-iPP.

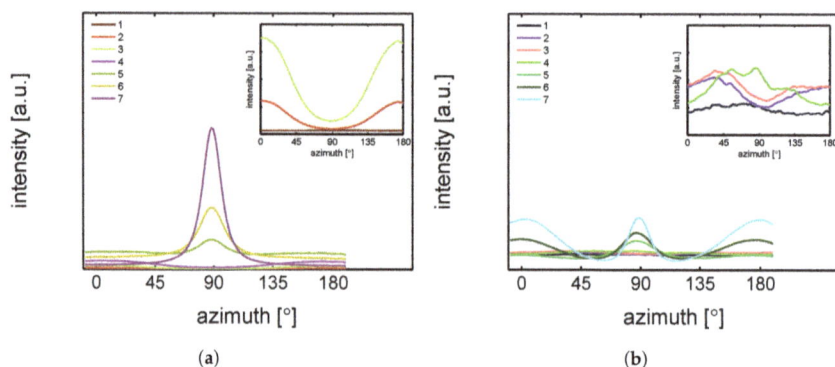

Figure 26. Azimuthal intensity of the lamellar scattering at various strains of γ-iPP. (**a**) uni-axial stretching at 25 °C and (**b**) 110 °C. The numbers in the legend correspond to the 2D patterns in Figure 25.

To quantify the void fraction, the 2D patterns are integrated according to Equation (22), which holds for the assumption of cylindrical symmetry, and by substitution of the result into Equation (23), the evolution of void fraction as a function of the strain is obtained. This procedure was applied on the α-, β- and γ-iPP for all the experimental conditions, and the result is shown in Figure 27. The increase of the void fraction becomes particularly clear at the strain where macroscopic yielding takes place. Based on the 2D images this can be linked to the transition of perpendicularly oriented voids towards voids parallel to the tensile direction, which coincides with a large increase in the scattering intensity. In the case of α-iPP deformed at 25 °C, ϕ_v increases simultaneously with the macroscopic softening. This increase stops at the end of the softening after which the volume fraction decreases. Since a volume fraction is considered, two possible explanations can be given for this observation: (1) The voids grow (or coalesce) to larger dimensions and therefore are no longer in the detectable size domain; (2) Due to the extension of the voids and the resulting cylindrical dimensions, together with the fibrillar material morphology in between, the negative hydrostatic stresses reduce severely, causing the voids to collapse. This also leads to a reduction in volume.

Quantitative analysis of the void size and shape could confirm the aforementioned hypothesis. To this extent, three different approaches to obtain such information are considered. Unfortunately, due to experimental restrictions none of them turned out to give reliable results. First of all, the approach presented in the work of Lode et al. [64] is considered. This approach was originally used to obtain craze dimensions in amorphous polycarbonate. To successfully fit the parameters of the model of Lode et al., the craze fibril scattering on the equator should show a maximum. Since in the situation at hand, the mean fibrillar spacing is too large, this intensity maximum disappears in the beam stop, making a fit of the parameters highly unreliable. If one wants to use this approach for iPP, USAXS or SALS could possibly be a good alternative for SAXS. Secondly, void dimensions could also be obtained by means of the radius of gyration, as is for example demonstrated by Zafeiropoulos et al. [65], Na et al. [19] or Pawlak et al. [61]. The deformed iPP in this work contains voids of polydisperse dimensions (different populations) and shapes (cylinders, ellipsoids, spheres, etc.). For this reason, an approach similar to Na et al. [19] and Pawlak et al. [61] had to be chosen, in which the total void volume is divided over three populations, each with their own dimension. Unfortunately, the dimensions where again to large to obtain reliable results. Moreover, the void volume does not consist of three discrete void sizes (in practice it is a continuous distribution), making the approach rather sensitive for the arbitrary constraints applied in the optimization routine. Finally, a method introduced by Ruland [66] is considered. This method is successfully used by Lu et al. [20], who obtained dimensions of cavities in iPP. The applicability of this approach depends on the presence of fiber-like entities, which is only the case at very high strains, possibly even higher than the maximum strains reached in this work. Indeed, the application of this method resulted in erroneous void dimensions. In addition to the possible lack of sufficient fiber-like shapes, the setup is aligned such, that the focus is on the sample, rather than on the detector. To summarize, dimensions and shapes of the cavities could not straightforwardly be determined in a reliable way, with either one of the discussed approaches.

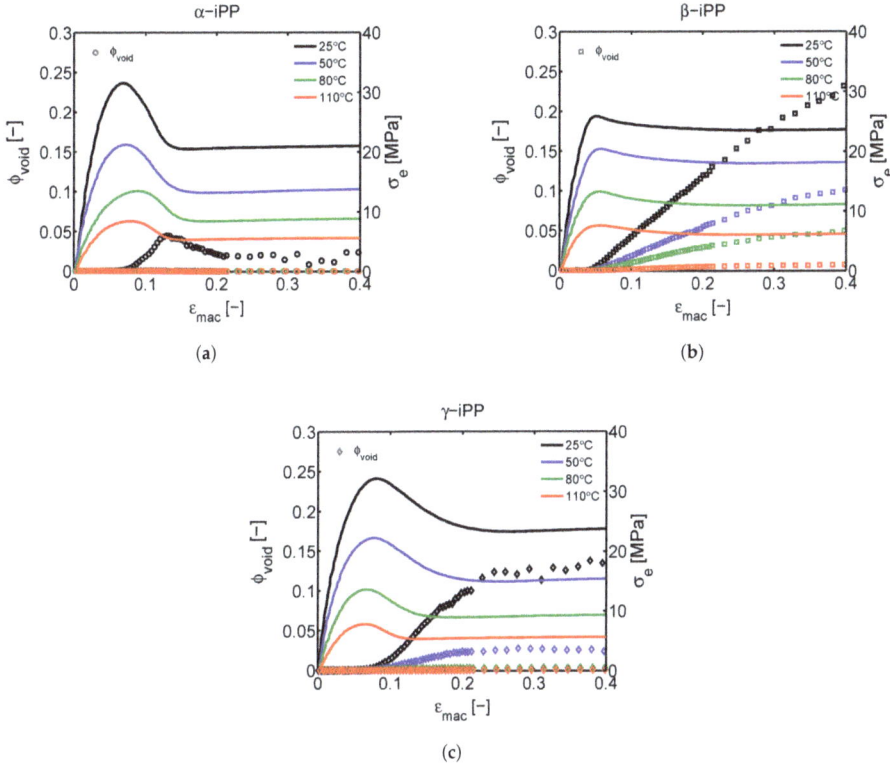

Figure 27. Void volume fraction as a function of the macroscopic strain for α-iPP (**a**), β-iPP (**b**) and γ-iPP (**c**), elongated at 25, 50, 80 and 110 °C.

When considering the α-iPP elongated at high temperatures, no notable void fraction can be measured although the 2D-patterns clearly show the presence of voids, evidenced by the non-circular scattering close to the beam center. The intensity in these 2D images is given on a logarithmic scale, and thus, the early stage of void initiation immediately becomes clear. Apparently these effects are not strong enough to cause a scattered intensity increase, significantly large to be reflected in the void fraction. In the case of β-iPP the void fraction starts to increase at the yield point and continues to grow throughout the entire experiment. The two phenomena that cause the ϕ_v in α-iPP to decrease seem not to be present in the β-sample. With increasing temperature, the volume percentage of voids decreases. At 25 and 50 °C, the γ-iPP shows an increase in the scattered intensity that is sufficiently large to be reflected in the void fraction. This starts to develop at the yield point, and continues to grow with increasing strain. However, after softening the void fraction reaches a plateau. The yield stress of β-iPP is found at the lowest strains. Therewith, the onset of voiding in the β samples starts the earliest. This observation is also reported by Pawlak [17]. Although in the work of Pawlak the void volume fraction is not quantified, the 2D SAXS images show that not only the voiding starts at lower macroscopic engineering strain for β-iPP compared to α-iPP, but also that the scattered intensity is much stronger. Since this is directly related to the volume fraction of voids (within the detectable length scale), it can be concluded that the findings in this work are in good agreement with the work of Pawlak. Na et al. [19] investigated the volumetric strain upon stretching α- and γ-iPP at room temperature. At low strains they first found a volumetric increase in γ-iPP, subsequently followed by an increase of the α-iPP volume at slightly higher strains. This seems to disagree with the findings reported here, however, volumetric strain does not necessarily match the void fraction obtained from SAXS experiments in which only part of the void fraction can be assessed, i.e., the part in the detectable length scale. It is very likely that in the case of α- and γ-iPP the void dimensions quickly grow out of the detectable range. Furthermore, determination of the volumetric strain at the early stages of deformation, where nano-scale voids are formed, is a less sensitive technique compared to X-ray scattering. Moreover, crystal phase transitions as demonstrated in the previous sections lead to a change in material density and, therefore, also affect the volumetric strain obtained from optical techniques. To conclude, as far as a good comparison can be made, no insurmountable differences have been found between the results presented in this work and in other studies. In fact, general agreement between the trends observed for different polymorphs is found.

3.3.2. Lamellar Morphology

The crystal phase transformations are accompanied by changes in the lamellar morphology. For structural information on the lamellar thickness, the long period and the amorphous layer thickness, in both the tensile and transverse direction, the equatorial and meridional regions were integrated separately, see Figure 6. The integrated 1D intensity was Lorentz corrected and the methods described in Section 2.3.3 were applied. In Figures A2–A4 of the Supporting Information, the results are presented as a function of the macroscopic strain, in the range where the analysis could be applied, i.e., before crystals were destroyed too much. The most interesting observation was noted during stretching γ-iPP at 110 °C, see Figure 28. As expected, the lamellar thickness obtained using Bragg's law is lower than the ones obtained by using the auto-correlation function. In the latter case, the dense amorphous transition zones in the vicinity of the crystals add up to the crystal thickness. In that case, self evidently, the amorphous layers are thinner. In the elastic deformation regime, the long period in the equatorial region clearly increases. The evolution of the long period in the meridional region behaves opposite and even displays a decreasing trend. Based on the results obtained from Bragg's law, it follows that the amorphous regions increase in the initial stages of deformation, however, also the crystalline regions slightly thicken. After yielding, the associated destruction of γ-iPP and the transformation to α-iPP, a sudden increase in long period is observed. This is dominated by the thickening of lamellae, that increase on average from approximately 8–9 nm, to 12–13 nm, see Figure 28. This strong increase is only observed in the γ-samples after yielding at high temperature; see the Supporting Information.

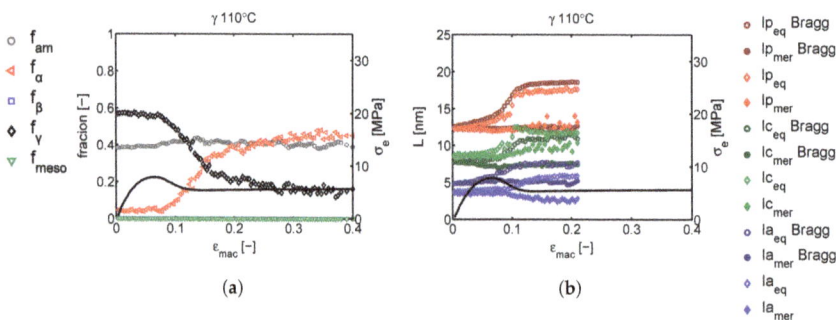

Figure 28. (a) Crystal phases in γ-iPP upon stretching at 110 °C; and (b), The evolution of the long period, lamellar thickness, and amorphous layer thickness.

If we consider the γ-crystal lattice and compare it with the newly formed α-lattice, the schematically depicted transformation in Figure 29 is obtained. This transition was investigated in detail, using WAXD experiments, by Auriemma et al. [14], who reported the transformation of γ-iPP (prepared in a low stereo regularity iPP) into α-iPP upon stretching. Their findings and interpretations of the different mechanisms involved at the unit cell level, combined with the increase of the long period simultaneously with the phase transition, suggest that the helical chain conformation is maintained during the transformation. During the destruction of γ-iPP at 110 °C, the ternary helical chain conformation seems to be maintained, and directly incorporated in the newly formed α-lattice. The observation of the increasing lamellar thickness could be worked out further, and offers a special way to obtain α-iPP with large crystal thickness obtained from hot drawing of γ-iPP.

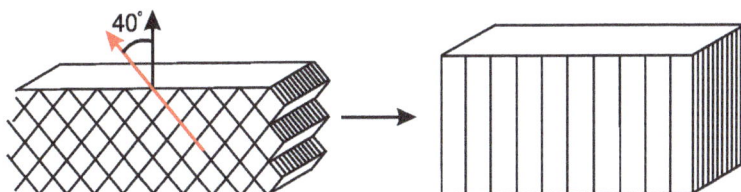

Figure 29. Suggested deformation of a γ-crystal with the orthorhombic unit cell structure at a temperature of 110 °C. The newly formed lamellar crystal is comprised of a monoclinic alpha unit cell structure.

4. Conclusions

Tensile and compression experiments were combined with in situ SAXS and WAXD measurements to reveal deformation-induced structural evolution phenomena at multiple length scales for the three well-known polymorphs of iPP. WAXD experiments are used to obtain information about phase transitions, selective melting and orientation of crystal planes. The orientation of lamellae, their thickness and the thickness of the intermediate amorphous layers are derived from the SAXS experiments, as well as the appearance and growth of voids. The findings, with respect to the structural evolution, are linked to the macroscopic intrinsic mechanical response.

Based on our previous work and the intrinsic behavior presented in Sections 3.1 and 3.2.2, we hypothesize that the typically amorphous phenomenon of softening is mainly due to the constrained amorphous regions in the vicinity of the crystals, rather than deterioration of the crystals.

With respect to the crystallographic structures, the WAXD experiments revealed that at low temperatures, all polymorphs undergo a (partial) phase transition to the oriented mesophase or amorphous phase at large strains. At elevated temperature, the newly formed structure is predominantly the thermodynamically most stable oriented α form. From the compression experiments, it is found that the true strain at which these transitions take place is similar for all polymorphs and, therefore, not the cause for the different macroscopic behavior.

The crystallinity of all polymorphs decreases at similar true strain, independent of the loading conditions, whereas softening is mainly observed in α- and γ-iPP. These are the crystalline forms with the highest density. The extent to which crystals constrain the amorphous phase is dominated by density (secondary interactions) and lamellar thickness or crystal defects. Apparently, the higher density of α and γ-iPP outweighs the influence of thicker lamellae in β-iPP. The crossed configuration of chains in γ-iPP lamellae seems to act as an additional constraint, whereby a rejuvenation treatment is needed in order to obtain the formation and growth of a stable neck in tensile deformation.

To obtain a rejuvenated constrained amorphous phase, mobility has to be created in constrained amorphous layers by either thermal treatments or mechanical deformation. The WAXD experiments show that the β-crystals deform at relatively low strains, clearly prior to yielding. This allows the constrained amorphous domains to gain mobility and soften. Since for β-iPP this happens far before yielding, it might partially explain the absence of softening.

Based on the observation of softening in α- and γ-iPP at a testing temperature of 110 °C, which is above the α_c-relaxation temperature, it is hypothesized that this is a result of the cross-hatched structure. Break-down of these structures deteriorates the structural integrity, causing a reduction of the stress. In the absence of cross-hatched structures, as is the case for β-iPP, softening is not observed at this temperature.

The evolution of the lamellar orientation suggests that upon stretching, α- and γ-iPP behave differently from β-iPP. In the latter one, the lamellae show no clear orientation while stretching at high temperatures. From the intrinsic behavior, it is found that the strain hardening modulus is the highest in β-iPP. The crystal shear starts the earliest, and also the onset of voiding, which is the most intense in β-iPP, is observed at the lowest strains. The observation that the crystal planes seem to slip before voiding starts suggests that in β-iPP, the critical shear stress is exceeded before the critical cavitation stress is reached.

The γ-iPP, with the crossed chain configuration in the crystal lattice, interestingly shows a strong increase in long period and lamellar thickness while stretching at high temperature. In the softening regime, where the γ-crystals transform to α-crystals, the lamellar thickness in the equatorial regions increases to such an extent that it is plausible to assume that the part of the chain originally incorporated in the crystal maintains the ternary helical conformation during this transition.

The low strain hardening modulus of α-iPP causes the initial disk-like voids to transform into an ellipsoid with a larger shape, oriented in the tensile direction, at relatively low apparent macroscopic strains, compared to the γ- and β-iPP, with the higher strain hardening modulus. In fact, for β- and γ-iPP, this transition is still not completely fulfilled at large macroscopic strains.

Where the crystal structure and topology determine to a large extent the pre-yield behavior and indirectly contribute to the level of the yield stress, the network is decisive in the post-yield behavior (strain hardening). The constrained amorphous phase in the vicinity of the crystalline material is of vital importance for the softening observed after yielding, and the strain hardening modulus determines when voids transform to other shapes.

Acknowledgments: SABIC is gratefully acknowledged to for the financial support of this project and the useful discussions. Furthermore, the staff at beamline BM26 of the European Synchrotron and Radiation Facility in Grenoble, France, is acknowledged for their help with the X-ray experiments.

Author Contributions: Harm J.M. Caelers, Enrico M. Troisi, Leon E. Govaert and Gerrit W.M. Peters conceived of and designed the experiments. Harm J.M. Caelers and Enrico M. Troisi performed the experiments. Harm J.M. Caelers and Enrico M. Troisi analyzed the data. Harm J.M. Caelers, Enrico M. Troisi, Leon E. Govaert and Gerrit W.M. Peters contributed reagents/materials/analysis tools. Harm J.M. Caelers wrote the paper.

Conflicts of Interest: The authors declare no conflict of interest. The founding sponsors had no role in the design of the study; in the collection, analyses or interpretation of data; in the writing of the manuscript; nor in the decision to publish the results.

Abbreviations

The following abbreviations are used in this manuscript:

iPP	isotactic polypropylene
1D	one-dimensional
2D	two-dimensional
SAXS	small angle X-ray scattering
WAXD	wide angle X-ray diffraction

Appendix A

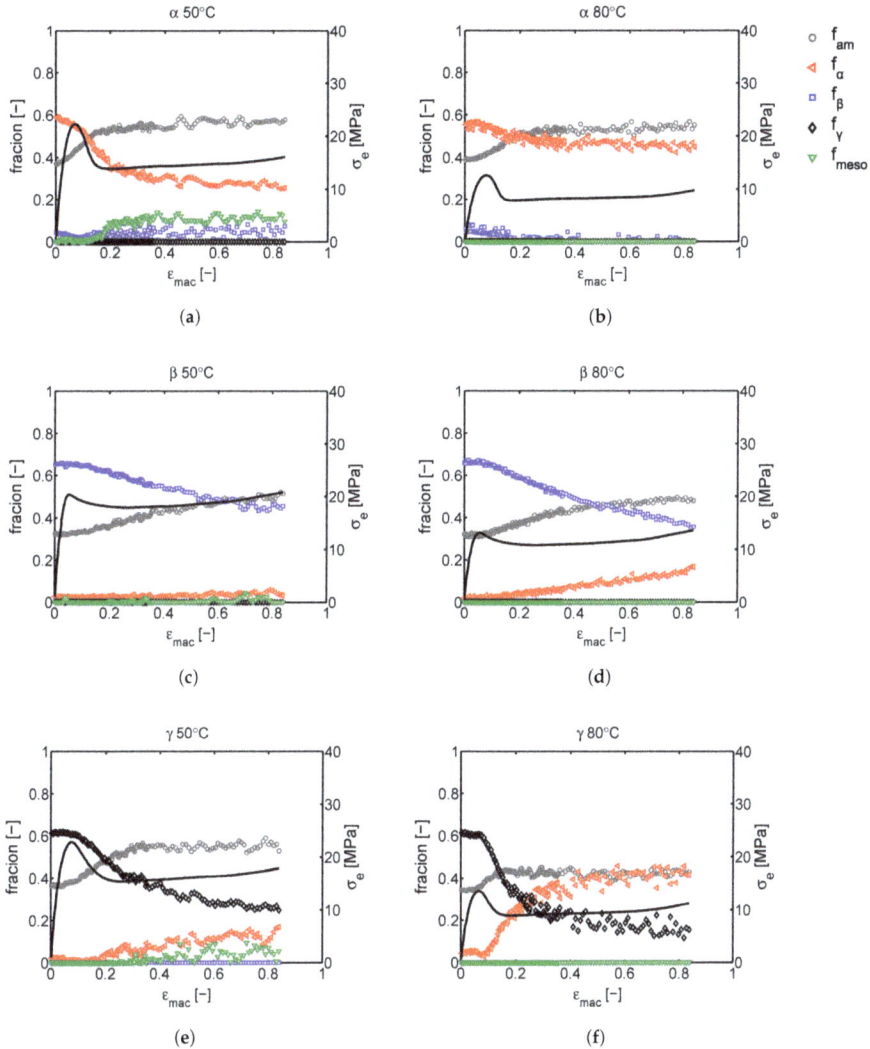

Figure A1. The crystal phase fractions obtained from the WAXD experiments. In (**a,c,e**), we see the evolution of the crystallinity and phase fractions upon stretching at 50 °C. Figure (**b,d,f**) show the result of the tensile experiments performed at 80 °C. From top to bottom, we see α, β and γ-iPP.

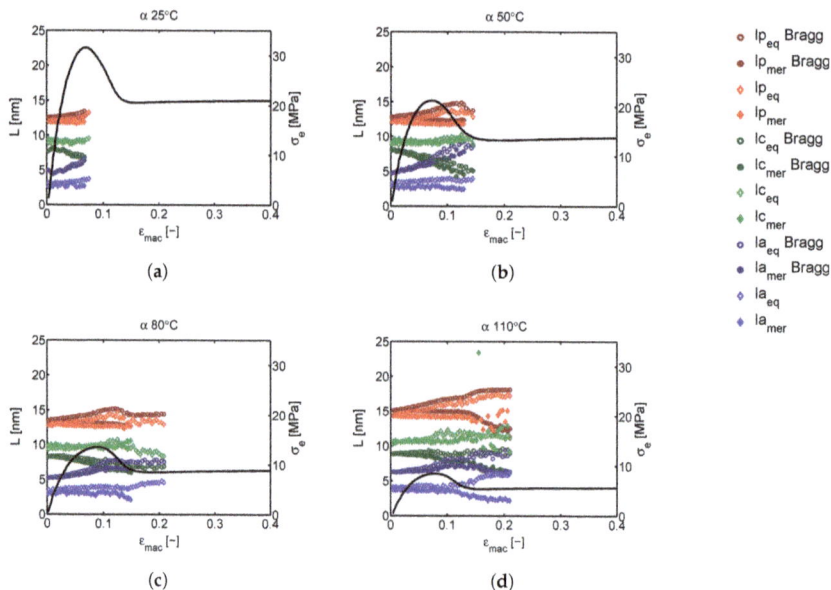

Figure A2. Long period, amorphous layer thickness and lamellar thickness determined via Bragg's law and the 1D autocorrelation function versus the macroscopic strain for α-iPP, elongated at 25, 50, 80 and 110 °C.

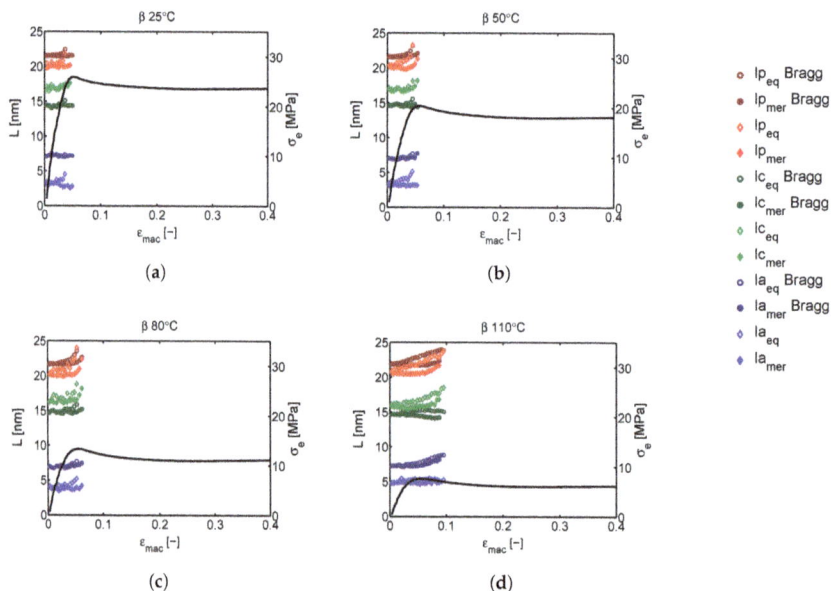

Figure A3. Long period, amorphous layer thickness and lamellar thickness determined via Bragg's law and the 1D autocorrelation function versus the macroscopic strain for β-iPP, elongated at 25, 50, 80 and 110 °C.

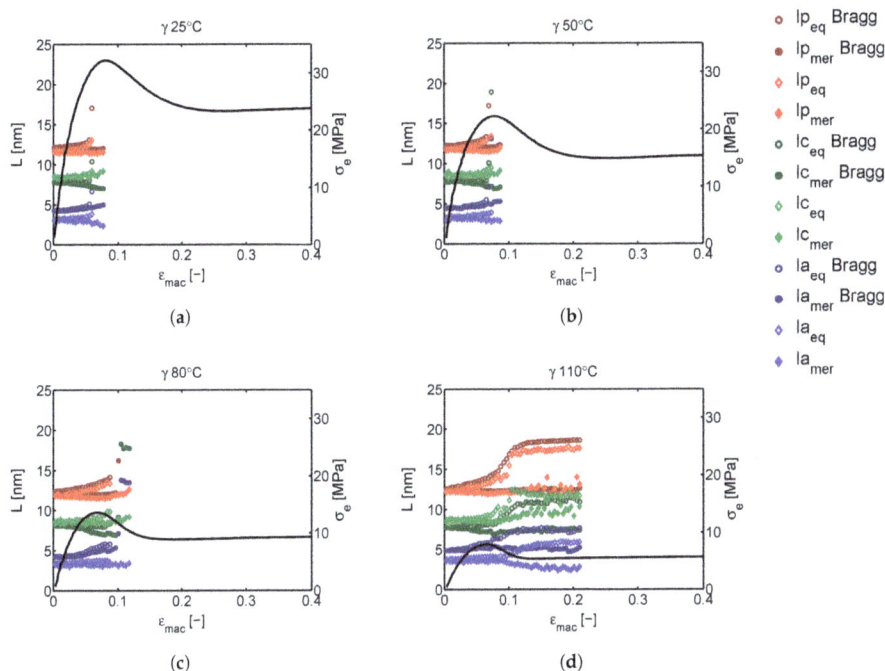

Figure A4. Long period, amorphous layer thickness and lamellar thickness determined via Bragg's law and the 1D autocorrelation function versus the macroscopic strain for γ-iPP, elongated at 25, 50, 80 and 110 °C.

References

1. Brückner, S.; Meille, S.; Petraccone, V.; Pirozzi, B. Polymorphism in isotactic polypropylene. *Prog. Polym. Sci.* **1991**, *16*, 361–404.
2. Lotz, B.; Wittmann, J.; Lovinger, A. Structure and morphology of poly(propylenes): A molecular analysis. *Polymer* **1996**, *37*, 4979–4992.
3. De Rosa, C.; Auriemma, F.; Di Capua, A.; Resconi, L.; Guidotti, S.; Camurati, I.; Nifant'ev, I.; Laishevtsev, I. Structure-property correlations in polypropylene from metallocene catalysts: Stereodefective, regioregular isotactic polypropylene. *J. Am. Chem. Soc.* **2004**, *126*, 17040–17049.
4. Varga, J. β-modification of isotactic polypropylene: Preparation, structure, processing, properties, and application. *J. Macromol. Sci. Phys.* **2002**, *41*, 1121–1171.
5. Roozemond, P.; Van Erp, T.; Peters, G. Flow-induced crystallization of isotactic polypropylene: Modeling formation of multiple crystal phases and morphologies. *Polymer* **2016**, *89*, 69–80.
6. Meille, S.; Ferro, D.; Brückner, S.; Lovinger, A.; Padden, F. Structure of β-isotactic polypropylene: A long-standing structural puzzle. *Macromolecules* **1994**, *27*, 2615–2622.
7. Varga, J.; Karger-Kocsis, J. Rules of supermolecular structure formation in sheared isotactic polypropylene melts. *J. Polym. Sci. Part B Polym. Phys.* **1996**, *34*, 657–670.
8. Varga, J.; Mudra, I.; Ehrenstein, G. Highly active thermally stable β-nucleating agents for isotactic polypropylene. *J. Appl. Polym. Sci.* **1999**, *74*, 2357–2368.
9. Valdo Meille, S.; Brückner, S. Non-parallel chains in crystalline γ-isotactic polypropylene. *Nature* **1989**, *340*, 455–457.
10. Mezghani, K.; Phillips, P. The γ-phase of high molecular weight isotactic polypropylene: III. The equilibrium melting point and the phase diagram. *Polymer* **1998**, *39*, 3735–3744.

11. De Rosa, C.; Auriemma, F.; Di Girolamo, R.; De Ballesteros, O.; Pepe, M.; Tarallo, O.; Malafronte, A. Morphology and mechanical properties of the mesomorphic form of isotactic polypropylene in stereodefective polypropylene. *Macromolecules* **2013**, *46*, 5202–5214.

12. Caelers, H.; Parodi, E.; Cavallo, D.; Peters, G.; Govaert, L. Deformation and failure kinetics of iPP polymorphs. *J. Polym. Sci. Part B Polym. Phys.* **2017**, *55*, 729–747.

13. Ma, Z.; Shao, C.; Wang, X.; Zhao, B.; Li, X.; An, H.; Yan, T.; Li, Z.; Li, L. Critical stress for drawing-induced α crystal-mesophase transition in isotactic polypropylene. *Polymer* **2009**, *50*, 2706–2715.

14. Auriemma, F.; De Rosa, C. Stretching isotactic polypropylene: From "cross-β" to crosshatches, from γ form to α form. *Macromolecules* **2006**, *39*, 7635–7647.

15. Zuo, F.; Keum, J.; Chen, X.; Hsiao, B.; Chen, H.; Lai, S.Y.; Wevers, R.; Li, J. The role of interlamellar chain entanglement in deformation-induced structure changes during uniaxial stretching of isotactic polypropylene. *Polymer* **2007**, *48*, 6867–6880.

16. Xiong, B.; Lame, O.; Chenal, J.M.; Rochas, C.; Seguela, R.; Vigier, G. In-situ SAXS study of the mesoscale deformation of polyethylene in the pre-yield strain domain: Influence of microstructure and temperature. *Polymer* **2014**, *55*, 1223–1227.

17. Pawlak, A. *Plastic Deformation and Cavitation in Semi-Crystalline Polymers*; American Institute of Physics Inc.: Ischia, Italy, 2014; Volume 1599, pp. 118–121.

18. Ran, S.; Zong, X.; Fang, D.; Hsiao, B.; Chu, B.; Phillips, R. Structural and morphological studies of isotactic polypropylene fibers during heat/draw deformation by in situ synchrotron SAXS/WAXD. *Macromolecules* **2001**, *34*, 2569–2578.

19. Na, B.; Lv, R.; Xu, W. Effect of network relaxation on void propagation and failure in isotactic polypropylene at large strain. *J. Appl. Polym. Sci.* **2009**, *113*, 4092–4099.

20. Lu, Y.; Wang, Y.; Chen, R.; Zhao, J.; Jiang, Z.; Men, Y. Cavitation in Isotactic Polypropylene at Large Strains during Tensile Deformation at Elevated Temperatures. *Macromolecules* **2015**, *48*, 5799–5806.

21. Lezak, E.; Bartczak, Z.; Galeski, A. Plastic deformation behavior of β-phase isotactic polypropylene in plane-strain compression at room temperature. *Polymer* **2006**, *47*, 8562–8574.

22. Lezak, E.; Bartczak, Z. Plastic deformation behavior of β-isotactic phase isotactic polypropylene in plane-strain compression at elevated temperatures. *J. Polym. Sci. Part B Polym. Phys.* **2008**, *46*, 92–108.

23. Zhang, C.; Liu, G.; Song, Y.; Zhao, Y.; Wang, D. Structural evolution of β-iPP during uniaxial stretching studied by in situ WAXS and SAXS. *Polymer* **2015**, *55*, 6915–6923.

24. Jia, C.; Liao, X.; Zhu, J.; An, Z.; Zhang, Q.; Yang, Q.; Li, G. Creep-resistant behavior of β-polypropylene with different crystalline morphologies. *RSC Adv.* **2016**, *6*, 30986–30997.

25. Lezak, E.; Bartczak, Z.; Galeski, A. Plastic deformation of the γ phase in isotactic polypropylene in plane-strain compression. *Macromolecules* **2006**, *39*, 4811–4819.

26. Humbert, S.; Lame, O.; Chenal, J.; Rochas, C.; Vigier, G. New insight on initiation of cavitation in semicrystalline polymers: In-situ SAXS measurements. *Macromolecules* **2010**, *43*, 7212–7221.

27. Wang, Y.; Jiang, Z.; Fu, L.; Lu, Y.; Men, Y. Lamellar thickness and stretching temperature dependency of cavitation in semicrystalline polymers. *PLoS ONE* **2014**, *9*, e97234.

28. Xiong, B.; Lame, O.; Chenal, J.M.; Rochas, C.; Seguela, R.; Vigier, G. Temperature-Microstructure Mapping of the Initiation of the Plastic Deformation Processes in Polyethylene via In Situ WAXS and SAXS. *Macromolecules* **2015**, *48*, 5267–5275.

29. Xiong, B.; Lame, O.; Chenal, J.; Rochas, C.; Seguela, R.; Vigier, G. In-situ SAXS study and modeling of the cavitation/crystal-shear competition in semi-crystalline polymers: Influence of temperature and microstructure in polyethylene. *Polymer* **2013**, *54*, 5408–5418.

30. Xiong, B.; Lame, O.; Chenal, J.M.; Rochas, C.; Seguela, R.; Vigier, G. Amorphous phase modulus and micro-macro scale relationship in polyethylene via in situ SAXS and WAXS. *Macromolecules* **2015**, *48*, 2149–2160.

31. Humbert, S.; Lame, O.; Vigier, G. Polyethylene yielding behaviour: What is behind the correlation between yield stress and crystallinity? *Polymer* **2009**, *50*, 3755–3761.

32. Schrauwen, B.; Janssen, R.; Govaert, L.; Meijer, H. Intrinsic deformation behavior of semicrystalline polymers. *Macromolecules* **2004**, *37*, 6069–6078.

33. Men, Y.; Rieger, J.; Strobl, G. Role of the Entangled Amorphous Network in Tensile Deformation of Semicrystalline Polymers. *Phys. Rev. Lett.* **2003**, *91*, 955021–955024.

34. Ishikawa, M.; Ushui, K.; Kondo, Y.; Hatada, K.; Gima, S. Effect of tie molecules on the craze strength of polypropylene. *Polymer* **1996**, *37*, 5375–5379.

35. Aboulfaraj, M.; G'Sell, C.; Ulrich, B.; Dahoun, A. In situ observation of the plastic deformation of polypropylene spherulites under uniaxial tension and simple shear in the scanning electron microscope. *Polymer* **1995**, *36*, 731–742.

36. Labour, T.; Gauthier, C.; Séguéla, R.; Vigier, G.; Bomal, Y.; Orange, G. Influence of the β crystalline phase on the mechanical properties of unfilled and CaCO3-filled polypropylene. I. Structural and mechanical characterisation. *Polymer* **2001**, *42*, 7127–7135.

37. Labour, T.; Vigier, G.; Séguéla, R.; Gauthier, C.; Orange, G.; Bomal, Y. Influence of the β-crystalline phase on the mechanical properties of unfilled and calcium carbonate-filled polypropylene: Ductile cracking and impact behavior. *J. Polym. Sci. Part B Polym. Phys.* **2002**, *40*, 31–42.

38. Van Drongelen, M.; Van Erp, T.; Peters, G. Quantification of non-isothermal, multi-phase crystallization of isotactic polypropylene: The influence of cooling rate and pressure. *Polymer* **2012**, *53*, 4758–4769.

39. Haward, R. Strain hardening of thermoplastics. *Macromolecules* **1993**, *26*, 5860–5869.

40. Van Melick, H.; Govaert, L.; Meijer, H. Localisation phenomena in glassy polymers: Influence of thermal and mechanical history. *Polymer* **2003**, *44*, 3579–3591.

41. Bras, W.; Dolbnya, I.; Detollenaere, D.; Van Tol, R.; Malfois, M.; Greaves, G.; Ryan, A.; Heeley, E. Recent experiments on a combined small-angle/wide-angle X-ray scattering beam line at the ESRF. *J. Appl. Crystallogr.* **2003**, *36*, 791–794.

42. Zhang, H.; Scholz, A.; De Crevoisier, J.; Vion-Loisel, F.; Besnard, G.; Hexemer, A.; Brown, H.; Kramer, E.; Creton, C. Nanocavitation in carbon black filled styrene-butadiene rubber under tension detected by real time small angle X-ray scattering. *Macromolecules* **2012**, *45*, 1529–1543.

43. Vonk, C.; Kortleve, G. X-ray small-angle scattering of bulk polyethylene-II. Analyses of the scattering curve. *Kolloid Z. Z. Polym.* **1967**, *220*, 19–24.

44. Ruland, W. The evaluation of the small-angle scattering of lamellar two-phase systems by means of interface distribution functions. *Colloid Polym. Sci. Kolloid Z. Z. Polym.* **1977**, *255*, 417–427.

45. Debye, P.; Bueche, A. Scattering by an inhomogeneous solid. *J. Appl. Phys.* **1949**, *20*, 518–525.

46. Porod, G. Die Röntgenkleinwinkelstreuung von dichtgepackten kolloiden Systemen-I. Teil. *Kolloid Z.* **1951**, *124*, 83–114.

47. Koberstein, J.; Morra, B.; Stein, R. The determination of diffuse-boundary thicknesses of polymers by small-angle X-ray scattering. *J. Appl. Crystallogr.* **1980**, *13*, 34–45.

48. Jansen, B.; Rastogi, S.; Meijer, H.; Lemstra, P. Rubber-modified glassy amorphous polymers prepared via chemically induced phase separation. 2. Mode of microscopic deformation studied by in situ small-angle X-ray scattering during tensile deformation. *Macromolecules* **2001**, *34*, 4007–4018.

49. G'Sell, C.; Jonas, J. Yield and transient effects during the plastic deformation of solid polymers. *J. Mater. Sci.* **1981**, *16*, 1956–1974.

50. Smit, R.; Brekelmans, W.; Meijer, H. Prediction of the large-strain mechanical response of heterogeneous polymer systems: Local and global deformation behaviour of a representative volume element of voided polycarbonate. *J. Mech. Phys. Solids* **1999**, *47*, 201–221.

51. Caelers, H.; Govaert, L.; Peters, G. The prediction of mechanical performance of isotactic polypropylene on the basis of processing conditions. *Polymer* **2016**, *83*, 116–128.

52. Meijer, H.; Govaert, L. Mechanical performance of polymer systems: The relation between structure and properties. *Prog. Polym. Sci.* **2005**, *30*, 915–938.

53. Mezghani, K.; Phillips, P. The γ-phase of high molecular weight isotactic polypropylene. II: The morphology of the γ-form crystallized at 200 MPa. *Polymer* **1997**, *38*, 5725–5733.

54. Yamada, K.; Matsumoto, S.; Tagashira, K.; Hikosaka, M. Isotacticity dependence of spherulitic morphology of isotactic polypropylene. *Polymer* **1998**, *39*, 5327–5333.

55. Von Compostella, M.; Coen, A.; Bertinotti, F. Fasern und Filme aus isotaktischem Polypropylen. *Angew. Chem.* **1962**, *74*, 618–624.

56. Lotz, B.; Wittmann, J. The molecular origin of lamellar branching in the α (monoclinic) form of isotactic polypropylene. *J. Polym. Sci. Part B Polym. Phys.* **1986**, *24*, 1541–1558.

57. Xu, W.; Martin, D.; Arruda, E. Finite strain response, microstructural evolution and β → α phase transformation of crystalline isotactic polypropylene. *Polymer* **2005**, *46*, 455–470.

58. Lezak, E.; Bartczak, Z. Plastic deformation of the γ phase isotactic polypropylene in plane-strain compression at elevated temperatures. *Macromolecules* **2007**, *40*, 4933–4941.

59. Plummer, C.; Kausch, H.H. Micronecking in thin films of isotactic polypropylene. *Macromol. Chem. Phys.* **1996**, *197*, 2047–2063.

60. Butler, M.; Donald, A.; Ryan, A. Time resolved simultaneous small- and wide-angle X-ray scattering during polyethylene deformation-II. Cold drawing of linear polyethylene. *Polymer* **1998**, *39*, 39–52.

61. Pawlak, A. Cavitation during tensile deformation of high-density polyethylene. *Polymer* **2007**, *48*, 1397–1409.

62. Pawlak, A.; Galeski, A. Plastic deformation of crystalline polymers: The role of cavitation and crystal plasticity. *Macromolecules* **2005**, *38*, 9688–9697.

63. Men, Y.; Rieger, J.; Homeyer, J. Synchrotron ultrasmall-angle X-ray scattering studies on tensile deformation of poly(1-butene). *Macromolecules* **2004**, *37*, 9481–9488.

64. Lode, U.; Pomper, T.; Karl, A.; Von Krosigk, G.; Cunis, S.; Wilke, W.; Gehrke, R. Development of crazes in polycarbonate, investigated by ultra small angle X-ray scattering of synchrotron radiation. *Macromol. Rapid Commun.* **1998**, *19*, 35–39.

65. Zafeiropoulos, N.; Davies, R.; Roth, S.; Burghammer, M.; Schneider, K.; Riekel, C.; Stamm, M. Microfocus X-ray scattering scanning microscopy for polymer applications. *Macromol. Rapid Commun.* **2005**, *26*, 1547–1551.

66. Ruland, W. Small- angle scattering studies on carbonized cellulose fibers. *J. Polym. Sci. Polym. Symp.* **1969**, *28*, 143–151.

polymers

MDPI

Article

Mixed Rigid and Flexible Component Design for High-Performance Polyimide Films

Xiaohui Yu, Weihua Liang, Jianhua Cao and Dayong Wu *

Technical Institute of Physics and Chemistry, Chinese Academy of Sciences, Beijing 100190, China; xhyu@mail.ipc.ac.cn (X.Y.); lwh@mail.ipc.ac.cn (W.L.); caojh@mail.ipc.ac.cn (J.C.)
* Correspondence: dayongwu@mail.ipc.ac.cn; Tel.: +86-010-82543772

Received: 25 July 2017; Accepted: 14 September 2017; Published: 15 September 2017

Abstract: To develop the polyimide (PI) which is closely matched to the coefficient of the thermal expansion (CTE) of copper, a series of PIs are prepared from 5,4′-diamino-2-phenyl benzimidazole (DAPBI), 4,4′-diaminodiphenyl ether (ODA), and 3,3′,4,4′-benzophenonetetracarboxylic dianhydride (BTDA) using a sequential copolymerization, blade coating, and thermal imidization process. The physical properties of the PIs are effectively regulated and optimized by adjusting the ratio of the rigid DAPBI and flexible ODA components. By increasing the DAPBI content, thermal stability, dimensional stability, and mechanical properties, the resultant polymer is enhanced. PI-80 exhibits an excellent comprehensive performance, a glass transition temperature of 370 °C, and a tensile strength of 210 MPa. Furthermore, the CTE as calculated in the range 50–250 °C is ca. 19 ppm/K, which is equal to that of copper. A highly dimensionally stable, curl-free, and high T-style peel strength (6.4 N/cm) of copper/PI laminate was obtained by casting the polyamic acid onto copper foil (13 μm) and thermally curing at 360 °C, which indicates that it has the potential to be applied as an electronic film for flexible displays and flexible printed circuit boards. A structural rationalization for these remarkable properties is also presented.

Keywords: polyimide film; linear coefficient of thermal expansion (CTE); copper clad laminate; structure and properties

---

1. Introduction

Aromatic polyimides (PIs) exhibit outstanding thermal stability, mechanical properties, solvent resistance, and electric insulation performance [1–6]. Aromatic PI films are widely used in the aerospace, automobile manufacture, and microelectronics fields owing to their facile fabrication by casting methods [7–16]. For example, PI films can be used as high-temperature-resistant insulating materials, flexible printed circuits, stress-relieving buffer layers, particle-blocking layers, dielectric and flexible connecting materials for multichip model systems, substrates for flexible solar cells or flexible displays, and lithium-ion battery separators [17–21].

Multichip modules require dielectric interlayers with low dielectric properties, excellent mechanical properties, and excellent adhesion. It is especially important that the coefficient of thermal expansion (CTE) of the interlayer should match that of the substrate (e.g., Si wafer, copper foil, aluminium foil) [22]. The use of PIs as dielectric layers can effectively enhance the safety and reduce the weight of electronic devices. However, typical commercial PI films can't tolerate the high temperatures associated with welding fabrication. Furthermore, there is a significant difference in CTE between the PI films and common substrates. For example, the CTEs of copper, silicon wafer, and glass are 19, 6, and 15 ppm/K, respectively [23,24]. Additionally, the CTEs of the commercial PI films Kapton and Upilex without stretching are ca. 47 and 45 ppm/K, respectively [25].

The CTEs of polymers can be tuned by modulating their chemical structures and aggregation properties. There are three strategies commonly used to reduce the CTEs of PI films. The first approach is to subject the PAA or PI film to uniaxial or biaxial stretching at a high temperature (>150 °C) [26–28]. Upon stretching, the orientation of the molecular chains is improved along the stretching direction by promoting the aggregation of the PIs, and the regularity of the molecular chains is enhanced. However, the reduction of the CTE data by this method is limited owing to the small stretching ratios possible without causing permanent damage. Moreover, the PIs suitable for stretching treatment are typically thermoplastic and the glass transition temperature (T_g) values of PIs are lower than 300 °C.

The second approach is to introduce inorganic nanoparticles that exhibit very low CTEs into the PI system in order to obtain the PI composite films with low CTEs [29–32]. The sol-gel method has been adopted, in which the precursors of inorganic particles are introduced into the polyamic acid (PAA) solution. However, macroscopic phase separation occurs in PI systems when the inorganic-particle content is higher than 5 wt %, and the properties of the resultant polymers are affected [33–36]. Geng et al. introduced reactive groups that can be linked with inorganic components into the PI backbone and adopted the end group (3-aminopropyl)triethoxysilane (APTES) to mitigate this macroscopic phase separation. However, this method had little impact on the CTEs of these inorganic-organic composite PI films.

The third approach is to import the rigid rod-like, planar structures or short-chain diamines and dianhydride monomers into the PI unit structure to enhance the proportion of rigid benzene rings and amide rings. This method can reduce the CTEs of PI films because it promotes the ordered arrangement and degree of molecular orientation in the polymer chains. Previous researches have suggested that designing polymers with rigid and ordered molecular chains is the best way to improve their thermal and dimensional stability [37–40].

Significant research efforts have been devoted to developing high-performance PIs comprising imidazole and oxazole rings owing to the outstanding heat resistance, high strength, and high moduli of polybenzimidazoles and polybenzoxazoles [41–43]. However, the flexibility of these films is poor owing to the high rigidity of the rod-like chains. Furthermore, the CTEs of these films are often negative, which mismatches those of common substrate materials [37,43].

To obtain a PI film with an equal CTE to that of copper over a wide temperature range, a series of different structural PIs were synthesized and optimized. The introduction of both rigid and flexible molecular units into polymer chains is an effective way to control and optimize their properties [37]. Consequently, in this study, PI films were prepared using the structurally rigid diamine 5,4'-diamino-2-phenyl benzimidazole (DAPBI), the structurally flexible diamine 4,4'-diaminodiphenyl ether (ODA), and 3,3',4,4'-benzophenonetetracarboxylic dianhydride (BTDA). PIs containing different percentages of ODA–BTDA and DAPBI–BTDA chain segments were synthesized by copolymerizing the diamines ODA and DAPBI with BTDA. ODA and BTDA, which exhibit flexible structures, were chosen to provide flexibility to the polymers. The rigid structural diamine DAPBI provides heat resistance and excellent mechanical properties to the polymers. By optimizing the ratio of flexible ODA–BTDA segments and rigid DAPBI–BTDA segments, the flexibility, CTE, mechanical properties, and thermal properties of PIs were tuned. When the ratio of ODA-BTDA and DAPBI-BTDA is 8:2, the optimized PI has a CTE value close to that of copper, and it also exhibits outstanding heat resistance and mechanical properties.

2. Materials and Methods

2.1. Materials

N,N-Dimethylacetamide (DMAc) was obtained from Tianjin Chemical Agent Factory (Tianjin, China) and was vacuum-distilled after drying over calcium hydride. BTDA and ODA were purchased from Forsman Scientific Company (Beijing, China). DAPBI was purchased from Nanjing Chemlin Chemical Industry Company (Nanjing, China). The dianhydride was dried in a vacuum oven at 150 °C for 12 h and the diamines were dried at 80 °C for 12 h. To improve the relative molecular weight, all the reaction vessels, solvents, and reactants must be anhydrous.

2.2. Methods

2.2.1. Infrared Spectroscopic Analysis

The inherent viscosities (η) of the PAAs were measured to evaluate the polymerization degree of the PIs. η was obtained by dissolving 0.5 g of PAA in 100 mL DMAc at 25 °C, and the flow of the PAA solution was measured using an Ubbelohde viscometer (Zhongwang Technology, Hangzhou, China). The η value was calculated using the following equation:

$$\eta = \frac{\ln(t/t_o)}{0.5 \ \mathrm{g/dL}} \tag{1}$$

where t and t_0 are the determination times for the PAA solution and a blank DMAc solution, respectively. In order to evaluate the imidization degree of the PI films, their Fourier transform infrared (FTIR) spectra were obtained on a Excalibur 3100 spectrometer (Varian, Palo Alto, CA, USA) from 600 to 4000 cm^{-1}.

The equilibrium water absorbability was determined by the weight difference of vacuum dried film specimens before and after immersion in deionized water at 25 °C for 24 h. Dry films which had been weighed (W_1) were immersed in water at room temperature for 24 h. The surfaces of the films were wiped to remove the water and weighed again (W_2). Water absorption was calculated by the equation: $(W_2 - W_1)/W_1 \times 100\%$.

2.2.2. Morphology and Structural Analysis

The microstructures of the polymers were characterized on Rigaku wide-angle and small-angle X-ray diffractometers (Rigaku Corporation, Tokyo, Japan). The 2θ scan data were collected at 0.02° intervals over ranges of 5°–60° and 0.6°–5°. The scan speed was 0.1°/min. The average distance (d) between the molecular chains can be obtained from Bragg's equation:

$$d = \lambda/2\sin \theta \tag{2}$$

where 2θ is the X-ray diffraction angle and λ is the wavelength of the X-rays. The cross-sectional micromorphologies of the PI films were sputtered with gold and then investigated using scanning electron microscopy (SEM) (Hitachi Limited, Hitachi, Japan) with HITACHI S-4300-4300 apparatus at an acceleration voltage of 10 kV. Moreover, the surface morphology and the element mapping of the PI layer and copper substrate after the T-peel test were measured with SEM and Energy Dispersive X-ray Spectrometric microanalysis (EDX) (Hitachi Limited, Hitachi, Japan) with HITACHI S-4300 apparatus.

2.2.3. Thermal Analysis

In order to investigate the thermal properties, the glass transition temperature (T_g), thermal decomposition temperature (T_d), and the CTEs of the films were characterized. The T_g values of the PI films were determined on a Mettler Toledo DSC 1 thermal analyzer (METTLER TOLEDO, Zurich, Switzerland). DSC measurements were carried out on 5–10 mg film samples heated under nitrogen atmosphere at a heating rate of 10 °C/min from 50 to 400 °C. The samples were first heated from 50 to 400 °C, then cooled at the same heating rate, and then heated again at the same conditions. The T_g of these PIs were determined in the second heating scan by the midpoint ASTM D3417-83 with the Mettler software. Dynamic mechanical analysis (DMA) was performed on thin film specimens (ca. $3 \times 0.65 \times X$ cm^3, the X represent the thickness of the film, which is about 3×10^{-3} cm) on a TA instrument DMA Q800 (TA Instruments, New Castle, PA, USA) at a heating rate of 5 °C/min and at a load frequency of 1 Hz under nitrogen atmosphere. The peak on the tan δ vs. temperature curve was regarded as the T_g of the film. T_d values for the PI films were obtained on a TA Q50 thermogravimetric analyzer (TA Instruments, New Castle, PA, USA). Measurements were carried out on 3–5 mg film samples heated at 5 °C/min in nitrogen atmosphere. Decomposition temperatures of 5% weight loss were recorded ($T_{d5\%}$) to evaluate the thermal stability of the films. The CTEs of the films were measured using a Mettler Toledo TMA/SDTA841^e (METTLER TOLEDO, Zurich, Switzerland). A constant force of 0.01 N was applied to film samples (ca. 1.5 × 0.4 × X cm^3). The samples were heated to 400 °C at a heating rate of 5 °C/min. The data were collected from the second heating run after the first run, which was performed up to the annealing temperature to eliminate the internal stresses in the films. Owing to the variation of the CTE with temperature, it is common practice to report the average CTEs determined over a specified temperature range. The average CTE is defined by:

$$\bar{a}_L = \frac{1}{L_0}\left(\frac{\Delta L}{\Delta T}\right) \tag{3}$$

where L_0 is the initial length of the sample between the grips at ambient temperature; $\Delta T = T_2 - T_1$, where T_1 and T_2 are the temperature limits; and $\Delta L = L_2 - L_1$, where L_1 and L_2 are the sample lengths at temperatures T_1 and T_2, respectively. All of the CTEs in this study were measured between 50 and 250 °C in the film plane direction and were derived using Equation (3).

2.2.4. Mechanical Analysis

The tensile properties of the films were examined with an Instron R5966 (INSTRON corproration, Canton, OH, USA). Tensile samples were machined to dimensions of ca. 4 × 0.5 × X cm^3 after imidization and tested with a crossed-head speed of 8 mm/min and a load capacity of 500 N. Each test was conducted on five samples, and the average tensile strength, modulus, and elongation at the break are reported.

2.2.5. Adhesion Properties

The adhesion strength between the PI layer and copper substrate was studied by the T-peel strength test method with an Instron R5966 (INSTRON corproration, Canton, OH, USA). The laminate samples were cut into 5 cm × 0.5 cm strips. The peeling speed was kept constant at 10 mm/min. The peel strength was expressed in N/cm and was determined from the peel load divided by the width of the PI/Cu laminate.

2.2.6. Synthesis of PAA Precursors

PAA precursors were prepared by copolymerization of the diamines ODA and DAPBI and the dianhydride BTDA at a 1:1 molar ratio of diamines and dianhydride. Six different PAAs were obtained by employing diamine mixtures with DAPBI molar contents of 0%, 20%, 40%, 60%, 80%, and 100%. The corresponding PI films were fabricated by coating, solvent drying, and thermal imidization, and denoted as PI-0, PI-20, PI-40, PI-60, PI-80, and PI-100.

The polymerization procedure for synthesizing the PAA for the PI-60 precursor is presented as an example: The molar ratio of ODA to DAPBI was 4:6, and the solid content was controlled at 15%. First, the ODA (1.602 g, 8 mmol) was dissolved in 60 mL DMAc at room temperature under N$_2$ atmosphere. After the ODA was dissolved, DAPBI (2.691 g, 12 mmol) was dispersed uniformly in the ODA solution, and a beige brown suspension was obtained. Then, a stoichiometric amount of BTDA (6.445 g, 20 mmol) was added gradually to the suspension, and the mixture was stirred continually for 6 h to form a viscous PAA precursor solution. The PAA was sealed and stored at 0 °C.

2.2.7. Thermal Imidization of PAA Films

PI films were obtained by casting the PAA solution onto clean, dry glass plates with a 300 μm depth blade. The solvent was then removed by drying the PAA films in an oven with a programmed procedure (50 °C for 1 h, 80 °C for 1 h, 100 °C for 1 h, 120 °C for 0.5 h, 150 °C for 0.5 h, and 180 °C for 0.5 h) at a heating rate of 10 °C/min. The PAA films were then cured in a vacuum oven at a heating rate of 5 °C/min to afford PI films. The PI-0 film was obtained with a curing program of 200 °C for

1 h, 250 °C for 1 h, 300 °C for 0.5 h, and 360 °C for 0.5 h. All the cured films were cooled to room temperature at a cooling rate of 3 °C/min. The PI films with thicknesses of 30 μm were easily peeled from the glass plate.

3. Results

3.1. Synthesis of PAAs and Fabrication of PI Films

The PI films containing benzimidazole moieties were obtained by the polymerization of the PAA precursor, blade coating, baking, and thermal imidization processes. The synthesis of the PIs, the molecular modeling of a homo-PI dimer simulated by ChemDraw (Cambridge Soft, Waltham, MA, USA), and a schematic diagram of the procedure for preparing the PI films and PI/copper laminates are shown in Scheme 1. PIs containing different ratios of the flexible ODA–BTDA segment and the rigid DAPBI–BTDA segment were prepared by adjusting the ratio of two diamines employed in the synthesis. Molecular models of the homo-PI ODA–BTDA and DAPBI–BTDA dimer structures were generated using ChemDraw software, and those with the lowest energy conformation are shown in Scheme 1b. The simulated results show that the ODA–BTDA dimer adopts a bent conformation and the DAPBI–BTDA dimer exhibits a planar construction. Hence the flexible ODA–BTDA segment governs polymer flexibility, and the DAPBI–BTDA segment endows the polymer chains with rigidity and regularity. Furthermore, the intermolecular hydrogen bonding formed between the N–H groups in DAPBI and the carbonyl groups of the imide ring can increase the intermolecular interaction strength. Thus, the DAPBI–BTDA segment was selected to improve the mechanical properties, dimensional stability, and thermal stability of the PIs [22]. According to the literature, benzimidazole moieties can react with copper oxide to form copper imidazolate inner complexes that provide active sites for coordinating interfacial bonding. Furthermore, the PIs containing the ODA diamine can exhibit improved copper adhesion owing to their flexible structures [44,45].

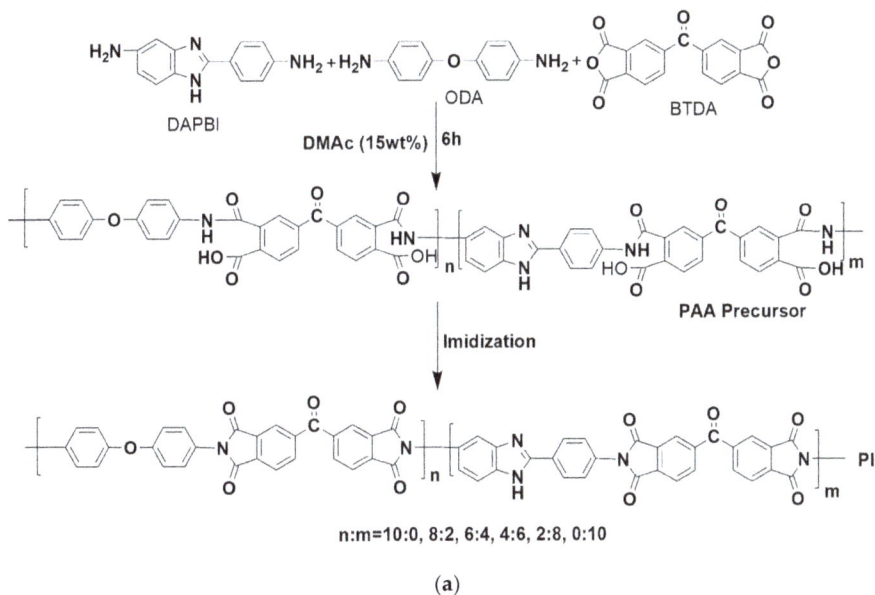

n:m=10:0, 8:2, 6:4, 4:6, 2:8, 0:10

(a)

Scheme 1. *Cont.*

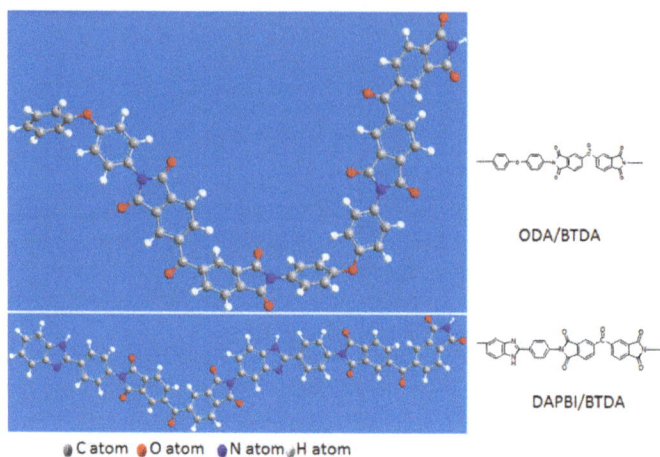

ODA/BTDA

DAPBI/BTDA

● C atom ● O atom ● N atom ● H atom

(b)

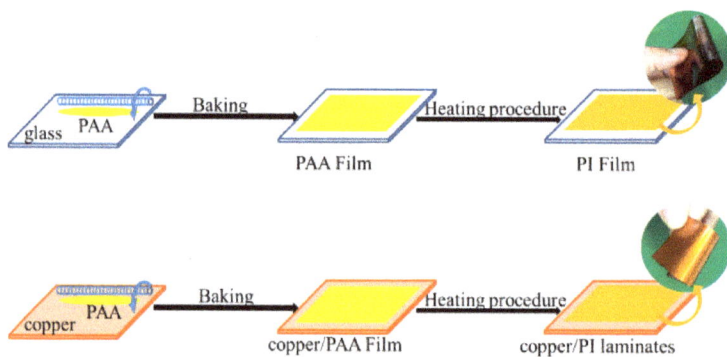

(c)

Scheme 1. (a) Synthesis of PAAs and thermal imidization of PIs. (b) Chemical structure of the homo-PI dimers ODA/BTDA and DAPBI/BTDA. (c) Fabrication of PI films and PI/copper laminates.

3.2. Viscosities of PAAs and FTIR Analysis of the PI Films

As the polymerization degree of the PAAs and imidization degree of the PIs affect the thermal properties and mechanical properties of the PI films, both polymerization and imidization were evaluated. First, the inherent viscosities of the PAAs were measured at 25 °C to assess their degree of polymerization, and the results are listed in Table 1. The high viscosities (1.8–2.3 dL/g) indicate that the PAAs obtained have moderately high molecular weights. The FTIR spectra of the PI films were recorded to assess the degree of imidization and they are shown in Figure 1. The characteristic absorption peaks of the symmetrical contraction vibration and the antisymmetric stretching vibration of the carbonyl group in the imide ring are observed at 1778 and 1716 cm^{-1}, respectively. Furthermore, the PI films present no absorption peaks characteristic of carboxyl and amide linkages, i.e., at 3500 cm^{-1} (COOH and NH), 1660 cm^{-1} (CONH), and 1550 cm^{-1} (C–NH), indicating that the PI films have undergone complete imidization. Upon increasing the molar percentage of benzimidazole stepwise from 0% to 100%, the carbonyl absorption peaks gradually shift from 1778.4 and 1716.6 cm^{-1} to 1774.5 and 1706.9 cm^{-1}, respectively. This indicated that hydrogen bonds form between the carbonyl group of the imide ring and the NH group of DAPBI. Thus, more hydrogen bonds may be formed with an increased imidazole ring content, and these intermolecular forces cause the carbonyl absorption peaks to move to a lower wavenumber [46,47]. A diagram showing this intermolecular hydrogen bonding is shown in Figure 2. All obtained films exhibited excellent chemical resistance and were insoluble in organic solvents, for example, NMP (*N*-Methyl-2-Pyrrolidone), m-cresol, DMAc, chloroform, etc. The water absorption results of these films are listed in Table 1. Owing to the existent intermolecular hydrogen bonding, the water absorption is a little higher than common polyimides.

Table 1. Inherent viscosities and thermal properties of the PI films.

Samples	η_{inh} [1] (dL/g)	T_g [2] (°C)		$T_{d5\%}$ [3] (°C)	R_{800} [4] (%)	Water absorption (%)
		DSC	DMA			
PI-0	1.85	278	296	523	54.2	2.0
PI-20	2.06	307	328	527	59.7	2.3
PI-40	2.20	314	334	522	61.7	2.5
PI-60	2.25	343	340	521	56.1	2.9
PI-80	2.30	NF [5]	371	523	62.7	2.5
PI-100	2.15	NF	397	529	65.6	3.0

[1] Inherent viscosity determined with a concentration of 0.5 g/dL in DMAc at 25 ± 0.1 °C. [2] Glass transition temperature (T_g) measured by DSC and DMA at a heating rate of 10 and 5 °C/min, respectively. [3] Temperatures at 5% weight loss were recorded by TGA at a heating rate of 5 °C/min under nitrogen atmosphere. [4] Residual weight retention at 800 °C under nitrogen. [5] Not found.

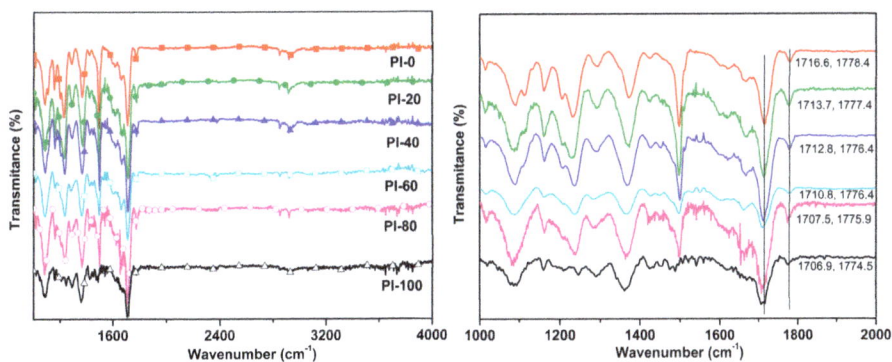

Figure 1. FTIR spectra of PI films.

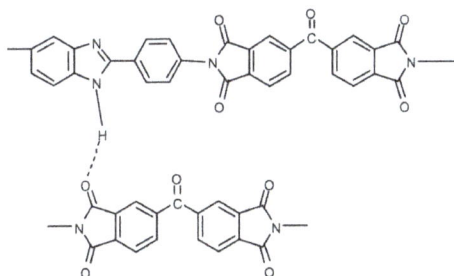

Figure 2. Interchain hydrogen bonding.

3.3. Morphology and Structures

The wide-angle X-ray diffraction (WAXD) patterns of the PI films are shown in Figure 3. Each presents a broad diffraction peak without obvious splitting peaks, indicating that the PIs are essentially amorphous. With an increase in the DAPBI–BTDA content, narrower and stronger intensity diffraction peaks with slight splitting are observed, indicating that the regularity of the molecular chains and the order degree of molecular chains are improved. According to Bragg's law (see Equation (2)), the distance between the molecular chains decreased from 4.8 to 4.3 Å. This demonstrates that the intermolecular interactions increase when the benzimidaole content increases.

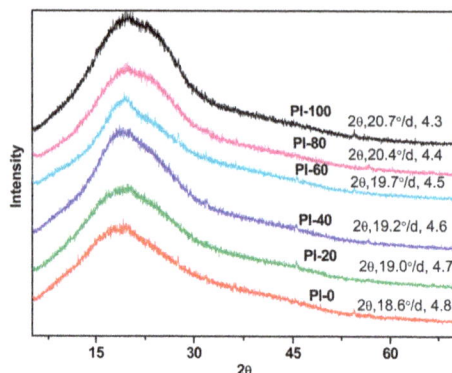

Figure 3. WAXD patterns of the PI films.

Figure 4 shows representative small-angle X-ray diffraction (SAXD) patterns for the PI-0, PI-40, and PI-80 films. The PI-0 sample presents no SAXD peaks because the polymer comprises a completely flexible and disordered molecular chain. Upon the introduction of DAPBI into the molecular chains, an SAXD diffraction peak appears at close to 1.2°, and the angle increases with increasing DAPBI content. According to Bragg's law (see Equation (2)), the long period of PI-40 and PI-80 is 75.4 and 65.4 Å, respectively. The SAXD results indicate that local structure ordering occurs in the PIs containing benzimidazole moieties, and the period diminishes with increasing benzimidazole content. This shows that the molecular chains containing benzimidazole rings possess a certain degree of planarity and rigidity, and thus, local structural ordering appears and the regularity of the molecular chain is improved as the DAPBI molar ratio increases.

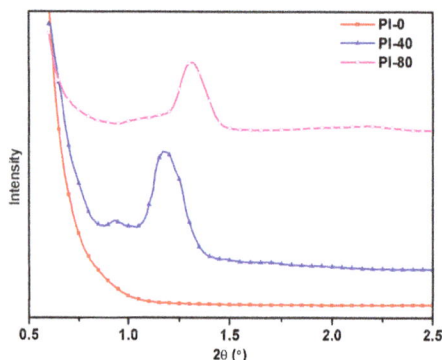

Figure 4. SAXD patterns of the PI films.

The cross-sectional morphologies of the PI films were observed by SEM, as shown in Figure 5. The morphologies of films PI-0 and PI-20 are smooth. However, when the DAPBI molar percentage is higher than 40%, densely packed layered ribbon-like morphological features appear. This indicates that the molecular chains of the polymers have poor regularity when the benzimidazole content is low. Conversely, the molecular chains of the polymer are arrayed more regularly, and the degree of orientation and local ordering improves with the influence of the benzimidazole component [41].

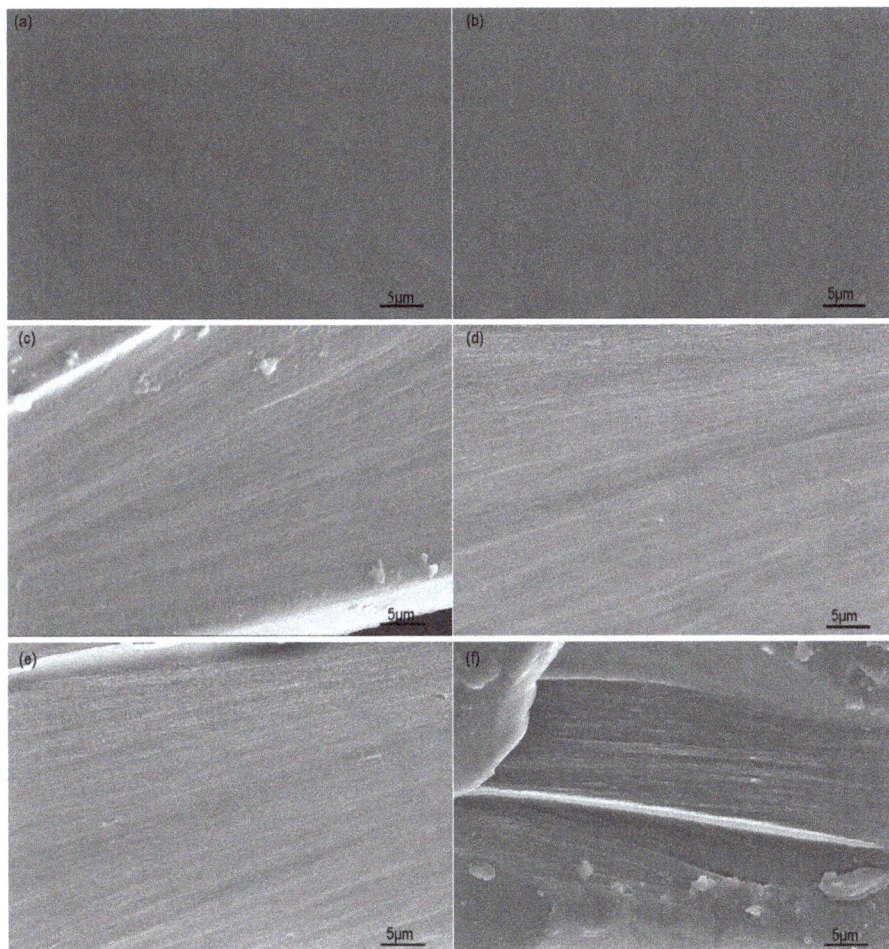

Figure 5. Cross-sectional SEM images of the PI films. (**a**) PI-0; (**b**) PI-20; (**c**) PI-40; (**d**) PI-60; (**e**) PI-80; (**f**) PI-100.

3.4. Thermal Properties

The T_g, T_d, and CTE data are given in Table 1. The DSC curves of films are shown in Figure 6. No endothermic peaks assigned to the melting peak are observed in the DSC curves, indicating that all the PIs are amorphous polymers. Owing to the poor molecular mobility introduced by the rigid structures, no well-defined T_g is detected for PI-80 and PI-100. The tan δ curves and storage modulus curves of films as obtained by DMA are shown in Figures 7 and 8. The T_g data (Table 1) obtained from

DSC and DMA are in the range of 278–400 °C, indicating that these PIs exhibit high heat resistance. With increasing DAPBI content, the rigidity of the polymer chains is enhanced, so the T_g of the PIs increases. The storage modulus curves show that all the films have large storage moduli (3–6 GPa) at 100 °C and still retain moduli of 10–100 MPa at temperatures above 300 °C. This demonstrates that the PIs with a higher content of DAPBI are very hard materials.

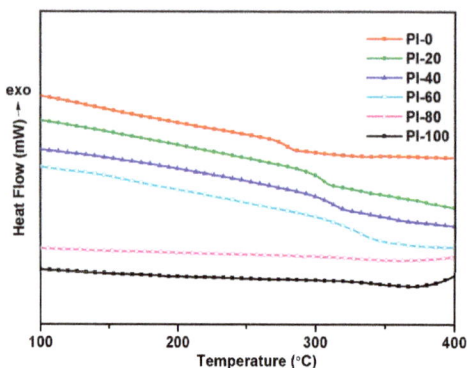

Figure 6. DSC curves of PI films.

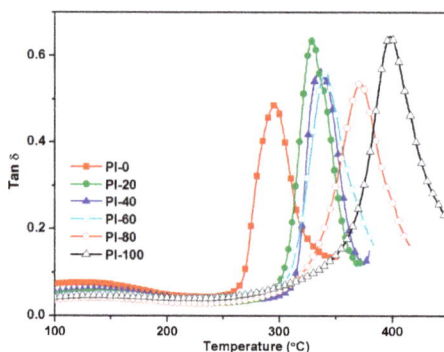

Figure 7. Tan δ vs. temperature for the PI films.

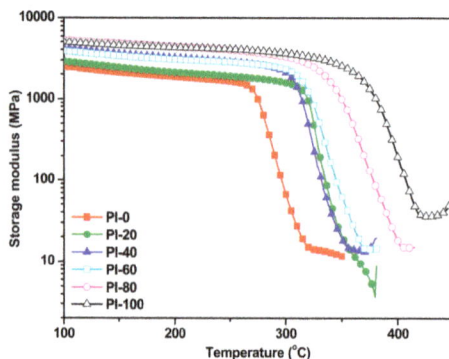

Figure 8. Storage modulus as a function of temperature for PI films.

The thermogravimetric analyses (TGA) of the PI films are shown in Figure 9 and show that the films present no weight loss below 480 °C. The excellent thermal stability indicates that these PI films are completely imidized. The films exhibit 5% weight loss in the temperature range of 521–529 °C in N_2, and the char yield at 800 °C reaches 65%.

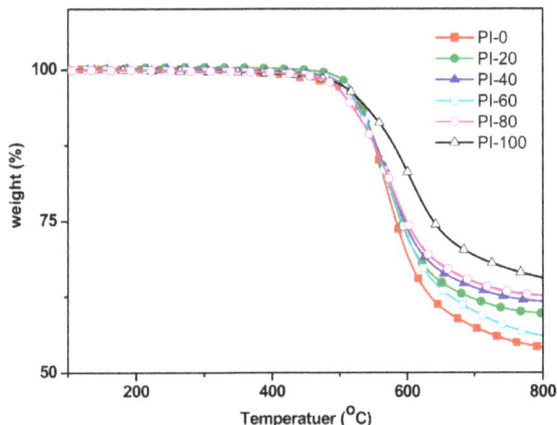

Figure 9. TGA curves of the PI films.

The temperature-strain curves of the films, commercial polyimides, and pure copper are shown in Figure 10. The average CTEs as calculated in the ranges 50–250, 50–300, and 100–200 °C are listed in Table 2. The curves in Figure 10 show that the strain of the films increases with an increasing temperature and the PI-0 and PI-20 increase more rapidly. This is mainly because the molecular chains of these two polymers are flexible, and the films deform far more rapidly at a high temperature. Films with a benzimidazole content over 40% exhibit increasingly excellent thermal dimensional stabilities than commercial polyimides, and the CTEs are less than 40 ppm/K. The temperature-strain curve of PI-80 is identical to the pure copper's temperature-strain curve. With the rigid structural DAPBI content increased, the regularity of molecular chains is enhanced. Thus, the CTE data of films decrease, namely, the dimensional stability is improved.

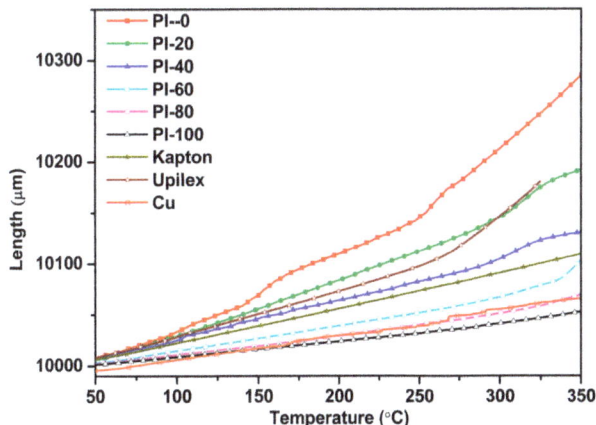

Figure 10. CTE curves of the PI films.

Table 2. CTE data for PI films in different temperature ranges.

Samples	CTE (50–250 °C) ppm/K	CTE (50–300 °C) ppm/K	CTE (100–200 °C) ppm/K
PI-0	67.9	82.0	76.2
PI-20	53.3	55.8	54.5
PI-40	37.8	38.6	38.9
PI-60	24.4	25.6	24.4
PI-80	19.0	19.6	18.8
PI-100	15.2	15.9	15.4
Cu	19.0	19.4	18.6
Kapton	33.5	34.2	33.0
Upilex	44.9	55.3	43.4

3.5. Adhesion Properties of PI-80/Copper Laminate

As shown in Figure 10 and Table 2, the PI-80 film has a comparable CTE to that of pure copper for a wide temperature range. The equivalent CTE of copper and the PI film can avoid the internal stresses of the aggregation generated during the processing. Consequently, the PAA precursor of PI-80 was coated onto 13-μm-thick pure copper (purchased from LingBao JinYuan ZhaoHui Copper Co., Ltd, Lingbao, China, and without any pretreatment) and then thermally imidized under N_2 atmosphere. The top-view and side-view photos of PI-80/copper laminates with PI layer thicknesses of 20 μm are shown in Figure 11. After being cured at 360 °C, the PI-80/copper laminates are flat and not warped. The adhesion strength between the PI layer and the copper substrate was tested by the T-peel strength test method with the PI-80/copper laminates. The load-displacement curve is shown in Figure 12. The peel strength is stabilization and the data is about 6.4 N/cm (3.2 N/0.5 cm), which indicated that a high enough adhesion laminate was obtained. The surface morphologies of the PI layer and copper substrate after peeling are shown in Figure 13. The surface of the pure PI-80 film without any treatment is smooth. The surface morphology of the PI-80 layer (b) after peeling becomes rough and displays a sheet structure. Also, the surface morphology of the copper substrate is different from the pure copper foil. It seems that there is one layer of polymers formed on the copper substrate. Further analysis of the interfacial interaction between the PI and copper substrate was conducted by Energy Dispersive X-ray Spectrometric microanalysis (EDX) with SEM. The element mapping and EDX spectrums of the peeled surface of PI-80 layer, copper substrate, and initial Cu foil are shown in Figures 14 and 15. As shown in Figure 14, there is a Cu element on the PI layer beyond the C, N, and O elements. Additionally, there are C, N, and O elements on the copper substrate except for the Cu element (Figure 15a–f). But there are no C, N, or other elements on the initial Cu foil (Figure 15g–i). All the SEM and EDX results declare that there is a strong interfacial interaction between the PI and copper. The N-H in the imidazole group of DAPBI can react with copper oxide at the interface and also the flexible structural ODA improves the adhesion force. Therefore, the PI-80/copper laminates show excellent peel strength. The element mappings and EDX spectrums of the peeled surface of the PI-60/copper laminate scanned for an approximately equal time with PI-80/copper are shown in Figure 16. The results show that there is an interfacial interaction between the PI-60 and Cu foil. However, the smaller peeled sheets of the PI-60 layer and lower C element on the copper substrate indicate that the interfacial interaction between the PI-60 and copper is smaller.

Figure 11. PI/copper laminates with PI layer thicknesses of 20 μm.

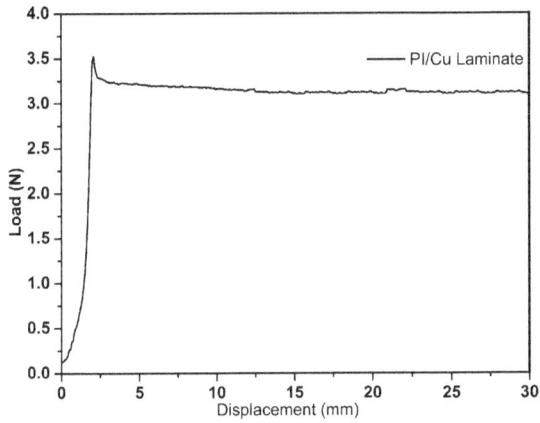

Figure 12. The load-displacement curve of the PI/Cu laminate.

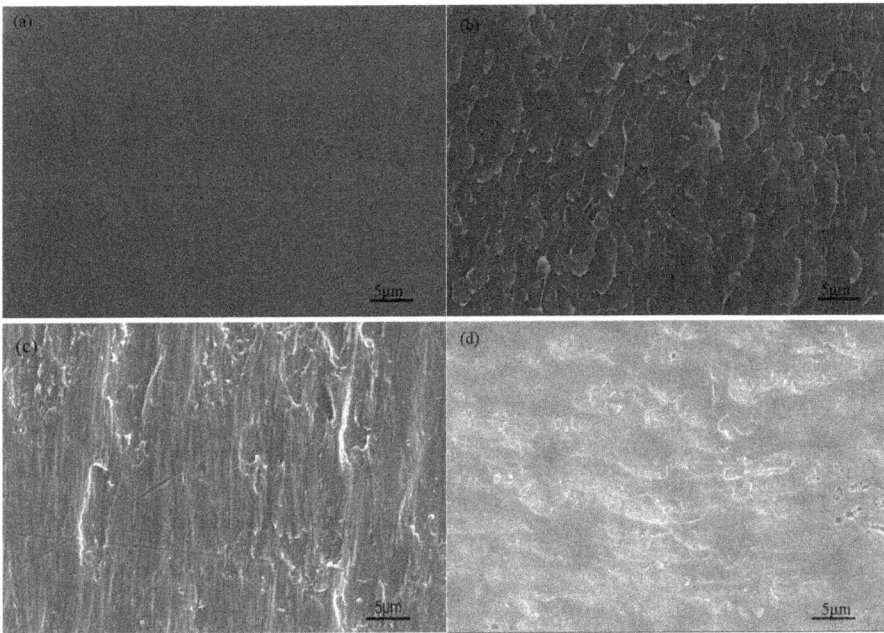

Figure 13. Surface SEM images of the PI-80 and copper substrate. (**a**) PI-80 film without any treatment; (**b**) PI film of PI-80/Cu laminate after peeling; (**c**) copper without any treatment; (**d**) copper substrate of the PI-80/Cu laminate after peeling.

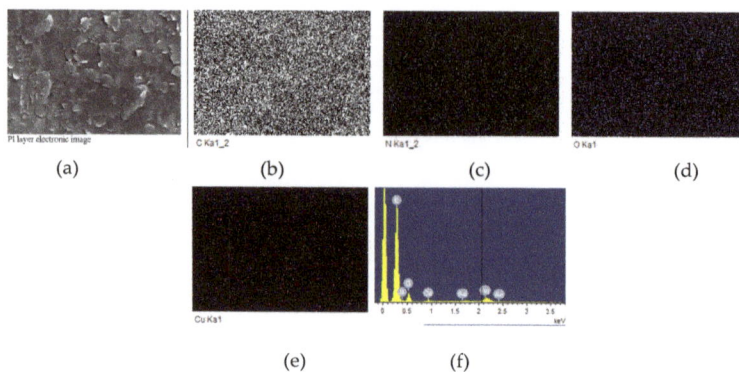

Figure 14. The element mappings of the peeled surface of PI-80 layer. (**a**) electronic image of PI layer; (**b**) C element in PI layer; (**c**) N element in PI layer; (**d**) O element in PI layer; (**e**) Cu element in PI layer; (**f**) EDX spectrum of peeled PI layer.

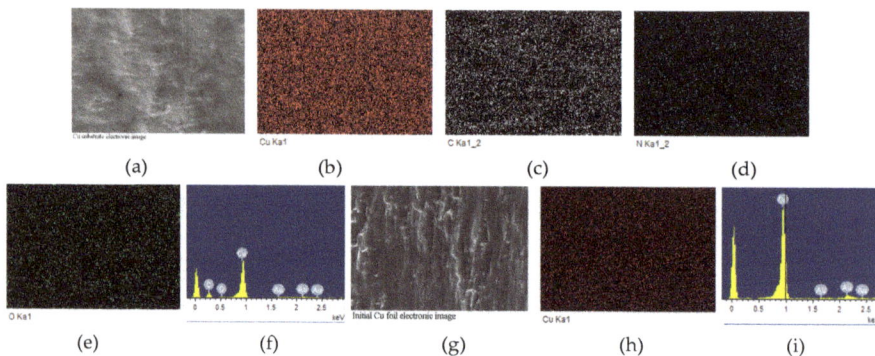

Figure 15. The element mappings of the peeled surface of copper substrate and initial Cu foil. (**a**) electronic image of Cu substrate; (**b**) Cu element in Cu substrate; (**c**) C element in Cu substrate; (**d**) N element in Cu substrate; (**e**) O element in Cu substrate; (**f**) EDX spectrum of peeled copper substrate; (**g**) electronic image of initial Cu foil; (**h**) Cu element on initial Cu foil; (**i**) EDX spectrum of pure Cu foil.

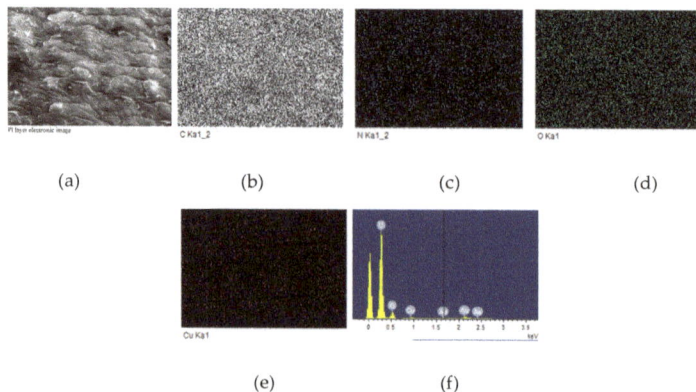

Figure 16. *Cont.*

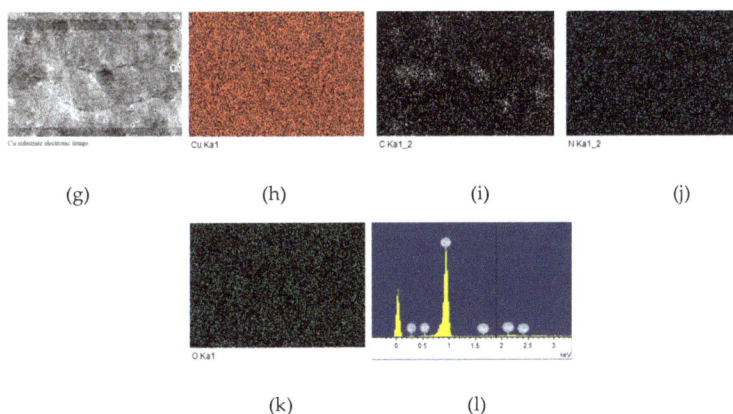

Figure 16. The element mappings of the peeled surface of PI-60/copper laminate. (**a**) electronic image of PI-60 layer; (**b**) C element in PI layer; (**c**) N element in PI layer; (**d**) O element in PI layer; (**e**) Cu element in PI layer; (**f**) EDX spectrum of peeled PI layer; (**g**) electronic image of Cu substrate; (**h**) Cu element in Cu substrate; (**i**) C element in Cu substrate; (**j**) N element in Cu substrate; (**k**) O element in Cu substrate; (**l**) EDX spectrum of Cu substrate.

3.6. Mechanical Properties

The mechanical properties of the PI films are listed in Table 3. All the films exhibit remarkable tensile strengths and moduli, which are in the ranges 146.0–220.8 MPa and 3.4–6.2 GPa, respectively. With increasing DAPBI content, the tensile strengths and moduli of the films increase, and their elongations decrease. The tensile strength of the PI-100 film is 1.5 times that of PI-0. The stress-strain curves of PI films are shown in Figure 17, showing that these PIs are amorphous polymers. The stress-strain curves of PI-0, PI-20, and PI-40 indicated these three polymers are hard and flexible materials. The stress-strain curves of PI-60, PI-80, and PI-100 declare that these three polymers are hard and tough materials. The results of tensile curves agree well with the DMA properties. The tensile results show that the rigid structural DAPBI could improve the rigidity and regularity of the polymer chains. According to the literature, the tensile strengths of the commercial films Kapton (PMDA-ODA) and Upilex-s (BPDA-ODA) without stretching are 130 and 140 MPa, respectively [25]. These PI films exhibit more outstanding mechanical properties, because of the rigid structure and stronger molecular interchain interaction.

Table 3. Mechanical properties of PI films.

Samples	Elastic (GPa)	Max stress (MPa)	Break strain (%)
PI-0	3.4 ± 0.05	146.0 ± 1.9	16.6 ± 4.9
PI-20	3.8 ± 0.14	157.8 ± 3.4	21.5 ± 8.7
PI-40	4.9 ± 0.16	173.3 ± 4.8	9.4 ± 3.8
PI-60	5.8 ± 0.07	205.6 ± 8.8	7.3 ± 1.0
PI-80	5.7 ± 0.48	209.0 ± 5.9	6.2 ± 0.4
PI-100	6.2 ± 0.37	220.8 ± 10.0	6.8 ± 1.2

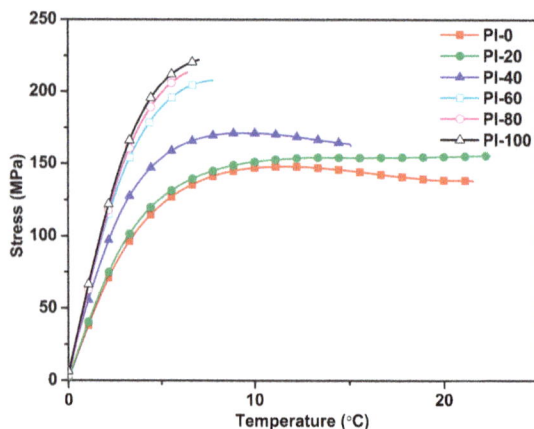

Figure 17. The stress-strain curves of PI films.

4. Conclusions

To obtain a PI film with an equal CTE to that of copper, PIs containing flexible ODA–BTDA segments and rigid DAPBI–BTDA segments were prepared from ODA, DAPBI, and BTDA. Six different PI films with different molecular compositions were obtained by adjusting the molar ratio of the two diamines. Owing to the inclusion of rigid structural benzimidazole moieties and hydrogen bonds forming between the carbonyl group of the imide ring and the NH group of DAPBI, the regularity of the molecular chain improves as the DAPBI molar ratio increases. Hence, the PIs containing benzimidazole moieties exhibit excellent thermal stability, mechanical properties and dimensional stabilities. The T_g of the PI was enhanced to 400 from 280 °C, the CTE (50–250 °C) was reduced to 15 from 67 ppm/K, and the tensile strength was increased to 220 from 146 MPa. Furthermore, the CTE of the PI-80 film was found to be equal to that of copper and was therefore used to prepare a PI/copper laminate that exhibited no distortion or delamination when heated. The SEM and EDX results of the peeled PI/copper laminate interface reveal that a strong interfacial interaction formed between the PI and copper owing to the chemical nature of DAPBI and flexible structural ODA. The excellent flatness, folding resistance, and high peel strength of this PI/copper laminate indicate that this PI film has potential for use in flexible printed circuit boards, flexible display substrates, and other next-generation electronic devices.

Acknowledgments: This work was financially supported by the "Strategic Priority Research Program" of the Chinese Academy of Sciences (Grant No. XDA09010105) and National Key R&D Program of China (No. 2016YFB0100105).

Author Contributions: Xiaohui Yu and Dayong Wu conceived and designed the experiments; Xiaohui Yu performed the experiments. Weihua Liang and Jianhua Cao analyzed the data; Dayong Wu contributed reagents/materials/analysis tools; Xiaohui Yu wrote the paper.

Conflicts of Interest: The authors declare no conflict of interest.

References

1. Liaw, D.J.; Wang, K.L.; Huang, Y.C.; Lee, K.R.; Lai, J.Y.; Ha, C.S. Advanced polyimide materials: Syntheses, physical properties and applications. *Prog. Polym. Sci.* **2012**, *37*, 907–974. [CrossRef]
2. Jia, M.C.; Li, Y.J.; He, C.Q.; Huang, X.Y. Soluble perfluorocyclobutyl aryl ether-based polyimide for high-performance dielectric material. *Appl. Mater. Interfaces* **2016**, *8*, 26352–26358. [CrossRef] [PubMed]

3. Peng, X.W.; Xu, W.H.; Chen, L.L.; Ding, Y.C.; Chen, S.L.; Wang, X.Y.; Hou, H.Q. Polyimide complexes with high dielectric performance: Toward polymer film capacitor applications. *J. Mater. Chem. C* **2016**, *4*, 6452–6456. [CrossRef]

4. Xu, L.L.; Jiang, S.D.; Li, B.; Hou, W.; Li, G.X.; Memon, M.A.; Huang, Y.; Geng, J.X. Graphene oxide: A versatile agent for polyimide foams with improved foaming capability and enhanced flexibility. *Chem. Mater.* **2015**, *27*, 4358–4367. [CrossRef]

5. Zhou, Y.Y.; Zhou, W.C.; Li, R.; Qing, Y.C.; Luo, F.; Zhu, D.M. Electroless plating preparation and electromagnetic properties of co-coated carbonyl iron particles/polyimide composite. *J. Magn. Magn. Mater.* **2016**, *401*, 251–258. [CrossRef]

6. Leventis, N.; Sotiriouleventis, C.; Mohite, D.P.; Larimore, Z.J.; Mang, J.T.; Churu, G.; Lu, H.B. Polyimide aerogels by ring-opening metathesis polymerization (ROMP). *Chem. Mater.* **2011**, *23*, 2250–2261. [CrossRef]

7. Li, X.D.; Zhong, Z.X.; Han, S.H.; Lee, S.H.; Lee, M.H. Facile modifications of polyimide via chloromethylation: II. Synthesis and characterization of thermocurable transparent polyimide having methylene acrylate side groups. *Polym. Int.* **2005**, *54*, 406–411. [CrossRef]

8. Sullivan, D.M.; Bruening, M.L. Ultrathin, gas-selective polyimide membranes prepared from multilayer films. *Chem. Mater.* **2003**, *15*, 281–287. [CrossRef]

9. Lim, J.; Yeo, H.; Goh, M.; Ku, B.C.; Kim, S.G.; Lee, H.S.; Park, B.; You, N.H. Grafting of polyimide onto chemically-functionalized graphene nanosheets for mechanically-strong barrier membranes. *Chem. Mater.* **2015**, *27*, 2040–2047. [CrossRef]

10. Zhuang, Y.B.; Seong, J.G.; Lee, W.H.; Do, Y.S.; Lee, M.J.; Wang, G.; Guiver, M.D.; Lee, Y.M. Mechanically tough, thermally rearranged (TR) random/block poly(benzoxazole-co-imide) gas separation membranes. *Macromolecules* **2015**, *48*, 5286–5299. [CrossRef]

11. Zhang, D.Z.; Chang, H.Y.; Li, P.; Liu, R.H.; Xue, Q.Z. Fabrication and characterization of an ultrasensitive humidity sensor based on metal oxide/graphene hybrid nanocomposite. *Sensor. Actuators B Chem.* **2016**, *225*, 233–240. [CrossRef]

12. Triambulo, R.E.; Cheong, H.G.; Lee, G.H.; Yi, I.S.; Park, J.W. A transparent conductive oxide electrode with highly enhanced flexibility achieved by controlled crystallinity by incorporating Ag nanoparticles on substrates. *J. Alloys Compd.* **2015**, *620*, 340–349. [CrossRef]

13. Lim, H.; Cho, W.J.; Ha, C.S.; Ando, S.; Kim, Y.K.; Park, C.H.; Lee, K. Flexible organic electroluminescent devices based on fluorine-containing colorless polyimide substrates. *Adv. Mater.* **2002**, *14*, 1275–1279. [CrossRef]

14. Park, J.S.; Kim, T.W.; Stryakhilev, D.; Lee, J.S.; An, S.G.; Pyo, Y.S.; Lee, D.S.; Mo, Y.G.; Jin, D.U.; Chung, H.K. Flexible full color organic light-emitting diode display on polyimide plastic substrate driven by amorphous indium gallium zinc oxide thin-film transistors. *Appl. Phys. Lett.* **2009**, *95*, 013503. [CrossRef]

15. Zhou, Z.X.; Zhang, Y.; Liu, S.W.; Chi, Z.G.; Chen, X.D.; Xu, J.R. Flexible and highly fluorescent aromatic polyimide: Design, synthesis, properties, and mechanism. *J. Mater. Chem. C* **2016**, *4*, 10509–10517. [CrossRef]

16. Wang, H.M.; Hsiao, S.H. Ambipolar, multi-electrochromic polypyromellitimides and polynaphthalimides containing di(*tert*-butyl)-substituted bis(triarylamine) units. *J. Mater. Chem. C* **2014**, *2*, 1553–1564. [CrossRef]

17. Zhang, Y.; Xiao, S.X.; Wang, Q.Y.; Liu, S.W.; Qiao, Z.P.; Chi, Z.G.; Xu, J.R.; Economy, J. Thermally conductive, insulated polyimide nanocomposites by AlO(OH)-coated MWCNTs. *J. Mater. Chem.* **2011**, *21*, 14563–14568. [CrossRef]

18. Lewis, J. Material challenge for flexible organic devices. *Mater. Today* **2006**, *9*, 38–45. [CrossRef]

19. Choi, M.C.; Kim, Y.; Ha, C.S. Polymers for flexible displays: From material selection to device applications. *Prog. Polym. Sci.* **2008**, *33*, 581–630. [CrossRef]

20. Zhang, B.; Wang, Q.F.; Zhang, J.J.; Ding, G.L.; Xu, G.J.; Liu, Z.H.; Cui, G.L. A superior thermostable and nonflammable composite membrane towards high power battery separator. *Nano Energy* **2014**, *10*, 277–287. [CrossRef]

21. Chen, J.L.; Liu, C.T. Technology advances in flexible displays and substrates. *Access IEEE.* **2013**, *1*, 150–158. [CrossRef]

22. Chen, H.L.; Ho, S.H.; Wang, T.H.; Chen, K.M.; Pan, J.P.; Liang, S.M.; Hung, A. Curl-free high-adhesion polyimide/copper laminate. *J. Appl. Polym. Sci.* **1994**, *51*, 1647–1652. [CrossRef]

23. Lee, Y.I.; Choa, Y.H. Adhesion enhancement of ink-jet printed conductive copper patterns on a flexible substrate. *J. Mater. Chem.* **2012**, *22*, 12517–12522. [CrossRef]

24. Long, K.; Kattamis, A.; Cheng, I.C.; Gao, Y.X.; Gleskova, H.; Wagner, S.; Sturm, J.C. High-temperature (250 °C) amorphous silicon TFT's on clear plastic substrates. *SID Dig.* **2005**, *36*, 313–315. [CrossRef]

25. Ding, M.X. *Polyimides: Chemistry, Relationship between Structure and Properties and Materials*; Science Press: Beijing, China, 2006; pp. 515–517, ISBN 7030165322.

26. Wang, L.N.; Yu, X.H.; Wang, D.M.; Zhao, X.G.; Yang, D.; Urrehman, S.; Chen, C.H.; Zhou, H.W.; Dang, G.D. High modulus and high strength ultra-thin polyimide films with hot-stretch induced molecular orientation. *Mater. Chem. Phys.* **2013**, *139*, 968–974. [CrossRef]

27. Nishino, T.; Miki, N.; Mitsuoka, Y.; Nakamae, K.; Saito, T.; Kikuchi, T. Elastic modulus of the crystalline regions of polyimide derived from poly(amic acid)–biphtalic dianhydride and p-phenylene diamine. *J. Polym. Sci. Polym. Phys.* **1999**, *37*, 3294–3301. [CrossRef]

28. Wang, H.Y.; Liu, T.J.; Liu, S.F.; Jeng, J.L.; Guan, C.E. Thermal and mechanical properties of stretched recyclable polyimide film. *J. Appl. Polym. Sci.* **2011**, *122*, 210–219. [CrossRef]

29. Ma, X.Y.; Ma, X.F.; Qiu, X.P.; Liu, F.F.; Jin, R.; Kang, C.Q.; Guo, H.Q.; GAO, L.X. Preparation and properties of imidazole-containing polyimide/silica hybrid films. *Chem. Res. Chin. Univ.* **2014**, *30*, 1047–1050. [CrossRef]

30. Jin, H.S.; Chang, J.H. Colorless polyimide nanocomposite films: Thermomechanical properties, morphology, and optical transparency. *J. Appl. Polym. Sci.* **2008**, *1*, 109–117. [CrossRef]

31. Huang, J.C.; Xiao, Y.; Mya, K.Y.; Liu, X.M.; He, C.B.; Dai, J.; Siow, Y.P. Thermomechanical properties of polyimide-epoxy nanocomposites from cubic silsesquioxane epoxides. *J. Mater. Chem.* **2004**, *14*, 2858–2863. [CrossRef]

32. Geng, Z.; Ba, J.Y.; Zhang, S.L.; Luan, J.S.; Jiang, X.; Huo, P.F.; Wang, G.B. Ultra low dielectric constant hybrid films via side chain grafting reaction of poly(ether ether ketone) and phosphotungstic acid. *J. Mater. Chem.* **2012**, *22*, 23534–23540. [CrossRef]

33. Jin, H.S.; Chang, J.H. Synthesis and characterization of colorless polyimide nanocomposite films containing pendant trifluoromethyl groups. *Macromol. Res.* **2008**, *16*, 503–509. [CrossRef]

34. An, L.; Pan, Y.Z.; Shen, X.W.; Lu, H.B.; Yang, Y.L. Rod-like attapulgite/polyimide nanocomposites with simultaneously improved strength, toughness, thermal stability and related mechanisms. *J. Mater. Chem.* **2008**, *18*, 4928–4941. [CrossRef]

35. Chang, H.C.; Sohn, B.H.; Chang, J.H. Colorless and transparent polyimide nanocomposites: Comparison of the properties of homo- and co-polymers. *J. Ind. Eng. Chem.* **2013**, *19*, 1593–1599.

36. Xenopoulos, C.; Mascia, L.; Shaw, S.J. Polyimide-silica hybrids derived from an isoimide oligomer precursor. *J. Mater. Chem.* **2002**, *12*, 213–218. [CrossRef]

37. Song, G.L.; Wang, S.; Wang, D.M.; Zhou, H.W.; Chen, C.H.; Zhao, X.G.; Dang, G.D. Rigidity enhancement of polyimides containing benzimidazole moieties. *J. Appl. Polym. Sci.* **2013**, *130*, 1653–1658. [CrossRef]

38. Chen, J.C.; Wu, J.A.; Lee, C.Y.; Tsai, M.C.; Chen, K.H. Novel polyimides containing benzimidazole for temperature proton exchange membrane fuel. *J. Mater. Chem.* **2015**, *483*, 144–154. [CrossRef]

39. Song, G.L.; Wang, D.M.; Zhao, X.G.; Dang, G.D.; Zhou, H.W.; Chen, C.H. Synthesis and properties of polyimides-containing benzoxazole units in the main chain. *High Perform. Polym.* **2013**, *25*, 354–360. [CrossRef]

40. Hasegawa, M. Semi-aromatic polyimides with low dielectric constant and low CTE. *High Perform. Polym.* **2001**, *13*, S93–S106. [CrossRef]

41. Wang, S.; Zhou, H.W.; Dang, G.D.; Chen, C.H. Synthesis and characterization of thermally stable, high-modulus polyimides containing benzimidazole moieties. *J. Polym. Sci. Polym. Chem.* **2009**, *47*, 2024–2031. [CrossRef]

42. Yin, C.Q.; Dong, J.; Zhang, Z.X.; Zhang, Q.H.; Lin, J.Y. Structure and properties of polyimide fibers containing benzimidazole and amide units. *J. Polym. Sci. Polym. Phys.* **2015**, *53*, 183–191. [CrossRef]

43. Song, G.L.; Wang, D.M.; Dang, G.D.; Zhou, H.W.; Chen, C.H.; Zhao, X.G. Thermal expansion behavior of polyimide films containing benzoxazole units in the main chain. *High Perform. Polym.* **2014**, *26*, 413–419. [CrossRef]

44. Xue, G.; Shi, G.Q.; Ding, J.F.; Chang, W.M.; Chen, R.S. Complex-induced coupling effect: Adhesion of some polymers to copper metal promoted by benzimidazole. *J. Adhes. Sci. Technol.* **1990**, *4*, 723–732. [CrossRef]

45. Chen, K.M.; Ho, S.M.; Wang, T.H.; King, J.S.; Chang, W.C.; Cheng, R.P.; Hung, A. Studies on the adhesion of polyimide coatings on copper foil. *J. Appl. Polym. Sci.* **1992**, *45*, 947–956. [CrossRef]

46. Ahn, T.K.; Kim, A.M.; Choe, S. Hydrogen-bonding strength in the blends of polybenzimidazole with btda- and dsda-based polyimides. *Macromol.* **1997**, *30*, 3369–3374. [CrossRef]

47. Song, G.L.; Zhang, Y.; Wang, D.M.; Chen, C.H.; Zhou, H.W.; Zhao, X.G.; Dang, G.D. Intermolecular interactions of polyimides containing benzimidazole and benzoxazole moieties. *Polymer* **2013**, *54*, 2335–2340. [CrossRef]

![polymers logo] *polymers*

MDPI

Article

Fabrication of Multi-Layered Lidocaine and Epinephrine-Eluting PLGA/Collagen Nanofibers: In Vitro and In Vivo Study

Fu-Ying Lee [1], Demei Lee [2], Tzu-Chia Lee [2], Jan-Kan Chen [3], Ren-Chin Wu [4], Kuan-Chieh Liu [2] and Shih-Jung Liu [2,5,*]

[1]   Department of Periodontics, Division of Dentistry, Chang Gung Memorial Hospital, Tao-Yuan 33305, Taiwan; fuying20@hotmail.com
[2]   Department of Mechanical Engineering, Chang Gung University, Tao-Yuan 33302, Taiwan; dmlee@mail.cgu.edu.tw (D.L.); joy820629@gmail.com (T.-C.L.); harobinca@hotmail.com (K.-C.L.)
[3]   Department of Physiology and Pharmacology, Chang Gung University, Tao-Yuan 33302, Taiwan; jkc508@mail.cgu.edu.tw
[4]   Department of Pathology, Chang Gung Memorial Hospital, Tao-Yuan 33305, Taiwan; renchin.wu@gmail.com
[5]   Department of Orthopedic Surgery, Chang Gung Memorial Hospital, Tao-Yuan 33305, Taiwan
*   Correspondence: shihjung@mail.cgu.edu.tw; Tel.: +886-3-2118166; Fax: +886-3-2118558

Received: 9 August 2017; Accepted: 3 September 2017; Published: 5 September 2017

Abstract: This study developed multi-layered lidocaine- and epinephrine-eluting biodegradable poly[(D,L)-lactide-*co*-glyco lide] (PLGA)/collagen nanofibers. An electrospinning technique was employed to fabricate the multi-layer biodegradable drug-eluting nanofibers. After fabrication, the nanofibrous membranes were characterized. The drug release characteristics were also investigated. In addition, the in vivo efficacy of nanofibers for pain relief and hemostasis in palatal oral wounds of rabbits were evaluated. Histological examinations were also completed. The experimental results suggested that all nanofibers exhibited good biocompatibility and eluted effective levels of lidocaine and epinephrine at the initial stages of wound recovery.

Keywords: biodegradable nanofibers; PLGA; collagen; epinephrine; lidocaine

---

1. Introduction

A free gingival graft (FGG) is a dental procedure that involves harvesting soft tissue from a distant site in the mouth and grafting it over a localized recession defect. FGG was first introduced in 1963 and found the clinical applications to treat mucogingival lesions one year later [1–3]. After the procedure, tissues at the excisional wound are replaced by fibrin and inflammatory cells. It is a surgical procedure widely employed to enhance the keratinized tissue encompassing a tooth and/or a dental implant. The technique has also been extensively employed to treat various periodontal mucogingival lesions including insufficient or lack of attached gingiva, the presence of high frenum attachments, shallow fornix, and denuded roots following gingival recession. However, the donor tissue is usually taken from the palate. The postoperative response at a palate wound site is generally uneventful. In addition, excessive bleeding and fierce post-surgery pain have been addressed after the FGG procedure [4–7].

For decades, science has recognized that wound dressings are extremely important components of wound care and are necessary for optimal wound healing. Open palatal wound recovery consists of a dynamic and intense process, which involves blood clot deposition, inflammation, and angiogenesis to constitute a vascularized granulation tissue. After that, an approximation of epithelium and tissue remodeling is completed to create a mature recovery process in four weeks [8]. Wound dressings perform three basic functions: protect wounds, help prevent infection, and maintain optimal moisture, consequently accelerating the healing process.

Lidocaine is a drug widely used to numb tissue in a specific area and may be applied directly to the skin for numbing [9]. Epinephrine, on the other hand, is primarily a medication used for a number of conditions including anaphylaxis, cardiac arrest, and superficial bleeding [10]. When lidocaine is mixed with a small amount of epinephrine, it allows larger doses to be used for numbing and increases the length of time it is effective [11]. Type I collagen is the most abundant collagen in the body that constitutes collagen fibers [12,13]. It is also one component of skin tissue that benefits all stages of the wound healing process. McGuire and Scheyer [14] employed a xenogeneic collagen matrix as a substitute to FGG for the augmentation of oral soft tissues. Their results showed that the collagen matrix may act as a suitable substitute for FGG in vestibuloplasty procedures designed to improve keratinized tissue around teeth.

This study developed biodegradable lidocaine, epinephrine, and collagen loaded nanofibrous membranes that offered a sustainable elution of hemostatic and analgesic drugs at the oral wound sites to decrease postoperative pain and accelerate palatal donor site wound healing. An electrospinning technique was employed to fabricate the multi-layer biodegradable drug-eluting nanofibrous membranes. The technique has been widely employed to fabricate nanofibers of various materials as well as nanofibrous membranes encapsulating living cells [15]. After fabrication, the nanofibrous membranes were characterized. The drug release characteristics were investigated. The efficacy of the drug-loaded membranes was assessed on a rabbit model. Histological analysis was completed as well.

2. Materials and Methods

2.1. In Vitro Release of Epinephrine and Lidocaine

2.1.1. Manufacture of Polycaprolactone Stent and Drugs Loaded PLGA Nanofibers

The polymeric materials included poly(ε-caprolactone) (PCL) (Sigma-Aldrich, St. Louis, MO, USA) and poly[(D,L)-lactide-*co*-glyco lide] (PLGA) polymer (lactide/glycolide: 50/50) (Sigma-Aldrich, St. Louis, MO, USA). Epinephrine and lidocaine hydrochloride were included, while type I collagen was from calf skin (Sigma-Aldrich, St. Louis, MO, USA).

A PCL stent with a dimension of 9 mm $\times$ 12 mm and an open window of 5 mm $\times$ 5 mm was sliced from a 0.15 mm thick film that is solvent-casted. Membranes with three various PLGA/drug combinations, namely 6:1, 3:1 and 2:1, respectively, were produced. To manufacture the multi-layered membranes with 6:1 polymer-to-drug ratio, PLGA/lidocaine (300 mg:50 mg), PLGA/epinephrine (300 mg:50 mg), and PLGA/collagen (300 mg:50 mg) were primarily dissolved in 1 mL of hexafluoroisopropanol (HFIP) (Sigma-Aldrich), respectively. The PLGA/lidocaine solution was spun and collected by the collection plate in a nonwoven form. The procedure was duplicated to manufacture subsequently the PLGA/epinephrine and PLGA/collagen nanofibers. All procedures were performed at room temperature. On the other hand, the preparation of nanofibrous membranes with 3:1 and 2:1 polymer-to-drug ratios followed the same electrospinning procedure, except that the polymer/drug used were 300 mg/100 mg and 300 mg/150 mg, respectively.

2.1.2. Characterization of Electrospun Nanofibrous Membranes

A field emission scanning electron microscope (FE-SEM) (JSM-7500F, Joel, Japan) was employed to quantify the electrospun nanofibers. The diameter compositions of nanofibers were acquired by characterizing the microphotos of 50 arbitrarily picked fibers for each specimen.

The spectra of electrospun nanofibrous membranes were characterized using a Fourier Transform Infrared (FTIR) spectrometry. FTIR analysis was completed on a Nicolet iS5 spectrometer (Thermo Fisher Scientific, Waltham, MA, USA) at a resolution of 4 cm^{-1} and 32 scans. Nanofibrous membranes were pressed as KBr discs, and spectra were analyzed over the 400–4000 cm^{-1} range.

The tensile characteristics of electrospun nanofibrous membranes were quantified on a Lloyd tensiometer (AMETEK, Berwyn, PA, USA).

The water contact angles of the nanofibers were assessed by a contact angle measuring apparatus (First Ten Angstroms, Portsmouth, VA, USA) (N = 5).

2.1.3. Drug Concentration Assessment

The elution behavior of lidocaine and epinephrine from the drug-loaded nanofibers with various PLGA/drug combinations, namely 6:1, 3:1 and 2:1, respectively, was evaluated. Specimens, with an approximate size of 2.0 mm × 3.0 mm and weight of 1 g, were sliced from the electrospun membrane and stored in glass tubes. Each tube contained one specimen along with 1 mL of buffered solution (0.15 mol/L, pH 7.4). All test tubes were maintained at 37 °C for 24 h. Then the eluent was collected for analysis, and fresh buffered solution (1 mL) was added. The procedure was repeated every 24 h until the specimen was fully dissolved. The experiment was triplicated (N = 3).

The lidocaine levels in the eluents were quantified using a high performance liquid chromatography (HPLC) assay. The HPLC analyses were carried out on a Hitachi L-2200R system, using an ATLANTIS dC18, 4.6 cm × 150 mm column (Waters Corp., Milford, MA, USA). Ammonium formate and methanol (Sigma-Aldrich; 20/80 (v/v)) were adopted as the mobile phase. The absorbency of monitoring was set at a wavelength of 210 nm, while the flow rate was maintained at 1.0 mL/min. To characterize epinephrine, a Discovery BIO Wide C18-5, 25 cm × 4.6 mm column was used. The mobile phase contained water, methanol, and acetic acid (Sigma-Aldrich) in a volume ratio of 85:10:5. The pH value of the mobile phase was adjusted by adding ammonium acetate (Sigma-Aldrich) to reach 3.1. The absorbency was maintained at 280 nm, while the flow rate was set at 1.0 mL/min. All specimens were assayed in triplicate (N = 3).

2.2. Cytotoxicity of Composite Nanofibers

Cytotoxicity of fabricate nanofibers was assessed by MTT assay (Roche, Berlin, Germany). Electrospun nanofibers were cut out with punch and placed onto 24-well culture plates. Human fibroblasts were seeded (5×10^3 cells/well) in DMEM at 37 °C under 5% CO_2/95% air conditions until cell confluence. Cell viability was observed on Days 1, 2, 3, and 7 by MTT assays and evaluated using an ELISA reader. The number of cells was also calculated under an optical microscope (Olympus IMT-2, Tokyo, Japan).

2.3. In Vivo Study of Animal

2.3.1. Animal-Related Procedure

Twelve New Zealand white rabbits were enrolled for the in vivo study. The average weight of the animals was 3.3 kg. All animal-related processes obtained the approval from the institution. The animals were taken care of following the guidelines of the Department of Health and Welfare, Taiwan.

The rabbit was sedated with 2% xylazine-HCl (5 mg/kg body weight) and Ketamine-HCl (Ketasol, 30 mg/kg body weight). The operative area was cleaned and sterilized. All surgical procedures were carried out using a #15 blade after the local injection of 0.5 mL 2% Xylestesin-A containing epinephrine at a concentration of 1:100,000. A rectangular shape incision of 6 mm × 8 mm (width × length) and 1–1.5 mm in thickness was created on the palate of rabbits. After eliminating the graft, wet gauze was applied onto the donor site for 60 s with moderate finger pressure.

The 12 rabbits were stochastically separated into two groups: In the control group, the excision wound was covered by a PCL stent without membrane and sutured with 6-0 polypropylene suture at the corners of the stent (Figure 1). In the test group, the donor site was covered by a PCL stent with multi-layered lidocaine/epinephrine/collagen loaded biodegradable nanofibrous membrane.

For simplicity reason, the membrane with the PLGA/drug combination of 3:1 was employed. After that, the stent was sutured with 6-0 polypropylene suture at the corners.

In the test group, the tissue fluid at the palatal excision wound was extracted using #30 standardized sterile paper point (DiaDent, Korea) on Days 1, 2, 4, 7, 10 and 14. Sampling was performed in situ for 30 s during each procedure. Immediately after collection, the paper point was eluted with 0.1 mL phosphate buffer solution and then stored at $-20\ ^{\circ}$C until it was analyzed. The HPLC analysis was employed to characterize the drug concentrations at each time period, using the same procedure as the in vitro study.

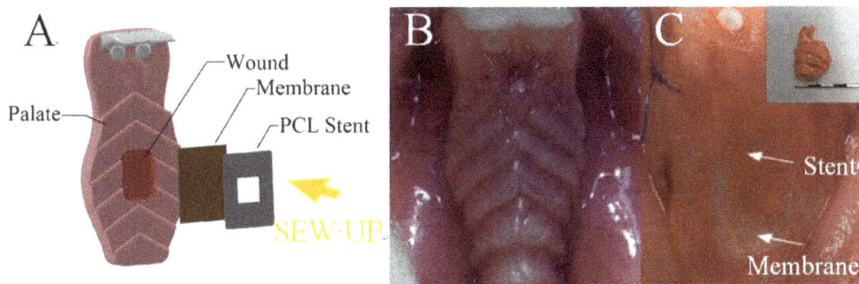

Figure 1. (**A**) Schematic of the deployment of biodegradable stent and nanofibrous membrane on the donor site; and (**B**) beforel and (**C**) after the deployment of stent and membrane.

2.3.2. Post-Surgery Evaluation

The daily changes of body weight, and food/water intakes were monitored post-surgery for 14 days. In addition, the hemostasis analysis was completed following the work of Saroff et al. [16].

2.3.3. Wound Healing and Histological Analysis

On Days 1, 3, 7 and 14 after operation, photos of the wound for each rabbit were taken. To evaluate and compare the tissue responses at a microscopically level, standard biopsy was done at the end of Days 1, 3, 7 and 14. The specimens were harvested and fixed in 10% formalin and embedded in paraffin, cut into 4 μm frontal section. The epithelial and connective tissue characteristics were examined by employing the Hematoxylin and Eosin (H&E) staining. The interpretation was completed at ×10 and ×20 magnification by an independent examiner.

2.4. Statistical Analysis

The differences between groups were assessed using a least significance difference test, by employing SPSS software (SPSS Inc., Armonk, NY, USA).

3. Results

3.1. In Vitro Characterization of Electrospun Drug-Loaded Nanofibrous Membranes

Figure 2 shows the SEM microphotos of the drugs loaded nanofibers and the diameter compositions of each fiber. The calculated diameters were 614 ± 213 nm, 623 ± 149 nm, and 659 ± 204 nm, respectively, for the 2:1, 3:1 and 6:1 lidocaine/epinephrine loaded PLGA/collagen nanofibers. The fiber diameter increased slightly with the percentage of polymers in the membranes. When subjected to an external force exerted by the electric field, the polymeric solution that has a higher polymer percentage showed a higher strength and was less extended. Electrospun nanofibers thus exhibited greater fiber diameters.

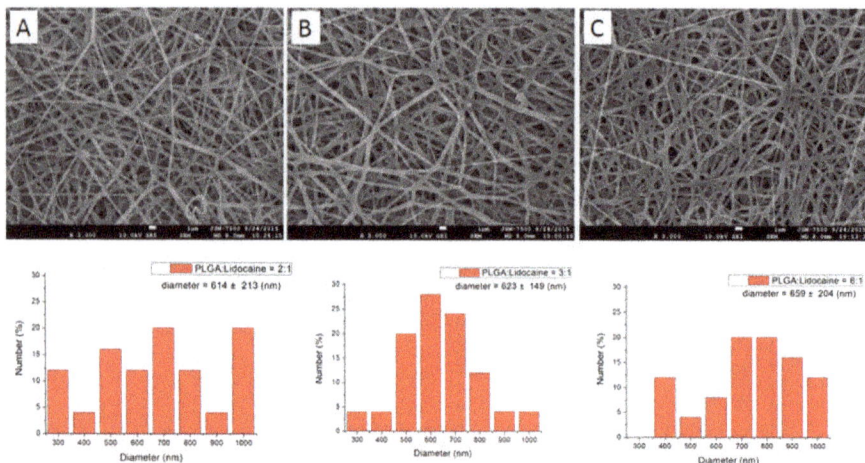

Figure 2. Scanning electron microscopy (SEM) photos and fiber diameter distribution of electrospun drugs loaded nanofibers. PLGA:Lidocaine (**A**) 2:1 (**B**) 3:1 (**C**) 6:1.

Figure 3 displays the calculated spectra of non-drug loaded PLGA nanofibers and drug incorporated PLGA nanofibers. The absorption peaks of epinephrine and lidocaine were identified in electrospun PLGA nanofibers. The vibration peak at 3400 cm^{-1} can be attributed to the N-H bonds of lidocaine and epinephrine, while the vibration at 3250 cm^{-1} of the O–H bond was improved with the presence of the drugs. In addition, the absorbance at the region of 3200 cm^{-1} could be resulted from the benzene of lidocaine [17] and epinephrine [18]. The results of the FTIR spectra demonstrated the incorporation of drugs in the PLGA matrix.

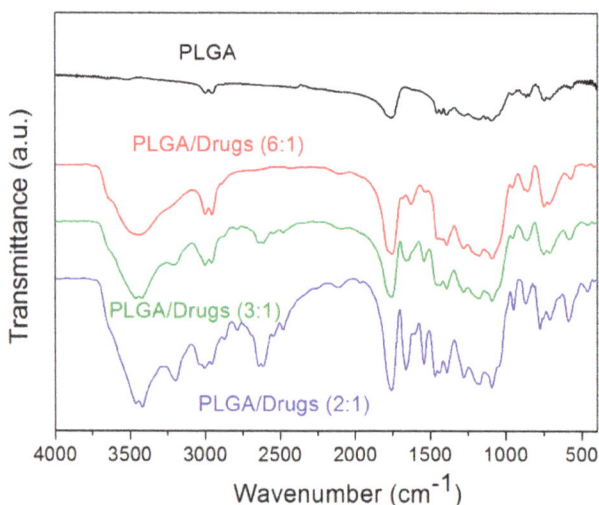

Figure 3. Fourier Transform Infrared (FTIR) spectra of electrospun nanofibrous membranes. The results of the FTIR spectra demonstrated the incorporation of drugs in the poly[(D,L)-lactide-*co*-glyco lide] (PLGA) matrix.

Measured water contact angles of pure PLGA, and 6:1, 3:1 and 2:1 PLGA:drug ratio loaded nanofibrous membranes were 127.4°, 118.5°, 87.6° and 54.6°, respectively. The addition of lidocaine and epinephrine increased the hydrophilicity of the PLGA nanofibers. Furthermore, the water contact angle (hydrophilicity) increased with the content of drugs in electrospun nanofibers.

The tensile test results demonstrated that the maximum strength (elongation at break) of the pure PLGA, and 6:1, 3:1 and 2:1 PLGA:drug ratio loaded nanofibers were 4.53 MPa (229.7%), 2.85 MPa (51.2%), 1.87 MPa (37.0%) and 1.40 MPa (26.1%), respectively. Obviously, the mechanical strengths decreased with the content of incorporated drugs.

Cytotoxicity experiments from MTT assays of the composite nanofibers were examined. The measured result in Figure 4 suggested that the fabricated nanofibers do not show any signs of cytotoxicity.

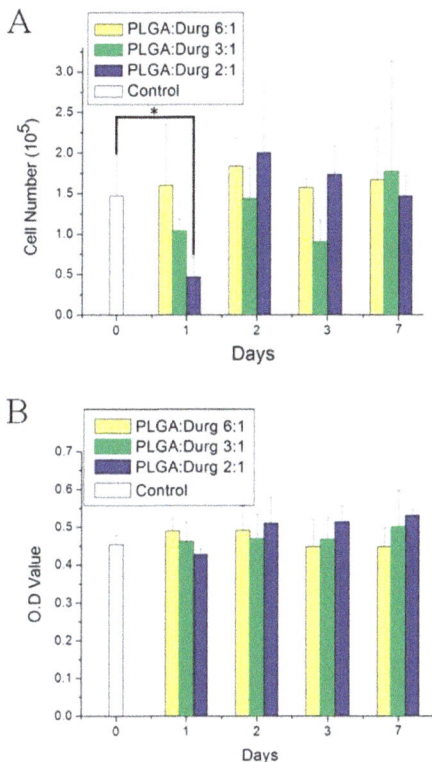

Figure 4. Toxicity of electrospun nanofibrous composite membranes ($p < 0.05$). The fabricated nanofibers show no signs of cytotoxicity. (**A**) cell number; (**B**) optical density (OD) value.

3.2. In Vitro Elution Characteristics of Epinephrine and Lidocaine

Figure 5A shows the elution behaviors of lidocaine, while Figure 5B displays the release patterns of epinephrine. The 6:1 polymer-to-drug ratio membrane showed a biphasic release profile, consisting of a burst elution during the first two days and a steady and gradually diminishing drug release. The elution pattern was comparable for the 3:1 and 2:1 polymer-to-drug ratios nanofibrous membranes, with a steadily decreasing release, indicating that the lidocaine and epinephrine were evenly encapsulated and distributed in the electrospun nanofibrous membranes. In addition, all biodegradable nanofibers released effective concentrations of epinephrine and lidocaine for over three weeks.

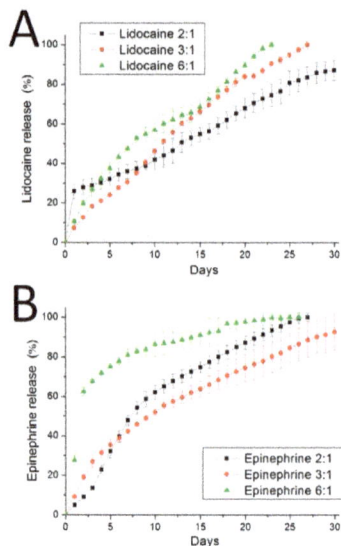

Figure 5. Release curves of: (**A**) lidocaine and (**B**) epinephrine from the nanofibrous membranes. All biodegradable nanofibers released effective concentrations of epinephrine and lidocaine for over three weeks.

3.3. In Vivo Elution of Lidocaine and Epinephrine

Figure 6 shows the in vivo elution profiles of lidocaine and epinephrine from the nanofibrous membranes. Release peaks were observed at Days 1 and 3, after which the drug concentration dropped significantly at Day 7 because both the stent and the membrane were not found at the wound sites by Day 7, possibly removed by the rabbit's tongue and swallowed by the rabbit. Measured drug level thus decreased accordingly.

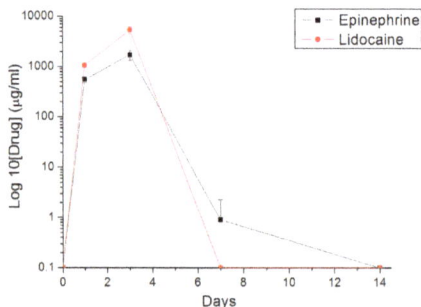

Figure 6. In vivo release of lidocaine and epinephrine from the nanofibrous membranes. Release peaks were observed at Days 1 and 3, after which the drug concentration dropped significantly at Day 7.

3.4. Efficacy of Released Drugs

3.4.1. Hemostasis Efficacy

The hemostasis efficacy of the drugs loaded nanofibrous membranes was evaluated [15]. For the initial hemostasis (1 min after the surgery), 67% (4/6) in the test showed hemostasis outcome, but only 17% (1/6) of wounds displayed hemostasis. Furthermore, while 100% (6/6) of the wounds in the test group exhibited effective hemostasis at 10 min, only 83% (5/6) of the wounds in the control showed hemostasis. The results here demonstrated the hemostasis capability of the drug-loaded collagen nanofibers.

3.4.2. Weight Changes

The preoperative body weights in the control group and the test group were comparable: 2.97 ± 0.43 kg in the control group A and 3.04 ± 0.68 kg in the test group B. No significant difference was found in the mean body weights of the rabbits between the control group and the test group by the end of the study.

3.4.3. Food Intake and Water Consumption

The food intake and water consumption were recorded. On Day 1, food intake by rabbits in the control group decreased significantly when compared to those in the test group. The amount of food intake increased gradually for all rabbits in both groups. On Day 14, rabbits of both groups regained their normal food intake (approximately 135 g). Throughout the study period, the rabbits in the test group exhibited greater food intake than did the rabbits in the control group. However, the difference did not reach significant level.

On the other hand, all rabbits reduced their water consumption by 50% at one day post-operation, after which the water consumption increased gradually. On Day 14, all rabbits regained their normal amount of water consumption (approximately 1100 mL). Again, the rabbits in the test group (composite membrane group) showed greater amount of water consumption than did those in the control group, although the difference is not statistically significant.

3.4.4. Wound Healing and Histological Analysis

Figure 7 shows the upper gingiva of the rabbits from each group on Days 1, 3, 7 and 14 after surgery. On Day 1, stents of both groups were in situ. The biodegradable composite membranes fully covered the palatal wound. From the window of the control site stent, formation of the granulation tissue could be observed. On Day 3, stents in the control group were sloughing, while the multiple layered nanofibrous membrane was broken at the wound site. Food debris was also found. All stents in both groups were completely disintegrated by Day 7. The gross wound appearance in the control group exhibited necrotic tissue along with new epithelialized gingiva tissues. In contrast, the wound dressed by the multiple layered nanofibrous membrane showed clear wound surface. On Day 14, the wounds in rabbits of both groups were fully recovered.

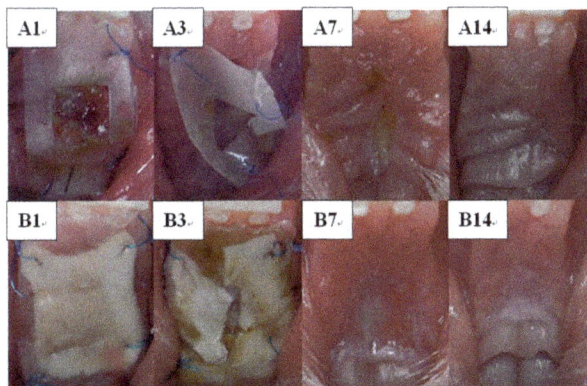

Figure 7. Wound healing in (**A**) control group and (**B**) test (nanofibrous composite membrane) group on Days 1, 3, 7 and 14 post-surgery.

Figure 8 shows the histological analysis result. On Days 1 and 3, necrosis with heavy Polymorphonuclear neutrophils (PMN) infiltrations were observed in both groups. Despite the fact that active fibrosis had been found in both groups, PMN infiltration was moderate in the drug-eluting membrane treated wounds, while the wounds in the control group showed heavy PMN infiltration on Day 7. On Day 14, fibrosis with few inflammatory cells was found on both groups and no significant difference was found between the groups.

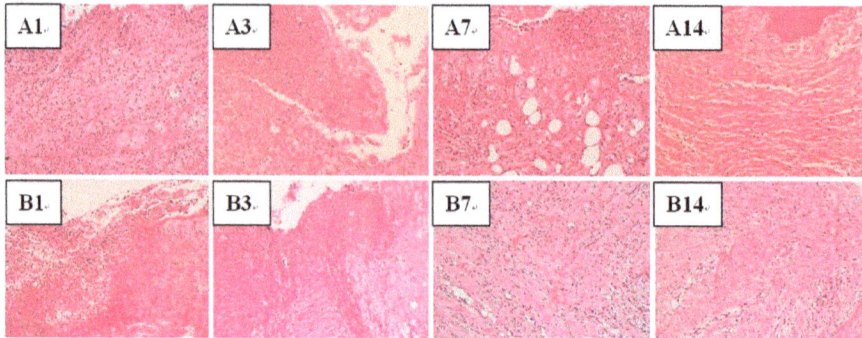

Figure 8. Histological images of (**A**) control group and (**B**) nanofibrous composite membrane group on Days 1, 3, 7 and 14. Polymorphonuclear neutrophils (PMN) infiltration was moderate in the drug-eluting membrane treated wounds, while the wounds in the control group showed heavy PMN infiltration on Day 7.

4. Discussion

Postoperative pain and bleeding are the most common complications following soft tissue grafting procedures. Pain control is important during the early phase in soft tissue surgery. In the early operative period, free gingival graft accounts for the greater incidence of donor site pain. The pain results in stress and impairs the neutrophil function. Furthermore, the stress affects the immune system and down regulates IL-1 gene expression and impairs wound healing [19]. After the procedures, tissues at excisional wound are substituted by fibrin and inflammatory cells. The palatal excision wound with large epithelial deficiencies cannot cure by initial closure of the wound edges. A secondary recovery with epithelial cell migration from the surrounding to the central region of the defect is usually required to heal the wound [20,21]. Ward was the first to advocate for the adoption of periodontal dressing after gingival surgery so as to decrease pain and infection at the wound site [22]. Various procedures for protecting and covering the donor site, including intraoral bandage or oral adhesive, interproximal wire ligation, surgical dressing by mattress sutures, and modified Hawley retainer, have also been proposed [23]. To facilitate the recovery process and to decrease the occurrence of complications, chemotherapeutic agents and hemostatic agents, ozonides and low-intensity pulse ultrasound, and low-energy laser irradiation were introduced [8,24–26].

This study developed multi-layered nanofibrous lidocaine/epinephrine-eluting PLGA membranes that provide good conformity to underlying gingiva, reduce postoperative pain and hemorrhage, and protect the clot from the forces applied during chewing. The drug release kinetics for the biodegradable drug-eluting nanofibrous membranes comprise of two different phases: an initial burst and a degradation-dominated phase. During electrospinning, most drugs are encapsulated and distributed in the nanofibers. Nevertheless, some drug compounds may be located on the nanofiber surface, thus resulting in the initial burst. After that, the drug-elution behavior is mainly decided by polymer degradation. The drugs are released as the polymers degrade with time [27]. The nanofibers thus exhibited a steady and diminishing release of epinephrine and lidocaine. The biodegradable multi-layered drug-eluting PLGA/collagen nanofibrous membranes could release high levels of epinephrine and lidocaine for over four weeks. This would provide advantages for pain and bleeding control and tissue regeneration.

Malmquist et al. [28] studied the HemCone dental dressing (HDD) and showed that it is a clinically effective hemostatic device that significantly reduces bleeding time after oral surgeries. HDD is fabricated from freeze-dried chitosan derived from shrimp shell chitin. Several authors demonstrated that HDD exhibits cell adhesion properties and releases growth factors from human palates stimulated by chitosan [29,30]. Despite the hemostatic capability, small amounts of unreacted residual acetic acid in wounds covered by the HDDs can cause temporary pain in the first two days after surgery [28]. This study employed PLGA as the delivery vehicle of the drugs. PLGA is a biocompatible and biodegradable polymeric material that shows a wide range of degradation times as well as adjustable mechanical properties. It has been widely investigated as delivery carriers for drugs, proteins and various other macromolecules [31,32].

Recent studies used platelet-rich fibrin (PRF) for promoting palatal donor wound healing. PRF is a fibrin matrix incorporating and releasing platelet cytokines, growth factors, and cells. The fibrin matrix is the natural guide of initial angiogenesis modulated by the binding to various growth factors. Fibrin plays an important role in modulating neutrophil activity and inducing epithelial cell migration to enhance the process of wound healing [33]. Kulkarni et al. [34] reported that PRF used for palatal wounds promotes complete wound closure at one week, reduces inflammation reaction at the periphery of the healing wound, and displays good control of bleeding at the time of surgery [32]. Although PRF has a positive effect on wound healing, patients may suffer from venous blood donation with another skin wound. In addition, as the preparation is strictly autologous, the amount of PRF obtained is limited.

Collagen is a natural substrate of extracellular matrix. The collagen dressing has the ability to realize hemostasis, chemotactic to fibroblasts and platelets, and induce the proliferation and differentiation of mesenchymal cells. Shanmugam et al. [35] demonstrated that collagen-based dressing offers significantly greater advantages over the traditional non-eugenol dressing in the healing of palatal wounds. As a scaffold, the nanofibrous membranes should be able to enhance cell proliferation and physiological function and retain normal states of cell growth. In this study, the electrospun multi-layered nanofibrous membranes show no signs of cytotoxicity. The possible effect of released acid from the hydrolysis of PLGA nanofibers on cell proliferation was also negligible. Furthermore, the animals dressed with the drug-eluting nanofibers show greater food intake and water consumption than did those without (the control). The measured in vivo release curve (Figure 6) showed that the multi-layered nanofibers release high concentrations of epinephrine and lidocaine during the first three days, which helps relieve pain [11]. This suggests that the fabricated nanofibers may provide effective analgesic effects at the wound site of free tissue grafts and promote wound healings.

It has been proposed that human skin fibroblasts exhibited higher growth on nanofibrous membranes possessing fiber diameters in the range of 350–1100 nm [36]. The measured diameters were 614 ± 213 nm, 623 ± 149 nm and 659 ± 204 nm, respectively, for the 2:1, 3:1 and 6:1 lidocaine/epinephrine loaded PLGA nanofibrous membranes. The fabricated membranes thus show positive cell proliferations in normal human fibroblasts. Despite the fact that lidocaine and epinephrine released from the nanofibers may delay cell growth and proliferation on Day 1, the negative impact of released drugs was found to be negligible starting from Day 2 and forward (Figure 4). On the other hand, the histological analysis images show no inflammation of the tissues. This further supports that the PLGA/drugs/collagen nanofibers can be an appropriate scaffold for tissue regeneration of free grafts.

5. Conclusions

This study developed multi-layered lidocaine- and epinephrine-eluting PLGA/collagen nanofibers and evaluated their efficacy for pain relief and hemostasis in palatal oral wounds of rabbits. All nanofibers exhibited good biocompatibility and eluted effective levels of lidocaine and epinephrine at the early stages of wound healing. Rabbits in the test group showed faster hemostasis as well as recovery of food and water intake postoperatively, compared with those in the control group. The experimental results demonstrate that the multi-layered biodegradable nanofibrous membranes offered adequate efficacy of hemostasis and sustainable pain relief for the initial healing of palatal oral wounds.

Acknowledgments: This work was supported in part by the Chang Gung Memorial Hospital (Contract No. CMRPD3D0153), and the Ministry of Science and Technology, Taiwan (Contract No. 104-2221-E-182-048-MY3).

Author Contributions: Fu-Ying Lee and Shih-Jung Liu conceived and designed the experiments; Tzu-Chia Lee performed the experiments; Jan-Kan Chen completed the cell culture; Ren-Chin Wu analyzed the histological data; Demei Lee and Kuan-Chieh Liu contributed reagents/materials/analysis tools; Shih-Jung Liu and Fu-Ying Lee wrote the paper.

Conflicts of Interest: The authors declare no conflict of interest.

References

1. Matter, J. Free gingival grafts for the treatment of gingival recession. A review of some techniques. *J. Clin. Periodontol.* **1982**, *9*, 103–114. [CrossRef] [PubMed]
2. Bjorn, H. Free transplantation of gingiva propria. *Odontol. Revy* **1963**, *14*, 323.
3. Pennel, B.M.; Tabor, J.C.; King, K.O.; Towner, J.D.; Fritz, B.D.; Higgason, J.D. Free masticatory mucosa graft. *J. Periodontol.* **1969**, *40*, 162–166. [CrossRef] [PubMed]
4. Griffin, T.J.; Cheung, W.S.; Zavras, A.I.; Damoulis, P.D. Postoperative complications following gingival augmentation procedures. *J. Periodontol.* **2006**, *77*, 2070–2079. [CrossRef] [PubMed]
5. Keceli, H.G.; Aylikci, B.U.; Koseoglu, S.; Dolgun, A. Evaluation of palatal donor site haemostasis and wound healing after free gingival graft surgery. *J. Clin. Periodontol.* **2015**, *42*, 582–589. [CrossRef] [PubMed]
6. Cairo, F.; Nieri, M.; Pagliaro, U. Efficacy of periodontal plastic surgery procedures in the treatment of localized facial gingival recessions. A systematic review. *J. Clin. Periodontol.* **2014**, *41* (Suppl. 15), S44–S62. [CrossRef] [PubMed]
7. Roccuzzo, M.; Bunino, M.; Needleman, I.; Sanz, M. Periodontal plastic surgery for treatment of localized gingival recessions: A systematic review. *J. Clin. Periodontol.* **2002**, *29* (Suppl. 3), 178–194; discussion 195–176. [CrossRef] [PubMed]
8. Wang, C.Y.; Tsai, S.C.; Yu, M.C.; Lin, Y.F.; Chen, C.C.; Chang, P.C. Light-emitting diode irradiation promotes donor site wound healing of the free gingival graft. *J. Periodontol.* **2015**, *86*, 674–681. [CrossRef] [PubMed]
9. Weibel, S.; Jokinen, J.; Pace, N.L.; Schnabel, A.; Hollmann, M.W.; Hahnenkamp, K.; Eberhart, L.H.; Poepping, D.M.; Afshari, A.; Kranke, P. Efficacy and safety of intravenous lidocaine for postoperative analgesia and recovery after surgery: A systematic review with trial sequential analysis. *Br. J. Anaesth.* **2016**, *116*, 770–783. [CrossRef] [PubMed]
10. Hartzell, T.L.; Sangji, N.F.; Hertl, M.C. Ischemia of postmastectomy skin after infiltration of local anesthetic with epinephrine: A case report and review of the literature. *Aesthet. Plast. Surg.* **2010**, *34*, 782–784. [CrossRef] [PubMed]
11. Sinnott, C.J.; Cogswell, I.L.; Johnson, A.; Strichartz, G.R. On the mechanism by which epinephrine potentiates lidocaine's peripheral nerve block. *J. Am. Soc. Anesthesiol.* **2003**, *98*, 181–188. [CrossRef]
12. Park, S.N.; Lee, H.J.; Lee, K.H.; Suh, H. Biological characterization of EDC-crosslinked collagen-hyaluronic acid matrix in dermal tissue restoration. *Biomaterials* **2003**, *24*, 1631–1641. [CrossRef]
13. Parry, D.A.D. Collagen fibril during development and maturation and their contribution to thr mechanical attributes of connective tissue. *Collagen* **1988**, *2*, 1–23.
14. McGuire, M.K.; Scheyer, E.T. Randomized, controlled clinical trial to evaluate a xenogeneic collagen matrix as an alternative to free gingival grafting for oral soft tissue augmentation. *J. Periodontol.* **2014**, *85*, 1333–1341. [CrossRef] [PubMed]
15. Townsend-Nicholson, A.; Jayasinghe, S.N. Cell Electrospinning: A Unique Biotechnique for Encapsulating Living Organisms for Generating Active Biological Microthreads/Scaffolds. *Biomacromolecules* **2006**, *7*, 3364–3369. [CrossRef] [PubMed]

16. Saroff, S.A.; Chasens, A.I.; Eisen, S.F.; Levey, S.H. Free soft tissue autografts. Hemostasis and protection of the palatal donor site with a microfibrillar collagen preparation. *J. Periodontol.* **1982**, *53*, 425–428. [CrossRef] [PubMed]

17. Fraceto, L.F.; Pinto Lde, M.; Franzoni, L.; Braga, A.A.; Spisni, A.; Schreier, S.; de Paula, E. Spectroscopic evidence for a preferential location of lidocaine inside phospholipid bilayers. *Biophys. Chem.* **2002**, *99*, 229–243. [CrossRef]

18. Wang, S.F.; Du, D.; Zou, Q.C. Electrochemical behavior of epinephrine at *l*-cysteine self-assembled monolayers modified gold electrode. *Talanta* **2002**, *57*, 687–692. [CrossRef]

19. Marucha, P.T.; Kiecolt-Glaser, J.K.; Favagehi, M. Mucosal wound healing is impaired by examination stress. *Psychosom. Med.* **1998**, *60*, 362–365. [CrossRef] [PubMed]

20. Kahnberg, K.E.; Thilander, H. Healing of experimental excisional wounds in the rat palate: (I) Histological study of the interphase in wound healing after sharp dissection. *Int. J. Oral. Surg.* **1982**, *11*, 44–51. [CrossRef]

21. Fejerskov, O. Excision wounds in palatal epithelium in guinea pigs. *Eur. J. Oral Soc.* **1972**, *80*, 139–154. [CrossRef]

22. Ward, A.W. Postoperative care in the surgical treatment of pyorrhoea. *J. Am. Dent. Assoc.* **1929**, *16*, 635–640.

23. Brasher, W.J.; Rees, T.D.; Boyce, W.A. Complications of free grafts of masticatory mucosa. *J. Periodontol.* **1975**, *46*, 133–138. [CrossRef] [PubMed]

24. Rossmann, J.A.; Rees, T.D. A comparative evaluation of hemostatic agents in the management of soft tissue graft donor site bleeding. *J. Periodontol.* **1999**, *70*, 1369–1375. [CrossRef] [PubMed]

25. Patel, P.V.; Kumar, S.; Vidya, G.D.; Patel, A.; Holmes, J.C.; Kumar, V. Cytological assessment of healing palatal donor site wounds and grafted gingival wounds after application of ozonated oil: An eighteen-month randomized controlled clinical trial. *Acta Cytol.* **2012**, *56*, 277–284. [CrossRef] [PubMed]

26. Maeda, T.; Masaki, C.; Kanao, M.; Kondo, Y.; Ohta, A.; Nakamoto, T.; Hosokawa, R. Low-intensity pulsed ultrasound enhances palatal mucosa wound healing in rats. *J. Prosthodont. Res.* **2013**, *57*, 93–98. [CrossRef] [PubMed]

27. Chen, D.W.C.; Liao, J.Y.; Liu, S.J.; Chan, E.C. Novel biodegradable sandwich-structured nanofibrous drug-eluting membranes for repair of infected wounds: An in vitro and in vivo study. *Int. J. Nanomed.* **2012**, *7*, 763–771.

28. Malmquist, J.P.; Clemens, S.C.; Oien, H.J.; Wilson, S.L. Hemostasis of oral surgery wounds with the HemCon Dental Dressing. *J. Oral. Maxillofac. Surg.* **2008**, *66*, 1177–1183. [CrossRef] [PubMed]

29. Cunha-Reis, C.; TuzlaKoglu, K.; Baas, E.; Yang, Y.; El Haj, A.; Reis, R.L. Influence of porosity and fibre diameter on the degradation of chitosan fibre-mesh scaffolds and cell adhesion. *J. Mater. Sci. Mater. Med.* **2007**, *18*, 195–200. [CrossRef] [PubMed]

30. Shen, E.C.; Chou, T.C.; Gau, C.H.; Tu, H.P.; Chen, Y.T.; Fu, E. Releasing growth factors from activated human platelets after chitosan stimulation: A possible bio-material for platelet-rich plasma preparation. *Clin. Oral Implant. Res.* **2006**, *17*, 572–578. [CrossRef] [PubMed]

31. Gentile, P.; Chiono, V.; Carmagnola, I.; Hatton, P.V. An overview of poly(lactic-*co*-glycolic) acid (PLGA)-based biomaterials for bone tissue engineering. *Int. J. Mol. Sci.* **2014**, *15*, 3640–3659. [CrossRef] [PubMed]

32. Makadia, H.K.; Siegel, S.J. Poly Lactic-*co*-Glycolic Acid (PLGA) as Biodegradable Controlled Drug Delivery Carrier. *Polymers* **2011**, *3*, 1377–1397. [CrossRef] [PubMed]

33. Weisel, J.W. Structure of fibrin: Impact on clot stability. *J. Thromb. Haemost.* **2007**, *5* (Suppl. 1), 116–124. [CrossRef] [PubMed]

34. Kulkarni, M.R.; Thomas, B.S.; Varghese, J.M.; Bhat, G.S. Platelet-rich fibrin as an adjunct to palatal wound healing after harvesting a free gingival graft: A case series. *J. Indian Soc. Periodontol.* **2014**, *18*, 399–402. [PubMed]

35. Shanmugam, M.; Kumar, T.S.; Arun, K.V.; Arun, R.; Karthik, S.J. Clinical and histological evaluation of two dressing materials in the healing of palatal wounds. *J. Indian Soc. Periodontol.* **2010**, *14*, 241–244. [PubMed]

36. Kumbar, S.G.; Nukavarapu, S.P.; James, R.; Nair, L.S.; Laurencin, C.T. Electrospun poly(lactic acid-*co*-glycolic acid) scaffolds for skin tissue engineering. *Biomaterials* **2008**, *29*, 4100–4107. [CrossRef] [PubMed]

Article

Preparation and Flame Retardance of Polyurethane Composites Containing Microencapsulated Melamine Polyphosphate

Shang-Hao Liu [1], Chen-Feng Kuan [2], Hsu-Chiang Kuan [2], Ming-Yuan Shen [3], Jia-Ming Yang [4] and Chin-Lung Chiang [4],*

[1] Department of Ammunition Engineering and Explosion Technology, Anhui University of Science and Technology, Huainan 232001, China; shliu998@163.com

[2] Department of Food and Beverage Management, Far East University, Tainan City 74448, Taiwan; cfkuan@mail.feu.edu.tw (C.-F.K.); hckuan@mail.feu.edu.tw (H.-C.K.)

[3] Department of Chemical and Materials Engineering, Southern Taiwan University of Science and Technology, Tainan City 71005, Taiwan; hbj678@gmail.com

[4] Green Flame Retardant Material Research Laboratory, Department of Safety, Health and Environmental Engineering, Hung-Kuang University, Taichung City 43302, Taiwan; young1362@gmail.com

* Correspondence: dragon@sunrise.hk.edu.tw; Tel.: +886-4-26318652-4008; Fax: +886-4-26525245

Received: 13 July 2017; Accepted: 29 August 2017; Published: 31 August 2017

Abstract: A new microencapsulated flame retardant containing melamine polyphosphate (MPP) and 4,4′-oxydianiline-formaldehyde (OF) resin as the core and shell materials, respectively, was synthesized by in situ polymerization. ^{29}Si NMR was used to measure the condensation density of polyurethane containing silicon compound (Si-PU). The structures and properties of the microencapsulated melamine polyphosphate (OFMPP) were characterized using X-ray photoelectron spectroscopy, scanning electron microscopy and water solubility. Thermal behavior of the OFMPP was systematically analyzed through thermogravimetric analysis. Flame retardance tests such as the limiting oxygen index and UL-94 were employed to evaluate the effect of composition variation on the MPP and OFMPP in polyurethane composites. The results indicated that the microencapsulation of MPP with the OF resin improved hydrophobicity and that the flame retardance of the Si-PU/OFMPP composite (limiting oxygen index, LOI = 32%) was higher than that of the Si-PU/MPP composite (LOI = 27%) at the same additive loading (30 wt %).

Keywords: microencapsulation; melamine polyphosphate; polyurethane; composite; flame retardant

1. Introduction

Polyurethane (PU) can be manufactured as per different requirements and is widely applied to fulfill daily needs and manufacture industrial goods. PU can be applied to coatings, adhesives, synthetic leather and elastomers [1,2]. Although its applications are broad, its thermal stability and flame retardance are relatively poor owing to its structure, limiting its application in certain areas. In recent years, many studies have been conducted on flame-retardant PU and considerably high flame retardance has been achieved [1–3]. In this study, silicon component was introduced into PU to increase thermal stability of polymer matrix, which was called as Si-PU. It was expected to expand above applications fields. Recently, commercial halogen-free flame retardants have gained increasing research attention. Several series of phosphorus and nitrogen compounds such as phosphates [4], melamine [5], ammonium polyphosphate (APP) [6], melamine polyphosphate (MPP) [7] and melamine phosphate (MP) [8] have been studied. The present study focused on MPP. Because MPP has two types of structures, APP and melamine, it can promote the formation of a char layer at the surface of the polymer via phosphorylation and promotion of cationic crosslinking during the flame retardation process, achieving an intumescent effect [9]. The effect of an intumescent flame retardant is thus related to char formation; however, char formation is also related to the level of intumescent flame retardant present and thermal degradation properties [5]. In recent years, microencapsulated flame retardants have been widely studied by scholars who believe that MPP in microcapsules can effectively improve flame retardance and thermal stability [10–15]. In this study, microencapsulation of MPP was utilized to obtain a surface with a char-forming compound. Furthermore, microencapsulated MPP (4,4′-oxydianiline-formaldehyde MPP (OFMPP)) was prepared and used. The thermal stability and flame retardance of the composites containing microencapsulated MPP were determined.

2. Materials and Methods

2.1. Materials

Isophorone diisocyanate (IPDI, purity 98%), ethylenediamine (purity 99%), 3-aminopropyltriethoxysilane (APTS, purity 99%), 4,4′-oxydianiline (ODA, purity 98%) and formaldehyde (37 wt %) were purchased from Acros Organics Co., Morris, NJ, USA. Anhydrous stabilized tetrahydrofuran (THF) was supplied by Lancaster Co., Morecambe, Lancashire, UK. Arcol polyol 1007 (polyether polyols 700) was purchased from Bayer Material Science Ltd., Kaohsiung, Taiwan. MPP (phase II, *n* > 1000) was purchased from San Jin Chemicals Corporation, Kaohsiung, Taiwan.

2.2. Preparation of Si-PU

First, IPDI (12.6 g) and polyether polyol (20 g) were placed into a four-necked flask. The contents of the flask were mechanically stirred in a nitrogen atmosphere at 80 °C (external oil bath). Next, a metal catalyst (DBTDL, 1 g) was placed into the flask. The mixture was stirred for 1.5 h to generate a prepolymer and the viscosity reached 5000 cp. The temperature was lowered to 50 °C to avoiding the escapting of solvent. THF solvent (50 mL), APTS (12.6 g) were added to the mixture and it was stirred for 0.5 h, following which H$_2$O (0.5 mL) was added. The temperature was increased to 70 °C and the viscosity increased due to increasing reaction extent. When the viscosity increased to 15,000 cp, the finished product (Si-PU) was removed from the reaction vessel, allowed to stand at room temperature for 6 h and then dried in an oven at 80 °C at a reduced pressure for 12 h. The reaction is shown in Scheme 1.

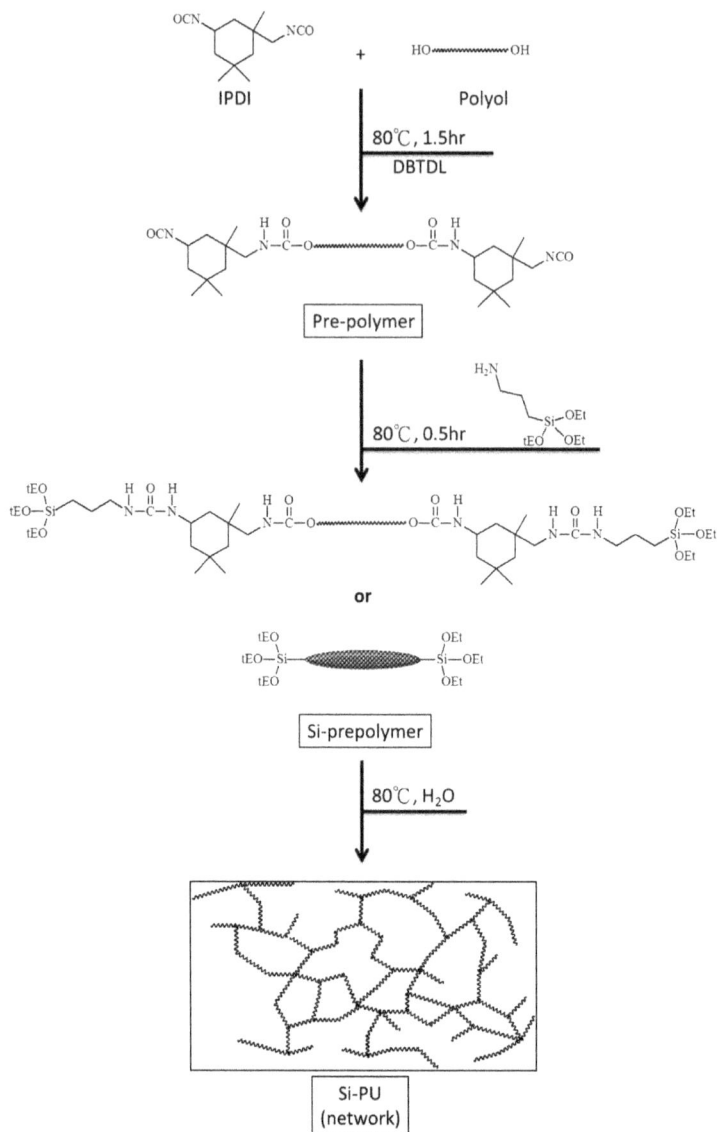

Scheme 1. Preparation of Si-PU composites.

2.3. Preparation of OFMPP

First, 4,4′-oxydianiline (10 g) and formaldehyde (5.99 g) were placed into a reaction vessel and then THF (40 mL) was added. Next, aqueous ammonia solution was added to adjust the pH to 8–9. The temperature was increased to 60 °C for 10 min and the mixture became transparent. Next, MPP (40 g) and ethanol (500 mL) were placed in another reaction vessel, preheated and stirred for 0.5 h. The transparent prepolymer was added into the mixture. The pH of the mixture was adjusted to be between 3 and 4 through the addition of aqueous hydrochloric solution and it was stirred at 80 °C for 3 h. The solid product was collected by filtration at reduced pressure, washed with ethanol and dried in a convection oven for 12 h (Scheme 2).

(a)

(b)

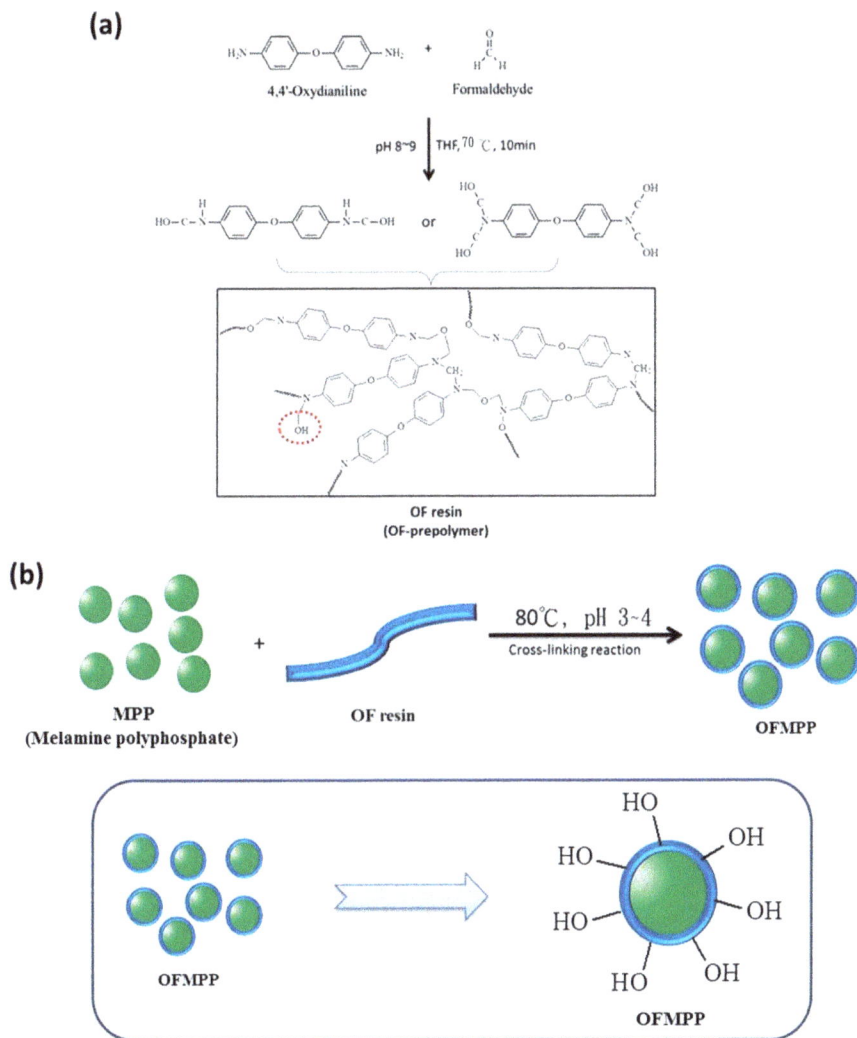

Scheme 2. (**a**) Preparation of OF resin (**b**) Preparation of OFMPP.

2.4. Preparation of Si-PU/OFMPP Composites

OFMPP flame retardant was added into the Si-prepolymer synthesized according to Scheme 3. Then, 0.5 mL of H$_2$O was added and the temperature was increased to 70 °C. The finished product (Si-PU/OFAPP) was removed after the viscosity increased and placed into a mold. It was allowed to stand at room temperature for 6 h and then placed in an oven, following which it was dried for 12 h at 80 °C and pressure was reduced.

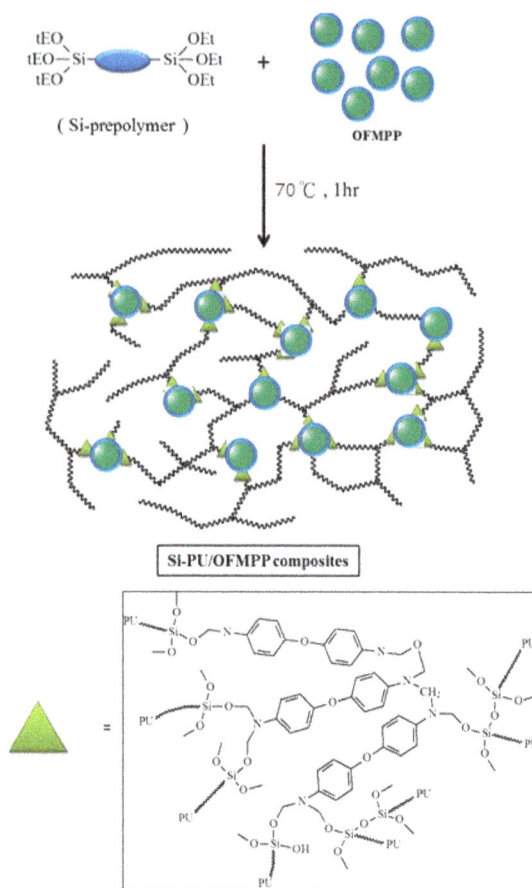

Scheme 3. Preparation of Si-PU/OFMPP composites.

2.5. Measurements

^{29}Si NMR was performed using a spectrometer (DSX-400WB, Bruker, Rheinstetten, Germany). The samples were treated at 180 °C for 2 h and then ground into fine powder. X-ray photoelectron spectra (XPS) were recorded using a PHI Quantera SXM/Auger with Al Ka excitation radiation (h_v = 1486.6 eV). The pressure in the analyzer was maintained at approximately 6.7 × 10^{-7} Pa. XPS data were processed using a DS 300 data system. MPP or microencapsulated MPP samples (10 g) was placed into 100 mL distilled water at a different temperature and stirred at that temperature for 60 min. The suspension was then filtered and 50 mL of the filtrate was collected and dried to a constant weight at 105 °C. The morphology of the burnt surface of the composites was examined using a scanning electron microscope (SEM) (JEOL JSM 840A, Osaka, Japan). Thermal degradation of the composites was examined using a thermogravimetric analyzer (TGA) (Perkin Elmer TGA 7, PerkinElmer Co., Waltham, MA, USA) from room temperature to 800 °C at a heating rate of 10 °C/min in nitrogen atmosphere. The measurements were performed for 6–10 mg of samples. LOI testing was performed following the ASTM D 2836 Oxygen Index Method (Atlas Fire Science Products Co., Kent, WA, USA), by using a test specimen bar that was 15 cm long, 6.5 ± 0.5 mm wide and 3.0 ± 0.5 mm thick. The sample bars were suspended vertically and ignited using a Bunsen burner. The flame was removed and the timer was started. The concentration of oxygen increased when the flame on the specimen was extinguished before it had burned for 3 min or burned 5 cm away from the bar. The oxygen content was adjusted until the limiting concentration was determined.

The UL94 test method based on the UL Standard for Tests for Flammability of Plastic Materials for Parts in Devices and Appliances was used. Each test specimen bar was 15 cm long, 6.5 ± 0.5 mm wide and 3.0 ± 0.5 mm thick. The vertical burning test was performed inside a fume hood. The samples were held vertically with tongs at one end and burned from the free end. The samples were exposed to an ignition source for 10 s, following which they were allowed to burn above cotton wool until both the samples and cotton wool were extinguished. Observable parameters were recorded to assess the fire retardance. The UL 94 test classified the materials as V-0, V-1 and V-2 according to the time period required before self-extinction and the occurrence of flaming dripping after the ignition source was removed. Each specimen was supported such that its lower end was 10 mm above the Bunsen burner tube. V-0 is the most ambitious and desired classification. A blue 20-mm-high flame was applied to the center of the lower edge of the specimen for 10 s and then removed. If burning ceased within 30 s, the flame was reapplied for an additional 10 s. If the specimen dripped, particles were allowed to fall onto a layer of dry absorbent surgical cotton placed 300 mm below the specimen. The specimens were not allowed to burn with flames for more than 10 s after either application of the test flame. The specimens that did not drip flaming particles that would ignite the surgical cotton were classified as V-0 level. The specimens that did not burn with glowing or flames up to the holding clamp for more than 30 s after either application of the test flame and did not drip flaming particles that would ignite the surgical cotton were classified as V-1 level. The specimens that did not burn with flames for more than 30 s after either application of the test flame and dripped flaming particles that ignited the surgical cotton were classified as V-2 level. The specimens that burned with flames for more than 30 s after either application of the test flame were classified as Fail.

3. Results and Discussion

3.1. ^{29}Si-NMR of Si-PU

The structure of the pure PU matrix was relatively weak and was easily destroyed when heated. After the structure was modified through the chemical reaction in Scheme 1, PU networks were formed that promoted thermal stability and flame retardance. During the structural formation of the Si-PU composite, a hydrolysis-condensation (sol-gel) reaction occurred. The degree of condensation was analyzed through solid-state ^{29}Si-NMR to obtain the percentage of condensation density.

As shown in Figure 1, the t-distribution and size of silicon spectra were explored. T^1 is a monosubstituted siloxane bond, T^2 is a disubstituted siloxane bond and T^3 is a trisubstituted siloxane bond, representing the number of bonds of the tri-alkoy group of APTS that underwent condensation. The figure clearly shows the area size of t-distribution at $T^2 > T^3 > T^1$. After the peak separation, the area size data were input into Equation (1) [16] to obtain a condensation density of 76.33%, as shown in Table 1, indicating that Si–O–Si functional group has a favorable network structure [17]. Network structure promoted the thermal stability of the composites.

$$D_c(\%) = \left[\frac{T^1 + 2T^2 + 3T^3}{3} + \frac{Q^1 + 2Q^2 + 3Q^3 + 4Q^4}{4} \right] \times 100 \tag{1}$$

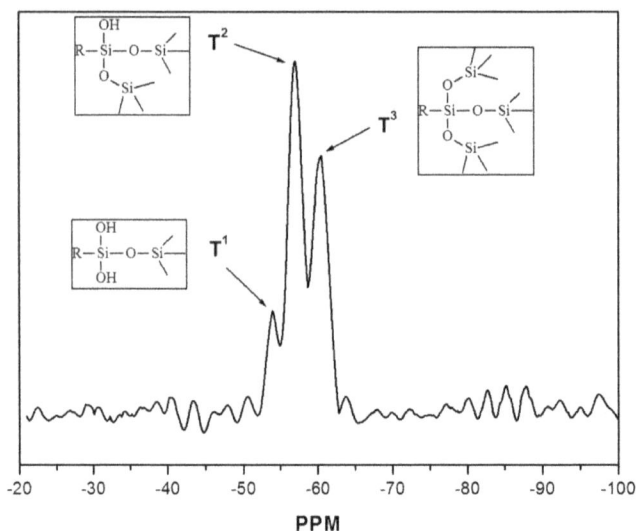

Figure 1. Solid-state ^{29}Si-NMR spectra of Si-PU composites.

Table 1. Area of the t-distribution.

Sample No.	Area (%)			Degree of condensation (%)
	T^1	T^2	T^3	
Si-PU composites	8	55	37	76.33

3.2. XPS of OFMPP

Figure 2 shows the peak chart produced by the XPS scan showing the microencapsulation of MPP. The P_{2S} and P_{2P} of MPP were 134 and 191 eV, respectively, clearly indicating the peaks. The intensities after microencapsulation were significantly reduced at P_{2S} and P_{2P}. O_{1s} and N_{1s} peaks were also slightly reduced. This may be because the outer layer of OFMPP had a shell. By contrast, at C_{1s}, the peaks of OFMPP were higher compared with those of MPP.

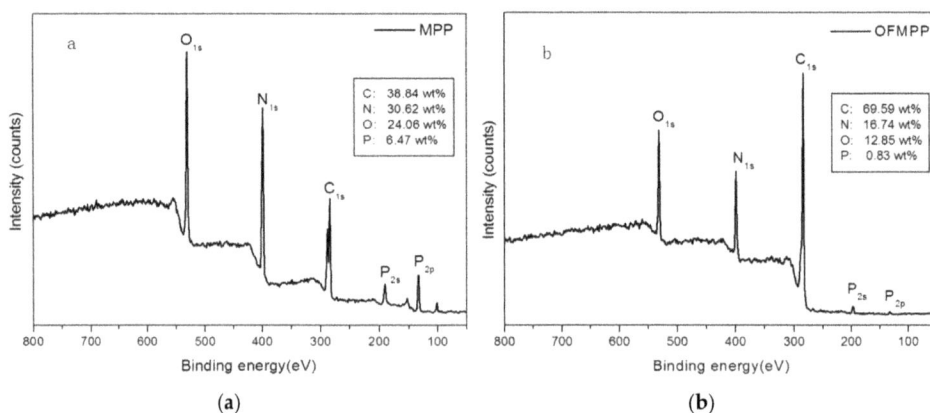

Figure 2. XPS spectra of: (**a**) MPP; and (**b**) OFMPP.

Thus, after microencapsulation, the outer surface of MPP has a layer of capsule shell, indicating that a new flame retardant, OFMPP, was successfully formed through the microencapsulation of MPP.

3.3. Water Solubility of OFMPP

The compatibility between fillers and polymer matrix was tested through water solubility. If the water solubility was poor, the fillers were hydrophobic. The compatibility between fillers and polymer matrix improved and the fillers were well dispersed in polymer matrix. Figure 3 and Table 2 present the weight change data of MPP and OFMPP after boiling in water solution. First, 1-g samples of each MPP and OFMPP were placed into four containers and 50 mL of DI water were added into each container. The temperature was increased to 25, 50, 75 and 100 °C (with 2 h of stirring), respectively. The samples were filtered and dried, following which the weight was measured to observe the change. If the weight remained constant, the compatibility with water was poor, indicating that the samples were hydrophobic [18–24]. If the weight was somewhat reduced, they were considered hydrophilic.

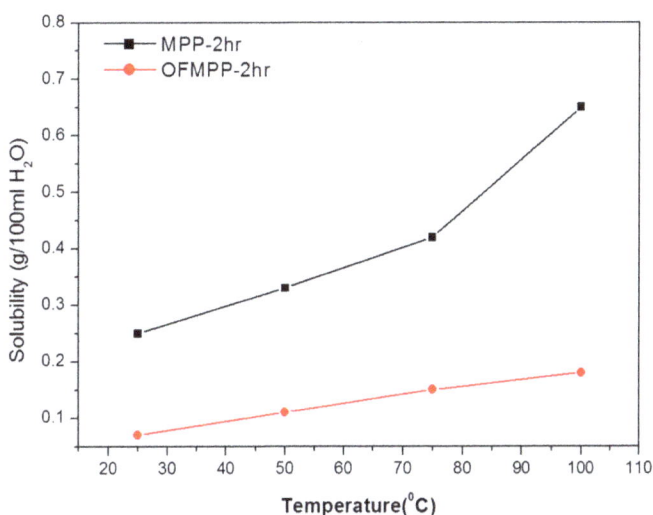

Figure 3. Water solubility of MPP and OFMPP.

Table 2. Water solubility data of MPP and OFMPP.

Sample code	25 °C (g/100 mL H_2O)	50 °C (g/100 mL H_2O)	75 °C (g/100 mL H_2O)	100 °C (g/100 mL H_2O)
MPP	0.25	0.33	0.42	0.65
OFMPP	0.07	0.11	0.15	0.18

According to Figure 3 and Table 2, at room temperature, MPP and OFMPP lost 0.25 and 0.07 g/100 mL H_2O, respectively, after 2 h of stirring. When the temperature was raised to 100 °C, they lost 0.65 and 0.18 g/100 mL H_2O, respectively. This indicates that MPP after water boiling is highly compatible with water, i.e., it is hydrophilic. The weight of OFMPP did not change much after being boiled in water, indicating an increase in its hydrophobicity. The aforementioned data show that OFMPP does not have high water solubility after microencapsulation, indicating that it was not affected for the storage at room temperature by moisture to experience deliquescence and that the stability increased during storage. This implies that the OFMPP within the composites did not easily migrate in the humid environment and the flame retardance of the composites was maintained to a satisfactory degree. By contrast, the hydrophobic nature of OFMPP enhanced its compatibility with the Si-PU matrix.

3.4. TGA of OFMPP

Figure 4 shows the thermogravimetric (TG) and derivative thermogravimetric (DTG) graphs of MPP and OFMPP, which indicate that MPP and OFMPP have two stages of thermal degradation. For MPP, the first stage occurred from 270 to 450 °C, wherein it released NH_3 and H_2O. Simultaneously, melamine was converted into melam, melem, and melon, and MPP was gradually converted into pyrophosphate and polyphosphate. At the second stage at more than 500 °C, melam pyrophosphates or melam polyphosphates began to be changed from MPP and the final product was char residue containing P–N [3,5,23–25]. For OFMPP, the TG curve at less than 350 °C was very similar to that of MPP, but OFMPP had high thermal stability after 350 °C, with the degradation temperature decreasing. The DTG clearly shows that the degradation rates in the two stages significantly decreased. This may be because of the outer shell (OF) of OFMPP, where OF mainly comprised the benzene ring, thus increasing the thermal stability and delaying the temperature for thermal degradation. The char yields at 800 °C for MPP and OFMPP were 2.9 and 6.7 wt %, respectively, where OFMPP improved the char formation by 3.8 wt %.

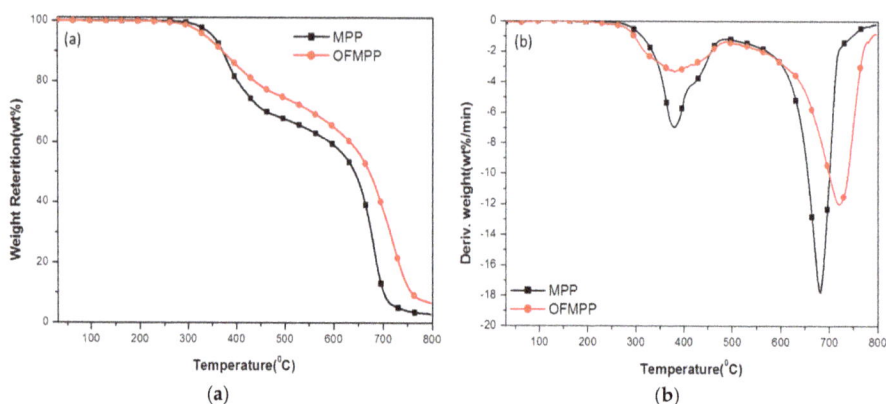

Figure 4. (a) TG; and (b) DTG curves of MPP and OFMPP.

The curves of TG and DTG show that the conversion of MPP into OFMPP through microencapsulation indicated high thermal stability, implying that OFMPP can have a strong effect.

3.5. LOI and UL-94 of Si-PU/OFMPP Composites

Figure 5 and Table 3 present the experimental results for pristine PU, Si-PU, Si-PU/MPP 30%, and Si-PU/OFMPP 10–40% after LOI and UL-94 tests. Figures 5 and 6 show the actual appearance after the composites were burned.

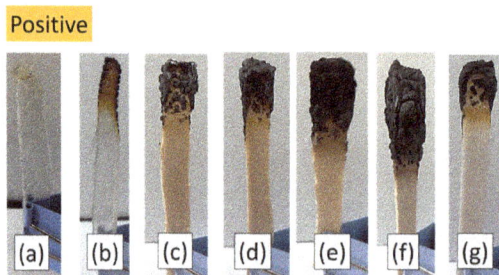

Figure 5. Residues after burning for 1 min: (a) pure PU; (b) Si-PU; (c) Si-PU/OFMPP 10%; (d) Si-PU/OFMPP 20%; (e) Si-PU/OFMPP 30%; (f) Si-PU/OFMPP 40% and (g) Si-PU/MPP 30%.

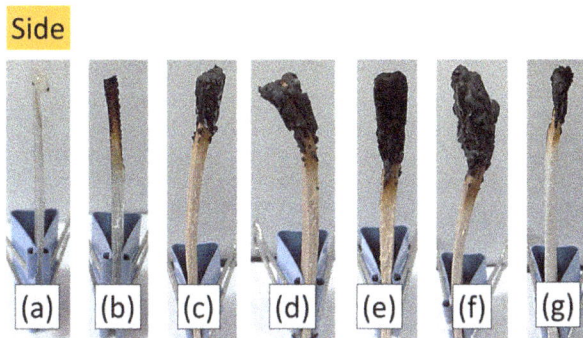

Figure 6. Residues after burning for 1 min: (**a**) pure PU; (**b**) Si-PU; (**c**) Si-PU/OFMPP 10%; (**d**) Si-PU/OFMPP 20%; (**e**) Si-PU/OFMPP 30%; (**f**) Si-PU/OFMPP 40% and (**g**) Si-PU/MPP 30%.

Table 3. Flame-retardant properties of pure PU, Si-PU, Si-PU/MPP 30% and Si-PU/OFMPP 10–40% composites.

Sample No.	LOI (%)		UL-94			
	Before soakage	After soakage	Before soakage		After soakage	
			Ranking	Dripping	Ranking	Dripping
Pristine PU	17	17	Fail	YES	Fail	YES
Si-PU	18	18	Fail	NO	Fail	NO
Si-PU/OFMPP 10%	19	19	Fail	NO	Fail	NO
Si-PU/OFMPP 20%	25	24	Fail	NO	Fail	NO
Si-PU/OFMPP 30%	32	30	V-0	NO	V-1	NO
Si-PU/OFMPP 40%	38	36	V-0	NO	V-0	NO
Si-PU/MPP 30%	27	24	Fail	NO	Fail	NO

Figure 5 and Table 3 show that the LOI value of pristine PU was 17%, which was improved to 18% for modified Si-PU. For the composites with added OFMPP at 10%, 20%, 30% and 40%, the LOI values were 19%, 25%, 32% and 38%, respectively; the LOI values were greatly improved after OFMPP was added. In terms of UL-94, pristine PU could not pass the flame retardant test. When the OFMPP concentration was 20 wt %, the UL-94 test result was still Fail. The specimens burned with flaming combustion for more than 30 s after either application of the test flame. However, at 30% concentration, the results of UL-94 showed significant changes and improvement to the highest V-0 grade. A comparison of the flame retardance between Si-PU/MPP 30% and Si-PU/OFMPP 30% showed that the LOI and UL-94 for Si-PU/MPP 30% were 27% and Fail, respectively, and 32% and V-0 for Si-PU/OFMPP 30%. The LOI of Si-PU/OFMPP 30% was 5% higher than that of Si-PU/MPP 30%; UL-94 was the most obvious, with Fail improved to V-0. This indicates that OFMPP offers outstanding performance compared with MPP in terms of flame retardance.

In Table 3, the LOI value for Si-PU/MPP 30% before hot water treatment was 27% and was reduced to 24% after hot water treatment (75 °C for 24 h). For Si-PU/OFMPP 30%, the LOI value was reduced to 30% from the original 32%. The LOI change for Si-PU/MPP 30% was more significant than that for Si-PU/OFMPP 30%. By contrast, in terms of the change for UL-94, the grade for Si-PU/OFMPP 30% changed from V-0 to V-1.

As for the appearance change after burning, the front view in Figure 5 shows that the top of pristine PU after burning was in a candle-like molten state without any char-forming effect, whereas Si-PU had char residue after burning. The composite with OFMPP showed that as the concentration was increased, the char layer became more obvious and the intumescent volume increased. By contrast, the char-forming effect for the composites with added MPP 30% was different from that of OFMPP 30%.

Si-PU/MPP 30% showed partial char, whereas the char for Si-PU/OFMPP 30% was more compact and wrapped around the entire polymer substrate.

The side view in Figure 6 shows the intumescent effect during the burning process. Both pristine PU and Si-PU showed no intumescent flame-retardant behavior, whereas the intumescent effect became more obvious for Si-PU/OFMPP 10–40% as the concentration increased, indicating higher flame retardance. Irrespective of the size or width of the intumescent area, Si-PU/OFMPP 30% had higher performance than Si-PU/MPP 30%. To summarize the results from LOI and UL-94, the comparison showed that OFMPP, a product after microencapsulation, was more effective than MPP in terms of flame retardance, implying that microcapsule formation improved the barrier effect. Furthermore, the composites with added OFMPP had satisfactory flame retardance, whereas those with added MPP had lower performance. Therefore, in terms of hot water treatment, OFMPP after microencapsulation offered higher water resistance. When it was added into the polymer matrix, the flame retardance did not change much, meaning that OFMPP was apparently more effective than MPP and that the microencapsulation technology is flame retardant to a certain extent.

3.6. SEM of Si-PU/OFMPP Composites after Burning

Figure 7 shows SEM images of the morphology of Si-PU, Si-PU/MPP 30% and Si-PU/OFMPP 30% after burning. Figure 7a shows intumescent pores and hollow voids on the surface of the Si-PU matrix after burning. The intumescent pores were formed by the char layer that was pushed when nonflammable gas from the polymer was released. The voids were originally the intumescent pores that burst after the char layer failed to withstand the strong release of the nonflammable gas. The formation of voids introduces oxygen into the polymer substrate, further burning the matrix so that an effective barrier cannot form to protect the polymer matrix. Figure 7b also shows voids in the char layer after MPP is added. This may be because MPP causes the release of nonflammable gases, for example, water vapor and NH_3. However, there are no pores, indicating that MPP's nitrogen ring in melamine sufficiently complements the compactness of the char layer, although the layer thickness is inadequate. Figure 7c shows that with OFMPP, the char density on the surface is sufficient and there are intumescent layers that do not burst. This indicates that the benzene ring structure on the outer shell of OFMPP has a sufficiently thick char layer to form a protective layer during burning [26–30]. This can support nonflammable air and the char layer stops the release of air while generating an intumescent flame-retardant layer.

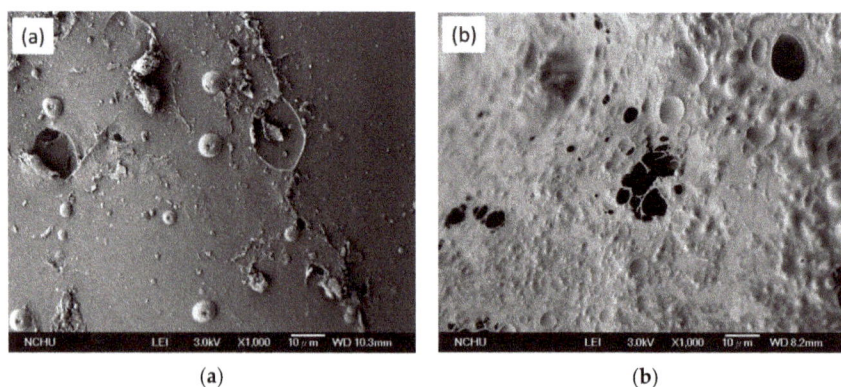

(a)   (b)

Figure 7. *Cont.*

Polymers **2017**, *9*, 407

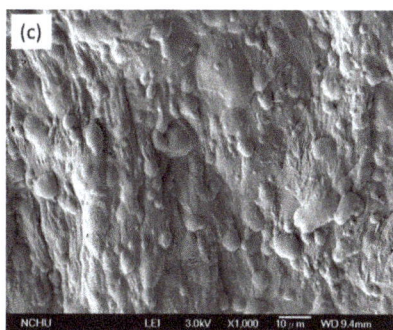

(c)

Figure 7. SEM images of char residues after burning: (**a**) Si-PU; (**b**) Si-PU/MPP 30% and (**c**) Si-PU/OFMPP 30%.

Thus, it can be concluded that Si-PU is a poor flame retardant, flame retardance of the composite is still inadequate only after MPP is added and the composite with MPP after microencapsulation can achieve compactness and sufficient barrier thickness. This means that the char layer of OFMPP is significantly more favorable than that of MPP. Thus, OFMPP can effectively enhance flame retardance and form an excellent thermal barrier.

4. Conclusions

The new generation of microencapsulation flame retardant developed in this study involved the OF resin wrapping around the surface of the MPP to form an OFMPP flame retardant through the characterization of XPS, ^{29}Si NMR and water solubility. The data from the tests showed that adding Si-PU/OFMPP 40% to polymer matrix could make the char yields reach approximately 26.9 wt % and flame retardance grades LOI-38% and UL-94 V-0, indicating high performance. The test curves of TGA and SEM also proved that OFMPP produced protective mechanisms at different temperatures, further improving the thermal stability and char yield. When MPP before and after microencapsulation are compared, the treatment clearly demonstrates high flame retardance and thermal stability. The new microencapsulated flame retardant successfully improved the shortcomings of flammable PU and could be developed for use in various applications.

Acknowledgments: The authors would like to express their appreciation to the National Science Council of the Republic of China for financial support of this study under grant MOST-105-2221-E-241-001-MY3 and MOST 106-3114-E-269-001.

Author Contributions: Shang-Hao Liu and Chin-Lung Chiang conceived and designed the experiments; Shang-Hao Liu, Jia-Ming Yang and Ming-Yuan Shen performed the experiments; Chin-Lung Chiang analyzed the data; Chen-Feng Kuan and Hsu-Chiang Kuan contributed materials; Chin-Lung Chiang wrote the paper.

Conflicts of Interest: The authors declare no conflict of interest.

References

1. Jeon, H.T.; Jang, M.K.; Kim, B.K.; Kim, K.H. Synthesis and characterizations of waterborne polyurethane-silica hybrids using sol-gel process. *Colloids Surf. A Physicochem. Eng. Asp.* **2007**, *302*, 559–567. [CrossRef]
2. Jin, J.; Dong, Q.X.; Shu, Z.J.; Wang, W.J.; He, K. Flame retardant Properties of Polyurethane/expandable Praphite Composites. *Procedia. Eng.* **2014**, *71*, 304–309. [CrossRef]
3. Thirumal, M.; Khastgir, D.; Nando, G.B.; Naik, Y.P.; Singha, N.K. Halogen-free flame retardant PUF: Effect of melamine compounds on mechanical, thermal and flame retardant properties. *Polym. Degrad. Stabil.* **2010**, *95*, 1138–1145. [CrossRef]

4. Liu, X.Q.; Wang, D.Y.; Wang, X.L.; Chen, L.; Wang, Y.Z. Synthesis of functionalized a-zirconium phosphate modified with intumescent flame retardant and its application in poly(lactic acid). *Polym. Degrad. Stabil.* **2013**, *98*, 1731–1737. [CrossRef]

5. Wang, Z.Z.; Lv, P.; Hu, Y.; Hu, K. Thermal degradation study of intumescent flame retardants by TG and FTIR: Melamine phosphate and its mixture with pentaerythritol. *J. Anal. Appl. Pyrolysis* **2009**, *86*, 207–214. [CrossRef]

6. Xia, Y.; Jin, F.F.; Mao, Z.W.; Guan, Y.; Zheng, A. Effects of ammonium polyphosphate to pentaerythritol ratio on composition and properties of carbonaceous foam deriving from intumescent flame-retardant polypropylene. *Polym. Degrad. Stabil.* **2014**, *107*, 64–73. [CrossRef]

7. Tai, Q.; Yuen, K.K.; Yang, W.; Qiao, Z.; Song, L.; Hu, Y. Iron-montmorillonite and zinc borate as synergistic agents in flame-retardant glass fiber reinforced polyamide 6 composites in combination with melamine polyphosphate. *Compos. Part A Appl. Sci. Manuf.* **2012**, *43*, 415–422. [CrossRef]

8. Yang, H.Y.; Song, L.; Tai, Q.; Wang, X.; Yu, B.; Yuan, Y.; Hu, Y.; Yuen, K.K. Comparative study on the flame retarded efficiency of melamine phosphate, melamine phosphite and melamine hypophosphite on poly(butylene succinate) composites. *Polym. Degrad. Stabil.* **2014**, *105*, 248–256. [CrossRef]

9. Naik, A.D.; Fontaine, G.; Samyn, F.; Delva, X.; Bourgeois, Y.; Bourbigot, S. Melamine integrated metal phosphates as non-halogenated flame retardants: Synergism with aluminium phosphinate for flame retardancy in glass fiber reinforced polyamide 66. *Polym. Degrad. Stabil.* **2013**, *98*, 2653–2662. [CrossRef]

10. Wang, Z.Z.; Wu, K.; Hu, Y. Study on flame retardance of co-microencapsulated ammonium polyphosphate and dipentaerythritol in polypropylene. *Polym. Eng. Sci.* **2008**, *48*, 2426–2431. [CrossRef]

11. Wu, K.; Wang, Z.Z.; Liang, H. Microencapsulation of ammonium polyphosphate: Preparation, characterization, and its flame retardance in polypropylene. *Polym. Compos.* **2008**, *29*, 854–860. [CrossRef]

12. Luo, J.; Wang, X.; Li, J.; Zhao, X.; Wang, F. Conductive hybrid film from polyaniline and polyurethaneesilica. *Polymer* **2007**, *48*, 4368–4374. [CrossRef]

13. Bocz, K.; Szolnoki, B.; Marosi, A.; Tábi, T.; Wladyka, P.M.; Marosi, G. Flax fibre reinforced PLA/TPS biocomposites flame retarded with multifunctional additive system. *Polym. Degrad. Stabil.* **2014**, *106*, 63–73. [CrossRef]

14. Chen, X.; Huo, L.; Jiao, C.; Li, S. TG-FTIR characterization of volatile compounds from flame retardant polyurethane foams materials. *J. Anal. Appl. Pyrolysis* **2013**, *100*, 186–191. [CrossRef]

15. Han, Y.H.; Taylor, A.; Mantle, M.; Knowles, K. Sol-gel-derived organic-inorganic hybrid materials. *J. Non-Cryst. Solids* **2007**, *353*, 313–320. [CrossRef]

16. Lee, T.M.; Ma, C.C.; Hsu, C.W.; Wu, H.L. Effect of molecular structures and mobility on the thermal and dynamical mechanical properties of thermally cured epoxy-bridged polyorganosiloxanes. *Polymer* **2005**, *46*, 8286–8296. [CrossRef]

17. Qian, X.; Song, L.; Bihe, Y.; Yu, B.; Yongqian, S.; Hu, Y.; Yuen, K.K. Organic/inorganic flame retardants containing phosphorus, nitrogen and silicon: Preparation and their performance on the flame retardancy of epoxy resins as a novel intumescent flame retardant system. *Mater. Chem. Phys.* **2014**, *143*, 1243–1252. [CrossRef]

18. Wang, X.; Pang, H.; Chen, W.; Lin, Y.; Ning, G. Nanoengineering core/shell structured brucite@polyphosphate@amine hybrid system for enhanced flame retardant properties. *Polym. Degrad. Stabil.* **2013**, *98*, 2609–2616. [CrossRef]

19. Hua, X.; Guo, Y.; Chen, L.; Wang, X.; Li, L.; Wang, Y. A novel polymeric intumescent flame retardant: Synthesis, thermal degradation mechanism and application in ABS copolymer. *Polym. Degrad. Stabil.* **2012**, *97*, 1772–1778. [CrossRef]

20. Salaün, F.; Huang, Z.; Zhang, Y. Preparation of CMC-modified melamine resin spherical nano-phase change energy storage materials. *Carbohydr. Polym.* **2014**, *101*, 83–88.

21. Poljansek, I.; Krajnc, M. Characterization of phenol-formaldehyde prepolymer resins by in line FT-IR spectroscopy. *Acta Chim. Slov.* **2005**, *52*, 238–244.

22. Salaün, F.; Vroman, I. Influence of core materials on thermal properties of melamine-formaldehyde microcapsules. *Eur. Polym. J.* **2008**, *44*, 849–860. [CrossRef]

23. Zheng, Z.; Yan, J.; Sun, H.; Cheng, Z.; Li, W.; Wang, H.; Cui, X. Microencapsulated ammonium polyphosphate and its synergistic flame-retarded polyurethane rigid foams with expandable graphite. *Polym. Int.* **2011**, *63*, 84–92. [CrossRef]

24. Wang, B.; Tang, Q.; Hong, N.; Song, L.; Wang, L.; Shi, Y.; Hu, Y. Effect of cellulose acetate butyrate microencapsulated ammonium polyphosphate on the flame retardancy, mechanical, electrical, and thermal properties of intumescent flame-retardant ethylene_vinyl acetate copolymer/microencapsulated ammonium polyphosphate/polyamide-6 blends. *Appl. Mater. Interfaces* **2011**, *3*, 3754–3761.

25. Wang, G.; Yang, J. Thermal degradation study of fire resistive coating containing melamine polyphosphate and dipentaerythritol. *Prog. Org. Coat.* **2011**, *72*, 605–611. [CrossRef]

26. Kang, C.; Huang, J.; He, W.; Zhang, F. Periodic mesoporous silica-immobilized palladium(II) complex as an effective and reusable catalyst for water-medium carbon-carbon coupling reactions. *J. Organomet. Chem.* **2010**, *695*, 120–127. [CrossRef]

27. Chiu, Y.C.; Ma, C.C.; Liu, F.Y.; Chiang, C.L.; Riang, L.; Yang, J.C. Effect of P/Si polymeric silsesquioxane and the monomer compound on thermalproperties of epoxy nanocomposite. *Eur. Polym. J.* **2008**, *44*, 1003–1011. [CrossRef]

28. Wu, C.S.; Liu, Y.L.; Chiu, Y.S. Epoxy resins possessing flame retardant elements from silicon incroporaed epoxy compouds cured with phosphorus or nitrogen containing curing agents. *Polymer* **2002**, *43*, 4277–4284. [CrossRef]

29. Liu, Y.L.; Hsu, C.Y.; Wei, W.L.; Jeng, R.J. Preparation and thermal properties of epoxy-silica nanocomposites from nanoscale colloidal silica. *Polymer* **2003**, *44*, 5159–5167. [CrossRef]

30. Wu, C.S.; Liu, Y.L.; Chiu, Y.C.; Chiu, Y.S. Thermal stability of epoxy resins containing flame retardant components: An evaluation with thermogravimetric analysis. *Polym. Degrad. Stabil.* **2002**, *78*, 41–48. [CrossRef]

![polymers logo] *polymers*

MDPI

Article

Octadecylamine-Grafted Graphene Oxide Helps the Dispersion of Carbon Nanotubes in Ethylene Vinyl Acetate

Li-Chuan Jia [1], Zhong-Han Jiao [1], Ding-Xiang Yan [2,*] and Zhong-Ming Li [1,*]

[1] College of Polymer Science and Engineering, State Key Laboratory of Polymer Materials Engineering, Sichuan University, Chengdu 610065, China; jialichuan@stu.scu.edu.cn (L.-C.J.); jiaozhonghan@stu.scu.edu.cn (Z.-H.J.)
[2] School of Aeronautics and Astronautics, Sichuan University, Chengdu 610065, China
* Correspondence: yandingxiang@scu.edu.cn (D.-X.Y.); zmli@scu.edu.cn (Z.-M.L.); Tel.: +86-028-8540-0211 (D.-X.Y. & Z.-M.L.)

Received: 11 August 2017; Accepted: 25 August 2017; Published: 27 August 2017

Abstract: In this paper, the dispersion of carbon nanotube (CNT) in ethylene vinyl acetate (EVA) is demonstrated to be significantly improved by the addition of octadecylamine (ODA)-grafted graphene oxide (GO) (GO–ODA). Compared to the CNT/EVA composite, the resultant GO–ODA/CNT/EVA (G–CNT/EVA) composite shows simultaneous increases in tensile strength, Young's modulus and elongation at break. Notably, the elongation at break of the G–CNT/EVA composite still maintains a relatively high value of 1268% at 2.0 wt % CNT content, which is more than 1.6 times higher than that of CNT/EVA composite (783%). This should be attributed to the homogeneous dispersion of CNT as well as the strong interfacial interaction between CNT and EVA originating from the solubilization effect of GO–ODA. Additionally, the G–CNT/EVA composites exhibit superior electrical conductivity at low CNT contents but inferior value at high CNT contents, compared to that for the CNT/EVA composite, which depends on the balance of CNT dispersion and the preservation of insulating GO–ODA. Our strategy provides a new pathway to prepare high performance polymer composites with well-dispersed CNT.

Keywords: carbon nanotube; homogeneous dispersion; ethylene vinyl acetate; mechanical performance; electrical conductivity

1. Introduction

Carbon nanotube (CNT) has been regarded as a very promising nanofiller for polymer matrices to achieve high performance and multifunction, due to its extraordinary mechanical, electrical, and thermal properties [1–8]. The CNT-based polymer composites are expected to make enormous technological and commercial impacts in view of extensive applications in sensing devices [9,10], electrical shielding and heating devices [11–15], as well as advanced composites for space and aircrafts [16,17]. Nevertheless, the resultant performances show much lower efficiency than expected though significant efforts have been made. This is mainly attributed to the fact that CNTs easily agglomerate in polymer matrices because of the high aspect ratio and strong van der Waals interactions, which would greatly limit their utilization [18,19].

Recently, there are growing efforts on controlled surface modification of CNT to realize well dispersion through covalent or noncovalent approaches [20–26]. The covalent functionalization lacks economic benefits and would introduce many defects that inevitably deteriorate the intrinsic performances of CNT, whereas the noncovalent functionalization is particularly attractive because it preserves nearly all intrinsic features of CNT. For traditional noncovalent approaches, homogeneous

CNT dispersion was always facilitated by surfactant and polymer wrapping based on van der Waals interactions or π–π stacking interactions. However, the surfactants may largely affect the properties of the obtained composites and the available polymers for CNT solubilization are limited [27,28]. Recently, graphene oxide (GO) was demonstrated as a high-efficient surfactant to disperse CNT due to the high solubility and adhesion of CNT onto the flat GO sheets through strong π–π stacking interaction [29–33]. For instance, Liao et al. reported that the incorporation of GO prominently improved the dispersion of CNTs in poly(vinyl alcohol) (PVA) and the resulted GO–CNT/PVA composite showed great improvements in the yield strength and Young's modulus compared to those of CNT/PVA composite [31]. Fu et al. demonstrated that the tribological performance of CNT/epoxy composite was significantly enhanced by the incorporation of GO that improves the dispersion of CNT [32]. Although significant progress was achieved in developing high-performance materials, it should be noted that the GO-dispersed CNT was mainly applied in the strong polar polymer matrices, such as PVA, epoxy, and polyurethane, owing to the presence of hydrophilic functional groups (hydroxyl, epoxide, and carboxylic groups). The highly hydrophilic feature of GO makes it hardly achieve the favorable dispersion quality of CNT in weak or non-polar polymer matrices, which greatly restricts the development of such type of composites for advanced functional materials. Recently, tremendous attentions have been paid to the oleophylic modification of GO by grafting long alkyl chains to improve its dispersion in non-polar organic solvents and enhances its compatibility with weak or non-polar polymer matrices [34–36]. Inspired by the high efficiency of GO to disperse CNT in strong polar polymers, it is reasonable to hypothesize that long alkyl chain grafted GO should be promising and ideal to realize the efficient dispersion of CNT in weak or non-polar polymers.

In the current study, octadecylamine (ODA) grafted GO (GO–ODA) was synthesized to assist the dispersion of CNT in ethylene vinyl acetate (EVA), because CNT-based EVA composites are widely used for various applications such as packaging films, adhesives, antistatic field and electromagnetic shielding in view of their excellent flexibility, durability, and chemical resistance [37–39]. Significantly improved dispersion of CNT was achieved because of the strong solubilization of GO–ODA to CNT. The effect of CNT dispersion on the mechanical and electrical properties of EVA composites was investigated thoroughly. This solubilization approach for CNT with the assistance of GO–ODA is confirmed to be an effective strategy to fabricate high-performance CNT-based polymer composites.

2. Materials and Methods

2.1. Materials

The commercially available EVA containing 28 wt % of vinyl acetate was kindly provided by Beijing Dongfang Petroleum Chemical Co. (Beijing, China). The density is 0.92 g/cm^3 and the melt flow rate is 25 g/(10 min) (190 °C, 21.6 N). Graphene oxide (GO) was prepared from expanded graphite, which was provided by Qingdao Haida Graphite Co., Ltd. (Qingdao, China) with an expansion rate of 200 mL/g, by the modified Hummers method, as described in our previous work [40]. CNT (NC 7000 series), with average diameter of 9.5 nm, length 1.5 μm, surface area 250–300 m^2/g, and 90% carbon purity, was supplied by Nanocyl S.A., (Sambreville, Belgium). ODA, xylene, deionized water and anhydrous ethanol were supplied by Chengdu Kelong Chemical Reagent Factory, Chengdu, China. All these chemicals were analytical reagent and used without further purification.

2.2. Preparation of GO–ODA

GO–ODA was prepared by facile refluxing of GO and ODA. Specifically, 0.6 g GO was initially dispersed in 300 mL deionized water via vigorous agitation and ultrasonic treatment for 60 min. Then the resulting uniform suspension was mixed with ODA/ethanol solution (0.9 g in 90 mL) in a three-neck flask. The mixture was refluxed with intensified mechanical stirring for 24 h at 85 °C and filtrated by a PTFE membrane (0.2 μm pore size). The filtrated powder was then washed thoroughly with ethanol and filtered to remove the remaining ODA. Finally, the mixture was dried in a vacuum oven overnight at 60 °C for 48 h.

2.3. Preparation of GO–ODA/CNT/EVA Composites

The schematic representation for the preparation of GO–ODA/CNT/EVA composites is illustrated in Figure 1. Firstly, GO–ODA was suspended in xylene *via* ultrasonication for 30 min, then CNT was added to GO–ODA dispersion (1:1 weight ratio) and further ultrasonicated for 60 min. Meanwhile, EVA was completely dissolved in hot xylene (75 °C) by mechanical stirring for 60 min. Subsequently, GO–ODA/CNT/xylene dispersion was poured into the EVA/xylene solution and the mixture was continuously stirred for 30 min at 75 °C. Afterwards, the mixture was recovered by flocculation with ethanol, followed by filtering and drying in a vacuum oven (60 °C) for 48 h. Finally, the dried mixture was compression molded under the pressure of 10 MPa at 160 °C, after preheating for 5 min. The resultant GO–ODA/CNT/EVA composites were marked as G–CNT/EVA composites. For comparison, the CNT/EVA composites were also prepared under the same processing conditions.

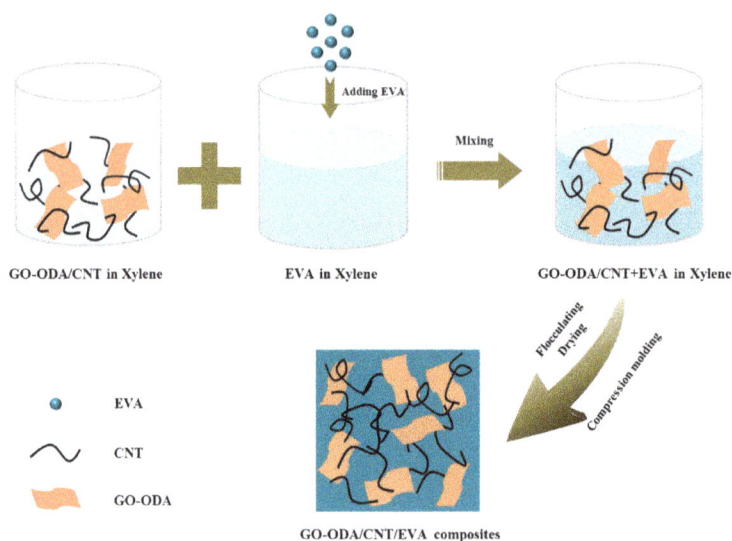

Figure 1. Scheme of the fabrication process of the G–CNT/EVA composites.

2.4. Characterizations

Fourier-transform infrared spectroscopy (FTIR) spectra were recorded with a 2 cm^{-1} spectral resolution on a Nicolet 6700 FTIR spectrometer (Thermo Nicolet, Waltham, MA, USA) in transmission mode. GO or GO–ODA was ground with KBr and pressed into KBr disks for FTIR measurements. X-ray diffraction (XRD) data was collected with a DX-1000 diffractometer (Dandong Fangyuan Instrument Co., Ltd., Liaoning, China) using CuKa irradiation at 40 kV in a scanning range from 2° to 30°. Water contact angle was measured by using a DSA 30 KRUSS drop shape analyzer (Kruss, Hamburg, Germany). Thermogravimetric analysis (TGA) was carried out to evaluate the thermal stability of GO and GO–ODA on a NETZSCH 209F1 (Netzsch, Bavarian, Germany), at a heating rate of 10 °C/min over 40–600 °C under nitrogen atmosphere. Scanning electron microscopy (SEM) images were taken by a field emission SEM (Inspect-F, FEI, Hillsboro, OR, USA) at an accelerating voltage of 20 kV. The specimens were cryo-fractured in liquid nitrogen and the fracture surfaces were coated with a thin layer of gold before morphological observation. The volume conductivity of the composite above 10^{-6} S/m was carried out on a four-point probe instrument (RTS-8, Guangzhou Four-Point

Probe Technology Co., Ltd., Guangzhou, China) and the volume conductivity below 10^{-6} S/m was measured with a Keithley electrometer Model 4200-SCS (Keithley, Beaverton, OR, USA). Tensile test was performed on an Instron universal test instrument (Model 5576, Instron Instruments, Norwood, MA, USA) at room temperature. The loading rate was 100 mm/min and the gauge length was 20 mm. More than five samples were tested to calculate the average value and standard deviations.

3. Results

3.1. Characterization of GO–ODA

FTIR spectra were performed to determine the chemical changes that occurred during the refluxing of ODA and GO. As shown in Figure 2a, the absorption bands of GO appear at 1710, 1642, 1420, and 1064 cm^{-1}, corresponding to C=O in carboxyl group, C=C in aromatic ring, C−OH stretching and C–O–C in epoxide, respectively [36]. The broad peak appearing at 3256 cm^{-1} is assigned to the hydroxyl groups. When it comes to GO–ODA, the existence of the octadecyl chain is clearly demonstrated because of the emerging peaks at 2919 cm^{-1} and 2850 cm^{-1} (–CH$_2$ stretching in the octadecyl chain) as well as the peak at 720 cm^{-1} [41,42]. Furthermore, the weakened intensity of C–O–C and disappearance of C=O in carboxyl group indicate the amidation reaction between the amine functionality of ODA with the epoxy and carboxyl functionality of GO, confirming the nucleophilic substitution between ODA with GO occurred during the refluxing [36]. The new peaks at 1576 cm^{-1} (N–H bending of amide) and 1470 cm^{-1} (C–N stretch of amide) in GO–ODA also indicate the presence of amide-carbonyl bond between ODA molecule and GO. All these results affirm the intercalation and chemical reaction of ODA with GO, which is in line with the reported in the literature [36,43]. The XRD curves of GO and GO–ODA are presented in Figure 2b. Compared to diffraction peak of GO at 9.6°, the peak for GO-ODA shifts to smaller angles of 5.3°, corresponding to the increase in the intra-gallery spacing from 0.89 nm to 1.7 nm. The enlarged intra-gallery space indicates the intercalation of the octadecyl chains between the GO nanosheets, supporting the reaction of ODA and GO. The influence of functionalization on the wetting property of GO was confirmed by the contact angle measurements. The contact angle of GO–ODA increases to 110.4° in comparison with that of hydrophilic GO (73.2°), revealing that the attached hydrophobic hydrocarbon chains make the GO–ODA hydrophobic (Figure 2c). TGA was further used to analyze the amination of ODA on GO surface and calculate the grafting ratio of ODA. As shown in Figure 2d, ODA shows a rapid weight loss starting at a low temperature (120 °C), and is almost exhausted when the temperature reaches 300 °C. In case of GO, two significant weight loss stages are observed, corresponding to the volatilization of water between 40 and 100 °C and the pyrolysis of the labile oxygen containing functional groups on GO between 170 and 270 °C, respectively. Compared to GO and ODA, the larger weight loss of GO–ODA in the range of 200–500 °C is mainly attributed to the decomposition of covalently bonded ODA on GO surface, demonstrating the successful graft modification of GO with ODA. The grafting ratio of ODA is calculated to about 9.1 wt % according to the yields of residual carbon at 600 °C [36].

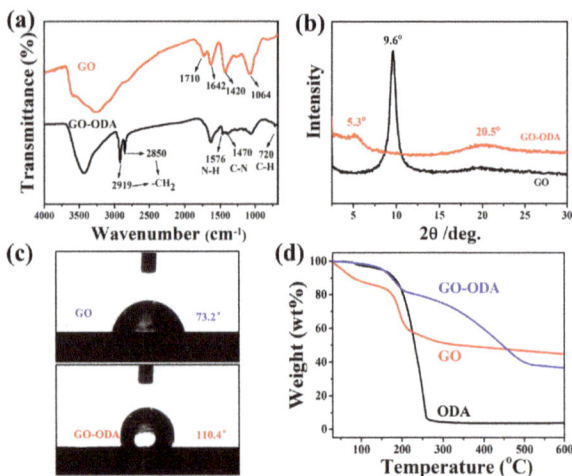

Figure 2. (a) Fourier-transform infrared spectroscopy (FTIR) spectra of graphene oxide (GO) and octadecylamine-grafted GO (GO–ODA); (b) X-ray diffraction (XRD) patterns of GO and GO–ODA; (c) Surface water contact angle of GO and GO–ODA; (d) TGA curves of ODA, GO and GO–ODA.

3.2. Dispersion of CNT in EVA

The dispersion of nanofillers has a direct correlation with the mechanical, electrical, and other properties of nanofiller/polymer composites. The fractured surface of the CNT-based EVA composite was observed with SEM to assess the CNT dispersion, as shown in Figure 3. For the G–CNT/EVA composites, CNTs are uniformly dispersed in EVA matrix without any visible aggregations (Figure 3a,b). While for the CNT/EVA composites, CNT agglomerations (labeled by red circle) are obviously observed (Figure 3c,d), mainly due to the intrinsic strong Van der Waals force of CNTs. These results indicate that GO–ODA plays a role of compatilizer for CNTs and EVA, which facilitates the dispersion of CNTs. Moreover, compared with the smooth fractured surface of 2.0 wt % CNT/EVA composite, the 2.0 wt % G–CNT/EVA composite shows a relatively rough fractured surface, demonstrating the excellent compatibility and strong interfacial adhesion of CNTs with EVA [36]. Based on the aforementioned analysis, the application of GO–ODA not only improves the CNT dispersion in EVA matrix but also enhances their interface adhesion, which should be attributed to the π–π stacking interaction between CNTs and GO–ODA and the interpenetration of GO–ODA chains with EVA chains. The well-dispersed CNTs in EVA matrix and high compatibility of CNTs with EVA are expected to impart the G–CNT/EVA composites with superior mechanical and electrical properties.

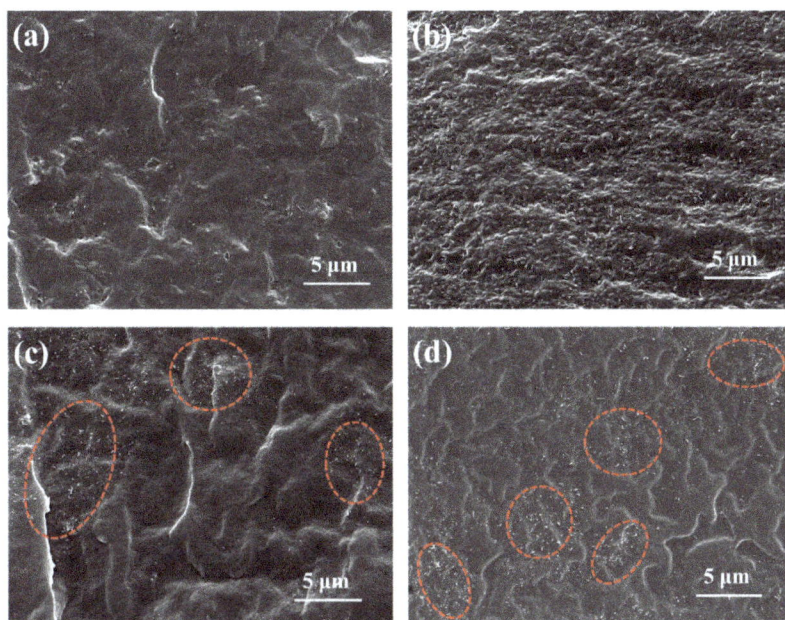

Figure 3. SEM images of the G–CNT/EVA composites with (**a**) 0.3 wt % and (**b**) 2.0 wt % CNT, respectively. SEM images of the CNT/EVA composites with (**c**) 0.3 wt % and (**d**) 2.0 wt % CNT, respectively. The red circles in (**c**) and (**d**) refer to the CNT agglomerations.

3.3. Effects of CNT Dispersion on Mechanical Properties

Figure 4a shows typical stress–strain curves of pure EVA, G–CNT/EVA, and CNT/EVA composites. With CNT incorporation, both G–CNT/EVA and CNT/EVA composites show increased tensile strength and Young's modulus, but decreased elongation at break, compared to those for pure EVA. The average Young's modulus, tensile strength, and elongation at break are calculated and illustrated in Figure 4b–d, respectively. The tensile strength and Young's modulus of the 0.3 wt % CNT/EVA composite are 22.7 and 13.5 MPa, which correspond to 56.6% and 53.4% increases from 14.5 and 8.8 MPa for pure EVA. Further increasing CNT content to 2.0 wt % in the CNT/EVA composite, the Young's modulus increases to 15.5 MPa due to the high stiffness of CNT, while the tensile strength drops to 15.3 MPa, mainly originating from the reduced elongation at break, which should be attributed to the heavy agglomeration of CNT in EVA matrix. This phenomenon is in line with previously reported CNT reinforced polymer composites [7,25]. When it comes to G–CNT/EVA composites, the improved CNT dispersion significantly boosts up the elongation at break as compared to the CNT/EVA composite. For instance, a more than 61.9% increase in elongation at break from 783% to 1268% is achieved at 2.0 wt % CNT content. Moreover, the G–CNT/EVA composites exhibit much higher tensile strength and Young's modulus than those of the CNT/EVA composites. The excellent mechanical performance of the G–CNT/EVA composites should be attributed to the improved dispersion of CNT and the high compatibility between CNTs and EVA molecular chain with the assist of GO–ODA, both of which facilitate effective stress transfer in the composites.

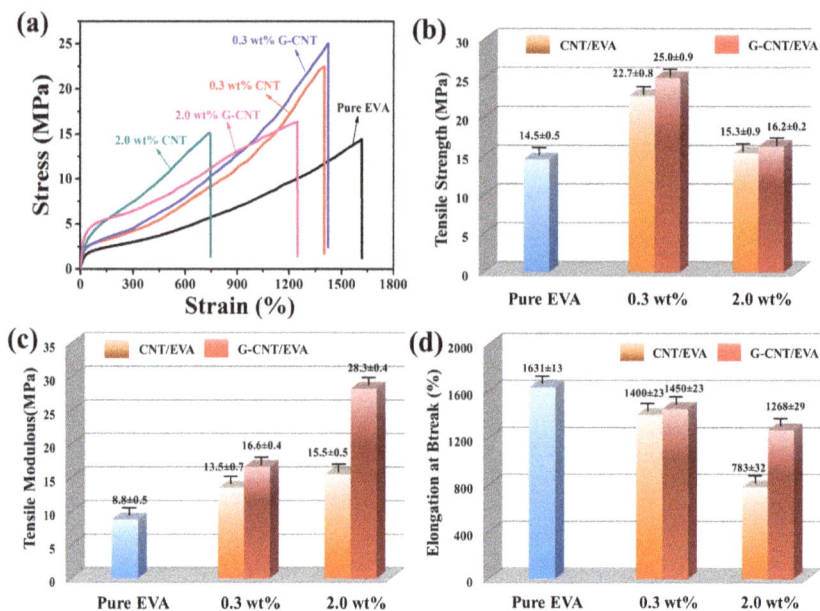

Figure 4. (**a**) Typical stress–strain curves of pure ethylene vinyl acetate (EVA) and its composites with various CNT loadings; (**b**) Tensile strength, (**c**) Young's modulus, and (**d**) elongation at break of pure EVA and its composites.

3.4. Effects of CNT Dispersion on Electrical Properties

Figure 5 shows the electrical conductivities of the G–CNT/EVA and CNT/EVA composites as a function of CNT content. Significant increases in the electrical conductivities of both the G–CNT/EVA and the CNT/EVA composites are observed with increasing CNT contents. Interestingly, the G–CNT/EVA composites exhibit superior electrical conductivities at low CNT contents, but inferior electrical conductivities at high CNT contents, compared to the CNT/EVA composites. At a very low CNT content of 0.2 wt %, the electrical conductivity of G–CNT/EVA composite already reaches 6.9×10^{-8} S/m, satisfying the antistatic criterion for commercial application [44]. Such value is about 4 orders of magnitude higher than that of the CNT/EVA composite (4.3×10^{-12} S/m). As the CNT loading rises to 0.5 wt %, the electrical conductivity of the G–CNT/EVA composites increases to 4.9×10^{-6} S/m, which is still higher than that of CNT/EVA composites. However, with further increasing CNT loadings to 1.0 wt % and 2.0 wt %, it can be seen that the electrical conductivities of CNT/EVA composites exceed that of the G–CNT/EVA composites. The main reasons for the interesting phenomenon can be found in the following two aspects. The first aspect is that the incorporation of GO–ODA can significantly improve CNT dispersion in EVA matrix due to π–π stacking interaction between GO–ODA and CNT. This contributes to the formation of inter-connected CNT networks. The second aspect is that the abundant presence of insulating GO–ODA between the adjacent CNTs will increase their contact resistance and thus hinder the electronic transport of CNTs in EVA. To provide visual demonstration of the conductive networks in the G–CNT/EVA and CNT/EVA composites, schematic representations are shown in Figure 5b,c and Figure 5d,e, respectively. At low CNT loading, the improved CNT dispersion contributes to the formation of effective conductive networks in the G–CNT/EVA composites (Figure 5b), while conductive networks are not formed in the CNT/EVA composites due to the agglomeration of CNTs (Figure 5d). Thus the first aspect plays a leading role in the electrical conductivity of the composites, which impart the G–CNT/EVA composites with higher electrical conductivity than that of the CNT/EVA composites. At high CNT loading, the CNT amount is enough to form inter-connected CNT networks in both the G–CNT/EVA composites (Figure 5c) and CNT/EVA composites (Figure 5e). The second aspect is dominant for the electrical conductivity of the composites and thus the G–CNT/EVA composites exhibit inferior electrical conductivity compared to that of CNT/EVA composites. A similar phenomenon was also reported for other CNT-based composites with the assistance of surfactants to improve CNT dispersion [45].

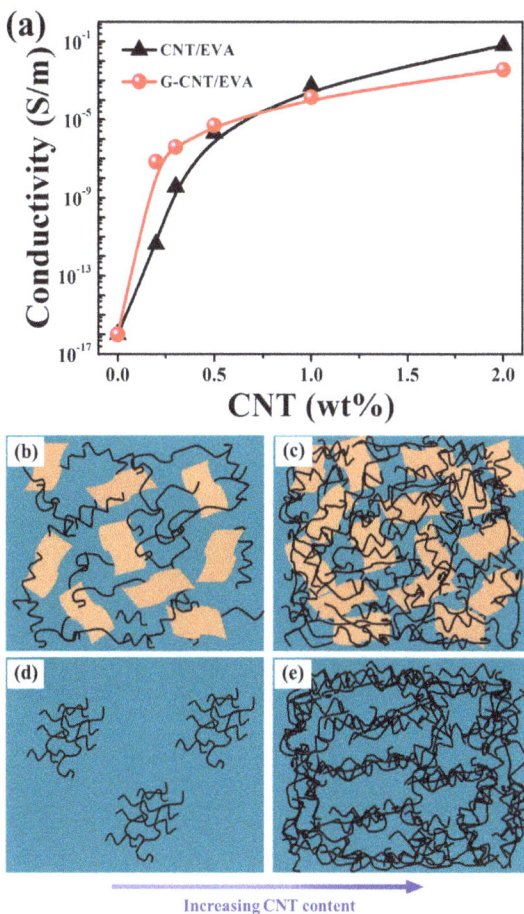

Figure 5. (**a**) Electrical conductivities of the G–CNT/EVA and the CNT/EVA composites. (**b**) and (**c**) The scheme of the conductive networks in the G–CNT/EVA composites at low and high CNT contents, respectively. (**d**) and (**e**) The scheme of the conductive networks in the CNT/EVA composites at low and high CNT contents, respectively.

4. Conclusions

We demonstrate a facile and effective approach to realize the homogeneous dispersion of CNT in EVA matrix, by using GO–ODA as a compatilizer. The resultant G–CNT/EVA composites show superior mechanical properties in comparison with the CNT/EVA composites, which could be attributed to the uniform dispersion of CNT in EVA matrix and the strong interfacial interaction between CNT and EVA, originating from the solubilization effect of GO–ODA. At a very low CNT content of 0.2 wt %, the electrical conductivity of G-CNT/EVA already reaches 6.9×10^{-8} S/m, satisfying the antistatic criterion for commercial application. The concept of dispersing CNT by using

Polymers **2017**, *9*, 397

hydrophobic GO–ODA in our work can be easily extended to other CNT-based polymer systems, which paves the way to develop such composites with excellent properties.

Acknowledgments: The authors gratefully acknowledge the financial support from the National Natural Science Foundation of China (Grant No. 51673134, 51421061, and 51473102,), the Science and Technology Department of Sichuan Province (Grant No. 2017GZ0412).

Author Contributions: Ding-Xiang Yan and Zhong-Ming Li conceived and designed the experiments; Li-Chuan Jia and Zhong-Han Jiao performed the experiments; Li-Chuan Jia, Ding-Xiang Yan and Zhong-Ming Li analyzed the data; Li-Chuan Jia wrote the paper; the paper was reviewed by all authors.

Conflicts of Interest: The authors declare no conflict of interest.

References

1. Chen, H.; Zeng, S.; Chen, M.; Zhang, Y.; Li, Q. Fabrication and functionalization of carbon nanotube films for high-performance flexible supercapacitors. *Carbon* **2015**, *92*, 271–296. [CrossRef]
2. Zhang, W.B.; Xu, X.L.; Yang, J.H.; Huang, T.; Zhang, N.; Wang, Y.; Zhou, Z.W. High thermal conductivity of poly (vinylidene fluoride)/carbon nanotubes nanocomposites achieved by adding polyvinyl pyrrolidone. *Compos. Sci. Technol.* **2015**, *106*, 1–8. [CrossRef]
3. Jia, L.C.; Yan, D.X.; Yang, Y.; Zhou, D.; Cui, C.H.; Bianco, E.; Lou, J.; Vajtai, R.; Li, B.; Ajayan, P.M.; et al. High strain tolerant emi shielding using carbon nanotube network stabilized rubber composite. *Adv. Mater. Technol.* **2017**, *2*, 1700078. [CrossRef]
4. Hayashida, K.; Tanaka, H. Ultrahigh electrical resistance of poly(cyclohexyl methacrylate)/carbon nanotube composites prepared using surface-initiated polymerization. *Adv. Funct. Mater.* **2012**, *22*, 2338–2344. [CrossRef]
5. Jia, L.C.; Yan, D.X.; Cui, C.H.; Jiang, X.; Ji, X.; Li, Z.M. Electrically conductive and electromagnetic interference shielding of polyethylene composites with devisable carbon nanotube networks. *J. Mater. Chem. C* **2015**, *3*, 9369–9378. [CrossRef]
6. Huang, J.; Mao, C.; Zhu, Y.; Jiang, W.; Yang, X. Control of carbon nanotubes at the interface of a co-continuous immiscible polymer blend to fabricate conductive composites with ultralow percolation thresholds. *Carbon* **2014**, *73*, 267–274. [CrossRef]
7. George, N.; Chandra, C.S.J.; Mathiazhagan, A.; Joseph, R. High performance natural rubber composites with conductive segregated network of multiwalled carbon nanotubes. *Compos. Sci. Technol.* **2015**, *116*, 33–40. [CrossRef]
8. De Volder, M.F.; Tawfick, S.H.; Baughman, R.H.; Hart, A.J. Carbon nanotubes: present and future commercial applications. *Science* **2013**, *339*, 535–539. [CrossRef] [PubMed]
9. Jia, L.C.; Li, M.Z.; Yan, D.X.; Cui, C.H.; Wu, H.-Y.; Li, Z.M. A strong and tough polymer-carbon nanotube film for flexible and efficient electromagnetic interference shielding. *J. Mater. Chem. C* **2017**. [CrossRef]
10. Zhou, J.; Yu, H.; Xu, X.; Han, F.; Lubineau, G. Ultrasensitive, stretchable strain sensors based on fragmented carbon nanotube papers. *ACS Appl. Mater. Interfaces* **2017**, *9*, 4835–4842. [CrossRef] [PubMed]
11. Cui, C.H.; Yan, D.X.; Pang, H.; Jia, L.C.; Xu, X.; Yang, S.; Xu, J.Z.; Li, Z.M. A high heat-resistance bioplastic foam with efficient electromagnetic interference shielding. *Chem. Eng. J.* **2017**, *323*, 29–36. [CrossRef]
12. Li, Y.; Zhang, Z.; Li, X.; Zhang, J.; Lou, H.; Shi, X.; Cheng, X.; Peng, H. A smart, stretchable resistive heater textile. *J. Mater. Chem. C* **2017**, *5*, 41–46. [CrossRef]
13. Jia, L.C.; Li, Y.K.; Yan, D.X. Flexible and efficient electromagnetic interference shielding materials from ground tire rubber. *Carbon* **2017**, *121*, 267–273. [CrossRef]
14. Chen, Y.; Zhang, H.B.; Yang, Y.; Wang, M.; Cao, A.; Yu, Z.Z. High-performance epoxy nanocomposites reinforced with three-dimensional carbon nanotube sponge for electromagnetic interference shielding. *Adv. Funct. Mater.* **2016**, *26*, 447–455. [CrossRef]
15. Zhang, K.; Yu, H.O.; Shi, Y.D.; Chen, Y.F.; Zeng, J.B.; Guo, J.; Wang, B.; Guo, Z.; Wang, M. Morphological regulation improved electrical conductivity and electromagnetic interference shielding in poly (L-lactide)/poly (ε-caprolactone)/carbon nanotubes nanocomposites via constructing stereocomplex crystallites. *J. Mater. Chem. C* **2017**, *5*, 2807–2817. [CrossRef]

16. Atar, N.; Grossman, E.; Gouzman, I.; Bolker, A.; Hanein, Y. Reinforced carbon nanotubes as electrically conducting and flexible films for space applications. *ACS Appl. Mater. Interfaces* **2014**, *6*, 20400–20407. [CrossRef] [PubMed]

17. Han, J.H.; Zhang, H.; Chen, M.J.; Wang, G.R.; Zhang, Z. CNT buckypaper/thermoplastic polyurethane composites with enhanced stiffness, strength and toughness. *Compos. Sci. Technol.* **2014**, *103*, 63–71. [CrossRef]

18. Oh, J.Y.; Jun, G.H.; Jin, S.; Ryu, H.J.; Hong, S.H. Enhanced electrical networks of stretchable conductors with small fraction of carbon nanotube/graphene hybrid fillers. *ACS Appl. Mater. Interfaces* **2016**, *8*, 3319–3325. [CrossRef] [PubMed]

19. Souier, T.; Maragliano, C.; Stefancich, M.; Chiesa, M. How to achieve high electrical conductivity in aligned carbon nanotube polymer composites. *Carbon* **2013**, *64*, 150–157. [CrossRef]

20. Niyogi, S.; Hamon, M.A.; Hu, H.; Zhao, B.; Bhowmik, P.; Sen, R.; Itkis, M.E.; Haddon, R.C. Chemistry of single-walled carbon nanotubes. *Accounts Chem. Res.* **2002**, *35*, 1105–1113. [CrossRef]

21. Hu, L.; Hecht, D.S.; Gruner, G. Carbon nanotube thin films: fabrication, properties, and applications. *Chem. Rev.* **2010**, *110*, 5790–5844. [CrossRef] [PubMed]

22. Choi, E.Y.; Roh, S.C.; Kim, C.K. Noncovalent functionalization of multi-walled carbon nanotubes with pyrene-linked nylon66 for high performance nylon66/multi-walled carbon nanotube composites. *Carbon* **2014**, *72*, 160–168. [CrossRef]

23. Roy, S.; Das, T.; Yue, C.Y.; Hu, X. Improved polymer encapsulation on multiwalled carbon nanotubes by selective plasma induced controlled polymer grafting. *ACS Appl. Mater. Interfaces* **2014**, *6*, 664–670. [CrossRef] [PubMed]

24. Daugaard, A.E.; Jankova, K.; Hvilsted, S. Poly(lauryl acrylate) and poly(stearyl acrylate) grafted multiwalled carbon nanotubes for polypropylene composites. *Polymer* **2014**, *55*, 481–487. [CrossRef]

25. George, N.; Bipinbal, P.K.; Bhadran, B.; Mathiazhagan, A.; Joseph, R. Segregated network formation of multiwalled carbon nanotubes in natural rubber through surfactant assisted latex compounding: A novel technique for multifunctional properties. *Polymer* **2017**, *112*, 264–277. [CrossRef]

26. Byrne, M.T.; Gun'ko, Y.K. Recent advances in research on carbon nanotube-polymer composites. *Adv. Mater.* **2010**, *22*, 1672–1688. [CrossRef] [PubMed]

27. Zhang, C.; Huang, S.; Tjiu, W.W.; Fan, W.; Liu, T. Facile preparation of water-dispersible graphene sheets stabilized by acid-treated multi-walled carbon nanotubes and their poly (vinyl alcohol) composites. *J. Mater. Chem.* **2012**, *22*, 2427–2434. [CrossRef]

28. Zhang, C.; Ren, L.; Wang, X.; Liu, T. Graphene oxide-assisted dispersion of pristine multiwalled carbon nanotubes in aqueous media. *J. Phys. Chem. C* **2010**, *114*, 11435–11440. [CrossRef]

29. Tian, L.; Meziani, M.J.; Lu, F.; Kong, C.Y.; Cao, L.; Thorne, T.J.; Sun, Y.P. Graphene oxides for homogeneous dispersion of carbon nanotubes. *ACS Appl. Mater. Interfaces* **2010**, *2*, 3217–3222. [CrossRef] [PubMed]

30. Dong, X.; Xing, G.; Chan-Park, M.B.; Shi, W.; Xiao, N.; Wang, J.; Yan, Q.; Sum, T.C.; Huang, W.; Chen, P. The formation of a carbon nanotube-graphene oxide core-shell structure and its possible applications. *Carbon* **2011**, *49*, 5071–5078. [CrossRef]

31. Li, Y.; Yang, T.; Yu, T.; Zheng, L.; Liao, K. Synergistic effect of hybrid carbon nantube-graphene oxide as a nanofiller in enhancing the mechanical properties of PVA composites. *J. Mater. Chem.* **2011**, *21*, 10844–10851. [CrossRef]

32. Shen, X.J.; Pei, X.Q.; Liu, Y.; Fu, S.Y. Tribological performance of carbon nanotube-graphene oxide hybrid/epoxy composites. *Compos. B* **2014**, *57*, 120–125. [CrossRef]

33. Wu, C.; Huang, X.; Wu, X.; Xie, L.; Yang, K.; Jiang, P. Graphene oxide-encapsulated carbon nanotube hybrids for high dielectric performance nanocomposites with enhanced energy storage density. *Nanoscale* **2013**, *5*, 3847–3855. [CrossRef] [PubMed]

34. Choudhary, S.; Mungse, H.P.; Khatri, O.P. Dispersion of alkylated graphene in organic solvents and its potential for lubrication applications. *J. Mater. Chem.* **2012**, *22*, 21032–21039. [CrossRef]

35. Ryu, S.H.; Shanmugharaj, A. Influence of long-chain alkylamine-modified graphene oxide on the crystallization, mechanical and electrical properties of isotactic polypropylene nanocomposites. *Chem. Eng. J.* **2014**, *244*, 552–560. [CrossRef]

36. Huang, H.D.; Zhou, S.Y.; Ren, P.G.; Ji, X.; Li, Z.M. Improved mechanical and barrier properties of low-density polyethylene nanocomposite films by incorporating hydrophobic graphene oxide nanosheets. *RSC Adv.* **2015**, *5*, 80739–80748. [CrossRef]
37. Jia, L.C.; Yan, D.X.; Cui, C.H.; Ji, X.; Li, Z.M. A unique double percolated polymer composite for highly efficient electromagnetic interference shielding. *Macromol. Mater. Eng.* **2016**, *301*, 1232–1241. [CrossRef]
38. Clancy, T.C.; Gates, T.S. Modeling of interfacial modification effects on thermal conductivity of carbon nanotube composites. *Polymer* **2006**, *47*, 5990–5996. [CrossRef]
39. Zhang, Z.; Zhang, Y.; Yang, K.; Yi, K.; Zhou, Z.; Huang, A.; Mai, K.; Lu, X. Three-dimensional carbon nanotube/ethylvinylacetate/polyaniline as a high performance electrode for supercapacitors. *J. Mater. Chem. A* **2015**, *3*, 1884–1889. [CrossRef]
40. Yan, D.X.; Pang, H.; Li, B.; Vajtai, R.; Xu, L.; Ren, P.G.; Wang, J.H.; Li, Z.M. Structured reduced graphene oxide/polymer composites for ultra-efficient electromagnetic interference shielding. *Adv. Funct. Mater.* **2015**, *25*, 559–566. [CrossRef]
41. Li, W.; Tang, X.Z.; Zhang, H.B.; Jiang, Z.G.; Yu, Z.Z.; Du, X.S.; Mai, Y.W. Simultaneous surface functionalization and reduction of graphene oxide with octadecylamine for electrically conductive polystyrene composites. *Carbon* **2011**, *49*, 4724–4730. [CrossRef]
42. Lin, Z.; Liu, Y.; Wong, C.P. Facile fabrication of superhydrophobic octadecylamine-functionalized graphite oxide film. *Langmuir* **2010**, *26*, 16110–16114. [CrossRef]
43. Yao, H.; Chu, C.C.; Sue, H.J.; Nishimura, R. Electrically conductive superhydrophobic octadecylamine-functionalized multiwall carbon nanotube films. *Carbon* **2013**, *53*, 366–373. [CrossRef]
44. Verma, M.; Verma, P.; Dhawan, S.; Choudhary, V. Tailored graphene based polyurethane composites for efficient electrostatic dissipation and electromagnetic interference shielding applications. *RSC Adv.* **2015**, *5*, 97349–97358. [CrossRef]
45. Kim, S.W.; Kim, T.; Kim, Y.S.; Choi, H.S.; Lim, H.J.; Yang, S.J.; Park, C.R. Surface modifications for the effective dispersion of carbon nanotubes in solvents and polymers. *Carbon* **2012**, *50*, 3–33. [CrossRef]